Lecture Notes in Physics

Edited by H. Araki, Kyoto, J. Ehlers, München, K. Hepp, Zürich
R. Kippenhahn, München, H. A. Weidenmüller, Heidelberg,
J. Wess, Karlsruhe and J. Zittartz, Köln

Managing Editor: W. Beiglböck

278

The Physics of Phase Space

Nonlinear Dynamics and Chaos, Geometric Quantization, and Wigner Function

Proceedings of the First International Conference on the Physics of Phase Space, Held at the University of Maryland, College Park, Maryland, May 20–23, 1986

Edited by Y. S. Kim and W. W. Zachary

Springer-Verlag

Berlin Heidelberg New York London Paris Tokyo

Editors

Y. S. Kim
Department of Physics and Astronomy, University of Maryland
College Park, Maryland 20742, USA

W. W. Zachary
Naval Research Laboratory, Department of the Navy
Washington, D.C. 20375, USA

ISBN 3-540-17894-5 Springer-Verlag Berlin Heidelberg New York
ISBN 0-387-17894-5 Springer-Verlag New York Berlin Heidelberg

Library of Congress Cataloging-in-Publication Data. International Conference on the Physics of Phase Space (1st: 1986: University of Maryland). The physics of phase space. (Lecture notes in physics; 278) 1. Phase space (Statistical physics)–Congresses. 2. Nonlinear theories–Congresses. 3. Chaotic behavior in systems–Congresses. 4. Geometric quantization–Congresses. 5. Wigner distribution–Congresses. I. Kim, Y. S. II. Zachary, W. W., 1935-. III. Title. IV. Series.
QC174.85.P48I58 1986 530.1'3 87-9857
ISBN 0-387-17894-5 (U.S.)

Printed in Germany

Printing: Druckhaus Beltz, Hemsbach/Bergstr.;
Bookbinding: J. Schäffer GmbH & Co. KG., Grünstadt
2153/3140-543210

PREFACE

This volume consists of the papers presented at the First International Conference on the Physics of Phase Space, 20 - 23 May 1986, at the University of Maryland, College Park, Maryland, U.S.A.

The main motivation for holding this Conference was to bring together a diverse collection of researchers working in the general area of phase space to present their current research and exchange new ideas concerning various approaches to current problems in pure and applied physics. Indeed, the Conference covered many different fields. The papers presented at the Conference may be divided into the following general areas:

A. Classical Nonlinear Dynamics and Chaos
B. Quantum Chaos
C. Wigner Distributions
D. Other Semiclassical Theories (WKB Methods, Maslov Theory, etc.)
E. Symplectic Geometry and Quantization
F. Relativity, Quantum Mechanics, and Thermodynamics

Almost sixty years have passed since the present form of quantum mechanics was formulated. Yet, the study of classical mechanics remains one of the major lines of research in physics, and interest in this field has been growing steadily in recent years. The transition from classical to quantum mechanics remains an interesting subject, and the number of investigations in this area has been growing very rapidly. The concept of phase space plays the decisive role in this area of physics, for example in nonlinear dynamics and chaos, geometric quantization, and the Wigner distribution function.

The Wigner function is one of the tools with which one can study the boundary between classical and quantum physics, and the number of publications on this subject has been increasing in recent years. In addition, other semiclassical approaches such as generalized WKB methods have recently become very important. The theory of geometric quantization comprises another procedure for studying the transition from classical to quantum physics. Current studies of nonlinear dynamics and chaos involve both classical mechanics and quantum physics, and investigation of the boundary between them. The purpose of this Conference was to emphasize the interconnection between these different fields.

In nonlinear dynamics and chaos, the Hamiltonian formulation on phase space is very important, particularly in the analysis of conservative systems. For example, in the relatively new subject of quantum chaos, direct use is made of phase space formulations in that usually this phenomenon is studied in systems for which classical chaos is known to exist. Thus, this class of problem is one in which the transition between classical and quantum mechanics plays a fundamental role.

The subject of chaos, in both its classical and quantum aspects, is now being investigated by an increasing number of researchers. Quantum chaos is now just beginning to be investigated, and its definitive character has yet to be determined.

Geometric quantization has its roots in the representation theory of Lie groups, which yields procedures for studying quantization - the derivation of quantum theories from corresponding classical ones. It is based upon differential geometry and makes essential use of the mathematical theory of vector bundles. For this reason, until recent times, this approach had not had much impact on the research techniques employed by physicists. However, a recent development has been that many physicists, particularly those in general relativity and elementary particle theory, are also using and developing geometric quantization techniques.

Another major recent development in the physics of phase space has been, and still is, the application of the Wigner function to both pure and applied physics. The Wigner function was introduced by Wigner in 1932 in connection with his effort to understand the effect of quantum correlation in statistical mechanics. The Wigner function depends on both the position and momentum variables, and is defined on phase space. While this function has many properties similar to those of the classical distribution function and quantum probability distribution, it is clearly distinct from these distribution functions. For this reason, continuous efforts are being made to see what fundamental role the Wigner function plays in the interpretation of quantum mechanics. The Wigner function also generates some interesting mathematical physics.

The present volume reflects these current trends. The Conference provided a forum for researchers in these fields to exchange their views.

The Organizing Committee consisted of Y.S. Kim and W.W. Zachary. This Conference was supported in part by the Office of Naval Research, the National Science Foundation, the International Union of Pure and Applied Physics, the Naval Research Laboratory, and the University of Maryland. The Conference was one of the topical conferences of the American Physical Society. The Organizing Committee is indebted to Professors H. Araki, H. Doebner, J.R. Klauder, C.S. Liu, P. Mohr, M. Shlesinger, and E.P. Wigner for their support and encouragement.

December 1986

Y.S. Kim, University of Maryland
W.W. Zachary, Naval Research Laboratory

CONTENTS

A. Classical Nonlinear Dynamics and Chaos

D. Other Semiclassical Theories

E. Symplectic Geometry and Quantization

F. Relativity, Quantum Mechanics, and Thermodynamics

A. CLASSICAL NONLINEAR DYNAMICS AND CHAOS

Entropy and Volume as Measures of Orbit Complexity

by

Sheldon E. Newhouse

Mathematics Department
University of North Carolina
Chapel Hill, North Carolina 27514

Abstract: Topological entropy and volume growth of smooth disks are considered as measures of the orbit complexity of a smooth dynamical system. In many cases, topological entropy can be estimated via volume growth. This gives methods of estimating dynamical invariants of transient and attracting sets and may apply to time series.

1. Introduction.

A basic problem in the theory of dynamical systems is to understand chaotic motion. One wants to attach numerical invariants to a system which measure the amount of chaos in the system. A natural invariant of a continuous or discrete system is the so-called *topological entropy*. This is a non-negative number which gives a crude quantitative measure of the orbit complexity of the system. The definition of the topological entropy is not very amenable to its calculation. Recently, results in the theory of smooth dynamical systems have related the topological entropy to the maximum volume growth of smooth disks in the phase space. Preliminary numerical studies indicate that in many cases volume growth rates may be estimated easily, and, hence the entropy itself may be estimated.

A system with positive topological entropy may have no complicated attracting sets. That is, the entropy may be given by the orbit structure on non-attracting (i.e., transient) sets. Typical orbits may spend varying amounts of time near these transient sets and then wind toward periodic attracting orbits (we are here, of course, thinking of dissipative dynamics). The entropy can give information about transient behavior, but it is interesting to ask how relevant it is for understanding asymptotic behavior. In this connection a simple example will be useful.

Consider the mapping $f_r(x)=rx(1-x)$ from the unit interval [0,1] to itself, where r is a real number in [0,4]. It is known that for r=0.25 and r=3.83 almost all orbits are asymptotic to periodic sinks. In the first case the sink is a fixed point while in the second case it is a periodic point of period three. Suppose we ask how much of a movement in r is necessary for the mapping f_r to have a set of positive measure whose orbits are not asymptotic to sinks. The answer is that much more is required for r=0.25 than for r=3.83. Is there some way of knowing this from f_r itself? We suggest that the topological entropy provides a clue. Indeed, the entropy for r=0.25 is zero while the entropy for r=3.83 is $\log(\frac{1+\sqrt{5}}{2}) \approx 0.481$. If one can estimate the entropy and its modulus of continuity, then one can get a predictive tool for the appearance of chaotic attractors.

2. Topological Entropy.

Let M be a smooth manifold and let $f: M \to M$ be a smooth self-map. We get f as either a discrete dynamical system or as a time-t map of a flow on M. Let d be a distance function on M induced by a smooth Riemannian metric. Let $\varepsilon > 0$ and let n be a positive integer. A set E is (n,ε)-separated if whenever $x \neq y$ in E there is an integer $j \in [0,n)$ such that $d(f^j x, f^j y) > \varepsilon$. Letting $r(n,\varepsilon,f)$ denote the maximum possible number of elements of any (n,ε)-separated set E, it easy to show that $r(n,\varepsilon,f) \leq C e^{n\alpha}$ for some $C > 0$ and $\alpha > 0$. The best such α is

$$r(\varepsilon,f) = \limsup_{n\to\infty} 1/n \log r(n,\varepsilon,f).$$

The number $h(f) = \lim_{\varepsilon \to 0} r(\varepsilon,f)$ is the *topological entropy* of f.

Properties of h(f):

1. $h(f^n) = nh(f)$ for $n \geq 0$.
2. $h(\phi f \phi^{-1}) = h(f)$ if ϕ is a continuous change of coordinates (i.e., h(f) is a topological invariant).
3. $h(f) = h(f^{-1})$ if f is a homeomorphism.
4. $h(f) = \sup \{h_\mu(f) : \mu \in \mathcal{M}(f)\}$, where $\mathcal{M}(f)$ is the set of f-invariant probability measures on M and $h_\mu(f)$ is the measure-theoretic entropy of f with respect to μ.

Note that if h(f) is positive, then f has invariant probability measures with positive entropy so f has some chaotic dynamics.

Examples:

1. Let $f(z)=P(z)/Q(z)$ be a rational function in

one complex variable z, where P and Q have no common factors.. Consider f as a mapping on the Riemann sphere S^2. It can be shown that

$h(f) = \log(\text{topological degree of } f)$
$\quad = \log(\max(\text{degree } P, \text{degree } Q))$

(see[L],[N1]).

2. Let A be an integer $N \times N$ matrix with determinant one, and let $\tilde{A}$ be the induced linear automorphism of the N-dimensional torus. Then, $h(f) = \sum_{\substack{\lambda \text{ is an eigenvalue} \\ \text{of } A \text{ with} \\ |\lambda|>1}} \log|\lambda|$ (see[B]).

3. Let $J=\{1,\ldots,N\}$ and let A be an $N \times N$ matrix of zeroes and ones. Let $\Sigma^{\mathbb{Z}} = J$ and let $\Sigma_A = \{\underline{a} \in \Sigma: A_{\underline{a}(i)\underline{a}(i+1)}=1$ for all $i\}$. Let $\sigma:\Sigma_A \to \Sigma_A$ be the shift map. Then $h(\sigma)$ is the logarithm of the largest modulus of the eigenvalues of A.

For more information on topological entropy, see [DGS].

3. Volume growth and its relation to topological entropy.

Let D^k be the closed unit ball in $\mathbb{R}^k$. A C^k disk in M is a C^k map $\gamma: D^k \to M$. . For such a C^k disk γ with $k \geq 1$, let $|\gamma|$ denote its k-dimensional volume with multiplicities. This is defined by

$$|\gamma| = \int_{D^k} |\Lambda^k T\gamma| \, d\lambda,$$

where $T\gamma$ is the derivative of γ, $\Lambda^k T\gamma$ is the kth exterior power of $T\gamma$, and $d\lambda$ is Lebesgue volume on D^k. When k=1, $|\gamma|$ is the length of the curve γ. When k=2, it is the surface area of γ, etc.

Given C^k $f:M \to M$ with $k > 1$ and γ as above, let

$$\bar{G}(\gamma,f) = \limsup_{n\to\infty} 1/n \log^+ |f^{n-1}\circ\gamma|.$$

Here, $\log^+$ is the positive part of the natural logarithm function. Thus, $\bar{G}(\gamma,f)$ is the volume growth rate of γ by f.

Let $G(\gamma,f) = \lim_{n\to\infty} 1/n \log^+ |f^{n-1}\circ\gamma|$ when the limit exists.

Let $\bar{G}(f) = \sup\{\bar{G}(\gamma,f): \gamma$ is any smooth disk in M$\}$, and let

$G(f) = \sup \{G(\gamma, f): \gamma$ is any smooth disk in $M\}$. We emphasize that the disks in the definitions of $\bar{G}(f)$ and $G(f)$ have their dimensions varying from 1 through dim M.

<u>Theorem 1</u> [N1]. Let $f: M \to M$ be a C^k self-map of the compact manifold M with $k > 1$. Then, $h(f) \le \bar{G}(f)$.

<u>Theorem 2</u> [Y]. Let $f: M \to M$ be a C^∞ self-map of the compact manifold M. Then, $h(f) \ge \bar{G}(f)$

Actually, the techniques in [N1] and [N2] can be combined with those in [Y] to prove the following sharper result.

<u>Theorem 3</u>.

1. Let $f: M \to M$ be a C^∞ self-map of the compact manifold M. Then, $h(f) = G(f)$ and the supremum in $G(f)$ is actually assumed by some disk γ.
2. The map $f \to h(f)$ is uppersemicontinuous on the space of C^∞ self-maps of M with the C^∞ topology.
3. For a fixed C^∞ map f, the mapping $\mu \to h_\mu(f)$ is uppersemicontinuous on the space of f-invariant measures on M. In particular, every C^∞ map has measures of maximal entropy.
4. The map $f \to h(f)$ is continuous on the space of C^∞ diffeomorphisms of a compact two-dimensional manifold M^2. (The lowersemicontinuity of $f \to h(f)$ for $C^{1+\alpha}$ diffeomorphisms of surfaces was proved by Katok [K].)
5. Let $f: M^2 \to M^2$ be a C^∞ diffeomorphism from the compact two-manifold with boundary M^2 into its interior. Assume that f is weakly dissipative in the sense that there is an integer $\tau > 0$ such that the Jacobian determinant of f^τ is less than one at each point in M^2. Let ∂M^2 denote the boundary of M^2. Then, $h(f) = G(\partial M^2, f)$.

○

Note that statement 5 of Theorem 3 applies to many forced oscillations. To compute the entropy, one only needs to compute the growth rate of the length of the boundary.

4. Numerical results.

We considered several Hénon mappings $x_1 = 1 + y - a x^2$, $y_1 = bx$ as a test for computing length growth for systems with two degrees of freedom.

Figure 1 below shows a plot of the log of the length of the n-th iterate of a certain line segment γ as a function of n for $5 \le n \le 2000$ with a=1.4, b=0.3. The x-units are in multiples of 5. The best least-squares line is also computed. The average of the entropies is actually the average of the quantities $1/j \log |f^j \gamma|$ for $5 \le j \le 2000$. We take the least-squares slope, LS, as an estimate of h(f). Note that LS is approximately 0.45 .

Figure 2 shows a similar plot for a=1.27, b=0.3. Note that there are two positive slopes. The first one is 0.35 and the second one is 0.09. This indicates the presence of a transient chaotic set with entropy ≈ 0.35 and a strange attractor with entropy ≈ 0.09. A plot of the iterates of a single orbit (not shown here) shows that the strange attractor has seven pieces. Its characteristic exponent is ≈ 0.084. Figure 3 shows the log-of-length versus length plot for a=1.28, b=0.3. The transient chaotic set seems to have merged with the strange attractor to produce a single attractor with entropy ≈ 0.30 and characteristic exponent ≈ 0.258.

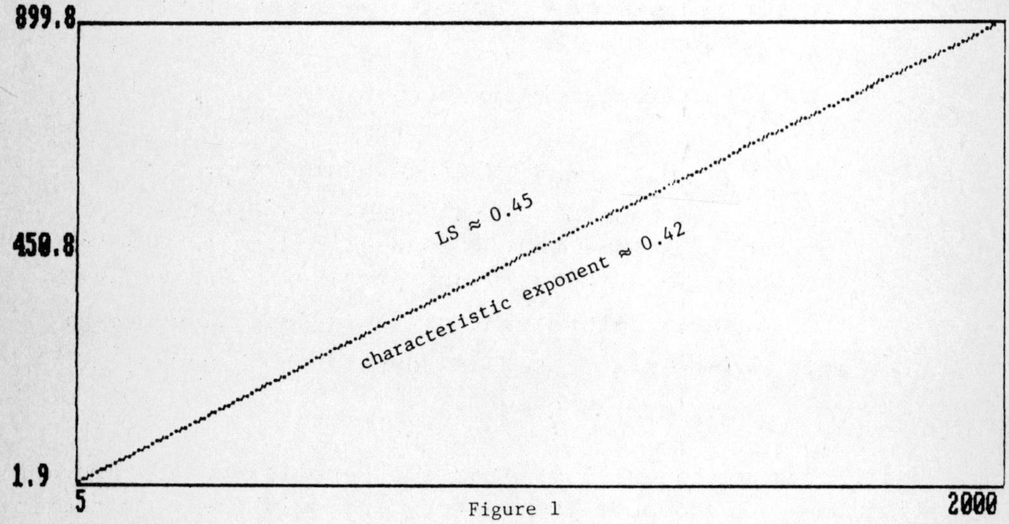

Figure 1

a = 1.4, b = 0.3, x - unit = 5, LS ≈ 0.45 average of entropies = 0.454

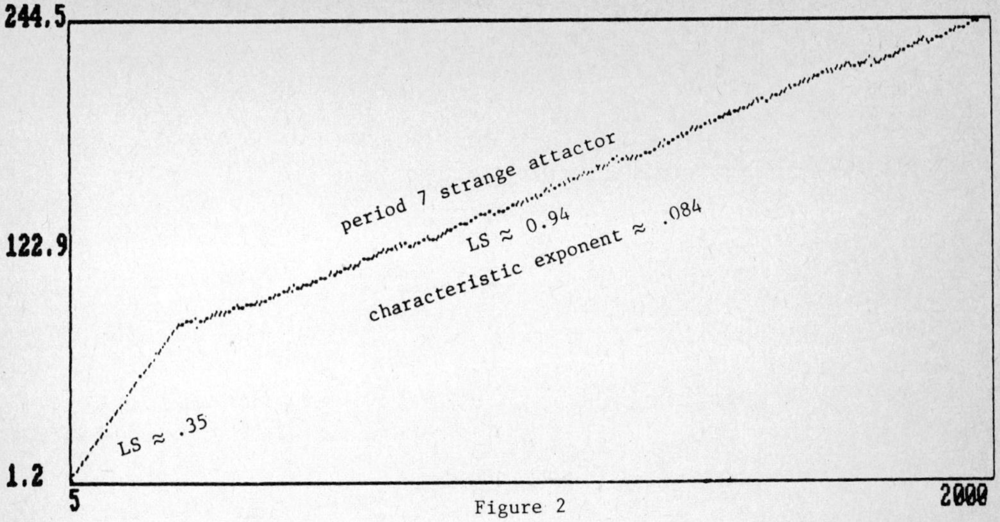

Figure 2

a = 1.27, b = 0.3, x - unit = 5

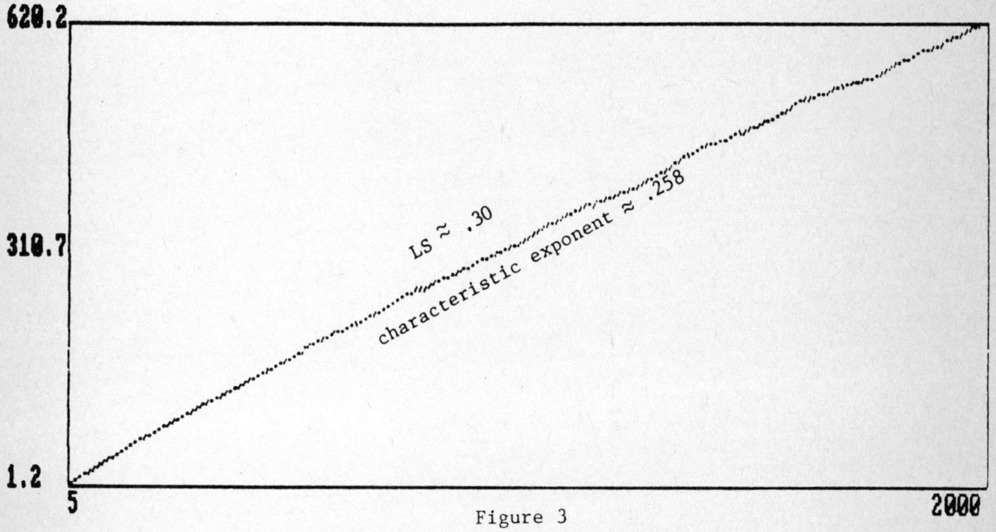

Figure 3

a = 1.28, b = 0.3, x - unit = 5

References

[B] R. Bowen, Entropy for Group Endomorphisms and Homogeneous Spaces, Trans. Amer. Math. Soc. 153(1971), 401-414, 181(1973), 509-510.

[DGS] M. Denker, C. Grillengerger, and K. Sigmund, Ergodic Theory on Compact Spaces, Lecture Notes in Math. 527 (1976). [K] A. Katok and L. Mendosa, to appear.

[L] M. Ljubich, Entropy properties of rational endomorphisms of the Riemann Sphere, Jour. Ergodic Theory and Dyn. Sys. 3(1983), 351-387.

[N1] S. Newhouse, Entropy and Volume, to appear in Jour. Ergodic Theory and Dyn. Sys.

[N2] S. Newhouse, Continuity Properties of Entropy, preprint, Mathematics Department, University of North Carolina, Chapel Hill, NC 27514, USA.

[Y] Y. Yomdin, Volume Growth and Entropy, and C^k-resolution of Semi-algebraic Mappings--Addendum to the Volume Growth and Entropy, to appear in Israel J. of Math.

A CHAOTIC 1-D GAS: SOME IMPLICATIONS

O.E. Rossler
Institute for Physical and Theoretical Chemistry,
University of Tubingen, 7400 Tubingen, West Germany

A 1-D classical gas with maximal chaos is described. It supports many simplified (color-coded) chemical reactions - including far-from-equilibrium dissipative structures. A proposed example generates a limit cycle. Its excitable analogue is a model observer. Gibbs symmetry invariably gives rise to a substitute Hamiltonian. The resulting pseudo-reversibility implies, for the model observer, that all external objects are subject to Nelson stochasticity and hence quantum mechanics.

The theory of classical solitons in 1 D is rich in implications. It was recently used to solve the relativistic no-interaction problem [1]. It might also help solve the problem of whether or not quantum mechanics can be reduced to classical mechanics.

In the following, classical nonrelativistic particles in 1 D are considered. Two types of particles are assumed. The "rods" present in a horizontal frictionless tube pass freely through each other. They interact only with the curved "bullets" that, while running in a vertical tube of their own, each may or may not protrude with their heads into the horizontal tube. The Hamiltonian in the simplest case (just 2 particles; $i_{max}=1$, $j_{max}=1$) becomes

$$H = \frac{p_i^{\,2}}{2} + \frac{p_j^{\,2}}{2} + \frac{\varepsilon}{x_i} + \frac{\varepsilon}{1-x_i} + \frac{\varepsilon}{y_j} + \frac{\varepsilon}{L-f_j(x_i)-y_j}\ , \qquad (1)$$

where f is a bowler-hat shaped function, being

$$f = \sqrt{(x-0.45)^2+10^{-6}} + \sqrt{(x-0.55)^2+10^{-6}} - 2\sqrt{(x-0.5)^2+10^{-4}}$$

in the simplest case, and $L = 0.5$ and $H = 1.8$, for example.

First, it is to be shown that this two degrees of freedom system is chaotic. This is easy if the bowler-hat function is ideal (flat zero except for a sharply protruding half circle). Then with $\varepsilon \to 0$ a classical collision problem, a point-shaped billiard on a 2-D table that sports a protruding half-circle on one side, results. Sinai's theorem (see [2]) which implies chaos applies. The above f ("smoothed tent") yields, with $\varepsilon = 0.002$, the same result numerically for non-selected initial conditions.

Second, the system has to be shown to remain "maximally chaotic" as more and more particles are added. Specifically, adding a second horizontal particle ($i = 1,2$) augments the right-hand side of H by 4 terms. The corresponding collision problem now is a point-shaped billiard on a 3-D table - with one side of the box sporting two protruding half-cylinders in the shape of a cross. Hence Sinai's theorem applies in two mutually independent directions, this time. And so forth. Hence the number of positive Lyapunov characteristic exponents remains $n-1$ ("maximal chaos").

Similarly if more vertical particles are added. Each new j augments H by $1+i(max)$ terms. The j types of f functions differ only in the positions of the protrusions on the x-axis. (To accomodate many vertical slots, smaller and smaller half circles are needed if the unit x-interval is retained. The y-lengths, L, then have to be decreased proportionally.) There is maximal chaos again: Each horizontal particle interacts chaotically with each vertical one.

Third, a first implication. The present 1-D gas can be made the basis of chemical interactions. Elskens [3] already considered reactions supported by an underlying deterministic dynamics - a 1-D gas of the quasiperiodic type. The simplest possibility is color coding: Colors (chemical identities) change in a lawful manner under collisions [3]. The reaction energies are hereby shielded from contributing to the mechanical ones. This is an admissible idealization (to be relaxed as more realistic molecular-dynamics Hamiltonians become available). Unlike bimolecular reactions, monomolecular ones require a special convention. Making these color changes contingent on arbitrary "supra-threshold" collisions is one possibility. Easier to implement is an artificial convention: There is a "color-changing position" in every unit interval L , both horizontally and vertically, and there is some clock (some - any - particle being in a certain color-specific interval, somewhere) determining whether or not a color change takes place. Another problem is that there are necessarily two subpopulations to each color, one among the bullets and one among the rods. It is conceivable that even with symmetric initial numbers and large n , their numbers might diverge under certain conditions. This unlikely situation has yet to be ruled out.

Again, chemical reactions relaxing toward equilibrium (cf. [3]) can be studied. The present gas has the asset of being strongly mixing so that some of the results can be expected to be even more realistic.

Fourth, it deserves to be stressed that far-from-equilibrium situations and even open conditions can be included. Such systems are able to generate nontrivial dissipative structures (like limit cycles or other attractors) [4]. They only have to obey mass conservation in the present context. A convenient example is the following simple 4-variable quadratic mass action system:

$$\begin{aligned}
\dot{a} &= 0.0011d - ab \\
\dot{b} &= ab - bc - 0.05b + 10^{-5}d \\
\dot{c} &= 0.002d - bc - 0.035c \qquad (2) \\
\dot{d} &= 0.05b + 2bc + 0.035c - \\
&\quad - (0.0011 + 10^{-5} + 0.002)d.
\end{aligned}$$

There is mass conservation (a+b+c+d = const.). One of the monomolecular reactions (that from d to a) actually has to be second-order in reality - involving a constant-concentration, energy-rich reaction partner. This "fifth" color is formally included in the above pseudo-collision convention for first-order reactions. The system of Eq.(2) produces, at the assumed parameter values and with a+b+c+d = 10, a deterministic limit cycle. It will be interesting to reproduce this limit cycle with the above 1-D molecular dynamics scheme - with small values of n like 100 to 1000, perhaps.

Fifth, a variant to the reaction system of Eq.(2) is bound to produce "excitable" behavior - stability toward very small-amplitude perturbations but autocatalytic instability toward somewhat larger ones (with subsequent re-excitability after a refractory period), cf. [4]. The system in this case will constitute a "formal neuron." Of course, if one such neuron can be implemented by Eq.(1), so can 10^{10}, say. That is, a full-fledged macroscopic observer (of well-stirred type) can be implemented - in principle.

Sixth, a new question can therefore be posed. How must the world appear to such a (fully transparent, in principle) observer?

The question can be approached using the present excitable system. (One neuron is as good as many in principle, especially so as arbitrary classical measuring devices may be provided to the system.) At first sight, nothing unusual is to be expected.

The situation changes if the fact that the observer contains equal-type classical particles is taken into account. Such particles, if really identical (that is, unlabellable), introduce a nontrivial symmetry. Note in this context that classical solitons - which provide the motivation for the present particles - are indeed unlabelable.

At first the simpler case of the reaction-free gas is to be considered. Here Gibbs's early finding of a "reduced phase space volume" [5] can be confirmed. N indistinguishable (as far as their material identities are concerned) particles reduce this volume by a factor of N! [5]. This is because the lack of knowledge about their material identities gives rise to N! equally eligible, mirror-symmetric trajectories once a single one is unambiguously defined in space-time. Mutual identification of all of them then leads to this reduction.

However, even more can be said. Pointwise identification, at corresponding instants in time, of all N! trajectories leaves certain subsurfaces of position space invariant. Along these hypersurfaces, 2 or more of the trajectories are already identical. These surfaces form natural boundaries. Each trajectory upon hitting such a surface possesses a "continuation" this side - so as if the surface were the boundary of an N-dimensional billiard table. Position space therefore is naturally divided up into "cells." Each cell contains a unique trajectory. As in crystallography, an irreducible unit exists for position space.

As an example, consider the horizontal subsystem of Eq.(1), that is, assume $j = 0$ and $i = 1, \ldots, N$. The position space of the mutually interpenetrating N rods is the unit N-cube. Indistinguishability leads to a "triangulation" of the latter - as is easy to verify for $N = 2$ and $N = 3$. In general, the so-called standard N!-triangulation of the N-cube applies. The unique trajectory, inside the standard triangle/simplex, is exactly the same as if the original Hamiltonian H had been replaced by

$$H^* = H + \frac{\varepsilon'}{x_2-x_1} + \frac{\varepsilon'}{x_3-x_2} + \ldots + \frac{\varepsilon'}{x_N-x_{N-1}} , \quad (3)$$

where $\varepsilon' \to 0$ and $x_1, \ldots, x_N$ are the sequentially ordered, at one moment in time, equal rods.

Equation (3) means that the N mutually transparent particles have become opaque: Each particle has seemingly acquired a point-shaped hard core as far as meeting with its own kind is concerned. This deterministic result is in accordance with recent measure-theoretic results of Bach [6] on statistical mechanical systems that involve classically indistinguishable particles. It explains these results in one dimension on a deterministic basis.

Seventh, the preceding result remains valid in the case of open, isothermal systems like the above model observer. The unique, dissipative trajectory of the observer (if chemical free energy is included as proper) once more becomes "N!-unique." Uniqueness is re-established for a subcell of position space, with substitute Hamiltonian H* as before. If N is large, the trajectory is effectively "curled up" in this cell in almost all directions. The observer thereby becomes pseudo-closed and pseudo-reversible [7]. More specifically, a "causal" and an "anti-causal subobserver" become inextricably interlaced [7]. As a consequence, external causality vacillates irreducibly for the observer. This vacillation of time's axis, in turn, every unit cell passage time τ, imposes a random diffusion, with the unit thermal noise energy E of the observer, on every external object of mass M. Therefore, Nelson's [8] diffusion coefficient, $D = E\tau/(2M)$, which implies quantum mechanics [8], governs all observations that the observer might perform.

To conclude, classical chaos theory and classical Gibbs symmetry can be combined.

I thank Joe Ford, Martin Hoffmann and Klaus Strecker for discussions.

References

[1] S.N.M. Ruijsenaars & H. Schneider (1986). A new class of integrable systems and its relation to solitons. Ann. of Phys. (in press).

[2] Ya.G. Sinai (1980). Apperdix. In: N. Krylov, Papers on Statistical Mechanics. Princeton Univ. Press.

[3] Y. Elskens (1984). Microscopic derivation of a Markovian master equation in a deterministic model of a chemical reaction. J.Stat.Phys. 37, 673-695.

[4] G. Nicolis and I. Prigogine (1977). Self-organization in Nonequilibrium Systems. Wiley, New York.

[5] J.W. Gibbs (1902). Elementary Principles of Statistical Mechanics. Yale University Press, New Haven. Last Chpt.

[6] A. Bach (1985). On the quantum properties of indistinguishable classical particles. Lett.Nuov.Cim. 43, 483-487.

[7] O.E. Rossler (1985). A possible explanation of quantum mechanics. Unpublished Manuscript.

[8] E. Nelson (1967). Deterministic derivation of the Schrodinger equation. Phys.Rev. 150, 1079-1085.
E. Nelson (1967). Dynamic Theories of Brownian Motion. Princeton University Press, Princeton.

SINGULAR APPROXIMATION OF CHAOTIC SLOW-FAST DYNAMICAL SYSTEMS

B. Rossetto, University of Toulon, F 83130 - LA GARDE (France)

1. We consider the autonomous dynamical system defined in R^3 :

$$(1)\qquad \begin{aligned} \varepsilon\dot{x} &= -(x^3 - ax + b) = -S(x,a,b) \\ \dot{a} &= -(0.1x + a - 1) = -F(x,a) \\ \dot{b} &= -(0.1x + a - 1)x + x + \lambda = G_\lambda(x,a) \end{aligned}$$

where $(\dot{\ }) = d/dt$, $0 < \varepsilon < 1$ and λ is a real parameter. This system is a metaphor set up by J. Argémi for a model of behavior of baryum-treated *Aplysia* neurons [1].

The first aim of this paper is to show for $\varepsilon \neq 0$, as for the singular approximation ($\varepsilon = 0_+$), the existence of slow trajectories for (1) ; in other words, we search a domain $D \subset R^3$ in which the solutions of (1) also verify :

$$(2)\qquad \dot{x} = h(x,a,b) + O(\varepsilon)$$

where h is a map of D into R^3, of degree 0 in ε, continuous and derivable, and $O(\varepsilon)$ is bounded in D and has a degree in ε greater than zero. The part of a trajectory that verifies (1) and (2) is called *slow trajectory of (1)*.

Multiplying (2) by ε and comparing with (1), one obtains the zero-order approximation for initial conditions of slow trajectories, $f^0(x,a,b) = O(\varepsilon)$:

$$(3)\qquad f^0(x,a,b) = S(x,a,b) = O(\varepsilon).$$

Therefore D, if it exists, is located in the ε - neighbourhood of the slow manifold S of the singular approximation, $S = 0$. So as to derive the zero-order approximation of slow velocity, we write $df^0/dt = O(\varepsilon)$, i.e., $(\partial f^0/\partial x)\,\dot{x} + (\partial f^0/\partial a)\,\dot{a} + (\partial f^0/\partial b)\,\dot{b} = O(\varepsilon)$:

$$(4)\qquad \dot{x} = -\frac{xF + G_\lambda}{P} + O(\varepsilon),\ \dot{a} = -F,\ \dot{b} = G_\lambda,\ S = O(\varepsilon),$$

where $P = 3x^2 - a$ must be different from zero. Thus, the slow motion of the singular approximation of (1) appears as the zero-approximation of slow trajectories, valid outside of the fold line, defined by $P = 0$. Yet there are some trajectories crossing the fold line, which satisfy (2) and which are consequently slow trajectories : those which cross the singular point of the singular approximation, Q_0, defined by $P = 0$, $S = 0$ and $xF + G_\lambda \equiv x + \lambda = 0$, and which, near Q_0, are given by the eigenvectors. We call them *pseudo-singular solutions* ($pss_0(Q_0)$) because such a point Q_0, which is not a singular point of the initial system (1), is called *pseudo - singular point* by J. Argémi.

Putting (4) into (1), we obtain the first approximation, $f^1_\varepsilon(x,a,b)$:

$$(5)\qquad f^1_\varepsilon(x,a,b) = SP - \varepsilon(x + \lambda) = O(\varepsilon^2),\ P \neq 0.$$

We derive the following successive approximations of slow trajectories in the same way that we have deduced f^1_ε from f^0 :

$$P^1_\varepsilon = P^2 + 6xS - \varepsilon$$

$$X^1_\varepsilon = P(x + \lambda) + SF$$

$$f^2_\varepsilon = SP^1_\varepsilon - \varepsilon X^1_\varepsilon = O(\varepsilon^3),\ P^1_\varepsilon \neq 0$$

$$P^2_\varepsilon = 6S^2 + 24x\,PS + P\,(P^2 - \varepsilon) - \varepsilon[\,P + 6x\,(x + \lambda) + PF + 0{,}1\,S\,]$$

$$X^2_\varepsilon = (x + \lambda)\,[\,P^2 + 12x\,S - \varepsilon(1 + 2F)\,] + 2\,PSF + \varepsilon SF$$

$$f^3_\varepsilon = SP^2_\varepsilon - \varepsilon\,X^2_\varepsilon = 0\,(\varepsilon^4)\;,\; P_\varepsilon \neq 0.$$

...

$$(6)\qquad f^m_\varepsilon = SP^{m-1}_\varepsilon - \varepsilon\,X^{m-1}_\varepsilon = 0\,(\varepsilon^{m+1})\;,\; P^{m-1}_\varepsilon \neq 0$$

with $P^{m-1}_\varepsilon = \dfrac{\partial f^{m-1}_\varepsilon}{\partial x}$, and $X^{m-1}_\varepsilon = -\dfrac{\partial f^{m-1}_\varepsilon}{\partial a}F + \dfrac{\partial f^{m-1}_\varepsilon}{\partial b}G_\lambda$.

Therefore, D is located in the ε^m - neighbourhood of the manifold $f^m_\varepsilon = 0$. The mth-order approximation of slow motion is solution of the dynamical system :

$$(7)\qquad f^m_\varepsilon = 0\;,\; \dot{x}^m = X^m_\varepsilon \operatorname{sgn} P^m_\varepsilon\;,\; \dot{a}^m = -F\,P^m_\varepsilon \operatorname{sgn} P^m_\varepsilon\;,\; \dot{b}^m = G_\lambda\,P^m_\varepsilon \operatorname{sgn} P^m_\varepsilon.$$

The manifold defined by the limit of (6) when $m \to \infty$, if it exists, is D. Any trajectory corresponding to an initial condition as close to D as one wants, has a velocity along the x-axis of degree -1 in ε and, if it is unstable with regard to the rapid motion (given by $\dot{x}^m = f^m_\varepsilon(x, \tilde{a}, \tilde{b})$, with $\tilde{a}$ and $\tilde{b}$ const.), will move off from D the more rapidly as ε is small : it is *sensitive to initial conditions* [4].

Let P_m be a pseudo-singular point satisfying $P^m_\varepsilon = 0$, $X^m_\varepsilon = 0$, and $f^m_\varepsilon = 0$. The *pss$_m$ (Q_m)*, defined as before, are not submitted to the restriction $P^\varepsilon_m \neq 0$ and their stability with regard to the rapid motion changes in Q_m. When $\varepsilon \to 0$, we have shown that Q_m tends to Q_{m-1} and a *pss$_m$ (Q_m)* to a *pss$_{m-1}$ (Q_{m-1})*. On the other hand, the map T of D into S, which connects each different *pss$_\infty$ (Q_∞)* to a *pss$_0$ (Q_0)*, is injective. It follows that a necessary condition for (1), or analogous models, to have an infinite number of those slow trajectories whose stability changes near the fold line, is for the singular approximation to have a pseudo singular node [5].

Now, we have shown, for the singular approximation of (1), the existence of periodic solutions, made up by a *pss$_0$ (Q_0)* closed-looped by a rapid trajectory. For the values of the bifurcation parameter λ for which Q_0 is a node, there is an infinite number of these periodic solutions and they are not separable. For the same values of λ and for $\varepsilon \neq 0$, a numerical integration provides, in the neighbourhood of a Hopf bifurcation, repetitive *doubling of period leading to chaos* [1].

2. Consider the electronic oscillator of S.V. Kiyashko, A.S. Pikovsky and M.I. Rabinovich [3]. The model has the form:

$$(8)\qquad \begin{aligned} \dot{x} &= 2hx + y - gz \\ \dot{y} &= -x \\ \varepsilon\dot{z} &= x - f(z) \end{aligned}$$

where $(\dot{\ }) = d/dt$. For a circuit made up with a field-effect transistor and a tunnel diode, we have the numerical values : $h = 0.12$, $g = 0.7$, $f(z) = 44\,\dfrac{z^3}{3} + 41\,\dfrac{z^2}{2} + \lambda z$, λ being a parameter, and $\varepsilon \simeq 0.01$. The authors have shown the existence of *chaotic solutions by re-injection* for this model, according to O.E. Rössler [6]. We study the singular approximation of (1), the solutions of which are in agreement with experimental results. Then, the use of a Poincaré map makes it possible to show an attractor and sequences of bifurcations accompanying transition to chaos. In particular, according as the re-injection takes place in one or another basin delimited by a pseudo-singular saddle, we observe a different *"elementary motif"*, in the sense used by R. Lozi [2]. On the other hand, the variety of motives observed depends on the position of the singular unstable focus of (8) ($x = 0$, $y = 0$, $z = 0$) with respect to the *cofold* line, which is, according to J. Argémi, the projection of the fold line on the slow manifold $x = f(z)$.

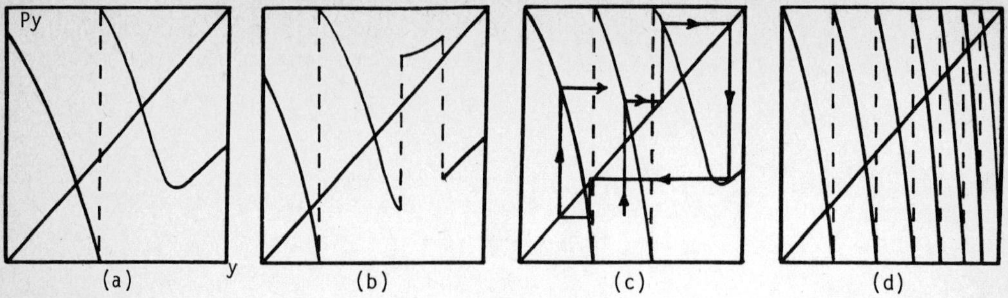

Fig. 1 - *Some aspects of a Poincaré map of the singular approximation of the chaotic oscillator (8) for different values of* λ : *(a)* : $\lambda = 6.94$; *(b)* : $\lambda = 6.95$; *(c)* : $\lambda = 7.0$; *(d)* : $\lambda = 7.165$.

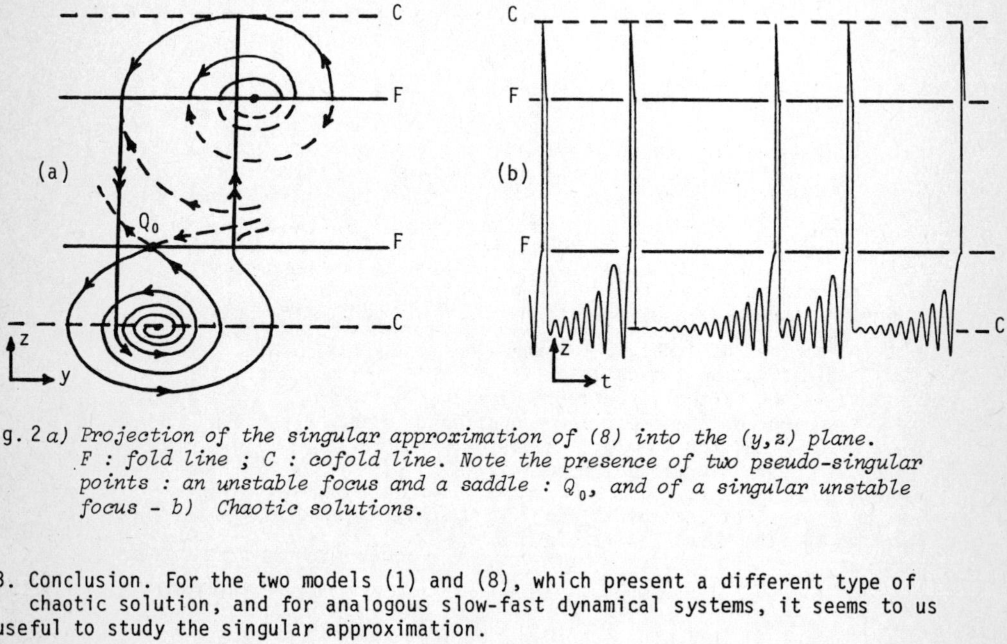

Fig. 2 *a) Projection of the singular approximation of (8) into the (y,z) plane. F : fold line ; C : cofold line. Note the presence of two pseudo-singular points : an unstable focus and a saddle :* Q_0, *and of a singular unstable focus - b) Chaotic solutions.*

3. Conclusion. For the two models (1) and (8), which present a different type of chaotic solution, and for analogous slow-fast dynamical systems, it seems to us useful to study the singular approximation.

Acknowledgements. I am indebted to J. Argémi, recently died, for very helpful discussions, and for his generous advice in this matter.

References.

[1] J. Argémi, B. Rossetto, J. Math. Biol., 17, 67-92 (1983).

[2] R. Lozi, Thesis of Doctorat d'Etat des Sciences, Nice (1983).

[3] M.I. Rabinovich, Ann. N.Y. Acad. Sci., 357, 435-452 (1980).

[4] B. Rossetto, to be published in 1986 in Japan J. Appl. Math.

[5] B. Rossetto, 7th Int. Conf. Analysis Optimization of Systems, INRIA, Antibes, France (1986).

[6] O.E. Rössler, Bull. Math. Biol., 39, 275-289 (1977) and Ann. N.Y. Acad. Sci., 316, 376-392 (1979).

DIMENSION CALCULATIONS IN A MINIMAL EMBEDDING SPACE: LOW-DIMENSIONAL ATTRACTORS FOR HUMAN ELECTROENCEPHALOGRAMS

A. M. Albano, L. Smilowitz
Bryn Mawr College
Bryn Mawr, PA 19010

P. E. Rapp, G. C. de Guzman, T. R. Bashore
The Medical College of Pennsylvania
3200 Henry Avenue, Philadelphia, PA 19129

1. Introduction

Some recent work has shown that under certain circumstances, human electroencephalograms (EEG's) can be described as motions on strange attractors of relatively low dimensionalities[1]. For EEG's, as for other chaotic signals, dimension calculations are, however, often rendered ambiguous by noise and complicated by the need to use high dimensional embedding spaces. We report here on the use of an orthonormal basis which removes some of these ambiguities. Calculations of the correlation dimension using this basis confirm previously reported results on changes in the EEG attractor dimension accompanying changes in the subject's cognitive state.

2. Dimension Calculations

Dimension calculations rely crucially on a reconstruction of the system's phase space from a single time series[2]. This is done by "embedding" the attractor in an n-dimensional space of "time-delay" vectors, $\mathbf{d}^{(n)}{}_k = (v_k, v_{k+1}, v_{k+2}), \ldots, v_{k+n-1}), \ k=1,\ldots,N$, where $v_k = v(k\tau)$ is the value of v at time $k\tau$. The correlation dimension, D_2, of the reconstructed attractor is determined by means of the correlation sum, or the fraction, $C_n(\epsilon)$, of those distances between embedding vectors that do not exceed ϵ. $C_n(\epsilon)$ scales as $\epsilon^{D2(n,\epsilon)}$, and D_2 is the limit of $D_2(n,\epsilon)$ as ϵ becomes small and n large. It is these limits that give rise to ambiguities, as small values of ϵ are most affected by noise, while large values of n introduce noise-like behavior at large ϵ's or large $C_n(\epsilon)$'s .

3. The Broomhead-King Basis.[3]

Broomhead and King have shown that the dimension of the smallest Euclidean space containing the attractor is given by the number of nonzero eigenvalues of the "covariance matrix", $C_{i,j} = N^{-1}\Sigma_{k=0,N-1}\,(v_{i+k}v_{j+k})$; $i,j = 1,\ldots,n$. The normalized eigenvectors, $\mathbf{e}_k$, of C, respectively corresponding to the eigenvalues, λ_k ; $k=1,\ldots,n$; $(\lambda_1 \geq \lambda_2 \geq \ldots \geq \lambda_n)$, constitute an orthonormal basis (the "Broomhead-King basis") such that λ_k is the mean square projection of the N embedding vectors on $\mathbf{e}_k$. Geometrical properties of the attractor are dominantly determined by directions corresponding to the largest eigenvalues, a fact that remains true even in the presence of noise. Use of this basis tends to alleviate the problems at the two extremes of ϵ mentioned earlier. This is dramatically illustrated by Fig. 1 which shows (a) a projection of an attractor into the v_k-v_{k+1} plane, and (b) a projection of the same attractor into the 1-2 plane of the Broomhead-King basis.

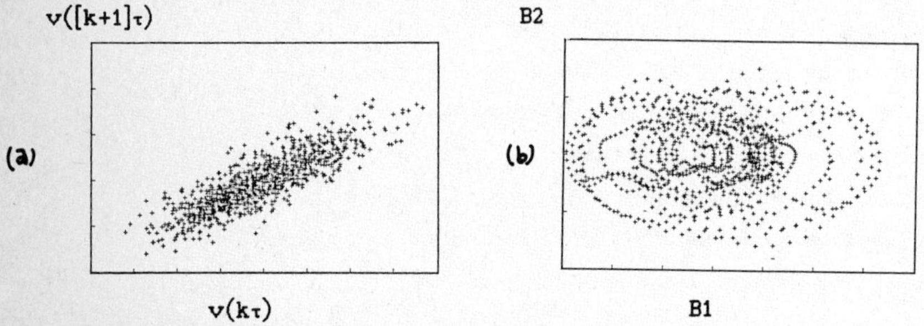

Figure 1. (a) Projection of an EEG attractor on the plane, $v(k\tau)$-$v([k+1]\tau)$, of time-delay vectors. (b) Projection of the same attractor on the 1-2 plane in the Broomhead-King basis

4. EEG Dimensions and Cognitive State.

We analyzed two sets of EEG measurements on the same subject: a 30-s epoch when the subject had his eyes closed and was resting, and another 30-s epoch with the subject counting backwards from 700 in steps of 7 ("serial 7's"). A simultaneous EEG recording was used to ensure that only EEG records uncontaminated by eye movement were used. Measurements were taken at the Oz location, digitized at 2-ms intervals.

Figure 2 shows $D_2(n,\epsilon)$ (the slope of log $C_n(\epsilon)$ vs. log ϵ) plotted vs. log $C_n(\epsilon)$, (a) for the resing case, and (b) for serial 7's. 1000 vectors in a 7-dimensional Broohmead-King basis were used, with a "window length", 7τ,of the order of the first zero of the signal's autocorrelation function. In both cases, the four graphs resulting from the use of 4-,...,7-dimensional subspaces are practically coincident implyimg that the high-n limit has been reached. Previously, it was necessary to go to an embedding dimension as high as 20 to reach this limit[1]. The value of D_2 is taken as the value of $D_2(n,\epsilon)$ at the "plateau" regions, where the $D_2(n,\epsilon)$ vs. log C_n graphs are flat. The "resting" case gives a D_2 of 2.20

± 0.05, while the "serial 7's" case gives 2.60± 0.05. This increase in D_2 agrees with results obtained earlier with another subject[1] using the time delay basis but for which, in the serial 7's case, the calculation did not converge sufficiently for an unambigous value of D_2 to be obtained.

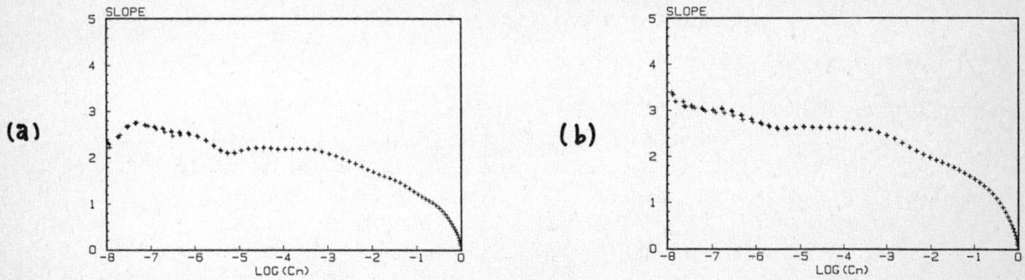

Figure 2. $D_2(n,\epsilon)$ = Slope of log $C_n(\epsilon)$ vs. logϵ for a human EEG using 1000 vectors in a 7-dimensional Broomhead-King basis, calculated in 4,5,6,and 7-dimensional subspaces. (a) Subject resting with eyes closed, (b) Subject with eyes closed and counting backward in steps of 7.

5. Conclusions

Reconstructing an attractor in the Broomhead-King basis makes dimension calculations less sensitive to noise and require smaller embedding spaces. Use of this procedure to study human EEG's confirm earlier results that EEG's may indeed be characterizable by small attrctor dimensions, and continue to offer the possibility that dimensions or similar quantitative measures of chaotic activity of EEG's might be used to track changes in cognitive activity or pathological conditions.

6. References

1. A.M .Albano, N.B. Abraham, G.C. de Guzman, M.F.H. Tarroja, D.K. Bandy, R.S. Gioggia, P.E. Rapp, E.D. Zimmerman, N.N. Greenbaun and T.R. Bashore: "Lasers and Brains: Complex Systems with Low-Dimensional Attractors"; A. Babyolantz: "Evidence of Chaotic Dynamics of Brain Activity During the Sleep Cycle", in Dimensions and Entropies in Chaotic Systems, G. Mayer-Kress, ed., (Springer-Verlag, Berlin, 1986); P.E. Rapp, I.D. Zimmerman, A.M. Albano, G.C. de Guzman, N.N. Greenbaun: "Experimental Studies of Chaotic Neural Behavior: Cellular Activity and Electroencephalographic Signals", in Nonlinear Oscillations in Chemistry and Biology, H.G. Othmer, ed. (Springer-Verlag, Berlin, to appear).
2. N.H. Packard, J.P. Crutchfield, J.D. Farmer, and R.S. Shaw: Phys. Rev. Lett. 45, 712 (1980); J.D. Farmer: Physica 4D, 336 (1982).
3. D.S. Broomhead and G.P. King, Physica D, (to appear); "On the Qualitative Analysis of Experimental Dynamical Systems", in Nonlinear Phenomena and Chaos, S. Sarkar, ed., (Adam Hilger, Bristol, 1986).
4. P. Grassberger and I. Procaccia: Physica 9D, 189 (1983); Phys. Rev. Lett 50, 349 (1983); Phys. Rev. A 28, 2591 (1983); Physica 13D, 34 (1984).

CHAOTIC SPACE-TIMES

John D Barrow
Astronomy Centre
University of Sussex
Brighton BN1 9QH
U.K.

It is now widely appreciated that very simple dynamical systems, notably iterated maps of the unit interval into itself, which possess regular initial data and deterministic evolution can exhibit behaviour which is for all practical purposes completely unpredictable. In this brief survey we shall highlight an example which shows how such chaotic unpredictability besets certain solutions of Einstein's equations which describe general relativistic cosmological models. Further details can be found in refs [1-3].

The Mixmaster universe is a spatially homogeneous anisotropic cosmological model first studied by Misner [4] and subsequently by many other authors [1,2,5-7]. The expansion scale factors a(t), b(t) and c(t) describe the evolution in time, t, of the three orthogonal directions of this expanding universe model. It expands from an initial Weyl curvature singularity at $t = 0$ (where $abc = 0$), attains a maximum value of the volume abc, and then probably contracts to a final singularity where $abc = 0$ again (the question of whether recollapse occurs is an interesting and unresolved one, see refs [8,9]). In the absence of matter the Einstein equations describing the time-evolution of the scale factors reduce to the following ordinary differential equations, [10],

$$(\ell n a^2)'' = (b^2 - c^2)^2 - a^4 \tag{1}$$

$$(\ell n b^2)'' = (c^2 - a^2)^2 - b^4 \tag{2}$$

$$(\ell n c^2)'' = (a^2 - b^2)^2 - c^4 \tag{3}$$

$$(\ell n a^2)'(\ell n b^2)' + (\ell n a^2)'(\ell n c^2)' + (\ell n b^2)'(\ell n c^2)' = a^4 + b^4 + c^4 - 2a^2b^2 - 2a^2c^2 - 2b^2c^2, \tag{4}$$

where $' \equiv d/d\Omega$ and the Ω-time coordinate is defined by

$$dt = -abc\, d\Omega\,. \tag{5}$$

Typically, $abc \propto t$ as the singularity is approached at $t = 0$ and so $\Omega \propto -\ell n t$. Hence the initial singularity is located at $\Omega = +\infty$. We shall be interested in the evolution as $\Omega \to +\infty$.

The qualitatitive behaviour of the system (1)-(5) is now fairly well-known to theoretical cosmologists. As $\Omega \to +\infty$ the evolution consists of an infinite sequence of stochastic oscillations. A four-dimensional Poincaré return mapping can be found for the dynamics described by (1)-(5). This discrete dynamical system is given by [3,11]

$$x_{n+1} = x_n^{-1} - [x_n^{-1}] \; ; \tag{6}$$

$$k_{n+1} = [x_n^{-1}] \; ; \; k \in Z^+ \tag{7}$$

$$y_{n+1} = \{1 + y_n(k_{n+1}x_n^{-1} - 1)\}/\{1 + y_n k_{n+1}(x_{n+1}^{-1} + x_n^{-1})\} \; ;$$

$$y_n \in (0,1), \tag{8}$$

where [x] denotes the integer part of the real number x. The mappings (6) and (7) were first found by the authors of ref [2], however the additional mappings given in [12] and recently studied by the authors of [13] are incorrect because they were derived from the false initial assumption that the variables are asymptotically independent.

An invariant measure can be found [11,2] for the system (6)-(8):

$$\mu(x,y,k) = \frac{\theta\{kx/(1+kx) < y < (k+1)x/(1+kx+x)\}}{x\{x(1-y) + yx^{-1}\}^2 \ln 2}, \tag{9}$$

where the θ-function is defined by

$$\theta(a < z < b) = 1 \quad \text{if } z \in (a,b)$$
$$= 0 \quad \text{otherwise.} \tag{10}$$

Integrating and summing over two of the three variables x,k,y in (9) yield the probability distributions for these variables alone:

$$\mu(x) = 1/\{(1+x)\ln 2\} \tag{11}$$

$$\mu(k) = \ln\{(k+1)^2/k(k+2)\}/\ln 2 \tag{12}$$

$$\mu(y) = 1/2(1-y)\ln 2 \quad \text{if } y \in (0,½]$$
$$= 1/2y\ln 2 \quad \text{if } y \in [½,1). \tag{13}$$

The probability distribution (12) is that of the integers appearing in the infinite continued fraction expansion of almost any real number [1,2,14].

It is advantageous to define two variables, u and v, which lie along the eigendirections of the non-linear mapping [3,11]

$$u = k+x \quad \text{and} \quad v = y(1+u)/(1-y). \tag{14}$$

The invariant measure (9) is

$$\mu(u,v) = 1/(1+uv)^2 \ln 2\,. \tag{15}$$

The Kolmogorov metric entropy h_μ of the map is non-zero and equal to

$$h_\mu = \pi^2/6(\ln 2)^2 = 3.4237..., \tag{16}$$

The system dynamics are thus chaotic.

It is interesting that there appears a close connection between the evolution of these mappings and the metric theory of numbers [1,2]. If initial data for the evolution towards t=0 are set by almost any irrational number $u_o = k_o+x_o$, then the sequence of iterates $\{k_1,k_2,k_3,...\}$ are just the partial quotients of the continued fraction expansion of u_o. Remarkably, for almost any u_o there exists a geometric mean value of k_n. By a theorem of Khinchin [15,1] we have that

$$\operatorname*{Lt}_{n\to\infty} (k_1k_2k_3....k_n)^{1/n} = \mathbf{K} \equiv 2.67.... \qquad (17)$$

The appearance of a continued-fraction structure is typical of hamiltonian dynamical systems with effectively closed confining potentials.

Recently there has been considerable interest in the behaviour of space-times possessing more than three spatial dimensions. In the spirit of the earlier ideas of Kaluza and Klein [16,17], it was hoped that gauge invariance might be explained as coordinate invariance in additional spatial dimensions. These higher-dimensional theories typically possess a space-time of the form $M^4 \times C^D$ where M^4 is the observed 4-dimensional space-time (or some approximation to it) and C^D is a D-dimensional compact space whose isometry group generates a low-energy quantum field theory of the Yang-Mills type. In the prototype of Kaluza-Klein, C^D was the circle S^1 and the associated isometry group the U(1) invariance of electromagnetism.

Various authors have investigated Mixmaster models with additional spatial dimensions [18-23]. In the case when the additional dimensions enter in the product form just described, it is found that chaotic behaviour is only exhibited by the spatially homogeneous models with three spatial dimensions. When inhomogeneous model universes are studied and the stipulation of a product structure for the space-time metric is removed then chaotic behaviour cannot occur generically when there are more than nine spatial dimensions. The situation when there are between four and nine spatial dimensions is not clearcut. The interpretation of the disappearance of chaotic behaviour when the spatial dimension is increased is straightforward. A necessary condition for chaotic behaviour to arise in vacuum spatially homogeneous cosmological models is that they possess effectively closed potentials when represented as hamiltonian systems. However, this is by no means a sufficient condition for chaos to ensue even if the walls of the confining potential exhibit hyperbolicity under relections. In the Mixmaster problem the walls of the potential expand outwards as $t \to 0$ and the singularity is approached. In order for chaos to occur it is also necessary that the maximum value of the velocity component of the motion normal to the walls exceed the velocity of the walls. If this is not so then the moving point will, after a few random bounces from the walls, enter a configuration in which it will never again catch up with the potential walls. No chaotic behaviour can exist in such a situation and the asymptotic behaviour is predictable. In the presence of matter the necessary and sufficient conditions for chaotic Mixmaster behaviour to occur are likely to be rather subtle unless a simplifying assumption (for example, that matter obeys an equation of state so the pressure is a continuous function of the density) is imposed. The reason is that, unless some condition is imposed to restrict the allowed form of the energy-momentum tensor, any space-time metric solves Einstein's equations. An interesting problem for future study is that of quantum chaotic behaviour in the Mixmaster universe. This presents a two-fold difficulty. On the one hand one must arrive at a good theory of quantum cosmology, whilst on the other, one is presented with the problem of defining what 'quantum chaos' is and determining whether it can exist.

Acknowledgements: I would like to thank D. Chernoff, J. Stein-Schabes, B.L. Hu and D. Brill for helpful discussions.

References

[1] Barrow, J.D., 1982. Phys. Reports 85,1.
[2] Belinskii, V.A., Lifshitz, E.M. & Khalatnikov, I.M., 1971. Sov. Phys. Usp. 13,745.
[3] Barrow, J.D., 1983, in *Classical General Relativity*, ed. W.Bonnor, J.Islam & M.A.H. MacCallum, (Cambridge U.P., Cambridge).
[4] Misner, C., 1969. Phys. Rev. Lett. 22,1071.
[5] Hu, B.L., 1975. Phys. Rev. D12,1551.
[6] Ryan, M. & Shepley, L.C., 1975. *Homogeneous Relativistic Cosmologies*, (Princeton U.P., Princeton, N.J.).
[7] Bogoiavlenskii, O.I., 1976. Sov. Phys. JETP 43,187.
[8] Barrow, J.D. & Tipler, F.J., 1986. Mon. Not. Roy. astr. Soc. 216,395.
[9] Barrow, J.D., Galloway, G. & Tipler, F.J., 1986. Mon. Not. Roy. astr. Soc. 000,000.
[10] Landau, L. & Lifshitz, E.M., 1974. *The Classical Theory of Fields*, (Pergamon, Oxford).
[11] Chernoff, D. & Barrow, J.D., 1983. Phys. Rev. Lett. 50, 134.
[12] Lifshitz, E.M., Lifshitz, I.M. & Khalatnikov, I.M., 1971. Sov. Phys. JETP 32, 173.
[13] Lifshitz, E.M., Khalatnikov, I.M. & Sinai, Y., 1984. Preprint.
[14] Gauss, C.F., 1812. Letter to Laplace. dated 30 Jan., *Werke* X, 371.
[15] Khinchin, A., 1934. Composito Math. 1,376.
[16] Kaluza, T., 1921. Sber. preuss. Akad. Math. Kl. 966.
[17] Klein, O., 1926. Z. Physik 37, 895.
[18] Barrow, J.D. & Stein-Schabes, 1985. Phys. Rev. D32, 1595.
[19] Furosawa, T. & Hosoya, A., 1985. Prog. Theo. Phys. 73, 467.
[20] Ishihara H., 1985. Prog. Theo. Phys. 74, 490.
[21] Halpern, P., 1986. Phys. Rev. D33,354.
[22] Demaret, J., Henneaux, M. & Spindel, P., 1986. Phys. Lett B000.000.

ON RELAXATION CHAOS : AN EXAMPLE FROM CELESTIAL MECHANICS

J.Koiller*, Instituto de Matematica, UFRJ
Caixa Postal 68530 Rio de Janeiro Brazil 21944

J.M.Balthazar and T.Yokoyama
Departamento de Matematica Aplicada, UNESP
Caixa Postal 178 Rio Claro Sao Paulo Brazil 13500

A new phenomenon in the dynamics of the asteroidal belt was discovered numerically and recently also explained theoretically by Wisdom: near the 3/1 resonance, the eccentricities can undergo sudden increases, typically from $e \sim 0.1$ to $e \sim 0.35$; thus, orbits become Mars (or even Earth!) crossing. We pursue here some developments of [1].

The model. The planar, elliptical, restricted and resonant three body problem is described by a two-degrees of freedom Hamiltonian (see [1a])

$$H = \alpha p^2/2 + \mu R(x,y,q), \tag{1}$$

where q is the critical angle (the resonant combination of Jupiter's and the asteroid mean motions), p its conjugate variable (function of the asteroid semimajor axis a, zero at exact resonance a^*); $\alpha < 0$ is a constant ; $(x,y) = \sqrt{2r}$ (cosw , sinw), where w is minus the longitude of the asteroid periapse and $2r \sim \sqrt{a_*}e^2$. The time scale t is such that Jupiter's period is 2π ; μ is the mass ratio of Jupiter and the Sun. High frequency terms of Jupiter disturbing function neglected

KAM theorem hypothesis do not hold [2,Appendix 8]. However, Wisdom noticed that approximate dynamics can be obtained from the "adiabatic principle". Indeed, in the intermediate time scale $\sigma = \varepsilon t$, $\varepsilon = \sqrt{\mu}$,

$$d^2q/d\sigma^2 = -\alpha R_q \tag{2}$$

$$dx/d\sigma = -\varepsilon R_y \quad , \quad dy/d\sigma = \varepsilon R_x. \tag{3}$$

Heuristically (2) can be thought as a 1-degree of freedom system with slowly varying "parameters" x,y . In the 3/1 resonance (2) is formally equal to a simple pendulum, since

$$R = F(x^2+y^2) + \bar{F}e_J x - A\cos q - B\sin q \tag{4a}$$

$$A = C(x^2-y^2) + De_J x + E e_J^2 \quad , B = 2Cxy + De_J y . \tag{4b}$$

Parameter values:

$\alpha = -12.98851$	$a^* = 0.4805968$	$e_J = 0.048$
$F = -0.2050694$	$\bar{F} = 0.1987054$	$\mu = 1/1047.355$
$C = 0.8631579$	$D = -2.656407$	$E = 0.3629536.$

* Visiting, under a CAPES/Brazil fellowship, the Department of Mathematics, Yale University.

Averaging the critical angle. A formalization of the canonical transformation method of [1c, Appendix] gives

$$H = \mu [h^o(I,\bar{x},\bar{y}) + \varepsilon h^1(\bar{x},\bar{y},I,\theta) + O(\varepsilon^2)] \qquad (5)$$

$$w = d\bar{x}\, d\bar{y} + \varepsilon\, dI\, d\theta \quad \text{(the symplectic form, see [2])}. \qquad (6)$$

Here (I,θ) are action-angle variables for a fixed regime (librations or circulations) of the "frozen pendulum"(FP) (2) with x,y fixed. More precisely, $(\bar{x},\bar{y},I,\theta) \mapsto (x,y,p,q)$, $(x,y) = (\bar{x},\bar{y}) + O(\varepsilon)$ is generated by $F = x\bar{y} + \varepsilon G(I,q,x,\bar{y})$, $y=F_x$, $\bar{x}=F_{\bar{y}}$, $\varepsilon\theta=F_I$, $p=F_q$ and G is the generating function reducing the FP to h^o. It follows that

$$h^1 = R_y G_x - h^o_x G_y \quad \text{(calculated at } (\bar{x},\bar{y},I,\theta)). \qquad (7)$$

In the σ-time scale (5,6) yield

$$\frac{d\bar{x}}{d\sigma} = -\varepsilon h^o_{\bar{y}} + \dots , \dots , \frac{dI}{d\sigma} = -\varepsilon h^1_\theta + \dots , \frac{d\theta}{d\sigma} = h^o_I + \dots . \qquad (8)$$

The averaging theorem [2,§52] implies that I is an adiabatic invariant and (2,3) is well approximated (up to time $O(1/\varepsilon)$ in σ-scale) by the averaged Hamiltonian εh^o , with symplectic form (6).

Classical Uncertainty Principle. Both the change of variables and averaging break down when (x,y) approaches a point in which the area inside the FP separatrix is $2\pi I$. Wisdom numerical studies show that the motion becomes uncertain for a while, eventually leaving the "uncertainty zone"(UZ) with a different action in one of the regimes. That is Wisdom's beautiful explanation for the eccentricity increases. We observe that the center curve of the UZ of energy h (h-UZ) is obtained by solving $R_q = 0$ for unstable FP equilibria $q^u(x,y)$ and setting

$$R^u(x,y) = R(x,y,q^u(x,y)) = h \quad . \qquad (9)$$

Why "relaxation-chaos". We show next that maximum eccentricities correspond to smallest possible librations I^* compatible with the constraint of reaching the UZ. The alternation of averaged arcs $h^o(I,\bar{x},\bar{y})$ with endpoints in $R^u=h$ vaguely resembles van der Pol's relaxation-oscillations.

Harmonic approximation for the librations. For the 3/1 resonance,

$$R^u = F(x^2 + y^2) + \bar{F}e_J x - P \quad , P^2 = A^2 + B^2. \qquad (10)$$

In the small librations regime, elliptic functions can be avoided:

$$\begin{aligned} & q = q^s + |\alpha/P|^{1/4} \sqrt{2I}\sin\theta \quad , \quad p = \varepsilon |P/\alpha|^{1/4} \sqrt{2I}\cos\theta \\ & h^o = R^s - |\alpha P|^{1/2} I \quad , \quad R^s = F(x^2+y^2)+\bar{F}e_J x + P = R(x,y,q^s(x,y)) \\ & \cos q^s = -A/P \quad , \quad \sin q^s = -B/P \\ & 2G = \theta + \sin\theta\cos\theta. \end{aligned} \qquad (11)$$

Maximum eccentricities. It is easily seen that circulation trajectories belong to the interior of (9). Symmetry considerations imply that the libration trajectory attaining maximum eccentricity departs the UZ at a point from the x-axis. The energy level $h=-1.93e_J^2$ is particularly interesting because (9) yields a "figure 8 contour" centered at the Sinclair's point $e \sim 0.11$ [1c,Figs.1-4,6,8,10,12]. Using (11) we get a libration amplitude $\sim 60^0$ and maximum $e \sim .35$ also in the x-axis, where $\sin w=0$. The harmonic approximation seems to give consistent results even beyond its range. We plan to attempt it for other resonances [3].

Singular horseshoes. Thin chaotic regions ($O(\exp(-\text{const}/\varepsilon))$) occur around separatrices of $h^o(I,x,y)$. Since the averaging method also fails at $I=0$, we expect the width to grow as $I \to 0$. This may have some interest, because for $I=0$ there is a figure 8 loop centered at Hill's point $e \sim 0.08$. Eccentricity increases to 0.24.

Reintroducing high-frequency terms. Their contribution for the chaotic zones is of even smaller order $O(\exp(-\text{const}/\mu))$. Adapting the "kick trick" (see [1a]), we get the following map for time $T=2\pi\mu$: break R^s into $R_{sec} = F(x^2+y\ )+ Fe_Jx$ and P; compose the partial flows. For an hydrodynamical analog, see Aref [4]. Notice $[R_{sec},P]\neq 0$ (Poisson bracket).

UZ motions. We estimate the UZ width by Melnikov's method [5]. It is of order ε with leading coefficient $m_h = \max|M|$ on $R^u=h$, where

$$M = 4[R_{sec},P]/|\alpha P|^{1/2} \text{ (Melnikov's function).} \tag{12}$$

Inside the UZ there is an unstable periodic orbit given by

$$h = R^u + \mu(q_x^u R_y^u - q_y^u R_x^u)^2/2\alpha + O(\mu^2). \tag{13}$$

The extensive chaotic region in (x,y) plane is the projection of the homoclinic orbits to it. Wisdom's criterion for chaos $[R_{sec},P] \neq 0$ is recovered and interpreted as transversal intersection of stable and unstable manifolds. The UZ reflects the local invariant manifolds. Trajectories stay in the UZ for a time $O(\varepsilon \log \varepsilon)$ in the slow scale $\tau = \mu t$. Let τ flow along $R^u=h$ and let $v=R^u-h$. Then (τ,v) parametrize the hUZ. The global dynamics is modelled by random composition of mappings U_θ, where θ is the phase at entrance point in the UZ, a random variable in the circle S^1. Details will appear elsewhere.

[1] J.Wisdom, a. Astr.J.87,577-593,1982. b.Icarus 56,51-74,1983. c. Icarus 63,272-286,1985.

[2] V.Arnold, Math.Meth.Classical Mechanics,Springer,1978.

[3] C.Murray, Icarus 65:1,70-82,1986.

[4] H.Aref, J.Fluid Mech. 143, 1-21,1984.

[5] C.Robinson, in Springer LNM 1007,1981.

COMPUTATION OF INVARIANT TORI AND ACCELERATION OF THE K.A.M. ALGORITHM*

R. L. WARNOCK
Lawrence Berkeley Laboratory
University of California
Berkeley, California 94720

R. D. RUTH
Stanford Linear Accelerator Center
Stanford University
Stanford, Calfornia 94305

We describe a method to compute invariant tori in phase space for classical non-integrable Hamiltonian systems. Our procedure is to solve the Hamilton-Jacobi equation stated as a system of equations for Fourier coefficients of the generating function. The system is truncated to a finite number of Fourier modes and solved numerically by Newton's method. The resulting canonical transformation serves to reduce greatly the non-integrable part of the Hamiltonian. Further transformations computed on progressively larger mode sets would lead to exact invariant tori, according to the argument of Kolmogorov, Arnold, and Moser (KAM)[1]. Our technique accelerates the original KAM algorithm, since each truncated Hamilton-Jacobi equation is solved accurately, rather than in lowest order. In examples studied to date, the convergence properties of the method are excellent, even near chaotic regions and on the separatrices of isolated broad resonances. One can include enough modes at the first step to get accurate results with only one canonical transformation. A second transformation gives an estimate of error. We propose a criterion for breakup of a KAM torus, which arises naturally in the Hamiltonian-Jacobi formalism. We verify its utility in an example with $1\frac{1}{2}$ degrees of freedom and anticipate that it will be useful in systems of higher dimension as well.

We present results for a system with one degree of freedom having a periodic time-dependent Hamiltonian. In angle-action variables the Hamiltonian is

$$H(\phi, J, \theta) = H_0(J) + V(\phi, J, \theta), \tag{1}$$

where V has period 2π in the time variable θ. We seek a canonical transformation $(\phi, J) \mapsto (\psi, K)$ in the form

$$J = K + G_\phi(\phi, K, \theta)\ , \tag{2}$$

$$\psi = \phi + G_K(\phi, K, \theta)\ , \tag{3}$$

such that the new Hamiltonian becomes a function of K alone. The Hamilton-Jacobi equation to determine the generator G is the requirement that the new Hamiltonian H indeed depend only on K; namely,

$$H_0(K + G_\phi) + V(\phi, K + G_\phi, \theta) + G_\theta = H_1(K)\ . \tag{4}$$

We seek periodic solutions of (4) with the Fourier development

$$G(\phi, K, \theta) = \sum_{m,n} g_{mn}(K) e^{i(m\phi - n\theta)}\ . \tag{5}$$

We rearrange (4) by adding and subtracting terms so as to isolate terms linear in G_ϕ and G_θ. We then take the Fourier transform for $m \neq 0$ to cast Eq. (4) in the form

$$g = A(g)\ , \tag{6}$$

* Work supported by the Department of Energy, contracts DE-AC03-76SF00098 and DE-AC03-76SF00515.

where $g = [g_{mn}]$ is a vector of Fourier coefficients and

$$A_{mn}(g) = \frac{i}{(\omega_0 m - n)} \frac{1}{(2\pi)^2} \int_0^{2\pi}\int_0^{2\pi} d\phi d\theta e^{-i(m\phi - n\theta)} \left[H(\phi, K + G_\phi, \theta) - H_0(K) - \omega_0 G_\phi\right] \ , \ m \neq 0 \ , \tag{7}$$

where $\omega_0(K) = \partial H_0/\partial K$. To truncate the system (6) for numerical solution we restrict (m,n) to some bounded set B of integers, with $m \neq 0$, and put

$$G_\phi = \sum_{(m,n)\epsilon B} img_{mn}(K)e^{i(m\phi - n\theta)} \ . \tag{8}$$

The set B is selected so that the only modes included are fairly close to resonance, and are driven by the perturbation V (directly or through harmonics).

We show results from solving (6) by Newton's iteration, starting from $g = 0$. The action variable K is changed at each iteration in such a way as to make the final frequency $\omega = \partial H_1/\partial K$ have a preassigned value. This is accomplished automatically by augmenting (6) with another equation to be iterated.

The example chosen is the non-integrable two-resonance Hamiltonian

$$H = \nu J + \frac{1}{2}\alpha J^2 + \varepsilon_1 J^{5/2}\cos(5\phi - 3\theta) + \varepsilon_2 J^2 \cos(8\phi - 3\theta), \tag{9}$$

where $\nu, \alpha, \varepsilon_1, \varepsilon_2$ are constants. For small $\varepsilon_1, \varepsilon_2$ we compute a KAM curve at a frequency equal to the golden mean $\omega_* = (\sqrt{5} - 1)/2$, which is between the two resonances, and explore its breakup as ε_1 and ε_2 increase to critical values. With $\nu = 0.5$, $\alpha = 0.1$, we find an apparently solid KAM curve for $\varepsilon_1 = 2\varepsilon_2 = 6 \times 10^{-5}$, for which case the resonance widths are $\Delta J_1 = 0.049$, $\Delta J_2 = 0.054$, as compared to the resonance separation $J_{r1} - J_{r2} = 0.25$. The curve $J(\phi, \theta = 0)$ shown in Fig. 1 was computed in 4 Newton iterations with 40 modes in the set B. It agrees well with results from direct integration of Hamilton's ordinary differential equations, and the corresponding canonical transformation leaves a very small residual perturbation. The average of the absolute value of the residual perturbation divided by a similar average of the original perturbation is 1.1×10^{-5}. Expanding the mode set to 77 modes and doing further iterations, we reduce this ratio to 6.4×10^{-8}.

To identify the breakup of the KAM curve ("transition to chaos") as the ε's are increased, we propose the criterion that the Jacobian of Eq. (3) vanish at some (ϕ, θ):

$$\partial\psi/\partial\phi = 1 + G_{K\phi} = \partial J/\partial K = 0 \ . \tag{10}$$

At such a point it may be impossible to solve uniquely for ϕ in terms of ψ. Fig. 2 shows $\partial\psi/\partial\phi$ corresponding to the case of Fig. 1. When $\varepsilon_1 = 2\varepsilon_2$ is increased to 1.4×10^{-4}, we get J and $\partial\psi/\partial\phi$ as shown in Figures 3 and 4, respectively. The anticipated zeros of $\partial\psi/\partial\phi$ appear in Fig. 4; however, the behavior of $\partial\psi/\partial\phi$ near transition is rather sensitive to the number of modes included. Judging from numerical integration of Hamilton's equations, we believe that the case of $\varepsilon_1 = 2\varepsilon_2 = 1.4 \times 10^{-4}$ is actually a little beyond transition.

As the transition to chaos is approached, it becomes more difficult to expand the mode set. If too many modes are included, convergence of the Newton iteration suffers, and there is little if any reduction in the residual perturbation beyond that obtained with about 100 modes. Nevertheless, with 100 modes the ratio of residual to original perturbation is small; even at $\varepsilon_1 = 2\varepsilon_2 = 1.2 \times 10^{-4}$ this ratio is 1.5×10^{-4}. This suggests that further canonical transformations, computed on progressively larger mode sets, would in fact yield an exact invariant torus. To date we have computed the second canonical transformation only in lowest order. The average absolute value of the torus distortion from the second transformation, divided by that from the first, varies from 2.8×10^{-6} at $\varepsilon_1 = 2\varepsilon_2 = 6 \times 10^{-5}$ to 4.1×10^{-3} at $\varepsilon_1 = 2\varepsilon_2 = 1.2 \times 10^{-4}$.

We conclude that the method provides a promising alternative to canonical perturbation theory and its modern variants. Unlike perturbation theory, its algebraic complexity does not increase as more accuracy is demanded, and the required computer programs are quite simple. The fact that the method is effective near chaotic regions is of great interest for applications. The generalization of (10) to higher dimensions, namely $\det(1 + G_{K\phi}) = 0$, may provide a useful criterion for the breakup of KAM surfaces in complicated systems of interest. We give an extended account of this work in Ref. 2.

REFERENCES

1. V. I. Arnold, "Mathematical Methods of Classical Mechanics", Springer, Berlin, 1978.
2. R.L. Warnock and R.D. Ruth, "Invariant Tori Through Direct Solution of the Hamilton-Jacobi Equation", SLAC-PUB-3865, LBL-21709, to be published.

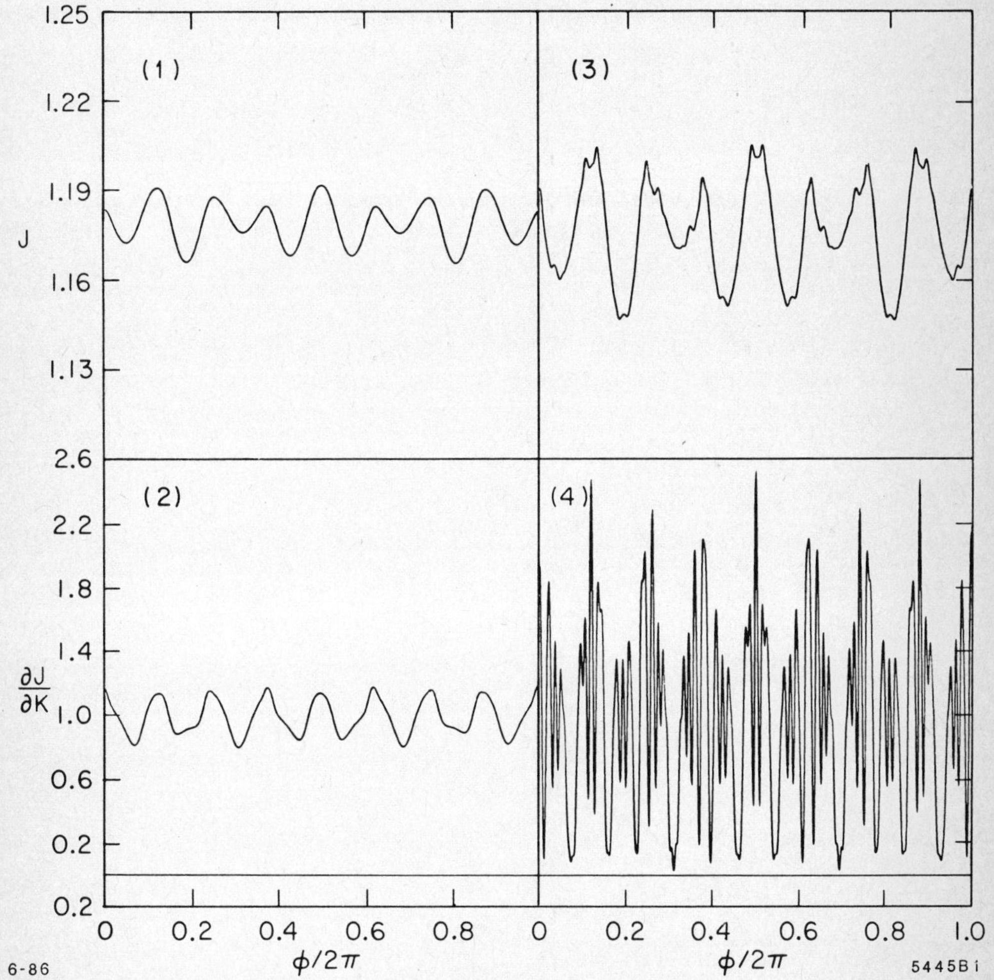

FRACTAL BASIN BOUNDARIES

Celso Grebogi, Edward Ott, and James A. Yorke
University of Maryland, College Park, MD

It is common for nonconservative systems to have more than one final time-asymptotic state (or "attractor"). In such cases, the attractor which eventually captures a given orbit is determined by the initial conditions of the orbit. The set of initial conditions which yield orbits going to a particular attractor is the basin of attraction for that attractor, and the boundary of the closure of that region is its basin boundary.

In order to illustrate the concepts of coexisting attractors, basins of attraction and basin boundaries, consider the simple case of a particle moving under the influence of friction in a potential V(x) as shown in Fig. 1(a). For almost any initial condition, the orbit will eventually come to rest at either of the two stable fixed points at $x = \pm x_0$. Figure 1(b) schematically depicts the phase space of the system and the basins of attraction of these two fixed point attractors. An initial condition chosen in the crosshatched region eventually comes to rest at $x = x_0$, while any initial condition in the blank region tends to $x = - x_0$. The boundary separating these basins is the smooth curve passing through the origin. It is a main point of this paper that a basin boundary need not be a smooth curve or surface. Indeed, for a wide variety of systems it is common for boundaries to exhibit a fractal structure and to be characterized by a noninteger dimension.

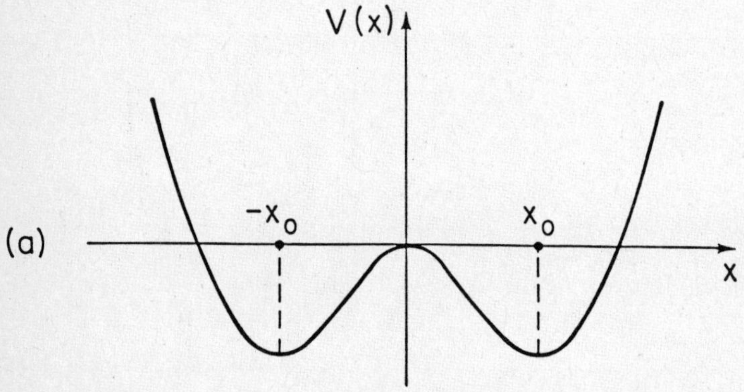

Figure 1(a)

Potential V(x) for a point particle moving in one dimension. With friction, almost every initial condition eventually comes to rest at one of the equilibrium points, x_0 or $-x_0$.

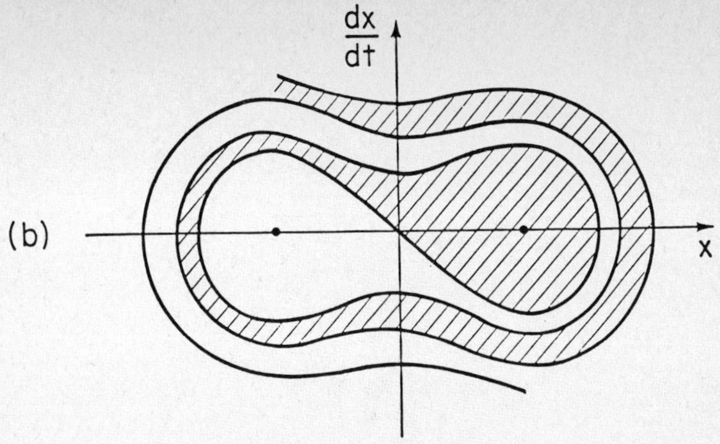

Figure 1(b)

Phase (velocity-position) space for the system in (a). The basin of attraction for x_0 (crosshatched) is separated from the basin of attraction for $-x_0$ (blank) by a smooth basin boundary curve.

The importance of studying the structure of basin boundaries is illustrated by the following example. Consider the simple two-dimensional phase space diagram schematically depicted in Fig. 2. There are two possible final states, or attractors, denoted by A and B. The region to the left (right) of the basin boundary Σ is the basin of attraction for attractor A (or B, respectively). In Fig. 2, points 1 and 2 represent two initial conditions with an uncertainty ε. While the orbit generated by initial condition 1 is definitely attracted to B, initial condition 2 is uncertain in that it may be attracted to either A or B. Now assume that initial conditions are chosen randomly with uniform distribution in the rectangular region shown in Fig. 2. We consider the fraction $f(\varepsilon)$ of initial conditions which are uncertain as to which attractor is approached when there is an initial error ε. For the simple case of Fig. 2, initial conditions within a strip of width 2ε centered on the boundary are uncertain; thus, $f(\varepsilon)$ is proportional to ε. It can, however, be demonstrated that systems with fractal boundaries are more sensitive to initial uncertainty and can obey[1-2]

$$f \sim \varepsilon^{\alpha}, \tag{1}$$

where α is <u>less than one</u>. We say that these systems possess <u>final state sensitivity</u>. We believe that many typical dynamical systems exhibit this behavior. In cases where α is significantly less than unity, a substantial reduction in the error in the initial condition, ε, produces only a relatively small decrease in the uncertainty of the final state as measured by f. Furthermore, it can be shown[2] that the scaling exponent α is the difference between the dimension of the phase space and the "capacity dimension" of the basin boundary. The increased sensitivity of final states to initial condition error when $\alpha < 1$ provides an important motivation for the study of fractal basin boundaries.

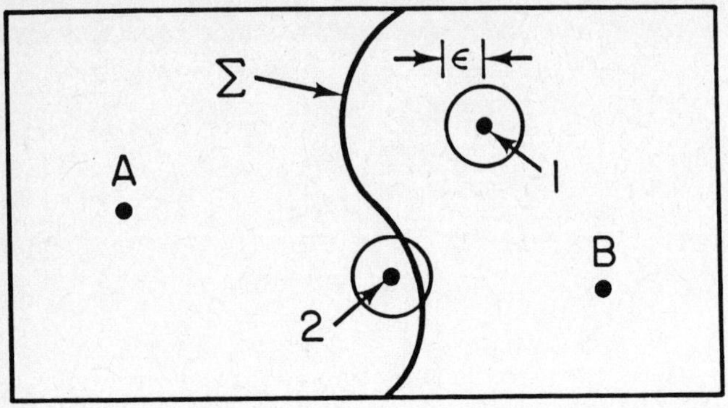

Figure 2

A schematic region of phase space divided by the basin boundary Σ into basins of attraction for the two attractors A and B. Points 1 and 2 represent two initial conditions with uncertainty ε.

In order to illustrate these concepts, consider the simple ordinary differential equation

$$\frac{d^2\theta}{dt^2} + \nu \frac{d\theta}{dt} + g \sin\theta = f \sin t , \qquad (2)$$

where ν, g, and f are parameters and θ is an angle variable (i.e., θ and $\theta \pm 2\pi$ are equivalent). This is just the equation of a simple pendulum with damping ν and a sinusoidally varying applied torque f sint. In addition to the pendulum, Eq. (2) also describes a number of other physical situations, including simple Josephson junction circuits, sliding change density waves, and voltage controlled phase locked loops.

We shall investigate the specific case where the damping, gravity and forcing parameters are set at $\nu = 0.2$, $g = 1.0$, and $f = 0.2$. For these values we numerically find that there are apparently only two attractors. The two attracting orbits, $\theta_-(t)$ and $\theta_+(t)$, represent solutions with average clockwise and counterclockwise rotation at the period of the forcing, $\theta_-(t + 2\pi) = \theta_-(t) - 2\pi$, $\theta_+(t + 2\pi) = \theta_+(t) + 2\pi$. (The existence of the $\theta_+(t)$ solution implies the existence of the $\theta_-(t)$ solution by the symmetry of Eq. (2), and vice versa.)

Figure 3 shows a representation of the basins of attraction for the θ_+ attractor (white region) and the θ_- attractor (black region). The figure is obtained by taking a grid of initial conditions, testing each initial condition on the grid to determine which attractor it goes to, and then plotting those which go to the θ_- attractor. (Since the grid is denser than the pixels used by the plotter, plotting the θ_- points blacks out the θ_- basin.)

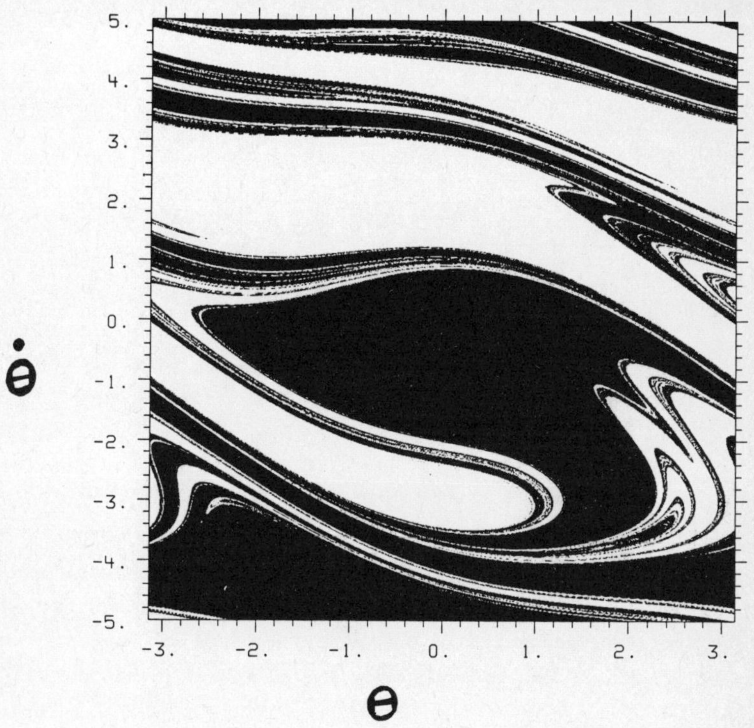

Figure 3

Basins of attraction for the forced damped pendulum.

In order to obtain the fractal dimension d we use the numerical technique discussed in Refs. 1 and 2. We use a random number generator to obtain N initial conditions uniformly distributed in the area shown in the figure. Then, for each of these randomly chosen initial conditions, $(\theta, d\theta/dt)$, we perturb the θ coordinate by a small amount ε, generating two subsidiary initial conditions, $(\theta + \varepsilon, d\theta/dt)$ and $(\theta - \varepsilon, d\theta/dt)$. We then integrate the pendulum equation forward in time to see which attractor each of the three initial conditions goes to. If all three do not go to the same attractor, then we say that the original initial condition is uncertain under ε perturbation of θ. We then repeat this procedure for each of the N randomly chosen initial conditions. Let $f(\varepsilon)$ be the fraction of initial conditions that are uncertain as $N \to \infty$. That is, if $N'(\varepsilon)$ of the initial conditions are uncertain, then

$$f(\varepsilon) = \lim_{N\to\infty} N'(\varepsilon)/N .$$

Numerically one uses a large N to obtain an approximation for $f(\varepsilon)$

$$f(\varepsilon) \cong N'(\varepsilon, S_i)/N ,$$

and the standard deviation in this approximation for f is $[N'(\varepsilon, S_i)]^{1/2}/N$. According to Eq. (1), to estimate d, we evaluate $f(\varepsilon)$ for a range of ε values, and then make a log-log plot of f versus ε. Thus, we can obtain an approximation to α and from it an approximation to d via $d = D - \alpha$. Here D is the dimension of the phase space; $D = 2$ for Fig. 3. We obtain $\alpha \cong 0.27$ and $d \cong 1.73$. Thus, for example, if an experimenter goes to a lot of trouble he might be able to increase his accuracy in specifying the system state by a factor of 10. This, however, buys him very little in the sense that the uncertain fraction of phase space only decreases by $10^{-0.27} \cong 0.5$, hardly worth the effort that is likely to be involved. Thus, we see that, from a practical point of view, fractal basin boundaries may effectively make reliable prediction of final states impossible.

We now indicate two recent lines of research that we have been pursuing on fractal basin boundaries:

A. Is the dimension of fractal basin boundaries unique? That is, if B is the boundary and S a subset of the phase space containing part of B in its interior, is the dimension of the set $S \cap B$ independent of S? The answer appears to be yes and no. In particular, there are common examples where the boundary dimension is unique (e.g., for Fig. 3), but there are also typical examples where it is not. In the latter case, it appears that the value of d can take on only a discrete number of values, say two values d_1 and d_2 (i.e., d will either be d_1 or d_2 depending on S). If this is the case, we also find that the regions of different dimension can be intertwined on arbitrarily fine scale. For example, if $d_1 > d_2$, then an S for which $d = d_1$ always has within it another set S' for which $d = d_2$, no matter how small the original S was.[3]

B. How do basin boundaries change as a system parameter is varied? We find that basin boundaries can undergo sudden discrete events as a system parameter is varied through a critical value. We call these basin boundary <u>metamorphoses</u>. In a metamorphosis a boundary can jump discontinuously in position, and its dimension also changes discontinuously.[4]

This work was supported by DOE, ONR, and DARPA (under NIMMP).

References

1. C. Grebogi, S. W. McDonald, E. Ott, and J. A. Yorke, Phys. Lett. 99A, 415 (1983).
2. S. W. McDonald, C. Grebogi, E. Ott, and J. A. Yorke, Physica 17D, 125 (1985).
3. C. Grebogi, E. Kostelich, E. Ott, and J. A. Yorke, to be published; and C. Grebogi, E. Ott, S.-T. Yang, and J. A. Yorke, to be published.
4. C. Grebogi, E. Ott, and J. A. Yorke, Phys. Rev. Lett. 56, 1011 (1986); and Physica D (to be published).

QUASI-PERIODIC SCHRÖDINGER EQUATIONS AND STRANGE NONCHAOTIC ATTRACTORS OF PENDULA AND JOSEPHSON JUNCTIONS

F. J. Romeiras*
Laboratory for Plasma and Fusion Energy Studies
University of Maryland, College Park, MD 20742, USA

Abstract: We discuss the existence and properties of strange nonchaotic attractors of differential equations forced at two incommensurate frequencies. One of the two equations we consider can be related to the Schrödinger equation; the other is the well-known model of the driven damped pendulum and of the current-driven resistively shunted Josephson junction. In particular, we show that these attractors are typical in the sense that they exist on a set of positive Lebesgue measure in parameter space, and also that they exhibit distinctive frequency spectra. These properties should make them experimentally observable.

Recently, attention has been called upon dissipative dynamical systems that typically exhibit a class of attractors that may be described as strange and nonchaotic [1]. The systems studied in Ref. [1] are particular maps of the form

$$\phi_{n+1} = g_\lambda(\phi_n,\theta_n), \qquad \theta_{n+1} = \theta_n + 2\pi\omega \ [\mathrm{mod}\ 2\pi] \ , \tag{1}$$

where g_λ is a 2π-periodic function of its second argument, ω is an irrational number, and λ is a parameter. Maps of this form can be obtained from ordinary differential equations forced at two incommensurate frequencies. It may therefore be conjectured that strange nonchaotic attractors will also be typical for these equations.

In order to verify this conjecture we have studied the equations

$$\frac{d\phi}{dt} = \cos\phi + \varepsilon\cos 2\phi + f(t), \tag{2}$$

$$\frac{1}{p}\frac{d^2\phi}{dt^2} + \frac{d\phi}{dt} + \sin\phi = f(t) \ , \tag{3}$$

where ε,ϕ are parameters and f is a two-frequency quasiperiodic function of t which was actually taken to be of the form $f(t) = K + V(\cos\omega_1 t + \cos\omega_2 t)$, where $\omega_1 = \frac{1}{2}(\sqrt{5} - 1)$ and $\omega_2 = 1$ were kept fixed while K,V were allowed to vary.

In the case $\varepsilon = 0$, Eq. (2) can be related by a transformation of both

---

*Permanent address: Centro de Electrodinamica, Instituto Superior Tecnico, 1096 Lisboa Codex, Portugal.

dependent and independent variables to the (time independent) Schrödinger equation; thus, the theory of the Schrödinger equation with quasiperiodic potential [2,3] can be used to aid in understanding Eq. (2) [4]. Equation (3) is the pendulum equation; it is also a useful model of the Josephson junction [5].

The following are the main results of our study:

(i) In the KV-plane (ε or p fixed) the diagram distinguishing negative and zero Lyapunov exponent (Λ) has a structure similar to the Arnold tongues of the circle map [See Fig. 1. All the figures shown in this paper refer to Eq. (2) with $\varepsilon = 0$; qualitatively similar figures are obtained for Eq. (2) with $\varepsilon \neq 0$ and Eq. (3).]

(ii) For a fixed value of V the curve giving the winding number W vs K is a "devil's staircase": a continuous, nondecreasing curve with a dense set of open intervals on which W is constant ($W = \ell\omega_1 + m\omega_2$ for Eq. (2), $\varepsilon = 0$; $W = \ell/n\ \omega_1 + m/n\ \omega_2$ for Eq. (2), $\varepsilon \neq 0$ and Eq. (3); ℓ,m,n are integers); between these intervals there is a Cantor set of apparently positive Lebesgue measure in which W increases with K. In the intervals the Lyapunov exponent Λ is always negative while in the Cantor set it is either negative (for small K) or zero (for large K). [See Fig. 2.]

(iii) In the intervals the equations exhibit two-frequency quasiperiodic attractors. In the Cantor set the equations exhibit either three-frequency quasiperiodic attractors (when $\Lambda = 0$) or strange nonchaotic attractors (when $\Lambda < 0$). The corresponding surface of section plots, obtained by plotting $\phi_n = \phi(t_n)$ [mod 2π] versus $\theta_n = \omega_1 t_n$ [mod 2π], where $t_n = t_0 + 2\pi n/\omega_2$, have qualitatively different characteristics. [See Figs. 3(A,B,C).]

(iv) The spectral distribution $N(\sigma)$, defined as the number of spectral components larger than some value σ, is different for the three types of attractor: $N(\sigma) \sim \sigma^{-\alpha}$ for strange nonchaotic attractors, $N(\sigma) \sim \log(1/\sigma)$ for two-frequency quasiperiodic attractors, and $N(\sigma) \sim \log^2(1/\sigma)$ for three-frequency quasiperiodic attractors. [See Fig. 4.]

(v) In the case of Eq. (3) with sufficiently small damping p, a transition from two-frequency quasiperiodic behavior to chaos is observed.

This work was supported by the U.S. Department of Energy, the Office of Naval Research, and the Portuguese Instituto Nacional de Investigacao Cientifica.

References

[1] C. Grebogi, E. Ott, S. Pelikan, and J. A. Yorke (1984) Physica 13D, 261.
[2] B. Simon (1982) Adv. Appl. Math. 3, 463.
[3] B. Souillard (1984) Phys. Rep. 103, 41.
[4] A. Bondeson, E. Ott, and T. M. Antonsen, Jr. (1985) Phys. Rev. Lett. 55, 2103.
[5] E. G. Gwinn and R. M. Westervelt (1985) Phys. Rev. Lett. 54, 1613.

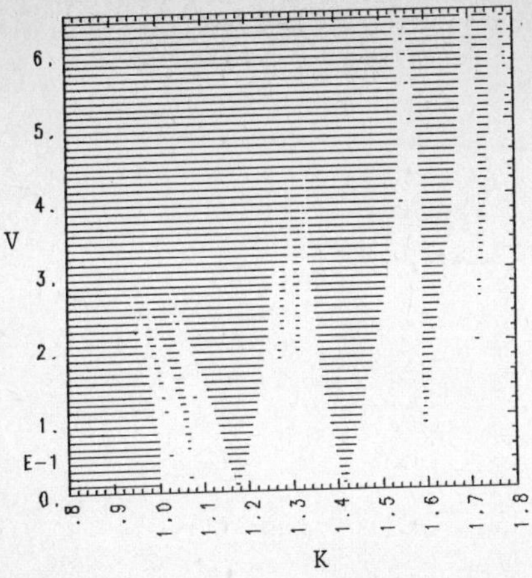

Fig. 1. Diagram of the KV-plane showing regions where $\Lambda<0$ (hatched) or $\Lambda=0$ (blank)

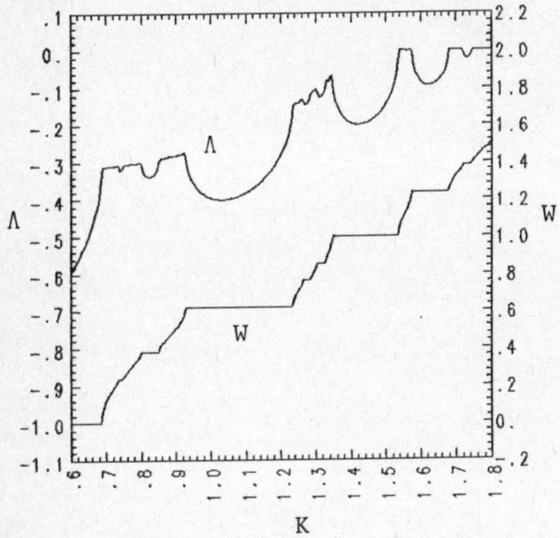

Fig. 2. Curves Λ vs. K and W vs. K for V=0.55.

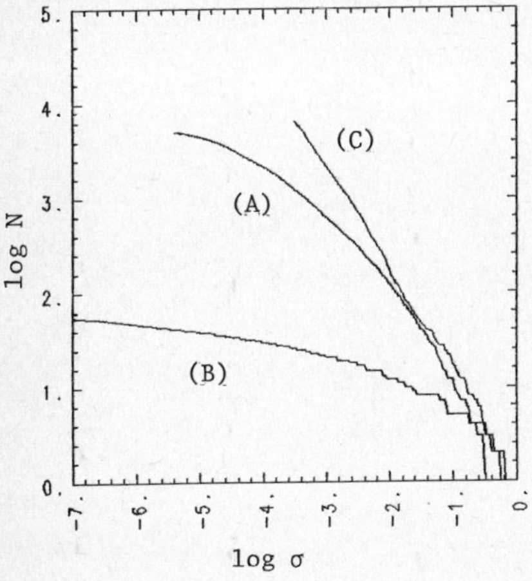

Fig. 4. Spectral distributions of the attractors of Fig. 3.

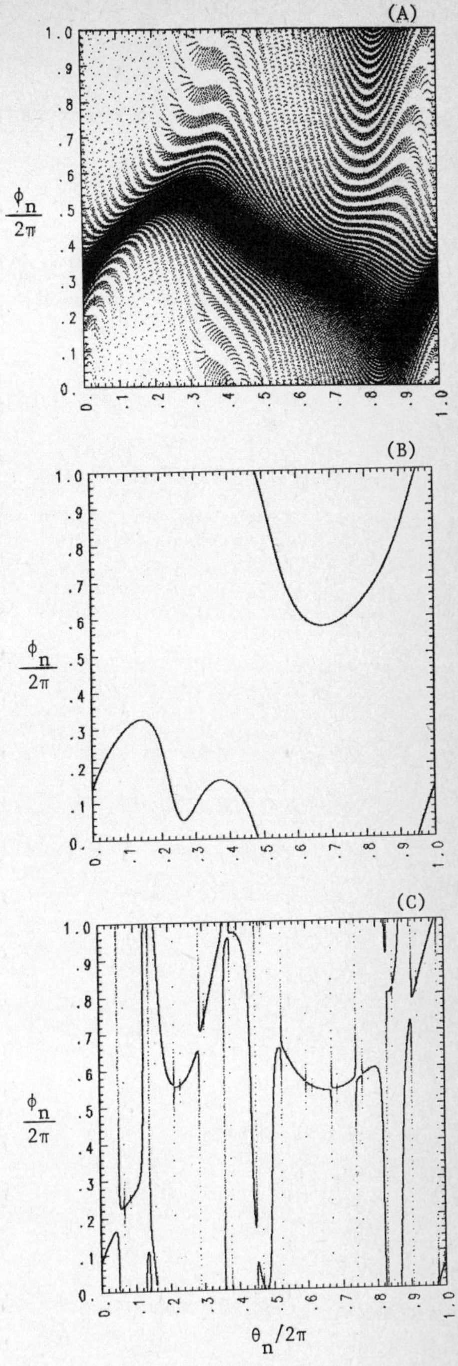

Fig. 3. Surface of section plots of (A) a three-frequency quasiperiodic attractor (K=1.54), (B) a two-frequency quasiperiodic attractor (K=1.39), (C) a strange nonchaotic attractor (K=1.34): V=0.55.

LONG-TIME CORRELATION IN CLASSICAL HAMILTONIAN SYSTEMS WITH TWO DEGREES OF FREEDOM

Haruyuki Irie
Department of Physics,
College of Science and Technology,
Nihon University, Tokyo 101, Japan
and
Tadatsugu Hatori
Plasma Physics Laboratory
Nagoya University, Nagoya 464, Japan

Recently, chaotic behavior for the conservative and the dissipative systems have been extensively studied in various fields. Long-time correlation for chaos is found for both systems. For the dissipative systems, e.g., the system described by the logistic map, the correlation function decays with power law at the critical point, which is the accumulating value of period doubling bifurcation and above which the orbit has a positive Lyapunov exponent. As for the conservative systems, several theoretical advances were made originating from stickiness to the island of a cycle or from self-similar structure of phase space (both for the chaotic orbit above the critical point)[1-7]. In this paper, it will be found theoretically in the classical Hamiltonian systems with two degrees of freedom that the correlation function for a orbit on and near the critical KAM surface, i.e., at and near the critical point for the orbit.

We consider the two-wave Hamiltonian

$$H = v^2/2 - M \cos x - P \cos[k(x-t)], \qquad (1)$$

which is studied by several authors[8,9]. Escande[9] has pointed out that (1) has an approximate self-similar structure of the Hamiltonian. That is, after killing the M-resonance or the P-resonance by use of a canonical transformation and using the primary two resonance approximation, he derived new Hamiltonian in new coordinates $(x',v',t';k',M',P')$, $H'=v'^2/2-M'\cos x'-P'\cos[k'(x'-t')]$, which has the same form as (1), where

$$\begin{aligned} &k'=(k+m)/(k+l),\ m=[z]-\lambda,\ l=[z]-1+\lambda, \\ &t'=\gamma t,\quad \gamma=(2\lambda-1)k/(k+m), \qquad (2) \\ &x'=(k+l)\theta-kt, \\ &v'=dx'/dt', \end{aligned}$$

with $\lambda=0(\delta z<1/2)$, $=1(1/2<\delta z)$, $z(=k/u-k+1\geqq 2)$ being the zoning number (or winding number), $u(=k/(k+z-1))$ the average velocity (or rotation number), $\delta z(=z-[z])$ and $[z]$ the fractional and integer part of the zoning number respectively, and θ the action variable in the Hamiltonian after killing the M-(or P-)resonance. M' and P' are somewhat complex functions of (z,k,M,P). Note that we observe the longer time behavior after this transformation because $|\gamma|$ is less than unity.

Next we introduce the velocity correlation function averaged over long time T

$$C(t;z,\underline{X}) \equiv \frac{1}{T}\int_0^T ds\ \delta v(s;z,\underline{X})\,\delta v(s+t;z,\underline{X}), \qquad (3)$$

where $\underline{X}=(k,M,P)$ and $\delta v(t;z,\underline{X})(=v(t;\underline{X})-u)$ is the deviation of velocity from its time average u. The dependence of the correlation function on the mean velocity u comes through the zoning number z defined below (2). Using the relation of old and new variables, we obtain

$$C(t;z,\underline{X})$$

$$= \beta^{-1} \frac{1}{T'} \int_0^{T'} ds' \delta v'(s';z',\underline{X}')\delta v'(s'+t';z',\underline{X}'),$$

$$= \beta^{-1} C'(t';z',\underline{X}'), \quad (4)$$

where

$$\delta v=[\gamma /(k+1)]\delta v',$$
$$\beta(z,k)=[(k+1)/\gamma]^2=[(k+1)\alpha(z,k)]^2, \quad (5)$$
$$\alpha(z,k)=1/|\gamma| = (k+m)/k > 1 .$$

In general, the form of function $C'(t')$ in the new coordinates is not equal to that in the original coordinates, $C(t)$. However, if self-similar structure exists, we can suppose both functions have equal form,

$$C(t;z,\underline{X}) = \beta^{-1} C(t/\alpha ;z',\underline{X}'), \quad (6)$$

where we used the fact that correlation function $C(t)$ is even with respect to t. Moreover, we obtain from (2) and (5)

$$\beta (z,k) = \alpha (z,k)^4 \rho (z,k)^2 , \quad (7)$$

with

$$\rho (a,k) = k/k' ,$$

so that

$$C(t;z,k,M,P) = \frac{1}{\alpha^4 \rho^2} C(\frac{t}{\alpha};z',\frac{k}{\rho},M',P') . \quad (8)$$

The two-wave renormalization transformation from (z,k,M,P) to (z',k',M',P') becomes chaotic map. Especially, the fractional part of the zoning number δz maps on itself and becomes chaotic, which reflects the chaotic sequence of quotients in the continued fraction representation of the zoning number. This one dimensional map has a denumerable set of unstable fixed points labeled by n and λ. The fixed points correspond to the zoning number whose continued fraction representation has a regular sequence. If we restrict ourselves to the map on the fixed points, we can discuss a orbit whose zoning number has a simple period in its continued fraction representation. Hereafter, we consider the restricted renormalization map $R(n,\lambda)$.

In the restricted map the zoning number is also fixed, so the map $R(n.\lambda)$ is on the three dimensional space $\underline{X}=(k,M,P)$. The map $R(n,\lambda)$ has an unstable fixed point $\underline{X}_*$ which has one unstable direction and two stable directions. The fixed point $\underline{X}_*$ corresponds to a critical KAM surface and depends on (n,λ) which fix the zoning number, so the location of the critical KAM surface.

We can solve the functional equation (8) for the orbit just on and in the vicinity of a critical KAM surface, i.e., just on and near the fixed point of the restricted renormalization map $R(n,\lambda)$. On critical KAM surfaces, the functional equation (8) becomes

$$C(t) = C(t/\alpha)/\alpha^4, \quad (9)$$

and its solution is $C(t)\propto 1/t^4$. The value of this exponent, 4, is independent of (n,λ), therefore the exponent of the power law decay of correlation for the orbit just on a critical KAM surface is independent of the location of the KAM surface.

In the vicinity of a critical KAM surface, assuming the correlation function has a form

$$C(t;\underline{X})=[S(\underline{X})-S(\underline{X}_*)]^q/t^p , \quad (10)$$

we obtain

$$p=\frac{d\ \ln\beta_*}{d\ \ln\alpha_*} = 2[1-\frac{k_*^2}{n-\lambda}], \qquad q=(4-p)\frac{\ln\alpha_*}{\ln\mu_*} . \quad (11)$$

In this case, the exponents p and q depend on $(n=[z],\lambda)$ and listed in Table 1 for some values.

	$\lambda = 0$		$\lambda = 1$	
[z]	p	q	p	q
2	0.000000	4.000000	1.236068	2.763932
3	0.868517	3.131483	1.464102	2.535898
4	1.236068	2.763932	1.582576	2.417424
5	1.431099	2.568901	1.656854	2.343146
6	1.549703	2.450297	1.708204	2.291796
7	1.628650	2.371350	1.745967	2.254033
8	1.684659	2.315341	1.774964	2.225036
9	1.726312	2.273688	1.797959	2.202041
10	1.758431	2.241569	1.816654	2.183346

Table 1

For more generic case, the renormalization map is chaotic and we can not say anymore.

In Hamiltonian systems with two degrees of freedom, the functional equation for the velocity correlation function is derived in the frame of two resonance approximation. By use of renormalization technique in the parameter space, the functional equation can be solved and the correlation function is found to decay with power law. We obtain the exponent on and near the critical KAM surfaces.

On the fixed point, i.e., on the critical KAM surface, we solve the functional equation for the correlation and show the correlation function decays with t^{-4}. The exponent is independent of the zoning number, i.e., the location of the KAM surface. Near the fixed points, using the tangent map of the renormalization we obtain the power law decay with the exponent being between 0 and 2. The value is determined according to the location of the KAM surfaces.

The power law decay comes from the self-similar structure of orbits in the phase space. The difference of exponents on and near a KAM surface may come from different self-similarities.

[1] J.D. Meiss, J.R. Cary, C. Grebogi, J.D. Crawford, A.N. Kaufman, and H.D.I. Abarbanel: Physica 6D, 375(1983).

[2] C.F.F. Karney: Physica 8D, 360(1983).

[3] B.V. Chirikov and D. L. Shepelyansky: Physica 13D, 395(1984).

[4] T. Hatori, T. Kamimura, and Y.H. Ichikawa, Physica D.

[5] T. Kohyama, Prog. Theor. Phys. Lett. 71, 1104(1984).

[6] Y. Aizawa, Prog. Theor. Phys. Lett. 71, 1419(1984).

[7] J.D. Meiss and E. Ott: Phys. Rev. Lett. 55, 2741(1985).

[8] A.B. Rechester and T.H. Stix: Phys. Rev. A19, 1656(1979).
A.B. Rechester and T.H. Stix: Phys. Rev. Lett. 36, 589(1976).

[9] D.F. Escande: Phys. Rep. 121, 165(1985).

Dimension density - an intensive measure of chaos in spatially extended turbulent systems

Thomas Kurz[a] and Gottfried Mayer-Kress

Center for Nonlinear Studies
Los Alamos National Laboratory, Los Alamos, New Mexico 87545

The determination of correlation dimensions by the Grassberger-Procaccia algorithm from an experimental time series has become a standard tool in the analysis of low dimensional chaotic systems [1]. Here we want to carry over this method to spatially extended systems which have a decaying spatial correlation. In these cases the total number of degrees of freedom or overall "dimension" grows with the size of the system. Then, in a finite-size system the dimension of the overall dynamics can be recovered already from a single point measurement, if the resolution is greater than some size-dependent treshold. Therefore, we expect that the measured dimension values will increase when smaller and smaller spatial structures are resolved. This feature is also observed in turbulence experiments [2]. Thus, the objective is to get an intensive (i.e., size-independent) measure which locally characterizes turbulent systems.

Following an idea of Y. Pomeau [3], we study the *dimension density* or density of degrees of freedom with the help of one-dimensional coupled map lattices described by a quantity $u(x_n, t_n)$ (see [4] for a review). When we consider a fixed resolution ϵ_0 then we expect that the dynamics $u(x_0, t)$ measured at one point x_0 will be influenced by the dynamics of points in a neighborhood $U(x_0, \epsilon_0)$. The dimension $D_2(x_0)$ measured for $u(x_0, t)$ should then be determined by the dynamics of $U(x_0, \epsilon_0)$.

The interdependence between two signals at x_1 and x_2 can be measured via the *two-point dimension* $D_2^{(2)}(x_1, x_2)$. It shall be defined as the correlation dimension of a combined time series with contributions from the two signals $u(x_1, t)$ and $u(x_2, t)$ (e.g., by interleaving the series or by adding them). If the dynamics at the two points is fully independent, we get $D_2^{(2)}(x_1, x_2) = D_2(x_1) + D_2(x_2)$ as one can easily see. On the other hand, we have $D_2^{(2)}(x_1, x_1) = D_2(x_1)$. Furthermore, we expect $D_2^{(2)}(x_1, x_1 + \Delta)$ to be a continuous, monotonically increasing function of the separation Δ. The dimension density $\rho(x_0)$ is then defined as the *rate of change* of the two-point dimension at x_0:

$$\rho(x_0) = \lim_{\Delta \to 0} \frac{D_2^{(2)}(x_0, x_0 + \Delta) - D_2(x_0)}{\Delta}. \tag{1}$$

The dynamics at each lattice point is generated by a "tent" map $h(u) = 1 - 2|u - 0.5|$. The coupling is diffusive with nearest-neighbor interaction:

$$u_{n+1}^{(i)} = h\left(u_n^{(i)} + \kappa(u_n^{(i-1)} - 2u_n^{(i)} + u_n^{(i+1)}) + \eta(i, n)\right) \tag{2}$$

where $u_n^{(i)}$ denotes the function value at the i-th lattice site at time step n, κ the coupling parameter and η an additional noise term. In our numerical computations we use a lattice of 100 maps with

[a] *Permanent address: Drittes Physikal. Institut, Universität Göttingen, D-3400 Göttingen, FRG.*

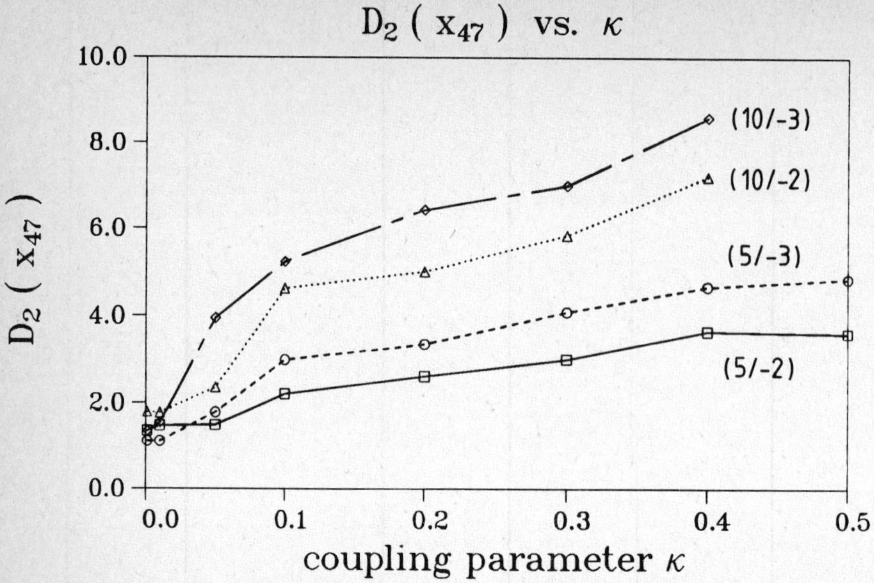

Fig. 1 Correlation dimension measured from single time series at reference point x_{47} at noise level $\eta = 10^{-6}$. The embedding dimension D and resolution of measurement $\log_2(\epsilon_0)$ are given in parentheses.

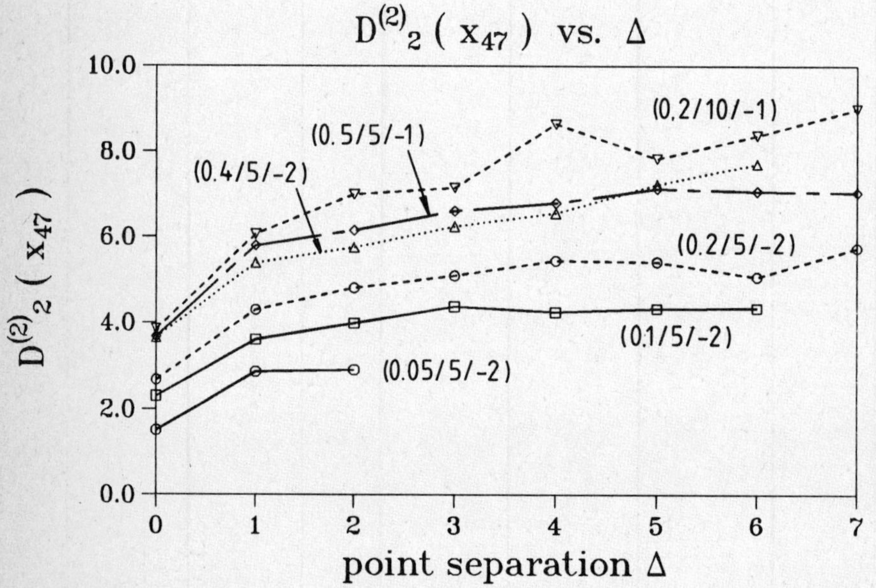

Fig. 2 Two-point dimension of points x_{47} and $x_{47+\Delta}$ as a function of separation Δ. The corresponding values of coupling strength κ, embedding dimension D and resolution $\log_2(\epsilon_0)$ are given in parentheses.

open boundary conditions and Gaussian distributed noise of maximally 0.1 % . The values of the coupling parameter κ vary between 10^{-4} and 0.5, beyond which the lattice gets unstable. While a single tent map causes numerical problems by hitting the origin after a few iterations, this has not been observed here. We compute $D_2^{(2)}$ of $\{u_n^{(i)}\}$ and $\{u_n^{(i+\Delta)}\}$ by analyzing the series of vectors $\binom{u_n^{(i)}}{u_n^{(i+\Delta)}}$ and embedding them in the corresponding $2D$-dimensional reconstructed phase space (typically we use about $n = 10000$ points and $D \leq 20$).

The numerical analysis of (2) shows that the standard methods have to be applied with great care. The correlation graphs have in general - as had to be expected - a slope which becomes steeper at small distances. Furthermore,they do not converge with increasing embedding dimension for small ϵ_0. Thus,the correlation dimension in a strict sense can not be proven to be finite. Therefore, we take the local slope of the correlation graphs for fixed embedding dimension and fixed ϵ_0 as a measure of correlation. It is well known that the method of Grassberger and Procaccia yields large errors for high correlation dimensions [1]. This imposes a limit on the accuracy of our results, especially if only small amounts of data are available. Since our model is spatially discrete, we replace the limes in definition (1) by a finite difference. To compute ρ at a reference point x_i, we determine the two quantities $D_2(x_i)$ and $D_2^{(2)}(x_i, x_i + \Delta)$, $\Delta = 0, 1, 2 \ldots$. Fig. 1 shows the slope of the correlation graphs for measurements of $D_2(x_{47})$ for various values of κ, embedding dimension D and resolution ϵ_0 (at a fixed noise level of 10^{-6}). It can be clearly seen that D_2 rises with increasing coupling strength, embedding dimension, and resolution (i.e., smaller ϵ_0). The given values for small ϵ_0 and high D are rather unreliable, however. In Fig. 2 we have plotted the two-point dimension versus point separation for different values of κ, D, and ϵ_0 and the same η. $D_2^{(2)}$ shows up as a generally increasing function of Δ which saturates for large distances at approximately $2D_2(x_{47})$. The fluctuations in the displayed curves can probably be explained with the large statistical errors of our measurements.

This result is in full agreement with the expectation that the two-point dimension should be the sum of the single-point dimensions if the two points are independent. Fig. 2 also shows that the two-point dimension increases more rapidly for small Δ than for larger ones. This makes a reliable estimation of the dimension density for high κ very difficult.
A more detailed discussion, which also includes the mutual information content between two points, separated in space and time,will be presented elsewhere [5].

ACKNOWLEDGEMENTS: One of us (G.M.-K.) would like to thank Y. Pomeau for stimulating discussions and U. Frisch and the Observatoire de Nice, where part of this paper was finished, for their hospitality.

REFERENCES:

[1] Proc. Pecos Conf. "Dimensions and Entropies in Chaotic Systems", G. Mayer-Kress (Ed.), Springer Series in Synergetics Vol. 32, Berlin,Heidelberg,New York 1986

[2] U. Frisch, "Fully Developed Turbulence: Where Do We Stand", in Proc. Peyresq Conf., World Publ. Comp. 1986

[3] Y. Pomeau,C.R. Acad. Sc. Paris, t. 300, Série II, no. 7, p. 239, 1985

[4] J.P. Crutchfield, K. Kaneko , to appear

[5] T. Kurz, G. Mayer-Kress, to be published

PROPERTIES OF THE MAXIMAL ATTRACTOR FOR THE LANDAU-LIFSCHITZ EQUATIONS

Tepper L. Gill*

Department of Physics
Virginia Polytechnic Institute and State University
Blacksburg, VA 24061

W. W. Zachary

Naval Research Laboratory
Washington, DC 20375-5000

The Landau-Lifschitz equations

$$\frac{d}{dt}\underset{\sim}{M} = \gamma\, \underset{\sim}{M} \times \underset{\sim}{H}^e - \frac{\lambda\gamma}{|\underset{\sim}{M}|}\underset{\sim}{M} \times (\underset{\sim}{M} \times \underset{\sim}{H}^e),\ \gamma, \lambda > 0, \tag{1}$$

describe the time-evolution of magnetization $\underset{\sim}{M}$ in classical ferromagnets and are of fundamental importance for the understanding of nonequilibrium magnetism. A general form for the effective magnetic field, $\underset{\sim}{H}^e$, is

$$\underset{\sim}{H}^e = \underset{\sim}{H}_{ext}(t) + \frac{C}{|\underset{\sim}{M}|}\Delta \underset{\sim}{M} - \frac{2A}{|\underset{\sim}{M}|}(\underset{\sim}{M} \cdot \underset{\sim}{n})\, \underset{\sim}{n} + \underset{\sim}{H}_d, \tag{2}$$

describing contributions from external fields and exchange, anisotropy, and demagnetization contributions, respectively. In the latter case we use the magnetostatic approximation together with the usual boundary conditions on the surface of the magnetic material to write [1]

$$\underset{\sim}{H}_d(\underset{\sim}{x}, t) = \underset{\sim}{\nabla}\int_\Omega \frac{\text{div}\, \underset{\sim}{M}(\underset{\sim}{x}', t)}{|\underset{\sim}{x} - \underset{\sim}{x}'|}\, dx' - \underset{\sim}{\nabla}\int_{\partial\Omega} \frac{M_n(\underset{\sim}{x}', t)}{|\underset{\sim}{x} - \underset{\sim}{x}'|}\, d\sigma(x'), \tag{3}$$

where Ω represents the volume of three-space occupied by the ferromagnet.

It is only in recent years that nonperturbative solutions of nonlinear versions of (1) have begun to be investigated. These studies deal with special cases in which the exchange term is replaced by a one-dimensional second derivative and special choices (usually zero) made for the other terms in (2). In these situations it was found that (1) is completely integrable [2]. On the other hand, for the general form (2) and (3) for $\underset{\sim}{H}^e$, it is expected that (1) has "chaotic" solutions which exhibit a strong dependence on the choice of initial conditions. This expectation is supported by numerical investigations of finite-dimensional models [3]. However, the only study of the system (1)-(3) without drastic simplifying assumptions known to the authors is the work of one of us [4], where the existence and uniqueness of periodic solutions is discussed. In the present work, we continue this program by proving that, under quite general conditions, the system (1)-(3) has attracting sets which are finite-dimensional in a suitable sense when $\lambda > 0$. Upper bounds are obtained for the Hausdorff and fractal dimensions of these sets.

The system (1)-(3) is very complicated and very little is known about it. Therefore, it is convenient to transform it to a form more amenable to analysis. To do this, we note that it follows from (1) that the absolute square of $\underset{\sim}{M}$ is conserved so that we can, after suitable normalization, consider $\underset{\sim}{M}$ on the unit two-sphere S^2. We then use the well-known result that the stereographic projection maps S^2 minus one point homeomorphically onto $\mathbb{R}^2$. Writing this transformation in terms of a complex quantity ψ,

$$\psi = \frac{m_x + i\, m_y}{1 + m_z},\ \underset{\sim}{m} = \frac{\underset{\sim}{M}}{|\underset{\sim}{M}|}, \tag{4}$$

where we have chosen the excluded point as the "south pole" $\underset{\sim}{m} = (0, 0, -1)$, we use the procedure described in [5] to write (1)-(3) in the form

$$\frac{d}{dt}\begin{pmatrix}\psi \\ \bar{\psi}\end{pmatrix} + \begin{pmatrix} B & O \\ O & B^* \end{pmatrix}\begin{pmatrix}\psi \\ \bar{\psi}\end{pmatrix} = \begin{pmatrix} f \\ \bar{f}\end{pmatrix}, \tag{5}$$

where $B = -i\,(1 - i\lambda)\,\gamma\, C\,\Delta$, $\gamma, C > 0$, and

$$f(\psi,\bar{\psi},t) = i\,(1 - i\lambda)\,\gamma\,\Bigg[-2C\bar{\psi}(\underset{\sim}{\nabla}\psi)^2(1 + |\psi|^2)^{-1}$$

$$+\frac{1}{2}h_+ - \frac{1}{2}h_-\psi^2 - h_z\psi$$

$$-2A\,(\psi\, n_- + \bar{\psi}\, n_+\,(1 - |\psi|^2)n_z)\,(1 + |\psi|^2)^{-1}\left(\frac{1}{2}n_+ - \frac{1}{2}n_-\psi^2 - n_z\psi\right)\Bigg].$$

We have grouped the external and demagnetization fields together in the quantity $\underset{\sim}{h} = \underset{\sim}{H}_{ext} + \underset{\sim}{H}_d$ and have used the combinations $h_\pm = h_x \pm ih_y$, $n_\pm = n_x \pm i\, n_y$. If $\partial\Omega$ is smooth, the negative Neumann Laplacian has compact resolvent on $X = L^2(\Omega)$ and so has a nonnegative discrete spectrum. Consequently, the spectrum $\sigma(B)$ of B is discrete with Re $\sigma(B) \geqslant 0$ so that B is a sectorial operator. Thus, (5) looks formally like a system of semilinear parabolic differential equations in a Hilbert space. Such systems have been extensively studied, and this is the advantage of (5) compared to the system (1)-(3). We investigate the properties of attractors for (5) and then relate these results to (1)-(3) via the stereographic projection (4).

We prove existence of a maximal attractor A for the system (1)-(3), i.e., a compact maximal invariant set which attracts all bounded sets, by first proving local existence and uniqueness of solutions of (5), and then extending these to global solutions for all nonnegative times and establishing a number of properties of the map

$$T(t) : \begin{pmatrix}\psi_o\\ \bar{\psi}_o\end{pmatrix} \mapsto \begin{pmatrix}\psi(t)\\ \bar{\psi}(t)\end{pmatrix}.$$

This is done in certain auxiliary spaces X^α defined, roughly speaking, as closures of the domain of B^α in suitable graph norms. The proofs involve the use of Sobolev imbedding theorems. We make the following assumptions.

I. Ω is a convex connected bounded open subset of $\mathbb{R}^3$ with C^∞ boundary $\partial\Omega$.

II. For the initial data of (5), assume that ψ_o, $\bar{\psi}_o$ are small in the sense that $\|\psi_0\|_\alpha = \|\bar{\psi}_0\|_\alpha < 1$, $\frac{7}{8} < \alpha < 1$, and $\psi_0, \psi_0 \in X^\eta$ for some $\eta > 2 + \alpha$.

III. The external fields can be decomposed in the form $\underset{\sim}{H}_{ext}(\underset{\sim}{x}, t) = \underset{\sim}{H}_s(\underset{\sim}{x}) + \underset{\sim}{H}_{rf}(\underset{\sim}{x}, t)$ into static and time-periodic parts, respectively, whose fourth derivatives are uniformly bounded on $\bar{\Omega}$, and H_{rf} are locally Hölder continuous and 2π-periodic in the time variable.

IV. We assume that

(1) if either $H_x^{ext} \neq 0$ or $H_y^{ext} \neq 0$, then $\max_{i=x,y} \|H_i^{ext}\|_{C^2(\bar{\Omega})}$ (which is time-independent) is sufficiently small, or

(2) if $H_x^{ext} = 0 = H_y^{ext}$, then $\|\psi_0\|_\alpha$ is sufficiently small.

Theorem 1. Assume I-IV *and suppose* $\frac{7}{8} < \alpha < 1$. *Then*:

(a) T(t) *is compact on* X^α *for* $t > 0$.

(b) *There exists a bounded absorbing set* $Y_0 \subset X^\alpha$, *i.e., for all bounded sets* $Y \subset X^\alpha$, *there exists* s > 0 *such that* $T(t)\, Y \subset Y_0$ *for* $t \geqslant$ s.

(c) *The maximal attractor* A *has the representation*

$$A = \bigcap_{\tau \geqslant 0} \overline{\bigcup_{s \geqslant \tau} T(s)\, Y_0} .$$

It can also be proved that the corresponding maximal attractor for the system (1)-(3) is $\tilde{A}$ = PA, where P denotes the transformation inverse to (4). Moreover, if A has finite topological dimension, then so does $\tilde{A}$ and these dimensions are equal.

We establish the finiteness of the Hausdorff and fractal dimensions of A, d_H (A) and d_F (A) respectively, by proving that T′(t), the Fréchet derivative of T (t), is a compact linear operator on X^α when $\frac{7}{8} < \alpha < 1$.

In order to obtain explicit upper bounds for d_H (A) and d_F (A), we employ a technique used by Constantin et al [6] in a study of the Navier-Stokes equations. We obtain:

Theorem 2. Assume I-IV with $\frac{7}{8} < \alpha < 1$. Then

$$d_H\,(A) \leqslant m \text{ and } d_F\,(A) \leqslant \frac{m + {}^{(m-1)}\left(\frac{k}{4C\lambda} - 1\right)}{m\left(m^{2/3} - \frac{k}{4C\lambda}\right)},$$

where $\frac{k}{4C\lambda} \in (1, m^{2/3})$ and $m \geqslant 2$ depend on the parameters in (2) as well as on a number of imbedding constants (and hence on the volume of Ω).

The second author wishes to thank C. Foiaş, H. Amann, S. Newhouse, J. A. Yorke, S. Antman, and R. Cawley for valuable remarks.

References

*On leave from the Department of Mathematics, Howard University, Washington, D.C. 20059.

1. A.I. Akhiezer, V.G. Bar'yakhtar, and S.V. Peletminskii, *Spin Waves*, North Holland, Amsterdam, 1968.
2. K. Nakamura and T. Sasada, Phys. Lett. *48A*, 321 (1974); M. Lakshmanan, Ibid. *61*A, 53 (1977).
3. K. Nakamura, S. Ohta, and K. Kawasaki, J. Phys. C*15*, L143 (1982); S. Ohta and K. Nakamura, Ibid, C*16*, L605 (1983); F. Waldner, D.R. Barberis, and H. Yamazaki, Phys. Rev. A*31*, 420 (1985); X.Y. Zhang and H. Suhl, Ibid. A*32*, 2530 (1985).
4. W.W. Zachary, "Executive and Uniqueness of Periodic Solutions of the Landau-Lifschitz Equations with Time-Periodic External Fields," submitted to Lett. Math. Phys.; "Some Approaches to the Study of Realistic Forms of the Landau-Lifschitz Equations," in 14th *International Colloqium on Group Theoretical Methods in Physics* (Y.M. Cho, ed.), World Scientific, Singapore, 1986, pp. 417-420.
5. M. Lakshmanan and K. Nakamura, Phys. Rev. Lett. *53*, 2497 (1984).
6. P. Constantin, C. Foias, and R. Temam, Memoirs Amer. Math. Soc., no. 314, 1985.

A COMPARISON OF THE FRACTAL DIMENSIONS OF CLOUD RADIANCE GRAPHS FOR TWO INFRARED COLOR BANDS

Charles Adler, Patricia H. Carter, and Robert Cawley

Naval Surface Weapons Center
White Oak, Silver Spring, MD 20903-5000

An experimental data file of infrared intensity from clouds is used to investigate the possibility of a fractal hypothesis. Intensity vs. angle is oscillatory and irregular, as would be the case for a coordinate of a chaotic process of a differentiable dynamical system. The obverse hypothesis is studied here in a brief follow-on to an earlier report, namely that the radiance dependence is non-differentiable. Measured values of fractal dimension for a few of the graphs are presented for the data in two color bands of the infrared, and the significance of the results is discussed.

The broad physical concept of a fractal is about fifteen or twenty years old now, and is due to Mandelbrot[1]. The first quantitative work on the fractal properties of clouds is that of Lovejoy[2], who combined infrared satellite cloud data with radar data for tropical rain areas to construct a plot of area vs. perimeter for perimeters ranging from 3 km to 3000 km. On a log-log plot the data fell along a remarkably straight line corresponding to a dimension value of 1.35. Evidence for scale lengths in the neighborhood of 1 km and less, depending upon wind speed, has been presented by Rys and Waldvogel[3] for hail clouds in severe convective storms.

In a recent publication[4] we have announced preliminary results of a new kind of experimental fractal dimension measurement, making use of a small part of a data file of infrared cloud radiance measurements. In contrast, the measurements of Refs. [2] and [3], together with many of the examples commonly discussed as illustrative of the natural occurrences of fractals, such as the shapes of islands, and most numerical simulation measurements involving fractal kinetic aggregations, all have been purely geometric in character. But there is another important way in which fractals can appear in physics, noted also by Mandelbrot[5], namely as the graph of a process, $t \to (t,V(t))$, which is a representation for the variation of a dynamical or physical quantity. See also the recent experiment of Allain and Cloitre[6]. The problem of measuring the dimension of the graph of V is more subtle since the units of t and V are no longer the same, and the dimension, in consequence, must obey an invariance against independent transformations of the scales of both. This property was proved in[4] for the infinite resolution limit of the graph covering procedure adopted for the numerical calculations reported there. From a theoretical point of view, the dimension of a compact set is the same as the dimension of the image of the set under any transformation that is differentiable and has differentiable inverse. Nevertheless, based on numerical studies of examples having, in the limit, known values of fractal dimension, more specifically of capacity D, we have developed significant improvements in covering and counting procedures which give much better values for dimension measurements.

The data we discuss here were taken under the Navy's background measurement and analysis program (BMAP), at Montauk Point, Long Island, in August 1983 and at Bedford, Massachusetts, in September 1984. The sensor consisted of two bore-sighted telescopes, one recording 3.8-5.0 μm (midwave) radiation and the other 7.3-11.9 μm (longwave) radiation. Each telescope contained a vertical focal plane array of 16 detectors, with each detector convering an instantaneous field of view (IFOV) of 0.33 mr x 0.33 mr. Azimuthal scanning was accomplished by means of a rotating mirror, covering about 2.2° at 36°/sec. The dwell time was 0.33 mr ÷ 36°/sec = 0.53 ms; the data were over-sampled, at the rate of 3.44 samples per dwell, giving a total of 61 ms for a full (400 point) scan of 38 mr. The total field of view of a single frame 16-channel scan was thus 38 mr x 5.3 mr; the angular diameter of the sun, for comparison, is about 9 mr.

A front panel switch permitted operation in either of two modes: (i) alternating, in which midwave frames, at one second intervals, were interspersed with longwave frames at the half-second marks, and (ii) 8/8, in which the first eight channels of every frame were longwave and the last eight channels were midwave, taken at the same time and at almost exactly equal, corresponding angles of elevations--the small discrepancy is due to a vertical offset of the two telescopes, which was about 10 cm.

In Figure 1 we show intensity vs. azimuthal angle for two corresponding elevations of a single sample frame in the 8/8-mode, taken at Montauk Point. The leading parts of the traces are from blue sky, and the first 50 points were discarded for the dimension measurements, which left the 320 usable points. The longwave tracing shows undershoot and droop distortions caused by quasi-dc response of the data collection circuits. This was the result of a 0.5-1000 Hz

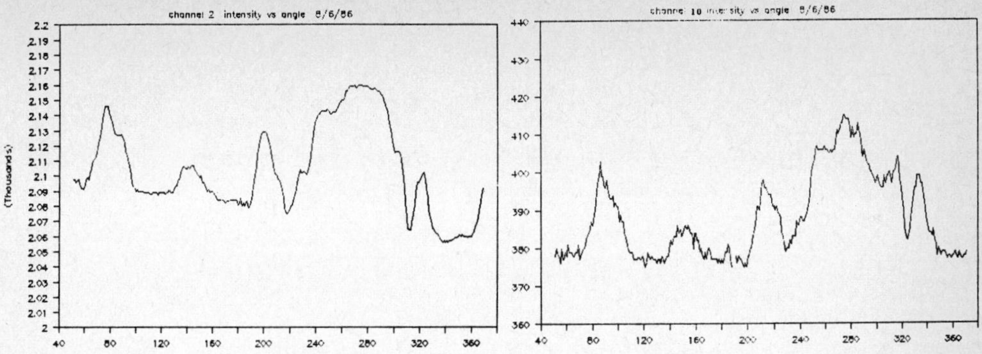

Figure 1. Intensity vs. azimuth for sample longwave (channel 2, left plot) and midwave (channel 10, right plot) color bands from data taken at Montauk Point. The curves represent data recorded simultaneously.

bandpass filter necessitated by the presence of 1/f-noise from the HgCdTe detector; the InSb detector used for the midwave radiance did not suffer the same disease. A gross similarity of the long wave and midwave radiance traces is evident, despite the generally different physical processes responsible. This is further reflected in the correspondence of dimension variations with elevation shown in Fig. 2.

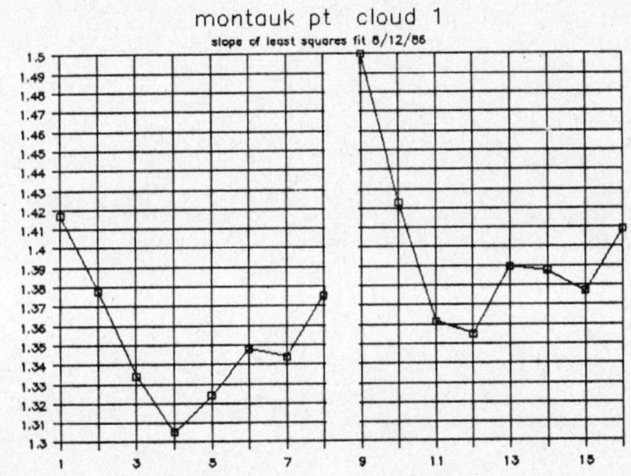

Figure 2. Capacities of radiance graphs for channels 1 to 16 of an 8/8-mode frame from Montauk Point. Channels 1 to 8 are longwave and channels 9 to 16 midwave, at the same angles of elevation. Graphs for channels 2 and 10 are shown in Fig.1

In Figure 3 we show sample Bedford intensity plots for channel 2 of frames 1, 3, 5, 7 and 2, 4, 6, 8 respectively mid- and long-wave, for an experimental run in the alternating mode. The striking difference between the two sets of curves is not due to the 0.5 sec delay between neighboring even and odd frames, for the variations observed with time, from second to second, for either long- or mid-wave separately, obviously is gradual, not abrupt. The longwave and midwave tracings typically were distinctly different in appearance at proximate times. Other Bedford data in the 8/8 mode show similar differences. Physically, the longwave radiance is dominated by thermal emissions;

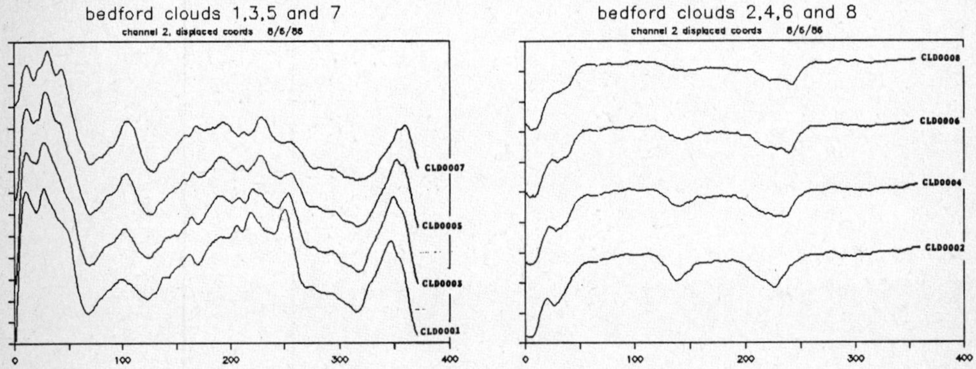

Figure 3. Intensity vs. azimuth for two sample longwave and midwave color bands from data taken at Bedford. The plots on the left are for channel 2 of frames 1, 3, 5, 7 (midwave) and those on the right are for frames 2, 4, 6, 8 (longwave). The frame 2 data were recorded 0.5 seconds after the frame 1 data, frame 3 is 0.5 seconds after frame 2, etc.

the black body peak for 300°K, for example, is at 9.9 μm, right in the middle of the 8-12 μm window. Solar scattering, on the other hand, is approximately comparable to thermal emissions in the midwave, 3-5 μm, region. Figure 4 shows the channel number dependence of the measured dimensions for frames 1 and 2. The value for the longwave channel 1 measurement is missing since detector no. 1 was dead. Despite the differences apparent in Figure 3, the curves of dimensions vs. channel number are approximately correspondent, being merely translated relative to one another just like the results shown in Figure 2.

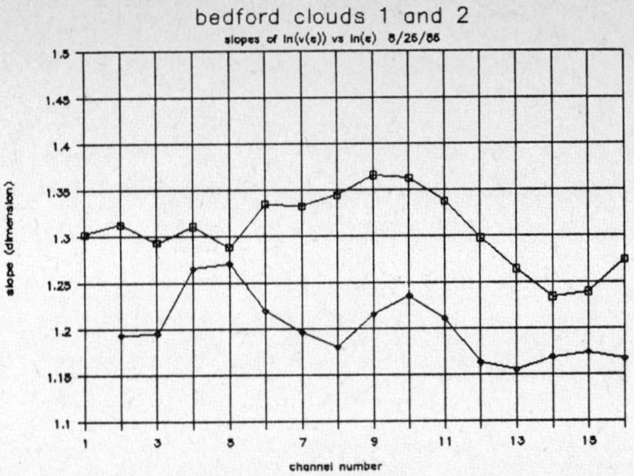

Figure 4. Capacities of radiance graphs for channels 1 to 16 of frames 1 and 2 of an alternating mode run at Bedford. Frame 1 is midwave and frame 2 longwave; the graphs for channel two are shown in Figure 3. The correspondences of channel number to elevation angle are the same for both frames.

Making use of time-dependences of measured dimension values, it is possible to detect the cloud motion correctly, as determined visually from an accompanying video, thereby lending broad and stronger support to a fractal hypothesis for the observed cloud radiance behavior. Detailed accounts of these results, as well as analyses of the numerical issues of the measurements, will be presented later.

The longwave dimension values are somewhat lower than midwave values, no doubt owing to relatively reduced effects of solar scattering and the dominance of blackbody equilibrium effects.

Acknowledgments

It is a pleasure to thank Bernie Kessler for providing us with the data and for his expert technical assistance in discussions about the experiment. This work was supported by ONR, the Naval Air Systems Command, the NSWC Independent Research Program and DARPA.

References

[1] B. Mandelbrot, "The fractal geometry of nature," W. H. Freeman, New York, 1977.

[2] S. Lovejoy, Science 216, 185 (1982). See also, _____, "The

statistical characterization of rain areas in terms of fractals," in Procs. 10th Conf. on Radar Meteorology, A.M.S., Boston, 1981.

[3] Franz S. Rys and A. Waldvogel, Phys. Rev. Lett. 56, 784 (1986).

[4] Patricia H. Carter, Robert Cawley, A. Lewis Licht, M. Susan Melnik and James A. Yorke, "Dimension measurements from cloud radiance," in "Dimensions and extropies in chaotic systems," G. Mayer-Kress, ed., Synergetics Series, Springer-Verlag, Berlin, 1986.

[5] B. Mandelbrot, Physica Scripta 32, 257 (1985).

[6] C. Allain and M. Cloitre, Phys. Rev. B33, 3566 (1986).

THE GENEALOGY OF PERIODIC TRAJECTORIES

M. A. M. de Aguiar, C. P. Malta, E. J. V. de Passos
Instituto de Física, Universidade de São Paulo
C.P. 20516, 01000 São Paulo, SP, Brazil

M. Baranger
MIT, Cambridge, MA 02139

and K. T. R. Davies
Oak Ridge National Laboratory
P.O. Box X, Oak Ridge, TN 37830

We have investigated numerically the periodic solutions of non-integrable classical Hamiltonian systems with two degrees of freedom. We obtained extensive numerical data and this was possible due to the development of new computational methods that are very fast and work very well independently of the periodic trajectory being stable or unstable. Our motivation for the present investigation was primarily to understand quantization as, in the study of many body nuclear systems, there are approximate methods that provide a classical description of collective modes. Therefore, quantization is required in order to describe bound-state spectrum or fluctuations. As is well known(1), the periodic trajectories form one-parameter families. Two convenient labelling parameters for a particular trajectory are its energy E or its period T. Most of our data are presented in the form E-T plots where each of the periodic families is represented by a line. The E-T plot provides a signature of the Hamiltonian H being studied, hence it is important to study the topology of the E-T plot.

The periodic trajectory is characterized by a matrix M called monodromy matrix(2). For two dimensions this is a 4x4 matrix having two unit eigenvalues. The other two have unit product. The trajectory is stable if the eigenvalues have magnitude 1, therefore the trajectory is stable if the trace of M lies between 0 and 4.

The topology of the E-T plot is determined by its branchings. At an isochronous branching, M has four unit eigenvalues and $\mathrm{Tr}\,M = 4$. At a period doubling branching, two eigenvalues must be -1 and $\mathrm{Tr}M=0$. Period-Triplings occur for $\mathrm{Tr}\,M=1$, period-quadruplings for $\mathrm{Tr}\,M = 2$, etc.. Obviously, the E-T plot at large T can become very dense and complicated, but it will never be everywhere dense. The families are discrete and the most important ones are found at small T. We present here the results for the following Hamiltonian

$$H = \frac{1}{2}(p_x^2 + p_y^2) + \frac{1}{2}x^2 + \frac{3}{2}y^2 - x^2y + \frac{1}{12}x^4 \quad .$$

It was chosen as a less symmetrical form of the Hénon-Heiles potential[3]. There are half dozen Hamiltonians under investigation but we do not expect the topological behaviour to depend essentially on the Hamiltonian used, thus we shall not mention results for the other Hamiltonians. Two of these families are obtained immediately: they correspond to harmonic oscillations of small amplitudes, around the equilibrium point, in the vertical and in the horizontal directions (normal modes). The vertical (V) family appears as a vertical line in the E-T plot (see fig. 1) because the potential V(0,y) is purely quadratic. The horizontal oscillation gives rise to the horizontal (H) family which has the period varying with amplitude. At low energy the H family is called boomerang (B) family (see fig. 1). The continuous line which represents a family on the E-T plot cannot begin or end except for two reasons: (1) it branches upon another family; (2) it becomes the family of small oscillations about an equilibrium point.

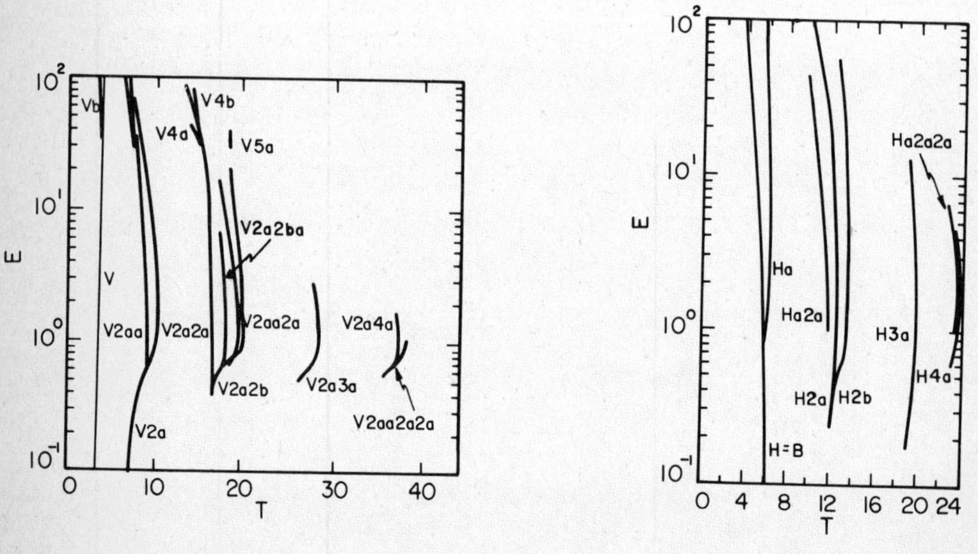

Figure 1 - V and H families and their branchings indicated by lower-case roman letters. The integer n before the letter indicates period n-pling (n=1 is omitted).

There are families that do not terminate anywhere: either they go on to infinity or they form closed curves (see fig. 2). The families S starting at the saddle points are always unstable with $TrM \to 4$ as $E \to \infty$. A family can exhibit more than one region of stability.

This happens for both the H and the V families.

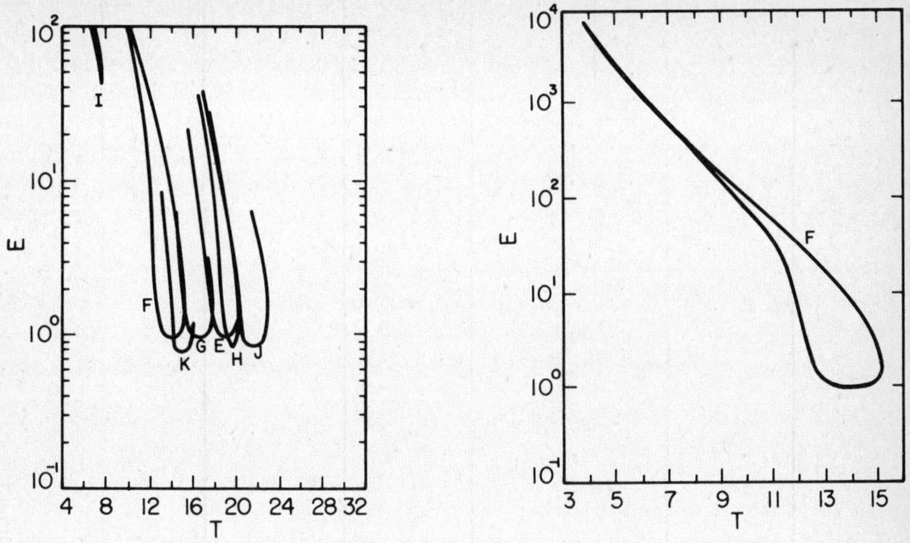

Figure 2 - Families that form closed curves.

Our main results are:

1) The vertical and the horizontal families are connected. For the potential above this connection happens via an isochronous branching (Vb = Hb) at E very high (23431).

2) At the points where TrM is tangent to zero or 4 there is a double branching, one stable and one unstable (see fig. 1).

3) For all the families that form closed curves TrM = 4 at the points where $\frac{dE}{dT} = 0$ and at these points there is no branching, the main trajectory switching simply from stable to unstable.

4) Period n-pling ($n \geqslant 3$) we believe gives rise to two distinct families, one stable and the other one unstable.

5) When two distinct families emerge at a branch point one of them is a libration and the other one is a rotation.

REFERENCES

1 - V. Arnold, Lés Méthodes Mathématiques de La Mécanique Classique, MIR, 1976.
2 - L. Pontriaguine, Équations Différentielles Ordinaires, MIR, 1975.
3 - M. Hénon and C. Heiles, Astronomical Journal 69 (1964), 73.

PERTURBATION THEORY AND THE SINGLE SEXTUPOLE

Leo Michelotti
Fermi National Accelerator Laboratory
Batavia, Illinois 60120

Perturbation theory plays at best an equivocal role in studying the behavior of a nonlinear dynamical system. Even the simplest systems possess complicated orbits, which makes the validity of a perturbative expansion doubtful. From a practical standpoint, however, convergence is seldom the real issue; for example, renormalized perturbative QED is certainly not assured to converge, yet its successes have been overwhelming. Rather, one would like to know whether the *first few* low order terms model the system's behavior "reasonably well" within the phase space region of interest. We shall consider this question for a very simple problem from accelerator theory: the single thin sextupole in one degree of freedom.

The design of a circular accelerator begins with the specification of a *central orbit.* Particles are constrained to remain close to the central orbit, to first order, by inserting quadrupole magnets to act as "lenses" which keep the beam focussed. Hill's equation describes the linearized transverse dynamics.

$$\frac{d^2x}{d\theta^2} + K(\theta)x = 0 \quad . \tag{1}$$

Here, x represents the horizontal, let us say, displacement of a particle from the central orbit; θ, the "independent variable," is an angular coordinate which labels points on the central orbit; K is a periodic function related to the transverse gradients of the quadrupoles' magnetic fields. The two independent Floquet solutions of this equation can be written

$$x(\theta) = \sqrt{\beta(\theta)}\, exp(\pm i\psi(\theta)) \tag{2}$$

where the *lattice functions* ψ and β are related by $d\psi = ds/\beta = Rd\theta/\beta$, s being arclength along the central orbit. [1] The function β is periodic, but ψ is not. Instead, it obeys the condition $\psi(\theta + 2\pi) = \psi(\theta) + 2\pi\nu$, where ν is the (horizontal) *tune* of the machine. It counts the number of times a particle oscillates about the central orbit in traversing the accelerator once.

Magnetic fields which vary nonlinearly with x are added to the accelerator either deliberately—to perform resonance extraction or to control certain dynamical effects, such as chromaticity—or accidentally—simply because we cannot build perfect dipoles and quadrupoles. In particular, inserting sextupole fields into the accelerator produces a force quadratic in the displacement variable. Eq. (1) then becomes

$$\frac{d^2x}{d\theta^2} + K(\theta)x + S(\theta)x^2 = 0 \tag{3}$$

where S is a periodic function which characterizes the strength and distribution of the sextupoles.

Now consider the case in which a single thin sextupole is inserted into the ring. "Thin" means that $S(\theta) \propto \delta(\theta)$, which in practice means that x remains unchanged in passing through the sextupole while a suitably defined "momentum", p, undergoes a kick, Δp, given by

$$\Delta p \;=\; -\lambda x^2 \tag{4}$$

$$\lambda \;\equiv\; -e\beta B'' l/2p_3 \tag{5}$$

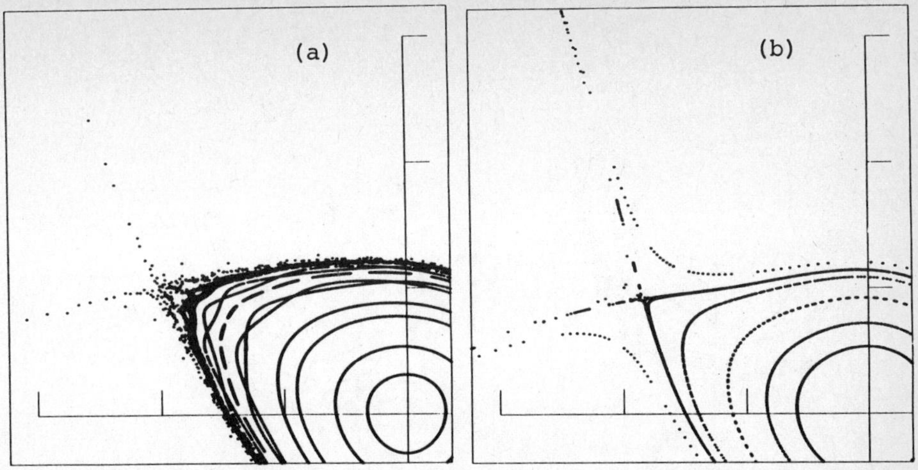

Figure 1: (a) Orbits of the sextupole mapping for $\nu = 0.15$. (b) Second order perturbation theoretic calculation of the stability boundary.

where e is the charge on a proton (the particle), p_3 is its longitudinal momentum, B'' is the (average) second derivative of the sextupole field, l is the length of the sextupole, and β, defined in Eq. (2), is evaluated at the position of the sextupole. The full Poincaré map then concatenates this with a phase space rotation through $2\pi\nu$, representing passage through the rest of the accelerator

$$\begin{pmatrix} x \\ p \end{pmatrix} \Leftarrow \begin{pmatrix} \cos 2\pi\nu & \sin 2\pi\nu \\ -\sin 2\pi\nu & \cos 2\pi\nu \end{pmatrix} \begin{pmatrix} x \\ p - \lambda x^2 \end{pmatrix} \quad . \tag{6}$$

We can set $\lambda \equiv 1$ without loss of generality by rescaling, $x \to x/\lambda$ and $p \to p/\lambda$. This is in keeping with Hénon's observation that any area preserving quadratic map can be put into a one-parameter form. [2]

We have studied this mapping in the tune range $0 < \nu < \frac{1}{2}$; Figures 1a and 2a illustrate a few orbits at the tunes $\nu = 0.15$, $\nu = 0.29$ respectively. The tic marks on the axes are separated by 0.5. The general features in these drawings are not surprising: (i) near the origin there are smooth (on the scale of the observations) KAM tori; (ii) as one gets farther in phase space a structure of islands and sub-islands develops; (iii) which finally breaks into a chaotic sea, nonetheless contains stable islands of its own.

It is hopeless to expect perturbation theory to say much about the rich fine-scale structure—which the figures exhibit rather poorly—of this mapping; it is, after all, the existence of this structure which makes us uneasy about the meaning of a perturbative expansion. However, the principal feature of interest is the *stability boundary*, and perturbation theory does enable us to calculate its position and shape surprisingly well. Figures 1b and 2b illustrate calculations done by applying Deprit's algorithm to the Hamiltonian associated with Eq.(6). [3] The dynamics in Figure 1 is dominated by a first order integer resonance, which must be put explicitly into the new Hamiltonian. With the appropriate distortion, also given by the perturbation expansion, *the separatrix of the resonance then can be associated with the stability boundary of the exact mapping.* By making this identification, we can compute the location of the latter to better than 10%.

Figure 2 is a remarkable case. Its most dramatic feature is the very large 2/7 resonance

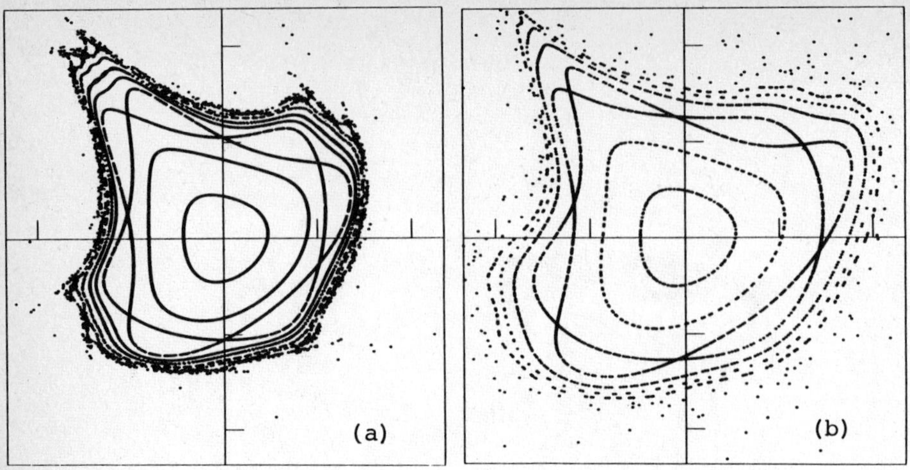

Figure 2: Same as Figure 1, but with $\nu = 0.29$.

which produces a system of seven islands. Seventh "order" resonances (i.e., resonances with winding number seven) should not appear until fifth order in the perturbation expansion, while the island chain is certainly more than a fifth order effect. In fact it is due to an *interference* between the 1/3 resonance, which appears at first order in the perturbation expansion, and the 1/4 resonance, which appears at second order. This is confirmed in Figure 2b which shows the perturbation theoretic prediction when those two resonances are explicitly taken into account.

Carrying out similar comparisons at other values of the tune,we have found that second order perturbation calculations can usually predict the stability boundary within 5-15% accuracy when the dominant resonances are put into the new Hamiltonian.

Of course, the real situation is far more complicated. At the minimum we must include both transverse directions in any realistic analysis of sextupole effects. This would change the horizontal force to something proportional to $x_1^2 - x_2^2$, where x_1 and x_2 represent the horizontal and vertical displacements from the central orbit, while introducing a vertical force proportional to $x_1 x_2$. The dynamics are in fact derivable from a Hamiltonian with a potential term of the form $g(\theta)(x_1^3 - 3x_1x_2^2)$. If g were constant we would recapture the Hénon-Heiles potential. In addition, more than sextupoles must be taken into account: octupoles produce cubic forces, decapoles produce quartic forces, and so forth. The "general" Hamiltonian representing transverse dynamics of a storage ring will possess harmonic polynomials in the transverse variables multiplying periodic functions of θ. The analysis of such Hamiltonian systems is a major challenge for accelerator theorists.

REFERENCES

1. Courant,E.D. and H.S.Snyder, Annals of Physics **3**(1),1(1958).
2. Hénon,M.,Quart. App. Math. **27**,291(1969).
3. Deprit, A.,Cel. Mech. **1**,12(1969).

STOCHASTIC INSTABILITY IN A SYSTEM WITH TWO DEGREES OF FREEDOM

K. Hizanidis and C. Menyuk
University of Maryland
Astronomy Program
College Park, MD 20742
USA

The physics of the interaction between charged particles and electromagnetic waves in the presence of an external magnetic field is a very important aspect in the study of several problems in plasma physics such as plasma heating and particle acceleration in the laboratory or in space. In this paper we consider the interaction between relativistic electrons and two externally imposed electromagnetic waves (1 and 2) of frequencies and wavenumbers (ω_1, ω_2), (k_1, k_2) respectively. We focus on the derivation of an approximate Hamiltonian which bears the resonance interaction to lowest order in the amplitude of the second (perturbing) wave.

We assume that the waves are circularly polarized. The total magnetic potential is given by: $\vec{A} = \hat{e}_x(A_1 \sin\psi_1 + A_2 \sin\psi_2) + \hat{e}_y(A_1 \cos\psi_1 + A_2 \cos\psi_2 + xB_o)$ where B_o is the external magnetic field, assumed constant, and $\psi_i = k_i z - \omega_i t$, $i = 1, 2$.

The Hamiltonian of the interaction is now:

$$H = mc^2 \left[1 + \left(\frac{P_x}{mc} + \alpha_1 \sin\psi_1 + \alpha_2 \sin\psi_2\right)^2 + \left(\frac{P_y}{mc} + \alpha_1 \cos\psi_1 + \alpha_2 \cos\psi_2 + \frac{x\Omega}{c}\right)^2 + \frac{P_z^2}{m^2c^2}\right]^{1/2} \qquad (1)$$

where P_x, P_y, P_z are the canonical momenta of the electron, $\alpha_i = eA_i/mc^2$, $i = 1, 2$ and $\Omega = eB_o/mc$ is the gyrofrequency.

In the presence of only one wave, Eq. (1) corresponds to an integrable system and has been thoroughly studied in the past.[1] However, integrating the equation of motion can only be formally done and, in the general case, the integration renders the energy as an implicit function of time. The presence of the second wave destroys the integrability by introducing an additional degree of freedom. The three degrees of freedom, that Eq. (1) apparently exhibits, can be reduced to two by three successive canonical transformations: one by employing the generating function $F = P_x'(X + c/\Omega\, P_y/mc)$, the second by introducing the action-angle variables (J, θ), $P_x = (2m\Omega J)^{1/2} \cos\theta$, $X = (2J/m\Omega)^{1/2} \sin\theta$, and the third $F_1 = (k_1 z - \omega_1 t + \theta) I_1 + (k_2 z - \omega_2 t + \theta) I_2$, to eliminate the time dependence. These successive transformations render a new Hamiltonian (we use the same symbol for convenience):

$$H = mc^2\{1 + \alpha_1^2 + \alpha_2^2 + (\frac{k_1I_1+k_2I_2}{mc})^2 + \frac{2\Omega(I_1+I_2)}{mc^2} + 2\alpha_1\alpha_2\cos(\theta_1-\theta_2)$$

$$+ 2\left[\frac{2\Omega(I_1+I_2)}{mc^2}\right]^{1/2}(\alpha_1\sin\theta_1 + \alpha_2\sin\theta_z)\}^{1/2} - \omega_1I_1 - \omega_2I_2 \ . \tag{2}$$

The four equations of motion derived from Eq. (3) are solved numerically. In Figures 1 and 2 representative sets of surfaces of section (I_1, θ_1) and (I_2, θ_z) are displayed, for N = 30 and 15 initial conditions, respectively, and for fixed values of H's. The actions are normalized to mc^2/Ω , and u_o's are the normalized (to mc) initial generalized momenta. The normalized (to c/Ω) initial x-position is 3 and the two waves (1, 2) have frequencies (3.33 MHz, 8.88 MHz) respectively. The external magnetic field is 0.35 Gauss and the ambient plasma density 10^3 cm^{-3}.

When the amplitudes α_1, α_2 are small compared to unity one can attempt a perturbative approach to Eq. (5). Since the implicit nature of any possible solution to the unperturbed ($\alpha_2 = 0$) problem makes the analysis extremely complex when $\alpha_2 \neq 0$, we define $\alpha_1 = \varepsilon$ and $\alpha_2 = \alpha\varepsilon^2$ where α is a parameter of order unity and $\varepsilon < 1$. This simplification is supported by the fact that most cases of practical interest, concerning either launched or naturally existing waves, have α's of order unity or less (usually, much less). Upon expanding Eq. (2) one obtains

$$H = H_o + \varepsilon H_1 + \varepsilon^2 H_2 + \cdots \tag{3}$$

with

$$H_o = mc^2\gamma - \omega_1I_1 - \omega_2I_2 \tag{4a}$$

$$H_1 = \frac{mc^2g}{\gamma}\sin\theta_1, \tag{4b}$$

$$H_2 = \frac{mc^2}{2\gamma}\left(1 - \frac{g}{2\gamma^2}\right) + \frac{mc^2g}{\gamma}\left(\frac{\cos 2\theta_1}{4\gamma^2} + \alpha\sin\theta_2\right), \tag{4c}$$

where $\gamma^2 = 1 + (k_1I_1 + k_2I_2)^2/m^2c^2 + 2\Omega(I_1 + I_2)/mc^2$ and $g = [2\Omega(I_1 + I_2)/mc^2]^{1/2}$. Since the second wave appears only to second and higher orders, we employ the Lie transformation method in power series in ε as developed by Deprit.[2] The transformed Hamiltonian $\overline{H}$, as a function of the new variables (we use the same symbols for convenience), becomes:

$$\overline{H} = \overline{H}_o + \varepsilon\overline{H}_1 + \varepsilon^2\overline{H}_2 + \cdots \tag{5}$$

where

$$\overline{H}_1 = \langle H_1\rangle = 0,\ \overline{H}_2 = \langle H_2\rangle + \frac{1}{2}\langle[w_1, \{H_1\}]\rangle + \{H_2\} \ . \tag{6}$$

The angle and curly brackets denote averaging over the angles and the oscillating part respectively of the functions involved. The generating function, w_1, inside the Poisson bracket, is $w_1 = mc^2 g \cos\theta_1/\gamma\Omega_1$, with $\Omega_1 = \partial H_o/\partial I_1$. The oscillating part $\{H_2\}$ in Eq. (6) is added on purpose since it is the one responsible for resonance interaction to lowest order in ε; this generating function w_2 which enters in the calculation of the Hamiltonian to third order in ε. Finally,

$$\overline{H} = mc^2\gamma - \omega_1 I_1 - \omega_2 I_2 + \varepsilon^2 \frac{mc^2}{2\gamma}\{1 - \frac{g}{2\gamma^2} - \frac{mc^2 g}{\gamma}[\frac{1}{\Omega_1}\frac{\partial}{\partial I_1}(\frac{g}{\gamma}) + \frac{\partial}{\partial I_1}(\frac{g}{\gamma\Omega_1}))]\} + \varepsilon^2 \frac{mc^2 g}{\gamma}(\frac{\cos 2\theta_1}{4\gamma^2} + \alpha \sin\theta_2) \quad . \tag{7}$$

This form of the Hamiltonian is amenable to the standard analysis that leads to the Chirikov criterion[3] for the onset of stochasticity. At a first glance, the widths of the resonances $\dot{\theta}_1 = 0$, $\dot{\theta}_2 = 0$, which lie on two orthogonal planes, are proportional to ε and $\varepsilon\sqrt{\alpha}$ respectively. The detailed analysis of Eq. (7) is presented elsewhere.

References

1. C. S. Roberts and S. J. Buchsbaum, Phys. Rev. A, 135, 381 (1964).
2. A. Deprit, Cel. Mech., 1, 12 (1969).
3. B. V. Chirikov, Phys. Reports, 52, 265 (1979).

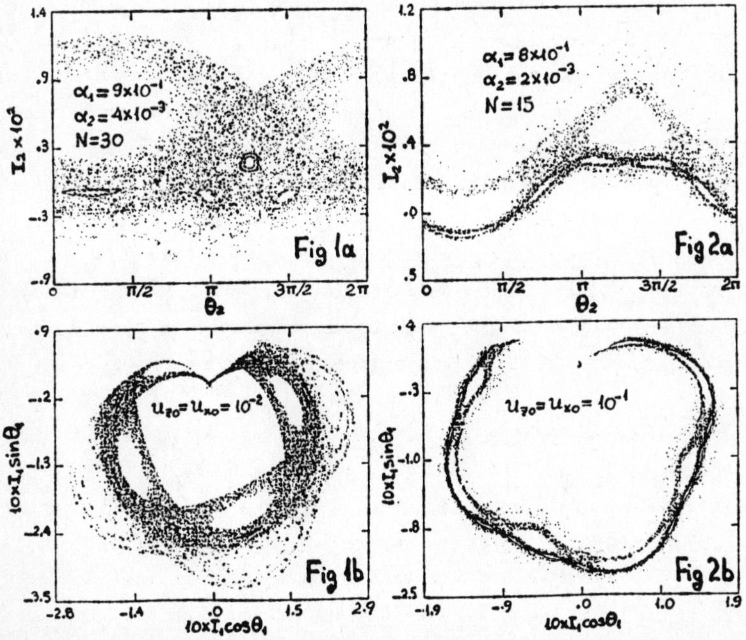

NONLINEAR STABILITY IN ANISOTROPIC MAGNETOHYDRODYNAMICS

John M. Finn and Guo-Zheng Sun
University of Maryland
College Park, MD 20742

Formal stability, i.e., positive energy linear stability, and nonlinear stability are defined. The generic way in which a system with infinite degrees of freedom may be nonlinearly unstable in spite of formal stability is illustrated. The "energy-Casimir" method of proof of nonlinear stability is applied to "relaxed states" of anisotropic magnetohydrodynamics (MHD). Such states are consistent with ergodic magnetic field lines and are formally stable under mild restrictions. Specifically, there is no free energy to drive interchange modes. Nonlinear stability is guaranteed under somewhat more restrictive conditions than is formal stability.

The subject of nonlinear stability of various physical systems has received a great deal of attention recently.[1] The method of investigation of stability used in most of this work is the so-called "energy-Casimir" (EC) method discussed in Ref. 1 and attributed to Arnold.[2] Our purpose in this note is to analyze and clarify (hopefully) nonlinear stability and the EC method and show how it can be applied to a relatively complex system, namely anisotropic MHD. In the process, we provide an admittedly personal critique of the usefulness of the method and the necessity of describing it in terms of noncanonical Poisson brackets and Casimir functions.

We begin with a review of formal stability, nonlinear stability, and the EC method. Take a Hamiltonian system of equations $\partial\phi_i/\partial t = [\phi_i, H]$, where $H[\phi]$ is the Hamiltonian and $[.,.]$ is the Poisson bracket. The variables ϕ_i need not be canonical variables, so that the bracket is not canonical. Therefore, there exist functions C called Casimirs that commute with anything $[.,C] = 0$, i.e., are in the null space of the bracket.[3] The first variation $\delta(H + C) = 0$ produces equations for an equilibrium; the more general the Casimir C, the more general the equilibrium allowed. A system is <u>formally stable</u> if the second variation $\delta^2(H + C)$ is positive definite. [A better sufficient condition for linear stability involves minimizing $\delta^2(H + C)$ subject to the constraint that C be conserved to first order, $\delta C = 0$.[4]] A system is formally stable if it is linearly stable with no negative energy modes.

Nonlinear stability is proved by the EC method by finding quadratic forms $Q_1(\Delta\phi)$, $Q_2(\Delta\phi)$ such that $Q_1(\Delta\phi) \leq \Delta H \equiv H(\phi_0 + \Delta\phi) - H(\phi_0) - (\delta H/\delta\phi)_0\Delta\phi$, $Q_2(\Delta\phi) \leq \Delta C \equiv C(\phi_0 + \Delta\phi) - C(\phi_0) - (\delta C/\delta\phi)_0\Delta\phi$, where ϕ_0 is the equilibrium state. Then, if $Q(\Delta\phi) \equiv Q_1(\Delta\phi) + Q_2(\Delta\phi)$ is positive definite, continuity of the functional $H_c = H + C$ with respect to the norm $\|\cdot\| = Q(\Delta\phi)^{1/2}$ implies nonlinear stability with respect to the same norm, because H and C are constants of motion.

To see how a system with <u>infinite</u> degrees of freedom can be formally stable, but nonlinearly unstable (this is impossible in finite degrees of freedom), consider an infinite collection of uncoupled nonlinear oscillators $d^2x_n/dt^2 = -\partial V_n/\partial x_n$, with $V_n(x_n) = \lambda_n x_n^2/2 - \mu_n x_n^4/4$. Initial conditions satisfying $|x_n(0)| \leq \Delta_n \equiv \sqrt{(\lambda_n/\mu_n)}$ remain bounded, and initial conditions with $|x_n(0)| > \Delta_n$ escape to infinity. If $\Delta_n \to 0$ as $n \to \infty$, (or, less restrictively, if the Δ_n are not bounded away from zero), then the system is unstable for $\|(x,p)\| = \Sigma(x_i^2 + p_i^2/2)$; for arbitrarily small $\delta > 0$, if $x_n(0) = \delta$, $x_n(t) \to \infty$ as $t \to \infty$ for n sufficiently large. See Figure 1.[7] Essentially, this comes about by an interchange of the limiting processes $n \to \infty$ and $\delta \to 0$.

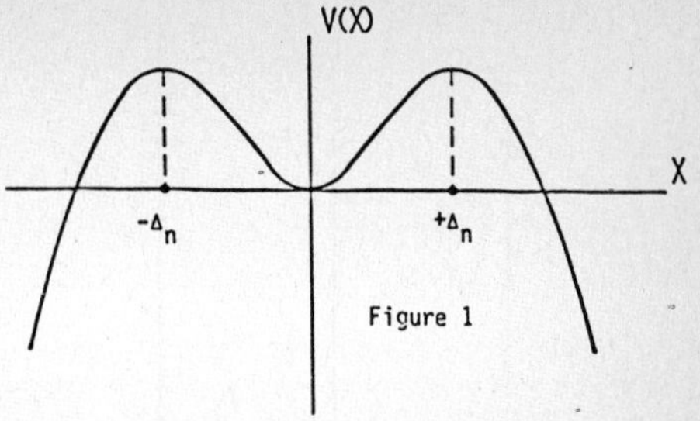

Figure 1

Notice that the proof of nonlinear stability by the EC method involves only the fact that H and C are constants of motion. It is not, strictly speaking, necessary to obtain a noncanonical bracket and its Casimirs; one may merely search for constants of motion of the original equations. In some cases, knowing the Poisson bracket may simplify the task of finding invariants by making it more systematic. However, in few cases is it trivial to find all the Casimirs. Moreover, equilibria can be described by $\delta(H+I) = 0$ for any invariant I, which can become a Casimir by the process of reduction. For example, in fluid systems whose boundary conditions are symmetric in z, the linear momentum $P_z = \int \rho v_z dV$ is a constant of motion which leads to equilibria with uniform v_z. However, P_z is not a Casimir for the usual bracket (e.g., the bracket for MHD given in Ref. 5.) Consider a two-body problem, with $H = p_1^2/2 + p_2^2/2 + V(q_1 - q_2)$. The total momentum $P = p_1 + p_2$ is a constant of motion, but not a Casimir with respect to the usual four-dimensional bracket. Since there are no Casimirs for the original bracket, $\delta H = 0$ would lead to the rather restricted class of equilibria consisting of both particles at rest in the lab frame. In the reduced variables P, $p = p_1 - p_2$, $q = (q_1 - q_2)$, when the center of mass position $Q = (q_1 + q_2)/2$ is ignored, P becomes a Casimir with respect to the three-dimensional degenerate bracket. Equilibria formed by considering P as a Casimir have both particles at rest in an <u>arbitrary</u> inertial reference frame.

To illustrate, we apply the EC method to anisotropic MHD. In addition to the energy, the magnetic helicity $K = \int \underset{\sim}{A} \cdot \underset{\sim}{B} dV$, the mass $M = \int \rho dV$, and two entropies $S_\perp = \int \rho \ell n\, \sigma_\perp\, dV$, $S_\parallel = \int \rho \ell n\, \sigma_\parallel\, dV$ are conserved, and are in fact Casimirs with respect to a noncanonical bracket that is a slight extension of that of Ref. 5 to the anisotropic case. Here, $\sigma_\perp = p_\perp/\rho B$ and $\sigma_\parallel = p_\parallel B^2/\rho^3$. The last three invariants are special cases of the general entropy functional $\int \rho F(\sigma_\perp, \sigma_\parallel) dV$. Therefore, the first variation produces a very restricted, special class of equilibria. The physical interest in such states is described in Ref. 6. The second variation of $H + C$ $(C = -\lambda K - \gamma M - \eta_{\perp} S_{\perp} - \eta_{\parallel} S_{\parallel})$ is

$$\delta^2(H+C) = \int dV\{\rho\delta \underset{\sim}{v}^2 + \sigma\delta \underset{\sim}{B}_{\perp}^2 + \tau\delta \underset{\sim}{B}_{\parallel}^2 - \mu\delta \underset{\sim}{A} \cdot \delta \underset{\sim}{B}$$

$$+ p_{\perp}(\delta p_{\perp}/p_{\perp} - \delta\rho/\rho)^2 + \frac{1}{2} p_{\parallel}(\delta p_{\parallel}/p_{\parallel} - \delta\rho/\rho)^2$$

$$+ p_{\parallel}[\delta\rho/\rho - (1 - T_{\perp}/T_{\parallel})\delta B/B]^2\} , \tag{1}$$

where σ and τ are the firehose and mirror parameters which are $1 + O(\beta)$, where $\beta \sim p/B^2$ is the plasma beta. These equilibria are stable for $\sigma,\tau > 0$ and for μ (proportional to the parallel current) small enough, even with unfavorable field line curvature. Applying the EC method, we find

$$\Delta(H+C) \geq Q \equiv \frac{1}{2} \int dV \{\rho_{min}\Delta \underset{\sim}{v}^2 + \sigma'\Delta \underset{\sim}{B}_{\perp}^2 + \tau'\Delta \underset{\sim}{B}_{\parallel}^2 - \mu\Delta \underset{\sim}{A} \cdot \Delta \underset{\sim}{B}$$

$$\eta_{\perp}\rho_{min}[\Delta\sigma_{\perp}/\sigma_{\perp max} + \sigma_{\perp max}\Delta B/\sigma_{\perp 0}B_0]^2$$

$$+ \eta_{\parallel}\rho_{min}[\Delta\sigma_{\parallel}/\sigma_{\parallel max} + 2\,\Delta\rho\sigma_{\parallel max}/\rho_{min}\sigma_{\parallel 0} - 2\rho_0\sigma_{\parallel\, max}\Delta B /$$

$$\rho_{min}\sigma_{\parallel 0}B_0]^2 + 2\eta_{\parallel}\rho_{min}\alpha[\Delta\rho/\rho_{min} + \frac{1}{\alpha}(2\rho_0\sigma_{\parallel max}^2/\rho_{min}\sigma_{\parallel 0}^2$$

$$-3g_{min} + \eta_{\perp}/2\eta_{\parallel})\Delta B/B_0]^2\} . \tag{2}$$

Here, $g = (\sigma_{\parallel 0} + \Delta\sigma_{\parallel})B_0^2/\sigma_{\parallel 0}(\underset{\sim}{B}_0 + \Delta\underset{\sim}{B})^2$, $\alpha = 3g_{min}(\rho_{min}/\rho_0)^2 - 2(\sigma_{\parallel max}/\sigma_{\parallel 0})^2$, and the modified firehose and mirror mode parameters σ' and τ' are still $1 + O(\beta)$. The quadratic form in (2) is positive definite if α, σ' and τ' are positive and if μ is small enough. As in virtually all of the examples of Ref. 1, extra conditions such as $0 < \rho_{min} < \rho < \rho_{max} < \infty$ are required to obtain a positive definite form Q in (2). For such complex systems, it is not known in general whether such caveats indicate the possibility of nontrivial nonlinear instability or indicate limitations of the method.

We wish to thank D. Holm and R. Littlejohn for useful discussions. This work was supported by the U.S. Department of Energy.

References

1. D. Holm, J. Marsden, T. Ratiu, and A. Weinstein, Phys. Rep. 123, 1 (1985).
2. V. I. Arnold, Am. Math. Soc. Trans. 19, 267 (1969).
3. See, for example, R. Littlejohn, AIP Conference Proceedings No. 88, 47 (1981).
4. J. Finn and T. Antonsen, Jr., Phys. Fluids 26, 3540 (1983).
5. P. Morrison and J. Greene, Phys. Rev. Lett. 45, 790 (1980).
6. G. Sun and J. Finn, to be submitted to Phys. Fluids.
7. C. Grebogi, E. Ott, J. Yorke, these proceedings, Fig. 1a.

ON RESONANT HAMILTONIANS WITH n FREQUENCIES.

Martin Kummer
Department of Mathematics
University of Toledo
Toledo, Ohio 43606

We study Hamiltonians of the form

$$H = \langle\omega,N\rangle + V(Z,\bar{Z}) \ , \tag{1}$$

where $\langle\omega,N\rangle := \sum_{n=1}^{n} \omega_k N_k$, $N_k = |z_k|^2$, and $V(Z,\bar{Z})$ is a convergent power series in the variables $Z = (z_k)_{k=1}^{n}$ and $\bar{Z} = (\bar{z}_k)_{k=1}^{n}$ which begins with a term of order three. The relation between our complex variables z_k and the more usual position and momentum variables (x_k,y_k) is

$$z_k = \left(\frac{\omega_k}{2}\right)^{\frac{1}{2}} x_k - i \left(\frac{1}{2\omega_k}\right)^{\frac{1}{2}} y_k \ . \tag{2}$$

Accordingly, the diff. equ. associated with the Hamiltonian (1) are

$$\dot{z}_k = i H,_{\bar{z}_k} \quad , \quad \dot{\bar{z}}_k = -iH,_{z_k} \quad (k=1,2,\dots,n). \tag{3}$$

Our goal is to study the flow that a Hamiltonian of type (1) induces close to the origin 0 of phase space under the assumption that the quadratic term describes no resonance of order < m ($m \in \mathbb{Z}$, $m \geq 3$) and precisely one resonance of order m, i.e.,

(i) $\langle p,\omega\rangle = 0$ for $p \in \mathbb{Z}^n \setminus \{0\}$ implies $\|p\| := \sum_{i=1}^{n} |p_i| \geq m$,

(ii) there exists a unique $g \in \mathbb{Z}^n$ with $\|g\| = m$ and $\langle g,\omega\rangle = 0$.

For related work see [1] - [8] and in particular [9]. Under our assumptions there is a canonical transformation which brings the Hamiltonian into the following normal form

$$H = \langle\omega,N\rangle + G(N) + A\, M_1 + O_{m+1}.$$

Here $M_1 = \mathrm{Re}(Z^g) := \mathrm{Re}\,(z_1^{g_1} z_2^{g_2} \dots z_n^{g_n})$, where $z_k^{g_k}$ has its usual meaning if $g_k \geq 0$. However, $z_k^{g_k} = \bar{z}_k^{-g_k}$, if $g \leq 0$. Also, $G(N)$ is a polynomial of degree $[\frac{m}{2}]$ in the variables $N := (N_k)_{k=1}^{n}$ and we assume $A \neq 0$. Actually, w.l.o.g. we may assume that A is real and positive and that $g_1 > 0$.

<u>Example</u>: $H = \omega_1 N_1 + \omega_2 N_2 + \omega_3 N_3 + BN_1N_2 + AM_1 + O_5$, $M_1 = \mathrm{Re}(z_1^3 \bar{z}_3)$, where $3\omega_1 - \omega_3 = 0$, and this is the only relation of type $p_1\omega_1 + p_2\omega_2 + p_3\omega_3 = 0$ with $(p_1,p_2,p_3) \in \mathbb{Z}^3 \setminus \{0\}$ and $|p_1| + |p_2| + |p_3| \leq 4$. Accordingly, $g_1 = 3, g_2 = 0, g_3 = -1$, $m = 4$.

Returning to the general case we first study the integrable approximation

$$K := \langle\omega,N\rangle + G(N) + A\, M_1 \tag{4}$$

to H and then invoke a theorem of J. Moser [10] to conclude that certain features of the flow of K carry over to the full Hamiltonian H. A complete set of "commuting" integrals of K is given by K, $R := (R_k)_{k=2}^{n}$, where $R_k := N_k - g_k\, g_1^{-1} N_1$. Removing from the total set of variation of R a set of measure zero, we obtain an open set $\mathcal{D} \subset \mathbb{R}^{n-1}$ with the following property:

Theorem : For each $R \in \mathcal{D}$ the level set $E_R := \{Z \in \mathbb{C} : R = \text{const.}\}$ fibers into (n-1)-tori over a two-dimensional base manifold S_R which can be realized as the following level surface in $\mathbb{R}^3$: $x^2 + (z - G_R(y))^2 = F_R(y)^2$. Here $(x,y,z) = (A\,\mathrm{Im}(Z^g), g_1^{-1} N_1, K - \langle R, \omega\rangle)$ are the coordinates of $\mathbb{R}^3$ and $G_R(y) := G(R+gy)$, $F_R(y) := F(R+gy)$ with $F(N) := A\, N_1^{|g_1|/2} N_2^{|g_2|/2} \ldots N_n^{|g_n|/2}$.

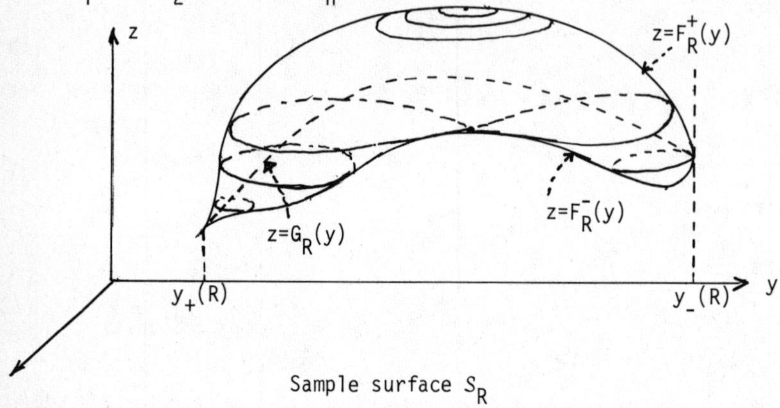

Sample surface S_R

The surface S_R typically looks like a "sausage" that extends over an interval $]\, y_+(R), y_-(R)[$ with $y_-(R) = \infty$ if all g_k's are non-negative. The flow lines of K on S_R are simply level lines of z. In particular, we have the

Theorem : Relative maxima (minima) of $F_R^+(y) := F_R(y) + G_R(y)$ and relative minima of $F_R^-(y) := F_R(y) - G_R(y)$ correspond to stable (unstable) invariant (n-1)- tori in phase space.

If $P_0 : (0, y_0, G_R(y_0))$ is an endpoint of S_R then $y_0 = R_j\, g_j^{-1}$ for some $j = 1,2,\ldots,n$ with $g_j \neq 0$ and the nature of such an endpoint is classified in the following

Theorem: $|g_j| = 1$: P_0 is an ordinary point of S_R and of the flow of K.

$|g_j| = 2$: P_0 is a corner of S_R and a c.p. of K which is stable/unstable depending on whether $F_R^+(y)$ and $F_R^-(y)$ are monotone in the same /opposite sense at $y = y_0$.

$|g_j| \geq 3$: P_0 is a cusp of S_R and a c.p. of K which is stable if only $G_R'(y_0) \neq 0$.

What are the implications of our analysis of the flow of K for the flow of the full Hamiltonian H? In order to summarize our results pertaining to this question we have to distinguish the cases n = 2 and n > 2. In the case n = 2 each non-degenerate c.p. of K on S_R gives rise to periodic orbits of H that fill a two-surface through the origin 0 of phase space. Under different resonance conditions analogous results have been obtained before. (See [2] - [5], [8]) This, however, seems not to be so in the case n > 2 to which we now turn. Here we make the additional assumption $G(N) \equiv 0$ if m is odd and $G(N)$ = homogeneous polynomial of degree $\frac{m}{2}$ if m is even. This condition guarantees that the resonance "reaches down" to the origin 0. It is complementary to any non-degeneracy condition on the polynomial $G(N)$ that via KAM - theory implies the existence of invariant n-tori in each neighborhood of 0. If it is satisfied, a non-trivial application of a theorem of J. Moser (see ref. [10]) leads to the prediction that to each non-degenerate c.p. of K on S_R there exist (n-1) - tori which, after inclusion of the term $\mathcal{O}_{m+1}$, persist in a suffiently small spherical shell about 0. These (n-1) - tori support quasiperiodic motions with frequency vectors that not only satisfy a KAM - like irrationality condition but whose endpoint must also lie in a very specific region of $\mathbb{R}^{n-1}$.

For the example introduced above, we find $F_R^{\pm}(y) = 3ByR_2 \pm A(3y)^{3/2}(R_3 - y)^{\frac{1}{2}}$ and assuming $A > 0$, $B > 0$, the surface S_R looks as in the figure below. The flow of K (= H without the term $\mathcal{O}_5$) on S_R always possesses the stable c.p. P_3. In addition, the unstable/stable pair (P_1, P_2) is present iff $3(2(3)^{\frac{1}{2}} - 3)^{\frac{1}{2}} > 2\,\frac{B}{A}\,R_2 R_3^{-1}$. These c.p. give rise to invariant two - tori of H with the same stability character.

Details of our analysis and proofs of our statements will be published elsewhere.

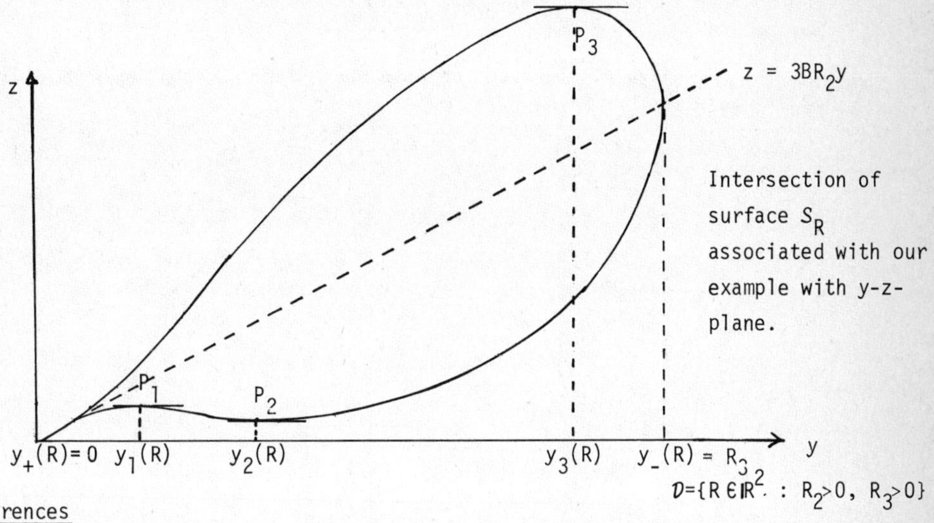

Intersection of surface S_R associated with our example with y-z-plane.

References

[1] Schmidt, D. & Sweet, D.: J. Diff. Eq. 14 (1973) 597 - 609.
[2] Brown, M.: J. Diff. Eq. 13 (1973) 300 - 318.
[3] Kummer, M.: Commun. Math. Phys. 48 (1976) 53 - 79.
[4] Cushman, R. & Rod, D.: Physica D 6 (1982) 105 - 112.
[5] Churchill, R. & Lee, D.: Lecture Notes in Pure and Appl. Math. 92 (1984) 239-286.
[6] Duistermaat, J.J.: Ergod. Th. & Dynam. Syst. 4 (1984) 533 - 562.
[7] Verhulst, F.: Siam J. Math. Anal. 15 (1984) 890 - 911.
[8] Churchill, R., Kummer, M. & Rod, D.: J. Diff. Eq. 49 (1983) 359 - 414.
[9] Augusteijn, M.F. & Breitenberger, E.: J. Math. Phys. 26 (1985) 1219 - 1227 and references given there.
[10] Moser, J.: Math. Annalen 169 (1967) 136 - 176.

SINGULAR PERTURBATION AND ALMOST PERIODIC SOLUTIONS OF NONLINEAR DYNAMIC SYSTEMS

Huang Yuanshi
Department of Mathematics, Fuzhou University
Fuzhou, Fujian, P. R. of China

In this paper, we consider the singularly perturbed nonlinear dynamic system

$$\frac{dx}{dt} = f(x,y,\varepsilon), \qquad \varepsilon\frac{dy}{dt} = g(x,y,\varepsilon), \tag{1}$$

where ε is a small real parameter, x and y are respectively real n-and m-dimensional vector functions. When $\varepsilon = 0$, we get the degenerate system

$$\frac{dx}{dt} = f(x,y,0), \qquad g(x,y,0) = 0. \tag{2}$$

Apart from some suitable hypotheses on smoothness for f, g and their Jacobian matrices, we assume that system (2) has a family of solutions

$$x = u(t,\alpha), \quad y = v(t,\alpha), \tag{3}$$

where $u(t,\alpha)$, $v(t,\alpha)$ are almost periodic in t, $\alpha = \mathrm{Col.}\ (\alpha_1, \alpha_2, \ldots, \alpha_k)$, and $\alpha_1, \alpha_2, \ldots, \alpha_k$ are k-independent parameters. Then the first-variation system of (2) with respect to (3) has the form

$$\frac{dz}{dt} = A(t,\alpha)z, \tag{4}$$

where $A(t,\alpha) = f_x(t,\alpha)-f_y(t,\alpha)g_y^{-1}(t,\alpha)g_x(t,\alpha)$. Here $f_x(t,\alpha)$ denote $\frac{\partial f}{\partial x}(u(t,\alpha), v(t,\alpha,0)$ and similar meanings are attached to $f_y(t,\alpha)$, $g_x(t,\alpha)$, and $g_y(t,\alpha)$. It is easy to verify that $\frac{\partial u(t,\alpha)}{\partial \alpha_j}$ $(j=1,2,\ldots,k)$ and $\frac{du(t,\alpha)}{dt}$ are all almost periodic solutions of system (4), and hence system (4) has at least (k+1) characteristic exponents equal to zero. Evidently, this is a critical case.

In this paper we shall use the definition of "characteristic exponents in the extensive sense" introduced by Lin Zhensheng [1]. We obtain the following theorem.

THEOREM 1 Suppose that

(I) system (2) has a family of almost periodic solutions (3) with k independent parameters, and $u(t,\alpha)$, $v(t,\alpha)$ are also almost periodic in each $\alpha_j (j=1,2,\ldots,k)$;

(II) the first-variation system (4) has exactly (k+1) zero characteristic exponents in the extensive sense;

(III) every eigenvalue of $g_y(t,\alpha)$ has nonzero real part for all t and α.

Then, for ε sufficiently small, system (1) has a unique family of almost periodic solutions

$$x = x(t,\alpha,\varepsilon), \quad y = y(t,\alpha,\varepsilon) \tag{5}$$

satisfying

$$||x(t,\alpha,\varepsilon)-u(t,\alpha)|| + ||y(t,\alpha,\varepsilon) - v(t,\alpha)|| \to 0 \text{ as } \varepsilon \to 0.$$

The main points of the proof for Theorem 1 are

(a) to show that there is an almost periodic matrix function $S(t,\alpha)$ such that $(\frac{du}{dt}, \frac{\partial u}{\partial \alpha_1}, \ldots, \frac{\partial u}{\partial \alpha_k}, S)$ is a regular matrix function;

(b) to perform the generalized normal coordinate transformation

$$\begin{aligned} x &= u(\theta,\alpha) + S(\theta,\alpha)\xi , \\ y &= v(\theta,\alpha) - g_y^{-1}(\theta,\alpha)g_x(\theta,\alpha)S(\theta,\alpha)\xi+\eta , \end{aligned} \tag{6}$$

which carries system (1) into the following system

$$\begin{aligned} \frac{d\theta}{dt} &= 1 + H(\theta,\xi,\eta,\alpha,\varepsilon) , \\ \frac{d\alpha}{dt} &= F(\theta,\xi,\eta,\alpha,\varepsilon) , \\ \frac{d\xi}{dt} &= B(\theta,\alpha)\xi + \beta_1(\theta,\alpha)\eta + G(\theta,\xi,\eta,\alpha,\varepsilon) , \\ \varepsilon \frac{d\eta}{dt} &= C(\theta,\alpha)\eta + H(\theta,\xi,\eta,\alpha,\varepsilon) , \end{aligned} \tag{7}$$

where θ is a real parameter, ξ,η are real (n-k-1) and m dimensional vectors respectively:

(c) to reduce system (7) into the following system

$$\begin{aligned} \frac{d\alpha}{d\theta} &= \overline{F}(\theta,\xi,\eta,\alpha,\varepsilon) , \\ \frac{d\xi}{d\theta} &= Bj(\theta,\alpha)\xi + B_1(\theta,\alpha)\eta + \overline{G}(\theta,\xi,\eta,\alpha,\varepsilon) , \\ \varepsilon \frac{d\eta}{d\theta} &= C(\theta,\alpha)\eta + \overline{H}(\theta,\xi,\eta,\alpha,\varepsilon) ; \end{aligned} \tag{8}$$

(d) to show that both system $\frac{d\xi}{d\theta} = B(\theta,\alpha)\xi$ and system $\varepsilon \frac{d\eta}{d\upsilon} = C(\theta,\alpha)\eta$

admit exponential dichotomies;

(e) to show that system (8) has a center integral manifold.

In the absence of α in Theorem 1, as a special case of Theorem 1, we obtain immediately an analogous result for noncritical case. Its direct proof has been given by Huang [2].

F. A. Howes [3] considered the existence and asymptotic behavior of periodic, almost periodic, and bounded solutions for the following singularly perturbed diagonal system

$$\frac{dx}{dt} = f(x,y,\varepsilon) ,$$
$$\Omega \frac{dy}{dt} = g(x,y,\varepsilon) , \qquad (9)$$

where $\Omega = \text{diag.}\ (\varepsilon^{h_1}, \varepsilon^{h_2}, \ldots, \varepsilon^{h_m})$ and $h_1, h_2, \ldots, h_m$ are integers, $0 < h_1 \ll h_2 \ll \ldots \ll h_m$. His results were got in noncritical cases. It is not difficult to see that our result can be applied to system (9) and that his result on almost periodic solutions can be generalized immediately to the critical case.

As a direct corollary of our results, the analogous problems on periodic or quasi-periodic solutions are also solved evidently. Moreover, this result may be applied to the initial value problems and the boundary value problems.

[1] Lin Zhensheng, Almost periodic linear system and exponential dichotomies, Chin. Ann. of Math., 3 (1982), 131-146.

[2] Huang Yuanshi, Almost periodic solitons of singularly perturbed autonomous systems, J. Fuzhou Univ., 3 (1933), 20-28.

[3] Howes, F. A., An application of Nagumo's lemma to some singularly perturbed systems, Int. J. Nonlinear Mech., 10 (1975), 315-325.

DIFFUSION IN A TURBULENT PHASE SPACE

Michael F. Shlesinger
Office of Naval Research
Physics Division
800 North Quincy Street
Arlington, Virginia 22217

Bruce J. West
Division of Applied Nonlinear Problems
La Jolla Institute
3252 Holiday Court, Suite 208
La Jolla, California 92037

Joseph Klafter
Corporate Research Science Laboratory
Exxon Research and Engineering Company
Annandale, New Jersey 08801

Abstract

We introduce a novel stochastic process, called a Lévy Walk, to provide a statistical description of motion in a turbulent fluid. The Lévy Walk describes random (but still correlated) motion in space and time in a scaling fashion and is able to account for the motion of particles in a hierarchy of coherent structures. When Kolmogorov's -5/3 law for homogeneous turbulence is used to determine the memory of the Lévy Walk, then Richardson's 4/3 law of turbulent diffusion follows in the Mandelbrot absolute curdling limit. If, as suggested by Mandelbrot, that turbulence is isotropic, but fractal, then intermittency corrections follow in a natural fashion.

We are all familiar with the Brownian motion of a pollen mote introduced into physics by Einstein and its connection to the diffusion equation. The mean square displacement of the trajectory of this pollen mote $\langle R^2(t)\rangle$ in three dimensions obeys the following law,

$$\langle R^2(t)\rangle = 6Dt\,, \tag{1}$$

where D is the diffusion constant and t is the time. Other laws of diffusion are also known. A random walk on a random walk path leads to

$$\langle R^2(t)\rangle \sim t^{1/2}\,, \tag{2}$$

and a random walk on a fractal (e.g., a percolating cluster at criticality) has[1]

$$\langle R^2(t)\rangle \sim t^{d_s/d_f} \tag{3}$$

where d_f and d_s are, respectively, the fractal and spectral dimensions of the cluster. For percolation in two dimensions , $d_s/d_f \sim 0.7$. For a random walk on the Brownian path $x(t)$ versus t one has $d_f = 2$, $d_s = 1$, and the recovery of eq. [1]. For a random walker which pauses for a random time between jumps (e.g., due to motion in a disordered system with a random distribution of activation barriers) one can have[2]

$$\langle R^2(t)\rangle \sim t^{\beta}, \quad 0<\beta<1 \qquad [4]$$

if the waiting time density $\psi(t)$ (which governs the pausing time between jumps) behaves asymptotically as the inverse power law $t^{-1-\beta}$ with $0<\beta<1$. So far all of these cases give a mean square motion slower than Brownian motion, i.e., the mean square displacement increases less rapidly than linearly in time. Motion faster than Brownian can be described by fractal Brownian motion denoted by $B_H(t)$. If $B_H(t)$ and $\lambda^{-H} B_H(\lambda t)$ are governed by the same probability distribution then[3]

$$\langle R^2(t)\rangle \sim t^{2H}, \; 0<H<1 \; . \qquad [5]$$

For $H > ½$ the motion is said to be persistent, and antipersistent (transient) for $H < ½$. The totally uncorrelated case of Brownian motion is recovered when $H = ½$. The $H = 1$ case gives $\langle R^2(t)\rangle \sim t^2$, which is the fastest motion allowed, and gives the same result as the relative motion of two particles moving in opposite directions on an expanding spherical wave.

After the above discussion some of you may be surprised to learn that for diffusion of a passive scalar in a fully developed turbulent flow[4-6] we have

$$\langle R^2(t)\rangle \sim t^{3+\gamma}, \qquad [6]$$

$\gamma \sim 0.27$ is called an intermittency correction, and the exponent 3 is called Richardson's law. How does one supercede the $\langle R^2\rangle \sim t^2$ result of the $H = 1$ fractal Brownian motion case? Up until now we have only considered the temporal behavior of the random walker. The statistical description of turbulence, however, needs to incorporate space-time (not necessarily Lorentz covariant) correlations in the motion.

Let us consider a random walker in a random vortex field, i.e., in turbulent flows there is a distribution of vortex sizes, energies, and locations. Larger vortices induce larger persistence lengths for the walker's motion. Also, the larger the vortex the larger will be the walker's velocity since vortex size scales directly with vortex energy.

We describe the motion of the random walker via a joint space-time probability density,

$$\Psi(\mathbf{R},t) = \psi(t|\,\mathbf{R})\,p\,(\mathbf{R}) \qquad [7]$$

where $p\,(\mathbf{R})$ is the probability that a jump (or correlated persistence length) of vector displacement **R** occurs and $\psi(t|\,\mathbf{R})$ is the *conditional* probability density that, *given* that the jump **R** occurs, it takes a time t to be completed. For simplicity we choose,

$$\psi(t|\,\mathbf{R}) = \delta\left(|\,\mathbf{R}|\, - \mathrm{V}(\mathbf{R})t\right) \qquad [8]$$

where we explicitly take into account that the velocity of the jump depends on the jump distance, i.e., vortex size. We calculate $\mathrm{V}(\mathbf{R})$ using Kolmogorov's scaling arguments on dissipation, as follows.[7] Let the average kinetic energy E_R associated with a scale R be $E_R \sim \mathrm{V}_R^2$. If the rate of energy dissipation ε_R across this scale is constant (for the inertial range of fully developed turbulence) then $\varepsilon_R \sim E_R \,/\, t_R \sim \mathrm{V}_R^3 \,/\, R$. Thus, $\mathrm{V}_R \sim R^{1/3}$. One can generalize this argument[8] to only allow points R which lie on a fractal of dimension d_f. Then $E_R \sim \mathrm{V}_R^2\, p_R$ where $p_R = (R \,/\, R_0)^{E-d_f}$, R_0 is an outer length scale, and E is the Euclidean dimension. This leads to

$$\mathrm{V}(\mathbf{R}) \sim R^{\frac{1}{3}+\frac{E-d_f}{6}} = R^{1/3+\mu/6} . \qquad [9]$$

A random walk analysis of this process yields[8]

$$\langle R^2(t)\rangle \sim \begin{cases} t^{3+\frac{3\mu}{4-\mu}}, & \beta \le \frac{1}{3}(1-\mu) \\ t^{2+\frac{6(1-\beta)}{4-\mu}}, & \frac{1}{3}(1-\mu) \le \beta \le (10-\mu)/6 \\ t & \beta \ge (10-\mu)/6 \end{cases} \qquad [10]$$

when the jump size probability density has the power-law form

$$p(R) \sim |\mathbf{R}|^{-1-\beta} \ . \qquad [11]$$

The first case in (10) recovers Richardson's law (with an intermittency correction) and corresponds to the mean time $\bar{t}$ of a jump being infinite, i.e., no characteristic jump time exists. The final case is asympototically equivalent to Brownian motion and occurs when the mean square time $\overline{t^2}$ to complete a jump is finite. *This shows that Kolmogorov's scaling does not necessarily imply Richardson's law.*

In one dimension, if $V(\mathbf{R}) = V$ independent of R then, it can be shown that[9]

$$\langle R^2(t)\rangle \sim \begin{cases} t^2 & -1 < \beta \le 0 \ , \quad \bar{t} = \infty \\ t^{2-\beta} & 0 < \beta < 1 \ , \quad \bar{t} < \infty \ , \ \overline{t^2} = \infty \\ t \ln t & \beta = 1 \ , \quad \overline{t^2} \textit{ log divergent} \\ t & \beta > 1 \ , \quad \overline{t^2} < \infty . \end{cases} \qquad [12]$$

These results have been useful for describing chaos in a Josephson junction where R is a parameter rather than a distance. In this latter case, $R > 0$ represents the number of times the voltage phase has rotated in a CW direction, and $R < 0$ the number of rotations in a CCW direction. The time to complete R rotations is proportional to R so a coupled space-time memory is needed. The t^2 case corresponds to chaos, or broad band voltage noise, and also to $\bar{t} = \infty$.

REFERENCES

1. S. Alexander and R. Orbach, J. Phys. (Paris) Lett. **43**: L-625 (1982).
2. M. F. Shlesinger, J. Stat. Phys. **10**, 421 (1974).
3. B. B. Mandelbrot, *The Fractal Geometry of Nature* (Freeman, New York 1983).
4. L. F. Richardson, Proc. Roy. Soc. London Ser A **110**, 709 (1926).
5. H. G. E. Hentschel and I. Procaccia, Phys. Rev. **A27**, 1266 (1983).
6. F. Wegner and S. Grossman, Zeits. für Physik **B59**, 197 (1985).
7. A. N. Kolmogorov, C. R. (Dokl.) Acad. Sci USSR **30**, 301 (1941).
8. M. F. Shlesinger, B. J. West, and J. Klafter, Phys. Rev. Lett (Submitted).
9. M. F. Shlesinger and J. Klafter, Phys. Rev. Lett. **54**, 2551 (1985).

Acknowledgement

This work was supported in part by the National Science Foundation Grant No. ATM-8509353.

INCREASE IN PHASE SPACE ACCESSIBLE TO PARTICLES WHEN THEIR ATTRACTIVE INTERACTIONS ARE SHORT-RANGED

John A. White
Department of Physics
American University
Washington, D.C. 20016

The volume of phase space accessible to a system of interacting particles depends on the distance over which they exert forces on one another. In particular, it is larger for attractive interactions of short-range than for interactions of infinite range which result in the same energy at uniform density. The reason is the increase in total kinetic energy which occurs for attractive forces of finite range, at given mean density $\overline{\rho}$, when $\overline{\rho^2}$ exceeds $\overline{\rho}^2$. This results in augmentation of the total volume of momentum space -and hence of phase space- accessible to the particles for nonuniform density.

Several consequences follow. The likelihood of density fluctuations increases, the entropy increases for given total energy, the free energy decreases for given temperature. Moreover, the familiar properties of extensivity (additivity) of entropy and free energies, and intensivity of the pressure, which are characteristic of Van der Waals' as well as ideal gases, cease to be true when attractive forces are short-ranged. This is especially the case within regions of linear dimension comparable to the attractive force range. It continues to be so for somewhat larger volumes, and even for very large volumes when close to the critical point for a phase transition.

For attractive forces as short-ranged as found in typical atomic systems, the augmentation of accessible phase space can result in substantial contributions to thermophysical properties, even when well away from the critical point. A simple illustration shows this.

Consider a gas of like particles. Suppose these repel one another as rigid spheres at small separations and attract at somewhat larger separations. Let the interactions be characterized by the following:

b = volume of repulsive core

c = volume of attractive region surrounding the repulsive core

a = energy-volume product for the attractive region.

When classical mechanics is used to describe this system, then essentially the theory of Van der Waals is recovered in the limit $c \to \infty$: b and a have the same meaning as in that theory and its successors; the introduction of $c \neq \infty$ is new.

A simple calculation then shows that for $c \neq \infty$, density fluctuations are enhanced for all wavelengths $\lambda > L = c^{1/3}$, with maximum enhancement for $\lambda \simeq 2L$. Their contribution to the entropy and free energy can be found by grouping fluctuations of increasingly long wavelength into "packets", each containing wavelengths in the range $2^{\pm 1/2}\lambda_n$, where $\lambda_n = 2^n L$. When a, b, c are assumed to be independent of density, though one or more of them may depend on temperature, the pressure of the gas in a region of linear dimension = $2^n L$ is found to be approximately, including the contribution of fluctuations when c is not too small:

$$\frac{P^{(n)}}{\rho} = -a\rho + k_B T \left(\frac{1 + \frac{b\rho}{4} + \left(\frac{b\rho}{4}\right)^2 - \left(\frac{b\rho}{4}\right)^3}{\left(1 - \frac{b\rho}{4}\right)^3} \right) + \rho \frac{\partial}{\partial \rho} \left(\rho^{-1} \sum_{\ell=1}^{n} \delta f^{(\ell)} \right) \tag{1}$$

where

$$\delta f^{(\ell)} = \frac{k_B T}{2^{3\ell+1} c} \ln \left| 1 - \frac{1 - 2^{1-2\ell}}{1 + (2a\rho)^{-1} \frac{\partial}{\partial \rho} P^{(\ell-1)}} \right| . \tag{2}$$

Here, the top line includes a simple approximation to the pressure of a hard sphere gas (second term) plus the contribution to the pressure from attractive interactions in the mean field approximation (first term). These two terms give the pressure for $n \leq 0$. The pressure is then nearly an intensive quantity and fluctuations are nearly as given by mean field theory. For $n \geq 1$, the increments of free energy density given on the second line contribute, each increment after the first decreasing as $\lambda_n^{-3} \ln (\ )$. Here the argument of the logarithm depends on the slope of the pressure isotherm for the next smaller region. Where that slope is small, higher derivatives along the pressure isotherm can be taken into account approximately, using a simple correction term given elsewhere.[1] Though the correction is needed when very close to the critical point, it can generally be ignored elsewhere without much error; in particular, it contributes only a little to the curves to be discussed below.

The calculation leading to the above expression for the pressure is a highly simplified one, but it predicts effects of the augmentation of accessible phase space which model closely those in some real systems. This is the case when a, b, and c are assigned the numerical values that give the experimentally measured density, temperature, and pressure at the critical point of some simple classical fluids like argon, krypton, methane. The contribution of fluctuations given by Eq.(2) then accounts, except at the lowest densities, typically for 10 - 30% or more of the pressure $P^{(n \to \infty)}$ over substantial portions of the gas-liquid phase diagram, including regions well away from the critical point.[2] Figure 1 shows this for two temperatures. The curves marked "mean field approximation" are given by the first two terms, those marked "theory" include the contributions of the third term in Eq.(1) for $n \to \infty$.

Comparison with experimentally measured pressures is indicated in Fig. 2. The solid curve is the same as the curve in Fig. 1 marked "theory" for temperature T_1 but extended to somewhat higher densities. The dots are composite experimental data at $T = T_c$ for several simple fluids, including Ar, Kr, CH_4, which all obey nearly the same empirical equation of corresponding states.[3] Agreement is to within approximately the width of the theoretical curve plotted in this figure, an agreement that cannot be achieved, to my knowledge, using any existing version of mean field theory.

Apart from showing the magnitude of the contribution from the increased accessible phase space when $c \neq \infty$, the illustration considered here indicates that, when this increase is taken into account, it is possible with a simple model to determine sufficiently well for some purposes the entire critical pressure isotherm from knowledge of a single point on it.

1. J.A. White, Bull. Am. Phys. Soc. 30, 714 (1985).
2. M.E. Pustchi and J.A. White, ibid. 30, 713 (1985).
3. K.S. Pitzer, et al., J. Am. Chem. Soc. 77, 3433 (1955).

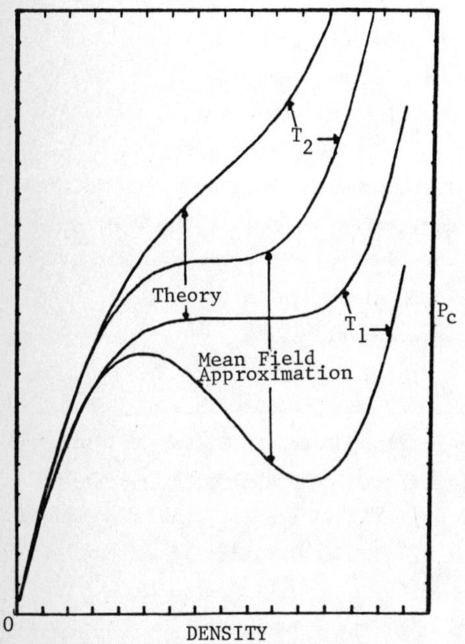

Fig. 1. Pressure isotherms at two temperatures.

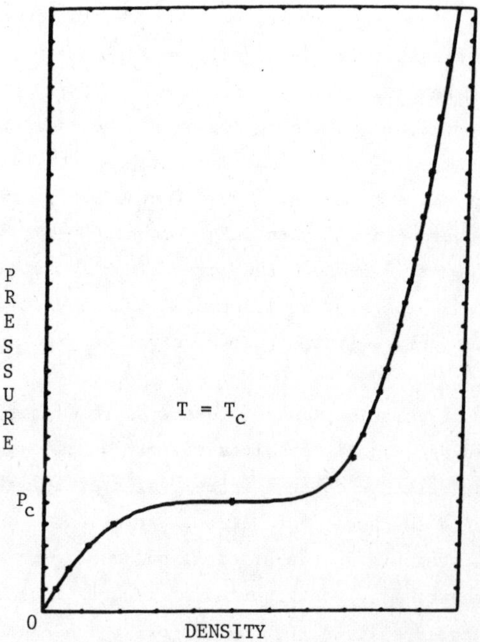

Fig. 2. Critical pressure isotherm, theory compared with experiment.

EXPERIMENTAL MEASUREMENTS OF PHASE SPACE

Roger McWilliams and Daniel Sheehan
Department of Physics
University of California
Irvine, California 92717

Many differential equations arising in physics, such as the Boltzmann equation, have solutions of the form $f = f(\underset{\sim}{x},\underset{\sim}{v},t)$, that is, the solutions are phase space density functions. Experimentalists commonly can measure number density along with electromagnetic field quantities, but a direct measure of f is difficult to achieve. Theorists have predicted wave-particle trapping, bunching, routes to turbulence, and phase space clumps.

The experiments reported here are of integrated phase space density involving one physical space dimension and one velocity space dimension, having integrated over the two velocity components perpendicular to the velocity space dimension displayed. For example, measurements were made of

$$f_i(x,v_y,t) \equiv \int f_i(x,y,z,v_x,v_y,v_z,t)dv_x dv_z \qquad (1)$$

with y and z fixed. A set of scans at various angles in velocity space allows the complete, non-integrated $f_i(\underset{\sim}{x},\underset{\sim}{v},t)$ to be obtained using tomographic reconstruction techniques.[1]

Phase space densities were measured via laser induced fluorescence (LIF) techniques[2] (Fig. 1). Resolution of the movable collimated laser beam and optics is 1 mm^3. The velocity resolution is 3 x 10^3 cm/sec and the time resolution is 2.0 μsec.

The experiments reported here were done in a Q-machine[3] (Fig. 1) which provided a low density ($n \sim 5 \times 10^9$ cm^{-3}), low temperature ($T_i \approx T_e \approx 0.2$ eV), nearly completely ionized barium plasma 1.0 m long and 5 cm in diameter. The con-

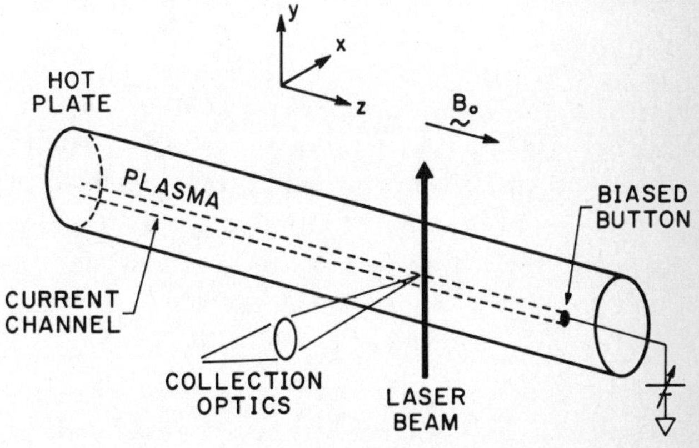

Fig. 1. Experimental apparatus.

fining magnetic field was 4 kG along the z-axis. The x-y plane is perpendicular to the magnetic field, with origin at r = 0 of the cylindrical plasma. All measurements shown in the accompanying figures were taken from scans along the x-axis, holding y and z constant, collecting v_y dependent information. Thus the displayed data are $f_i(x,v_y,t)$. An electrically conducting, variably biased button of 6 mm diameter placed in an x-y plane and centered at x = -0.5 cm, y = 0 may produce an electrostatic ion cyclotron instability which perturbs phase space. Fig. 2 shows two quadrants of the x-v_y plane at a fixed time when a small ampltiude, $e\phi/T < 1$, ion instability was generated.

From symmetry of the driving system and the ion Larmor orbit, the remaining two quadrants may be surmised by a mapping of $f_i(x,v_y,t) \rightarrow f_i(-x-1,-v_y,t)$. Contours of equal phase space density are drawn in Fig. 2. Peak value of the phase space density in the plot occurs along the $v_y = 0$ axis outside of the button region and has a value of 2.8×10^4 sec cm^{-4}. The contours follow fractions of this value and from large v_y to $v_y = 0$ (top to bottom in the figures) occur progressively at 5, 10, 15, 20, 30, 40, 50, 60, 70, 80, and 90 percent of the peak value. For a distribution function which is isotropic and homogeneous in physical space, contours parallel to the x-axis would result, as seen on the right. Fig. (2) also shows that ions whose orbits pass through the button current channel are affected by the instability. Hence, perturbations to the phase space density are observed only to the left in the figure. Ions whose paths may have $x > x_{Button}$ will display phase space changes for $v_y < 0$ only, consistent with the mapping among quadrants in the $x - v_y$ plane.

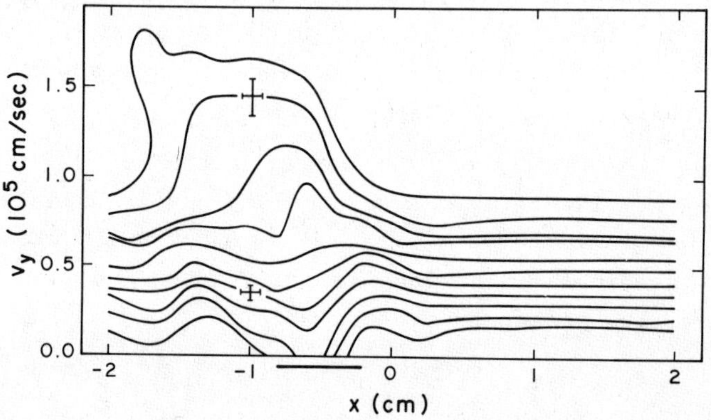

Fig. 2 Phase space density in two quadrants of x-v_y plane. Bar below horizontal axis indicates extent of biased button.

Time evolution of phase space density is shown in Fig. 3. The ion wave (with $e\phi/T < 1$) frequency was about $\omega = 3.2 \times 10^5$ sec^{-1}.

Shown are four plots of phase space taken at different phases during a wave period with $\delta\theta \sim \pi/2$ between plots. The ions show a coherent response tied to the wave phase. There is a "tongue" of phase space density oscillating at ω and reaching out in x and v_y. Additional observations on the distribution function measured parallel to $\underset{\sim}{B}_o$ show a phase resolved v_z-bunching of the ions in the z-direction. To generalize the concept of bunching, this tongue is a measurement of phase space bunching in four dimensions, three velocity dimensions and one spatial dimension. As an aside, properties such as the energy density u(x,t) and the velocity dependent energy density $u(x,v_y,t)$, momentum density, etc. may be calculated from this type of figure. When the wave was driven nonlinearly to large amplitudes ($e\phi/T \gg 1$) phase space density was found essentially to be independent of wave phase, and no coherent ion response could be found within the diagnostic time resolution. Hence, at some intermediate wave amplitude not studied in this experiment the ion response changed from coherent to turbulent.

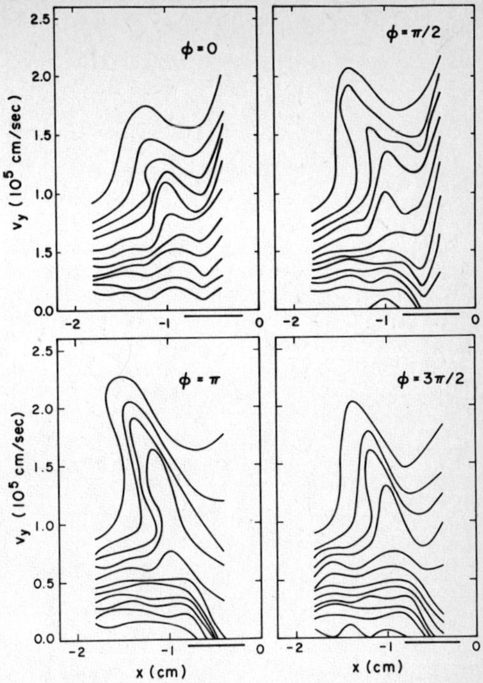

Fig. 3. Phase space density at increasing phases during one period of a small amplitude instability.

In summary, direct, non-perturbing measurements of phase space density integrated over two velocity components have been made in an experiment. Measurements were made of unperturbed phase space, along with linear and nonlinear wave effects. Coherent and incoherent responses showed linear particle responses and that a transition to turbulence occurred. Phase space particle bunching was seen.

Supported by National Science Foundation Grant #PHY-8306108.

References

1. R. Koslover and R. McWilliams, UCI Technical Report #86-18, submitted to Rev. Sci. Instrum, 1986.
2. D. N. Hill, S. Fornaca, and M. G. Wickham, Rev. Sci. Instrum. 54, 309 (1983).
3. N. Rynn, Rev. Sci. Instrum. 35, 40 (1964).

SIMULATION OF ARBITRARY ENSEMBLES BY EXTENDED DYNAMICS: A UNIFIED SCHEME

J. Jellinek and R. S. Berry
Department of Chemistry and the James Franck Institute,
The University of Chicago, Chicago, Illinois 60637

The classical problem of ergodicity , i.e., that of the relation between dynamical and statistical properties of physical systems, acquired a new, practical aspect as a result of a dramatical increase in our ability to actually inquire into these properties through extensive molecular dynamics and Monte Carlo simulations with computers. The statement that a system is ergodic means[1] that the trajectories describing its time evolution in the phase space are, for almost all initial conditions, "chaotic" enough to generate an invariant indecomposable measure in all of the relevant part of the phase space. The central question of the modern theory of dynamical systems, which deals with the ordered vs chaotic behavior and the transitions between the two, can be formulated as follows[1]: does a given dynamics generate an invariant measure, i.e., a distribution function, in the phase space of a system and, if it does, what is this measure? From the point of view of the relation between statistical physics and mechanics it is of interest to reverse the question and to ask how to generate a dynamics which would simulate a particular statistical mechanical ensemble. A number of thermodynamic quantities (e.g., heat capacity, isothermal compressibility, etc.) are directly related to fluctuations in specific ensembles and it would be of interest to correlate these quantities with the details of the corresponding dynamics.

A dynamics corresponding to a microcanonical ensemble can be generated using Hamilton's equations of motion (or their equivalents). Since these equations, when written for a physical system, conserve the energy of this system they cannot simulate any other ensemble in which the total energy is not fixed. A number of suggestions was put forward (see Refs. 2-10 and citations therein) regarding the possible dynamics for mimicking isothermal and isothermal-isobaric ensembles. The most satisfactory, and the only one which produces a continuous deterministic dynamics simulating a true canonical ensemble, is the procedure due to Nose[10] (see also Refs. 11-13). Below we generalize this procedure and point out the nature of the trajectories associated with a dynamics ergodic with respect to a canonical ensemble. The generalization includes clarification of the implications of the scalings involved in the procedure and, as a result of this, a reassessment of the whole approach. We conclude with formulating a general unified scheme for dynamical simulation of any statistical mechanical ensemble.

Consider an extended Hamiltonian

$$H(\vec{q},\vec{p},s,p_s)=\sum_{i,\lambda}(1/2m'_i)[p_{i\lambda}^2/h_{i\lambda}^2(s)]+\Phi(\{f_{i\lambda}(s)q_{i\lambda}\})+p_s^2/2Qu^2(s)+kTv(s), \quad [1]$$

where $\vec{q}\equiv\{\vec{q}_i\}$ and $\vec{p}\equiv\{\vec{p}_i\}$ $(i=1,\ldots,N;\ \lambda=x,y,z)$ are the coordinates and momenta--called "virtual" by Nose[10]--of a system of N particles; s and p_s are a (dimensionless) coordinate and its corresponding conjugate momentum representing a thermal "bath" at temperature T; m_i are the physical masses of the particles in the system and Q is the "mass" of the "bath". The scaling functions $h_{i\lambda}(s)$, $f_{i\lambda}(s)$, $u(s)$, as well as the function $v(s)$, are assumed to be differentiable and nonvanishing. (Nose[10] considered the special case of Eq. [1] with $h_{i\lambda}(s)=s$, $f_{i\lambda}(s)=1$ $(i=1,\ldots,N;\lambda=x,y,z)$, $u(s)=1$ and $v(s)=(3N+1)\ln s$). Introduce the primed quantities

$$q'_{i\lambda}=f_{i\lambda}(s)q_{i\lambda}\,,\quad p'_{i\lambda}=p_{i\lambda}/h_{i\lambda}(s),\ p'_s=p_s/u(s), \qquad [2]$$

where $\vec{q}'\equiv\{\vec{q}'_i\}$ and $\vec{p}\equiv\{\vec{p}_i'\}$ will be interpreted as the physical coordinates and momenta, respectively. Define in the extended phase space $\{\vec{q}_i,\vec{p}_i,s,p_s\}$ a weighted microcanonical (wμc) distribution function

$$\rho(\vec{q},\vec{p},s,p_s) = \delta(H(\vec{q},\vec{p},s,p_s)-E)/w(s), \qquad [3]$$

where w(s) is a continuous nonvanishing function (Nose[10] considered the special case w(s)=1), and calculate the corresponding partition function

$$Z_{w\mu c}=\int d\vec{p}\int d\vec{q}\int dp_s\int ds(H(\vec{q},\vec{p},s,p_s)-E)/w(s) \qquad [4]$$

(the constants unimportant in the present context are omitted). Using Eqs. [1] and [2] we can rewrite Eq. [4] as

$$Z_{w\mu c}=\int d\vec{p}'\int d\vec{q}'\int dp'_s\int ds[u(s)/w(s)]\prod_{i,\lambda}[h_{i\lambda}(s)/f_{i\lambda}(s)]\delta(\sum_i(\vec{p}_i'^2/2m'_i)+\Phi(\{\vec{q}'_i\})+ p_s'^2/2Q+kTv(s)-E)=$$

$$\int d\vec{p}'\int d\vec{q}'\int dp'_s\int ds(1/kT)[u(s)/w(s)v'(s)]\prod_{i,\lambda}[h_{i\lambda}(s)/f_{i\lambda}(s)]\delta(s-s_o), \qquad [5]$$

where v'(s) is the derivative of v(s) and s_o is the zero of the function $F(s)\equiv\sum_i(\vec{p}_i'^2/2m'_i)+\Phi(\{\vec{q}'_i\})+p_s'^2/2Q+kTv(s)-E$ (we assume that F(s) has only one real zero). In obtaining the right-hand side (r.h.s.) of Eq. [5], use was made of the identity $\delta(F(s))=\delta(s-s_o)/F'(s)$. Denoting the energy of the physical system $H_o(\vec{q}',\vec{p}')\equiv\sum_i(\vec{p}_i'^2/2m'_i)+\Phi(\{\vec{q}'_i\})$ and

$$G(s)\equiv[u(s)/w(s)v'(s)]\prod_{i,\lambda}[h_{i\lambda}(s)/f_{i\lambda}(s)] \qquad [6]$$

and requiring that

$$G(s_o)\equiv G(s)\Big|_{s=s_o=v^{-1}([E-\sum_i(\vec{p}_i'^2/2m'_i)-\Phi(\{\vec{q}'_i\})-p_s'^2/2Q]/kT)} = K(E,Q,T;p'_s)\exp[-H_o(\vec{q}',\vec{p}')/kT], \qquad [7]$$

where $K(E,Q,T;p'_s)$ is assumed to be an integrable function of p'_s and v^{-1} is the function inverse to the function v(s), we obtain from Eq. [5]:

$$Z_{w\mu c}=C(E,Q,T)\int d\vec{p}'\int d\vec{q}'\exp[-H_o(\vec{q}',\vec{p}')/kT]=C(E,Q,T)Z_c, \qquad [8]$$

where Z_c is the canonical partition function and C(E,Q,T) is a constant parametrically dependent on E,Q and T. An immediate consequence of Eq. [8] is that for any physical quantity $A(\vec{q}',\vec{p}')=A(\{f_{i\lambda}(s)q_{i\lambda}\},\{p_{i\lambda}/h_{i\lambda}(s)\})$

$$\langle A\rangle_c^{\{\vec{q}'_i,\vec{p}'_i\}} = \langle A\rangle_{w\mu c}^{\{\vec{q}_i,\vec{p}_i,s,p_s\}}, \qquad [9]$$

where < > stands for the corresponding ensemble average, and the superscripts refer to phase spaces in which the averages are calculated. Nose's[10] perception was that his was the only Hamiltonian which led to equalities [8] and [9]. The functional equation [7] admits, however, an infinite number of solutions with respect to the set of functions $\{f_{i\lambda}(s)\}$, $\{h_{i\lambda}(s)\}$, u(s), v(s) and w(s). In fact, for any fixed regular Hamiltonian [1], one can find a function w(s) such that Eq. [7] holds, and for any fixed w(s) there are infinitely many different Hamiltonians [1] implied by Eq. [7]. Each solution of Eq. [7] defines a new Hamiltonian [1] and a new distribution function [3] for which the equalities [8] and [9] hold. Each new Hamiltonian generates, through Hamilton's equations, a different dynamics in the space $\{\vec{q}_i, \vec{p}_i, s, p_s\}$:

$$
\begin{aligned}
&dq_{i\lambda}/dt = p_{i\lambda}/m_i' h_{i\lambda}^2(s), \\
&dp_{i\lambda}/dt = -[\partial\Phi/\partial(f_{i\lambda}(s)q_{i\lambda})] f_{i\lambda}(s), \qquad [10] \\
&ds/dt = p_s/Qu^2(s), \\
&dp_s/dt = \sum_{i,\lambda} \{[p_{i\lambda}^2/m_i' h_{i\lambda}^3(s)][dh_{i\lambda}(s)/ds] - q_{i\lambda}[\partial\Phi/\partial(f_{i\lambda}(s)q_{i\lambda})] \times \\
&\qquad \times [df_{i\lambda}(s)/ds] + [p_s^2/Qu^3(s)][du(s)/ds] - kT[dv(s)/ds].
\end{aligned}
$$

Those of the dynamics [10] which are ergodic with respect to the corresponding distribution functions [3] will, due to Eq. [9], produce time averages of physical quantities equal to their canonical ensemble averages. The distribution functions [3] imply dynamics confined to an energy shell in the space $\{\vec{q}_i, \vec{p}_i, s, p_s\}$ (Eqs. [10], of course, generate such dynamics) with the principle of equal a priori probability not satisfied, unless w(s)=1. This, however, does not cause any conceptual difficulty since the points on this shell do not represent states of a conservative physical system.

Note that those of the different dynamics [10] which are ergodic simulate the same canonical ensemble for the physical system. This suggests a diagnostic test for ergodicity: if two or more different dynamics produce the same time averages of physical quantities, we can infer with some confidence that each of these dynamics displays ergodicity.

Equations [10] can be rewritten in terms of any generalized extended space $\{\vec{P}_i, \vec{Q}_i, s, P_s\}$ without affecting the dynamics themselves if the usual uniform time t is used.[14] Introduction of a scaled time t', defined through the differential relation $dt' = dt/\alpha(s)$, where $\alpha(s)$ is a continuous nonvanishing function (Nose[10] considered the case $\alpha(s)=s$), however, has a nontrivial effect on the dynamics: although the trajectories remain unaltered they are traced out with a varying nonuniform rate.[13,14] In this sense, scaling of the time leads to new dynamics.[14] Each of these new dynamics produces, in general, new values for time averages and time correlation functions of physical quantities, and thus at most one of them will simulate the desired canonical ensemble. One of the implications of the first two of the transformations (2) which project the extended space $\{\vec{q}_i, \vec{p}_i, s, p_s\}$ onto the physical phase space $\{\vec{q}_i', \vec{p}_i'\}$ is that, although a trajectory in the space $\{\vec{q}_i, \vec{p}_i, s, p_s\}$ does not cross itself, its image in the space $\{\vec{q}_i', \vec{p}_i'\}$ does. This is a consequence of the fact that each point in the space $\{\vec{q}_i', \vec{p}_i'\}$ is an image of infinitely many different points of the space $\{\vec{q}_i, \vec{p}_i, s, p_s\}$. In fact, each point in the space $\{\vec{q}_i', \vec{p}_i'\}$ is a bundle of incoming and outgoing trajectories, and it is the repeated recrossing through which eventually the Boltzmann weighting $\exp[-H_o(\vec{q}',\vec{p}')/kT]$ in the space $\{\vec{q}_i',\vec{p}_i'\}$ is achieved. The microcanonical and canonical ensembles are the only traditional ensembles in statistical mechanics for which the distribution functions are defined in the mechanical phase space of a physical system. Other ensembles, e.g., the constant pressure or the grand canonical ensemble, involve additional state variables, such as volume or number of particles. A way to unification is effected through three steps. The first, "augmentation step", consists of adding to a mechanical phase space the extensive thermodynamic state variables of interest as additional "coordinates" and introducing "momenta" conjugate to these new "coordinates". Andersen[9] was the first to implement this idea for the particular case of an isoenthalpic-isobaric ensemble. The second, "extension step", is to include into the phase space also the coordinate s and momentum p_s of the "bath". Finally, the third step is to consider a general augmented extended Hamiltonian of the type of Eq. [1] and a general wμc distribution functon of the type of Eq. [3]. The Hamiltonian will be of the type of Eq. [1] with the extra kinetic and potential energy terms due to augmentation variables. Each of these added potential energy terms has the form of a product of an extensive thermodynamic variable and the corresponding intensive variable; the intensive variables are the additional parameters defining the new ensemble. The added kinetic and potential energy terms are written in general form in terms of the corresponding "virtual momenta" and "coordinates" scaled by functions of s. (A particular case of such a Hamiltonian for the canonical-isobaric ensemble was considered by Nose[10]). Repeating the steps which led to Eq. [7] one arrives at an analogue of this equation in which the Boltzmann factor in the r.h.s. should be replaced by the distribution function of the desired ensemble for the physical system. This functional equation has infinitely many solutions with respect to the scaling functions in the Hamiltonian and the weighting function w(s). Each solution furnishes a new realization of the Hamiltonian and thus a new dynamics in the extended, augmented phase space. Additional dynamics can be introduced through scaling of the time. Those of the dynamics which are ergodic in the wμc sense, i.e., with respect to the analogue of the distribution function [3], will simulate the desired ensemble for the physical system. Further details and proofs of the statements presented in this contribution can be found in Ref. 14.

References

1. J.-P. Eckmann and D. Ruelle, Rev.Mod.Phys. 57, 617 (1985).
2. L.V. Woodcock, Chem. Phys. Lett. 10, 257 (1971).
3. W.G. Hoover, A.J.C. Ladd and B. Moran, Phys. Rev. Lett. 48, 1818 (1982).
4. D.J. Evans and G.P. Morris, Chem. Phys. 77, 63 (1983).
5. D.M. Heyes, Chem. Phys. 82, 285 (1983).
6. J.M. Haile and S. Gupta, J. Chem. Phys. 79, 3067 (1983).
7. D. Brown and J.H.R. Clarke, Mol. Phys. 51, 1243 (1984).
8. H.J.C. Berendsen, J.P.M. Postma, W.F. van Gunsteren, A. DiNola and J.R. Haak, J. Chem. Phys. 81, 3684 (1984).
9. H.C. Andersen, J. Chem. Phys. 72, 2384 (1980).
10. S. Nose, J. Chem. Phys. 81, 511 (1984); Mol. Phys. 52, 255 (1984).
11. W.G. Hoover, Phys. Rev. A 31, 1695 (1985).
12. J.R. Ray and A. Rahman, J. Chem. Phys. 82, 4243 (1985).
13. D.J. Evans and B.L. Holian, J. Chem. Phys. 83, 4069 (1985).
14. J. Jellinek and R. S. Berry, in preparation.

SELF-ORGANIZED STRUCTURES IN THE FORCED BURGERS' TURBULENCE

Hiroshi Nakazawa
Department of Physics, Kyoto University, Kyoto 606/Japan

1. Problem We discuss the structure of the stationary state of randomly excited Burgers' fluid [1,2] between walls,

$$\partial u(x,t)/\partial t + u\partial u/\partial x = \nu\partial^2 u/\partial x^2 + \sigma(x)f(t), \quad 0\leqq x\leqq\pi,$$
$$\sigma(0) = \sigma(\pi) = u(0,t) = u(\pi,t) = 0. \tag{1}$$

It is assumed that f(t) is Gaussian and white, $\langle f(t)\rangle=0$ with $\langle f(s)f(t)\rangle =\delta(s-t)$. As a prototype of forced turbulence, our interest is in the inviscid limit $\nu\to+0$, with a given non-random $\sigma(x)$ that represents a large scale Fourier mode into which the energy is injected randomly.

2. General Aspects of (1) a) The Hopf-Cole transformation (cf. [1]) $u(x,t)=-2\nu\phi_x(x,t)/\phi(x,t)$ linearizes also (1) as follows:

$$\partial\phi/\partial t = \nu\partial^2\phi/\partial x^2 + \phi(x,t)[f(t)\textstyle\int_0^x\sigma(y)dy + C(t)]. \tag{2}$$

Here C(t) is an arbitrary function. Its solution is an infinite series of nonlinear functionals of $\{f(s);\ s\leqq t\}$, however, and the inverse transform to u meets difficulties as a quotient of such quantities. b) As ν tends to 0, almost all portions of a profile of u(x,t) is occupied by segments of the inviscid solution $\bar{u}$, $\bar{u}_t+\bar{u}\bar{u}_x=\sigma(x)f(t)$. On at most denumerable points [2], however, u(x,t) can have steep downward jumps called shock discontinuities in the limit $\nu=+0$. If a jump has a magnitude of D, at small $\nu\sim 0$ the steep transition occurs with x in the well-known structure, $u(x,t)=U-(D/2)\tanh[D(x-x_0)/\nu]$, $D>0$, within a thickness of $O(\nu D)$ in x. c) For all these complications, however, samples of u(x,t) converge [2] as ν tends to 0 in L^1 sense to a limit (denote it as u_0), $\lim_{\nu\to 0}\int_0^\pi|u(x,t)-u_0(x,t)|dx=0$, if $\sigma(x)$ is smooth enough. If the total energy exists at t=0, this fact also implies the convergence of averaged energy density etc. Numerical calculations suffice to be done, therefore, on a small but positive ν for our purpose.

d) Particular to the forced problem (1), the total momentum $P(t)=\int_0^\pi u(x,t)dx$ obeys a stochastic differential equation (SDE),

$$dP/dt = \nu[u_x(\pi,t)-u_x(0,t)] + \Sigma f(t), \qquad \Sigma = \int_0^\pi \sigma(x)dx. \tag{3}$$

If $\Sigma\neq 0$, especially if $\sigma(x)$ has a constant sign on $0<x<\pi$, $P(t)$ is almost a Wiener process $\Sigma\int_0^t f(s)ds$ for ν small. However, draining of the momentum (for positive P at $x\sim\pi$ and for negative P at $x\sim 0$) can take place for $\nu\sim+0$, at (and only at) wall shock discontinuities with $u_x=-O(1/\nu)$ and with a considerable time delay from the momentum injection. e) By (1) the total energy $E(t)=\int_0^\pi u^2(x,t)dx/2$ obeys a SDE (now in Itô's sense),

$$\begin{aligned} dE/dt &= \int_0^\pi [-\nu u_x{}^2(x,t) + \sigma^2(x)/2 + \sigma(x)u(x,t)f(t)]dx, \\ d\langle E\rangle/dt &= -\nu\int_0^\pi \langle u_x{}^2(x,t)\rangle dx + A/2, \qquad A = \int_0^\pi \sigma^2(x)dx. \end{aligned} \tag{4}$$

The average rate of energy injection is A/2, a constant. In the limit $\nu=+0$ the energy is dissipated at any shock discontinuities.

3. Mean Field Approximation For any Gaussian quantity v its Wick power $:v^n:$ is defined by $\exp(av)/\langle\exp(av)\rangle=\sum_{n=0}^\infty a^n:v^n:/n!$, as usual. Any linear functional $v(t)=\int K(t,s)f(s)ds$ of $f(t)$ is Gaussian. Wick polynomials of these $v(t)$'s of n-th degree span a closed subspace (V_n, say) of random variables with V_m and V_n orthogonal for $m\neq n$. The orthogonality refers to $(X,Y)=\langle X^*Y\rangle$. The Fock space $V=\bigoplus_{n=0}^\infty V_n$ form the totality of relevant random variables formed with $f(t)$ (Wiener-Itô decomposition). Let P_n be the projection into $\bigoplus_{k=0}^n V_k$. By (1) there holds for $u_n(x,t)=P_n u(x,t)$,

$$\partial u_n/\partial t + \tfrac{1}{2}\partial(P_n u^2)/\partial x = \nu\partial^2 u_n/\partial x^2 + \sigma(x)f(t), \qquad n\geqq 1. \tag{5}$$

The approximation adopted here is to take n=1 and replace $P_1u^2(x,t)$ by $P_1u_1{}^2(x,t)$. This gives, with the notation $v(x,t)=u_1(x,t)$,

$$\partial v/\partial t + \tfrac{1}{2}\partial(v^2 - :v^2:)/\partial x = \nu\partial^2 v/\partial x^2 + \sigma(x)f(t). \tag{6}$$

More explicitly, this is written in the following form of mean field approximation,

$$\partial v/\partial t + \tfrac{1}{2}\partial(2\langle v\rangle v + \langle v^2\rangle - 2\langle v\rangle^2)/\partial x = \nu\partial^2 v/\partial x^2 + \sigma(x)f(t), \tag{7}$$

which has the sense of the best linear prediction; cf. [3].

Though crude, this approximation enjoys solvability in the limit $\nu=+0$ w.r.t. its stationary solution [4]. The solution is obtained in the form of $v(x,t)=V(x)+\int_{-\infty}^t K(x,t-s)f(s)ds$, by observing equations satisfied by (non-random) functions $\{V(x),K(x,\tau)\}$ with the boundary layer analysis ··· $\{V,K\}$ for the inviscid version of (7) are first obtained and then they are matched to boundary layer solutions that enable the fulfilment of boundary conditions. The procedure is straightforward but lengthy. We therefore note only the essential features of the results.

The stationary solutions of (7) at $\nu=+0$ are not unique: For any $\sigma(x)$ there exist at least three of them. Though more can arise with special $\sigma(x)$'s, we describe only these three below which are relevant to our later discussions:

$$\langle v(x,t)\rangle = V(x) = 2^{1/2}\gamma_n A^{1/3}\sin\{\tfrac{1}{3}\,\mathrm{Sin}^{-1}[\,3\cdot 2^{-3/2}\gamma_n{}^{-3}A^{-1}\textstyle\int_a^x \sigma^2(y)\,dy]\},$$

$$\langle v^2(x,t)\rangle = e(x) + V^2(x) = \gamma_n{}^2 A^{2/3}, \qquad e(x) = \text{variance}, \tag{8}$$

$$n = 1/2 \text{ or } 1, \qquad a = \begin{cases} 0 \text{ or } \pi \text{ for } n = 1/2, \\ \text{given by } \int_0^a \sigma^2(x)\,dx = A/2 \text{ for } n = 1. \end{cases}$$

Profiles of these solutions are shown in Fig. 1 classified by (n,a) for $\sigma(x)=\sin(x)$; no other stationary solution exists for this $\sigma(x)$.

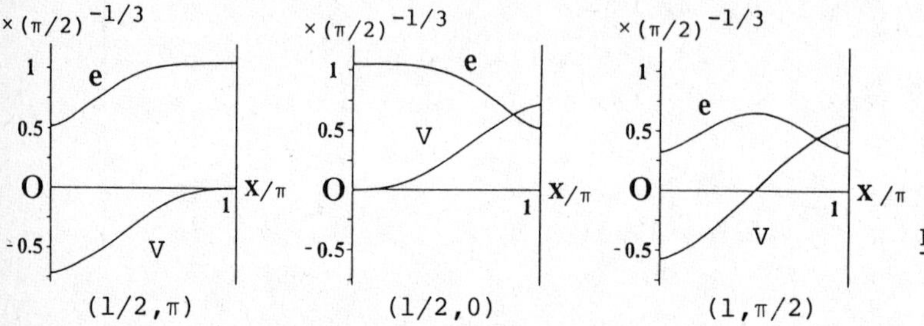

<u>Figure 1</u>

<u>4. Numerical Results</u> A realization of the Wiener process $B(t)=\int_0^t f(s)\,ds$ can remain small for $0\leq t\leq T$ with non-zero probability. Therefore, $u(x,t)$ of (1) for $\nu=+0$ has chances to experience during a finite time interval $0\leq t\leq T$ a nearly free decay, for which $u(x,t)=0$ is the sole fixed point by the persistence of dissipation. Therefore, the stationary state of $u(x, t)$ should be unique. The above result of (7) for $\nu=+0$ suggests that the status of the mean field approximation (7) to (1) will be the same as in the one-dimensional Ising problem. However, there are also reasons to expect quasi-stationary states of $u(x,t)$ of (1) which are not strictly stationary but can be long-lived or visited frequently, and (8) could represent these quasi-stationary states [4]. To clarify the circumstance, numerical simulations of (1) were performed.

The mesh size $\Delta x=\pi/500$ was chosen so that at least 10 points will be in a typical shock of width ν. The difference scheme was taken to be the explicit one of Lax [5] with $\Delta t=\pi/10^4$. These Δt and Δx correspond to $\nu=\pi/50$ and Reynolds number Re=58.1. Cases $\sigma(x)\equiv 0$, $=\sin(x)$ and $\propto\sin(2x)+0.3\sin(x)$ were taken up in double precision. Runs show the following.

a) The free decay of (1) for $\sigma(x)\equiv 0$ gives a half-life of $\sim 3000\Delta t=O(1)$ for the initial energy of $O(1)$.

b) The case $\sigma(x)=\sin(x)$ shows a correspondence without exception between the sign of $P(t)$ and the form of the profile $u(t)$, as depicted in Figure

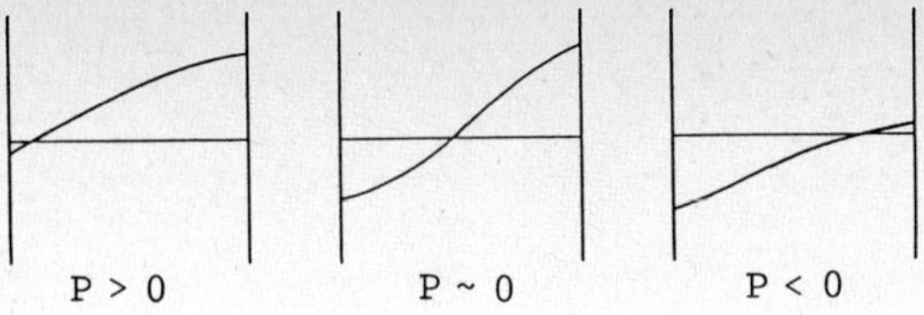

Figure 2

2. The distribution of momentum P shows peaks at P=0 and $P=\pm 0.2 \sim \mp 0.6$ which appear to form a caldera except for a very high but extremely narrow peak at P=0. Figure 3 shows the behavior of the momentum up to $12960 \times \Delta t \sim 40$ with the corresponding distribution of P.

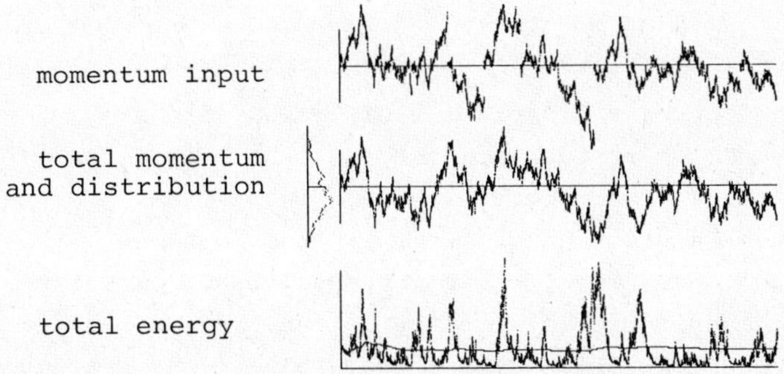

Figure 3

Summarizing for $\sigma(x)=\sin(x)$, we conclude the following from the correspondence of Figure 2. There exist clusters of profiles which are centered around asymmetrical ones and in which a sample of u(x,t) passes large portion of its history. Sojourn times of u(x,t) in a cluster, however, show dispersions, from short to very long, as seen in Figure 3. From a mechanical point of view, the evolution of u(x,t) of (1) is a sequence of processes of the nutrition that receives energy and momentum around $x \sim \pi/2$, of the formation (by using nourishments through its convection term) of structures with shock discontinuities at walls, and finally draining used momentum and energy there. The peaks of P(t) off the center and the mentioned clusters of profiles correspond to these structures which are transient and tottering, so to say. Though this type of self-organized structures are not so conspicuous, they seem to share universal features with many turbulence phenomena.

It is now a natural idea [4] that the solutions $(n=1/2, a=0)$, $(1/2, \pi)$ and $(1, \pi/2)$ of (7) might be taken to arise with relative probabilities P, P and P', respectively, to evaluate one-time (but not two- or more-time) average profiles of u(x,t) of (1). Here $2P+P'=1$ must hold. Figure 4 shows the outcome for choices A:{P=1/2, P'=0} and B:{P=0, P'=1}

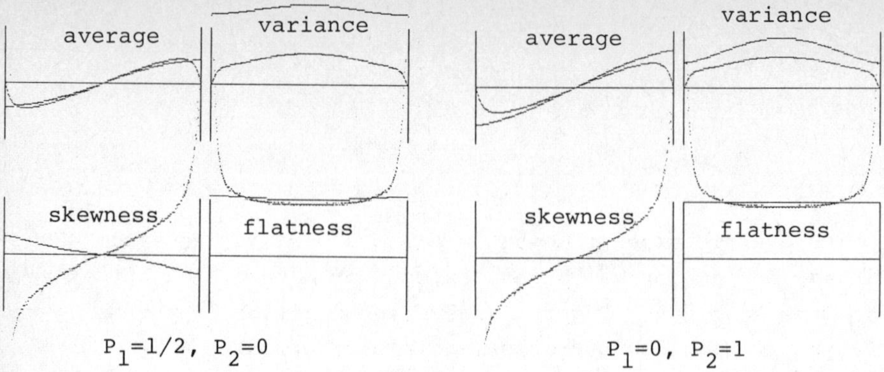

Figure 4

with lines that do not vanish at x=0 or π, together with numerical results. Choice A reproduces the average profile $\langle u(x,t)\rangle$ nicely, but shows departures from higher order moments. Choice B in contrast overestimates $\langle u(x,t)\rangle$, but gives more consistent higher order moments. Putting aside the non-Gaussian character exhibited in the skewness factor and the mentioned many-time problems, the solutions of (7) may be said to reproduce those of (1) in tendency to the order of magnitude. Doubtlessly, the approximation in (7) is too crude to estimate quantitatively the contributions of fluctuations. However, the closed solutions are there, offering ways to proceed by starting from them onto more quantitative assessments of one-time problems.

[1] J. M. Burgers (1974): *The nonlinear diffusion equation.* D. Reidel Publ. Co.

[2] H. Nakazawa (1982): *Stochastic Burgers' equation in the inviscid limit.* Adv. Appl. Math. 3, 18.

[3] H. Ito (1984): *Optimal Gaussian solutions of nonlinear stochastic partial differential equations.* J. Stat. Phys. 37, 653.

[4] H. Nakazawa (1977): *Inhomogeneous stationary state of Burgers fluid stirred randomly by an external mechanism.*

——————— (1981): *Structure of the stationary state in a class of forced Burgers turbulence with boundaries.*

[5] P. D. Lax (1954): *Weak solutions of nonlinear hyperbolic equations and their numerical computation.* Comm. Pure Appl. Math. 7, 159.

CRITICAL POINTS, CRITICAL EXPONENTS, AND STABILITY-INSTABILITY TRANSITIONS IN HAMILTONIAN SYSTEMS

F. T. Hioe
St. John Fisher College
Rochester, New York 14618
and
Z. Deng
Department of Chemistry
University of Rochester
Rochester, New York 14627

We will present two remarkably simple analytic results on stability-instability transitions for motions in Hamiltonian systems. The first gives the critical values of the coupling parameter at which the motion of a specific two-dimensional Hamiltonian system undergoes stability-instability transitions for a given set of initial conditions. The second gives a universal critical exponent which governs the behavior of the largest Lyapunov exponent near the stability-instability transition points for a general Hamiltonian system of n-dimensions.

1. A Two-Dimensional Nonlinearly Coupled Quartic Oscillator System.

We consider a two-dimensional nonlinearly coupled system whose Hamiltonian is given by

$$H = \frac{1}{2}(p_x^2 + p_y^2 + x^2 + y^2) + x^4 + 2Cx^2y^2 + y^4 \tag{1}$$

where the x, y, p_x, p_y represent the displacements and momenta of the oscillator, and C represents the coupling parameter. The equations of motion for x and y are

$$d^2x/dt^2 + x + 4(x^3 + Cxy^2) = 0, \tag{2a}$$

$$d^2y/dt^2 + y + 4(y^3 + Cx^2y) = 0. \tag{2b}$$

Depending on the initial values of displacements and momenta and the value of the coupling C, the resulting motion of the system may be regular or chaotic[1], and generally its character is a complicated function of these parameters and it will have to be studied by numerical analysis. It turns out that there is a remarkably simple result concerning the stable or unstable character of the motion if, as it is often the case, the system starts with an initial condition $x(0) = A$, $y(0) \simeq 0$, and $\dot{x}(0) = \dot{y}(0) = 0$. For this initial condition for which the energy of the system is given by $E = A^2(1 + 2A^2)/2$, we have found[2] that the motion as a function of the coupling parameter C undergoes an infinite number of stability-instability transitions at

$$C_n = \frac{1}{2}n(n+1), \qquad n = 0, 1, 2, \cdots, \tag{3}$$

independent of A so long as $A^2 \gg 1$. More precisely, the motion is unstable for $C < 0$, stable for $0 \leq C \leq 1$, unstable for $1 < C < 3$, stable for $3 \leq C \leq 6$, etc. The remarkable simplicity of the expression (3) for the transition points is a combination of a fluke and beauty, as eqs.(4)-(7) below explain. The critical points could not be so simply expressed if, for example, the quartic terms x^4 and y^4 are not present in the Hamiltonian in which case the critical points are solutions of a transcendental equation[3].

For an analytic proof of the above result, we have made use of a beautiful theorem concerning certain characteristic values of the Lamé equation with $k^2 = 1/2$ discovered by Ince[4] and proved by Erdélyi[5]. The Lamé equation

$$d^2u/d\tau^2 = [\nu(\nu+1)k^2 sn^2\tau - h]u \tag{4}$$

enters our problem because the linearized equations of motion for small perturbations Δx and Δy from x and y satisfy equations of this form:

$$d^2(\Delta x)/d\tau^2 = (6k^2 sn^2\tau - a)(\Delta x), \tag{5a}$$

$$d^2(\Delta y)/d\tau^2 = (2Ck^2 sn^2\tau - b)(\Delta y), \tag{5b}$$

where

$$k^2 = 2A^2/(1+4A^2), \qquad \tau = (1+4A^2)^{1/2}t, \tag{6}$$

$$a = 1 + 4k^2, \qquad b = 1 - 2k^2 + 2Ck^2. \tag{7}$$

For $A^2 \gg 1$, we see that $k^2 = 1/2$ which has the distinction of being complementary to itself, and for which the Lamé equation has been shown to possess an unusually simple set of characteristic values given by $n(n+1)/2$.

While the dependence on the energy and on the coupling parameters of the stability-instability transition points is generally very complicated for an arbitrary Hamiltonian system, we have found[6] a universal critical exponent which governs the behavior of the largest Lyapunov exponent in the neighborhood of any critical points, as we shall describe below.

2. General Hamiltonian Systems of n-Dimensions.

Consider a general Hamiltonian system whose Hamiltonian is given by

$$H = \frac{1}{2}\sum_{j=1}^{n} m_j \dot{x}_j^2 + V(x_1, x_2, \cdots, x_n) \tag{8}$$

for which the jth equation of motion is typically of the form

$$\ddot{x}_j + a_1^{(j)} x_{j_1}^{r_1} + a_2^{(j)} x_{j_2}^{r_2} + \cdots + C_1^{(j)} x_{k_1}^{p_1} x_{k_2}^{p_2} \cdots + C_2^{(j)} x_{l_1}^{q_1} x_{l_2}^{q_2} \cdots + \cdots = 0, \tag{9}$$

where the parameters of the system are the $a^{(j)}$'s and the $C^{(j)}$'s. We again consider an initial condition often encountered in practice given by

$$x_j(0) = A, \quad x_k(0) \simeq 0 \ \ for \ \ k \neq j, \quad and \ \ \dot{x}_l(0) = 0 \ \ for \ \ all \ \ l, \tag{10}$$

for which the motion of the system is a simple periodic motion if $x_k(0)$ are exactly equal to zero for all $k \neq j$. However, any small deviations from this initial condition as specified by $x_k(0) \simeq 0$ in (10) may result in a stable or unstable motion, and as a function of any chosen parameter of the system, the system generally undergoes many stability-instability transitions. We have found and proved analytically[6] that as a parameter C, say, of the system approaches one of its critical values C_ρ from the unstable region, the behavior of the largest Lyapunov exponent μ is given by

$$\mu = const.|C - C_\rho|^\beta \tag{11}$$

with $\beta = 1/2$ independent of the transition point, type of transitions, or the dimensionality of the system.

We shall only briefly outline here one of the crucial steps leading to our analytic proof of this result. It involves a study of the analytic or nonanalytic nature of certain relevant quantities appearing in the characteristic equation whose roots determine the character of the motion. In the stability-instability transitions of types I and III, the relevant roots are given by $s_j = [\alpha_j \pm (\alpha_j^2 - 4)^{1/2}]/2$ as α_j, which remains real in the neighborhood of the transition point, is an analytic function of ϵ and can be written as $\alpha_j = \pm 2 \pm \epsilon$, where ϵ =positive const.$|C - C_\rho|$, and where the positive and negative signs refer to transitions of types I and III respectively. In the stability-instability transitions of type II, on the other hand, we need to consider a complex conjugate pair α_j and α_{j+1} which in the unstable region close to the transition point are not analytic functions of ϵ and can be shown to be given by $(A \pm i\epsilon^{1/2})/2$, where A is real. In all cases, we find $|s_j|^2 = 1 + \epsilon^{1/2}$, and hence the behavior of the largest Lyapunov exponent μ which is related to s_j by $\mu = const. \ln s_j$ is given by eq.(11).

In summary, we have presented two exceptionally simple results, eqs.(3) and (11), which clearly stand out in their simplicity among many difficult and complicated results in the studies of stable, unstable, regular and chaotic motion.

Acknowledgement

This research is supported in part by the U. S. Department of Energy, Division of Chemical Sciences, under Grant number DE-FG02-84ER13243.

References

1. Z. Deng and F. T. Hioe, Phys. Rev. Lett. **55**, 1539 (1985), **56**, 1757 (1986).
2. F. T. Hioe and Z. Deng, An analytic solution of stability-instability transitions in a two-dimensional Hamiltonian system, submitted for publication.
3. Z. Deng and F. T. Hioe, Phys. Lett. A **115**, 21 (1986).
4. E. L. Ince, Proc. Royal Soc. Edinburgh, **60**, 47 (1940).
5. A. Erdélyi, Phil. Mag. **31**, 123 (1941).
6. F. T. Hioe and Z. Deng, Stability-instability transitions in Hamiltonian systems of n-dimensions, submitted for publication.

MAXIMUM LIKELIHOOD METHOD FOR EVALUATING CORRELATION DIMENSION

Robert Cawley and A. Lewis Licht*

Naval Surface Weapons Center
White Oak, Silver Spring, Maryland 20903-5000

1. INTRODUCTION

The correlation dimension [1],[2] has been used widely to characterize time-series or data-strings. It is one among several possible geometrical indicators of the time-asymptotic portion of phase space visited by system orbits of a dynamical system; others include [3] the pointwise dimension, capacity, information dimension, and also the higher order correlation and Renyi entropy-based dimensions [4],[5]. Historically, the correlation dimension was welcomed as an alternative to the capacity as a "physical observable" since the latter typically has involved computational effort which increases exponentially with embedding dimension [6], while the former does not suffer from this disease. That correlation dimension measurement has been widely accepted by the physics community is no doubt also due in part to familiarity of the theoretical concept for physicists accustomed to analogous ideas from many-body theory; and interesting further generalizations motivated by these parallels to statistical mechanics have also been introduced [4],[5],[7]. As a result however, the older notion of pointwise dimension, useful in dynamical systems theory and better known among mathematicians, has been largely neglected by physicists even though it also does not involve exponential computing cost increase with embedding dimension. But the dream of painless data analysis to extract dimension values encounters important numerical problems associated with the fact that the number of points in a time-series is inherently finite. There are some indications that these practical matters may cause less trouble in matters of accuracy for the pointwise dimension than for the correlation dimension [8].

*On assignment from Department of Physics,
University of Illinois at Chicago,
Box 4348, Chicago, IL 60680

In this paper, we examine a method for extracting correlation dimension which is due to Takens [9]. The method gives a simple algorithm for finding the correlation dimension of a data set. It also has an easy extension, which we do not consider here, to the pointwise dimension. The method is based on Fisher's [10] maximum likelihood rule. We implement the method numerically and examine the results. We also investigate the "boundary problem", i.e., the problem of estimating the largest separation in the data set for which scaling should hold.

We begin by reviewing Takens' analysis. The numerical application of the method is discussed. It is applied to the Hénon map [11], and to a model system based on a randomized 2-dimensional lattice.

2. MAXIMUM LIKELIHOOD METHOD

We begin by recalling the definition of a data-state vector [12-14] and of correlation dimension.

From the sequence V_i: $V_i \in R$, $i = 1,2,3..$, which can be assumed to form the time-history of a coordinate function for a dynamical system, we form a data-state vector of dimension d,

$$\vec{X} = \{V_i, \ldots V_{i+(d-1)\Delta}\}, \tag{1}$$

where d, Δ are positive integers, denoting respectively, the embedding dimension and delay parameter. Defining

$$C_N(\varepsilon) = \frac{1}{N_p} \sum \Theta\,(\varepsilon - | \vec{X}_i - \vec{X}_j |),\ 1 \leq i < j \leq N \tag{2}$$

where N_p is the number of pairs, the correlation function is

$$C(\varepsilon) = \lim_{N \to \infty} C_N(\varepsilon) \tag{3}$$

The correlation dimension ν, is defined as

$$\nu = \lim_{\varepsilon \to 0} \frac{\log(C(\varepsilon))}{\log(\varepsilon)} \quad , \tag{4}$$

if it exists.

The quantity $C(\varepsilon)$ is the fraction of points separated by distances less than ε, and the function $P(\varepsilon)$, where

$$C(\varepsilon) = \int_0^{\varepsilon} P(\varepsilon')\ d\varepsilon' \tag{5}$$

is the probability density of pair separations.

From Eq. (4), we expect for small ε, if the limit exists, that

$$C(\varepsilon) = \varepsilon^{\nu} + \text{higher order terms}, \tag{6}$$

for an appropriate normalization of ε. We assume for some fixed $\varepsilon_m > 0$, $\varepsilon < \varepsilon_m$,

$$\bar{C}(\varepsilon/\varepsilon_m) = \frac{C(\varepsilon)}{C(\varepsilon_m)} = (\varepsilon/\varepsilon_m)^{\nu}, \tag{7}$$

exactly. The corresponding probability density is

$$\bar{p}\ (\varepsilon/\varepsilon_m) = \nu(\varepsilon/\varepsilon_m)^{\nu-1}\ . \tag{8}$$

Now, let $\varepsilon_1, \ldots \varepsilon_M$ be a sequence of distances between randomly chosen pairs ($\vec{X}_i$, $\vec{X}_j$), disregarding pairs for which the separation is greater than ε_m. The sequence of sample values, assuming statistical independence,

$$\vec{r} = \{ r_i \ldots r_M : r_i = \varepsilon_i/\varepsilon_m ; i = i .. M \} \tag{9}$$

has "likelihood"

$$L(\underset{\sim}{r}) = \prod \bar{p}(r_j)\ , \tag{10}$$

$$= \nu^M \prod r_j^{\ \nu-1}\ .$$

We adjust ν to maximize this likelihood, requiring

$$\frac{\partial\ L(\nu)}{\partial \nu} = 0\ . \tag{11}$$

This leads to

$$\nu = -\ M\ (\ \Sigma\ \ln(r_i)\ )^{-1} \tag{12}$$

3. IMPLEMENTATION FOR REAL DATA

The customary analysis of a data string involves measuring the slope of ln C(ε) vs. ln(ε). Using

$$C(\varepsilon) = \int_0^{\varepsilon} d\varepsilon' P(\varepsilon') \qquad (13)$$

we get an expression for the local slope,

$$S(\varepsilon) = \frac{d \ln C(\varepsilon)}{d \ln(\varepsilon)} = \frac{\varepsilon P(\varepsilon)}{C(\varepsilon)} ; \qquad (14)$$

while the continuum version of (12) provides a "running mean" log estimate for ν, viz.

$$\nu(\varepsilon) = - \frac{\int_0^{\varepsilon} d\varepsilon' P(\varepsilon')}{\int_0^{\varepsilon} d\varepsilon' \ln(\varepsilon'/\varepsilon) P(\varepsilon')} . \qquad (15)$$

Equations (14) and (15) define ε-dependent values for ν and S, which we employ because the scaling region, $\varepsilon_0 < \varepsilon < \overline{\varepsilon}_1$ is not known precisely. Thus, if (7) held for all ε, we would have $S(\varepsilon) = \nu(\varepsilon)$.

In the data reduction, the quantities actually evaluated are

$$\nu(\varepsilon_n) = C(\varepsilon_n) / \{\Sigma_{i,j} \ln(\varepsilon_n / |\vec{X}_i - \vec{X}_j|) \theta(\varepsilon_n - |\vec{X}_i - \vec{X}_j|)\} , \qquad (16)$$

$$S(\varepsilon_n) = \frac{\varepsilon_n P(\varepsilon_n)}{C(\varepsilon_n)} , \qquad (17)$$

$$P(\varepsilon_n) = C(\varepsilon_n) - C(\varepsilon_{n-1}) . \qquad (18)$$

$P(\varepsilon_n)$ and $S(\varepsilon_n)$ are "binned" quantities giving average frequency and slope over the nth ε-bin, $\varepsilon_{n-1} < \varepsilon < \varepsilon_n$.

To facilitate comparisons between different data sets, it is convenient to normalize to

$$\vec{X}_i \to K \{V_i, V_{i+\Delta}, \ldots, V_{i+(d-1)\Delta}\} / (V_{max} - V_{min}) , \qquad (19)$$

where

$$V_{max} = \text{Max} \{V_i\} , \qquad V_{min} = \text{Min} \{V_i\} , \qquad (20)$$

and where K is an integer. The maximum value of ε is then bounded by
$\varepsilon \leqslant \sqrt{d}\,K$.

4. THE HÉNON MAP

The Hénon map,

$$\xi_{n+1} = 1 + \eta_n - a\,\xi_n^2 \tag{21}$$
$$\eta_{n+1} = b\,\xi_n\ , \tag{22}$$

where the standard values, a = 1.4 and b = 0.3, were chosen, was run through 50,000 steps. The sequence containing the next 1200 successive values of ξ_n was then taken as a sample. Using the normalization index K = 50, $P(\varepsilon)$, $\nu(\varepsilon)$, and $S(\varepsilon)$ were constructed for embedding dimensions d = 1,2,3,4.

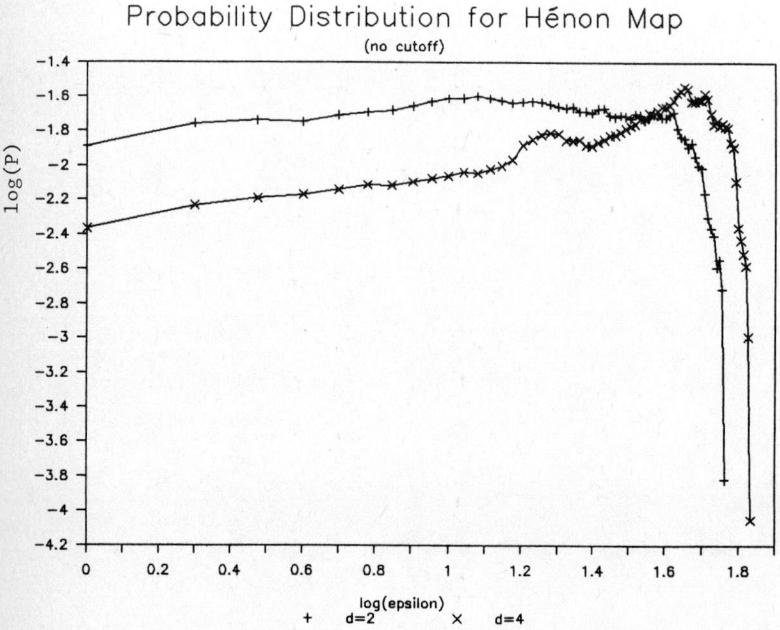

Figure 1:Probability distributions for the Hénon map. All logs are to base 10. Number of data points N=1200, normalization index K=50, and time delay Δ =1.

Figure 1 shows the Hénon curves for $P(\varepsilon)$ for d = 2 and 4. In Figures 2 and 3, the functions $\nu(\varepsilon)$ and $S(\varepsilon)$ are plotted for d = 2 and 4. The curves for d = 1 and 3 were similar.

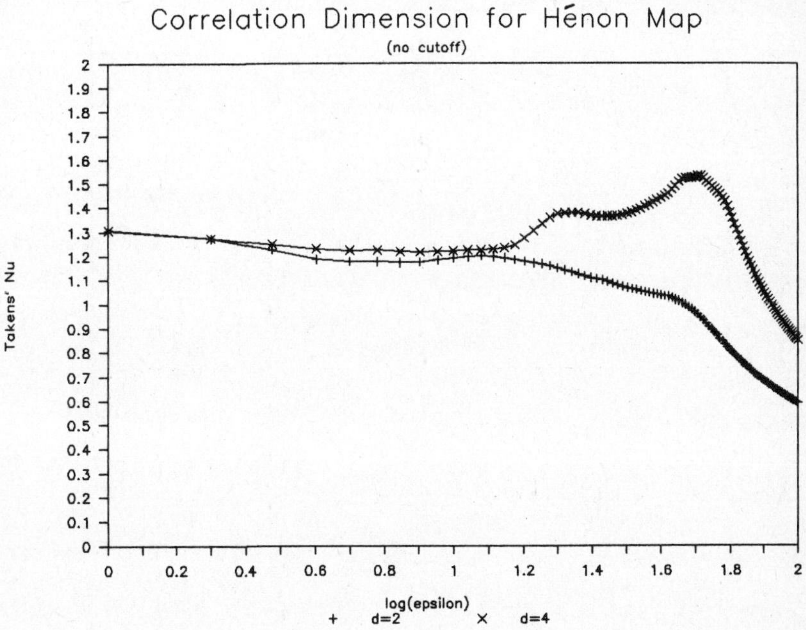

Figure 2: Henon map correlation dimension $\nu(\varepsilon)$ vs $\log_{10}(\varepsilon)$.

The best value for the correlation dimension may be $\nu = 1.22$. That there are pitfalls in dimension calculations has been shown by Caswell and Yorke [15]. The present determinations will not improve that situation, so we just take this value for a guide. The slopes of $\log P(\varepsilon)$ in the plots of Fig. 1, correspondingly, are expected to be 0.22. The lines in the figure do have slope about 0.25 each, for ε values up to about 10. The curves for $S(\varepsilon)$ in Fig. 3 show that the slope function for $C(\varepsilon)$ is $1.1\pm.1$ for $\varepsilon < 10$. Comparing Fig. 2 and 3 indicates that Takens' $\nu(\varepsilon)$ looks very much like a smoothed version of $S(\varepsilon)$, with a cleaner plateau, but with a tendency also to be somewhat higher at larger ε.

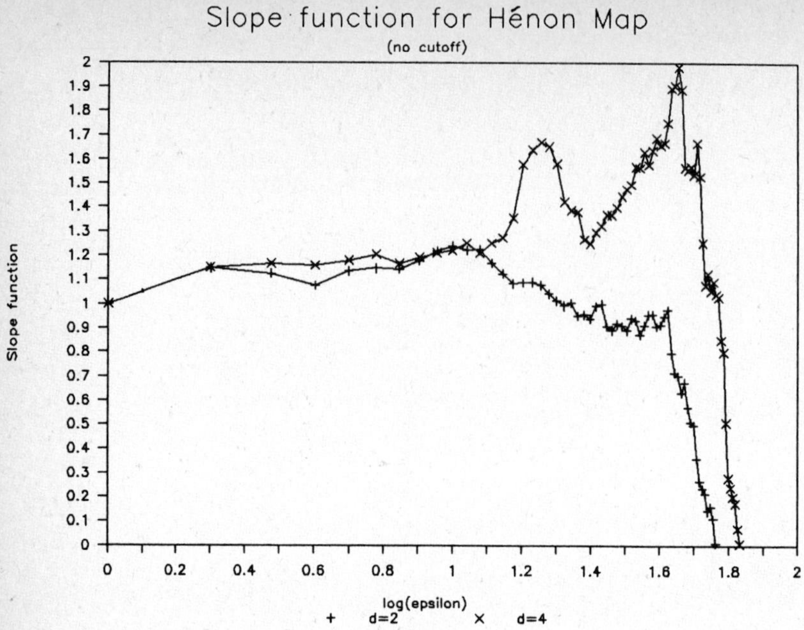

Figure 3: Hénon map slope function, $S(\varepsilon)$ vs $\log_{10}(\varepsilon)$.

5. THE BOUNDARY PROBLEM

Our plots based on the Hénon map show departures from scaling for large ε. It seems likely that much of the non-scaling has a geometrical cause, due to the finite extent of the data set.

Let $\vec{X}_c$ denote the "center" of the data set, defined in a d-dimensional embedding space as

$$\vec{X}_c = X_c \; (1,1,1,\ldots,1) \quad , \tag{23}$$

where

$$X_c = \frac{1}{2} \left(V_{max} + V_{min}\right) \quad . \tag{24}$$

Let R_0 denote the "radius" of the data set, defined as

$$R_0 = \mathrm{Max} \; |\vec{X} - \vec{X}_c| \; . \tag{25}$$

The data set is then contained in the sphere of radius R_0 centered at $\vec{X}_c$,

$$U(R_0,\vec{X}_c) = \{\vec{X}:|\vec{X} - \vec{X}_c| < R_0\} . \tag{26}$$

Those points $\vec{X}_r$ that are in the "ε-rim"

$$W_r(R_0, \varepsilon, \vec{X}_c) = U(R_0, \vec{X}_c) \setminus U(R_0-\varepsilon, \vec{X}_c) , \tag{27}$$

have ε-spheres $U(\varepsilon, \vec{X}_r)$ that intersect the boundary

$$\partial U(R, \vec{X}_c) = \{\vec{X}:|\vec{X} - \vec{X}_c| = R\} . \tag{28}$$

They will contribute less to the sums for $U(\varepsilon)$, $P(\varepsilon)$ and $C(\varepsilon)$ than will the points in the "ε-core" $U(R_0-\varepsilon,\vec{X}_c)$, whose ε-spheres lie entirely within $U(R_0,\vec{X}_c)$.

One way to avoid this undercounting is to sum over only those pairs which have at least one member in the core. Then the correlation integral becomes truncated,

$$C_T(\varepsilon) = \frac{1}{N_{\bar{\varepsilon}}} \sum_{ij} \theta(\varepsilon - |\vec{X}_i - \vec{X}_j|) \;\Big|\; \text{for } \vec{X}_j \in U(R_0-\varepsilon, \vec{X}_c) \tag{29}$$

where we choose a limiting $\bar{\varepsilon}$ with $\varepsilon<\bar{\varepsilon}$, and $N_{\bar{\varepsilon}}$ is the number of pairs that have at least one member in $U(R_0-\varepsilon, \vec{X}_c)$. Similar expressions hold for $P_T(\varepsilon)$ and $\nu_T(\varepsilon)$. In the data reduction, we arbitrarily take

$$\bar{\varepsilon} = .3\ K , \tag{30}$$

$$= 15 \text{ with } K = 50 .$$

We also approximate R_0 by K/2, thus, here $R_0 = 25$.

This truncation will not work for all data sets. Some sets have a shell structure, with few points in the core. This is the case for the Hénon map with d = 3 and 4.

6. THE LATTICE

The effects of truncation are well demonstrated by a data set constructed from the points of a 2-dimensional lattice.

Let
$$x_n = i_n + r_n ,$$
$$y_n = j_n + r_n' , \tag{31}$$

where i_n, j_n = -L, -L+1,...,+L, integers,and r_n, r_n' are random numbers, $|r_n|$, $|r'_n| < .3$. Let n = 1,2,3... be an ordering of the points of a lattice. We take as the data set the first N/2 points for which

$$x_n^2 + y_n^2 \leq L^2 , \tag{32}$$

in the form $\{x_1, y_1, x_2, y_2, x_3, y_3 \ldots\}$.

From the geometry, we expect that $\nu = 2$. Figure 4 shows log P(ε) vs. log (ε) for the lattice, for embedding dimension d = 2 and 4. The curves have slopes that are almost constant and near the expected value of 1 for $\varepsilon < 10$, but with a slight negative curvature that steepens rapidly as ε increases above 10.

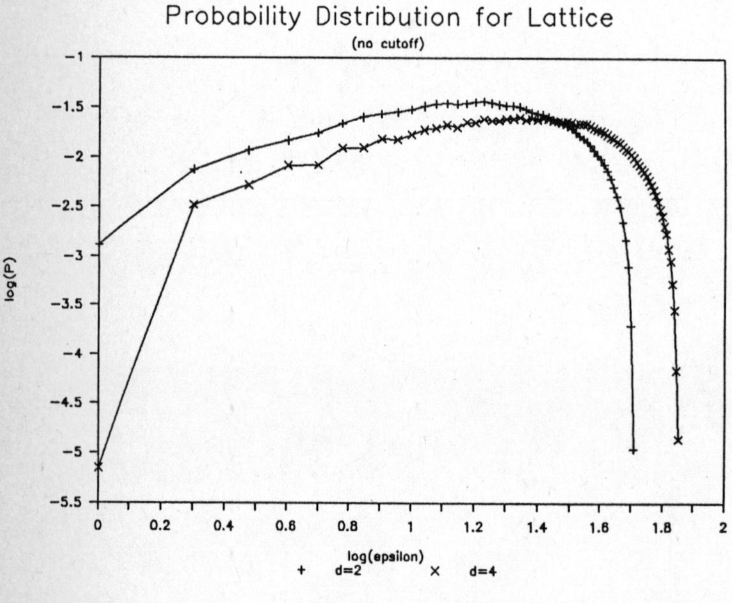

Figure 4: Probability distributions for the lattice. Same N, Δ ,K as in Figure 1.

Figure 5 shows the effect of truncation. The truncated and untruncated log P(ε) curves are shown for d = 2. The truncated curve continues rising beyond ε = 10, where the untruncated curve begins to flatten out. Similiar effects are shown in Fig. 6, for Takens' ν(ε).

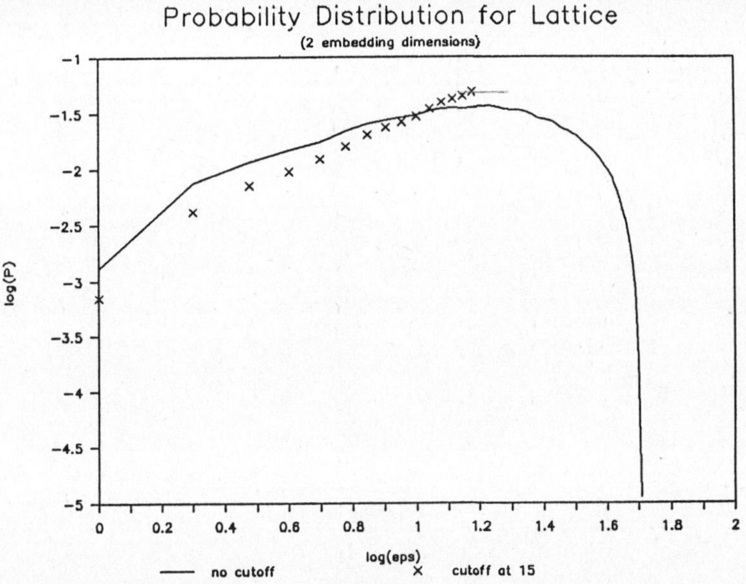

Figure 5: Probability distributions for the lattice. Truncated, with cutoff ε_{max} = 15 , X ; and untruncated ____________. Embedding dimension d = 2.

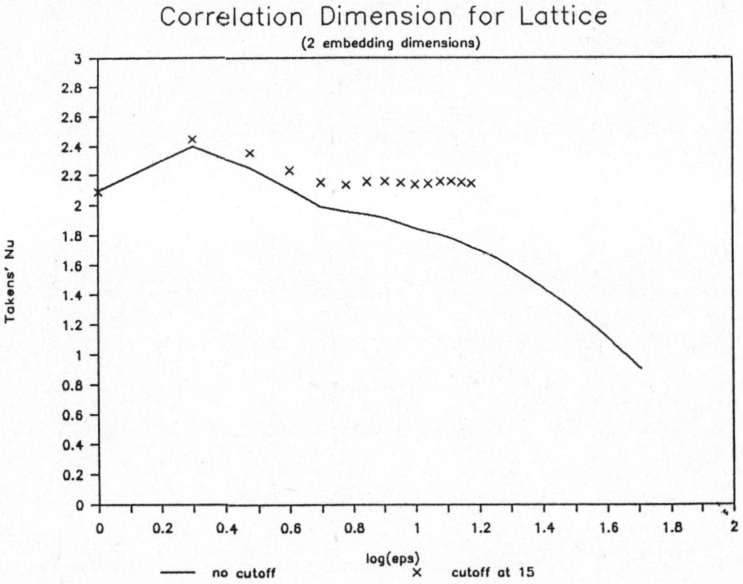

Figure 6: Lattice correlation dimension function ν(ε) vs log(ε) . Embedding dimension d=2. Truncated, X ; untruncated, __________.

The effect of truncation is thus to straighten out the log $(P(\varepsilon))$ curve. This is illustrated further by the curves in Figure 7. The truncated $\nu(\varepsilon)$ and $S(\varepsilon)$ reach a plateau above the point $\log(\varepsilon_c) = 0.8$, $\varepsilon_c = 6.3$, where the untruncated functions are distinctly falling.

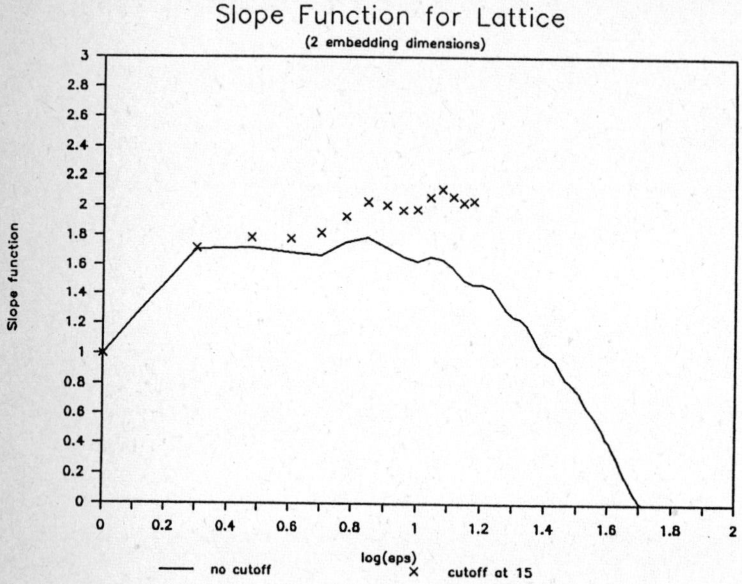

Figure 7: Lattice slope function $S(\varepsilon)$ vs $\log(\varepsilon)$. Embedding dimension d=2. Truncated, X ; untruncated, __________.

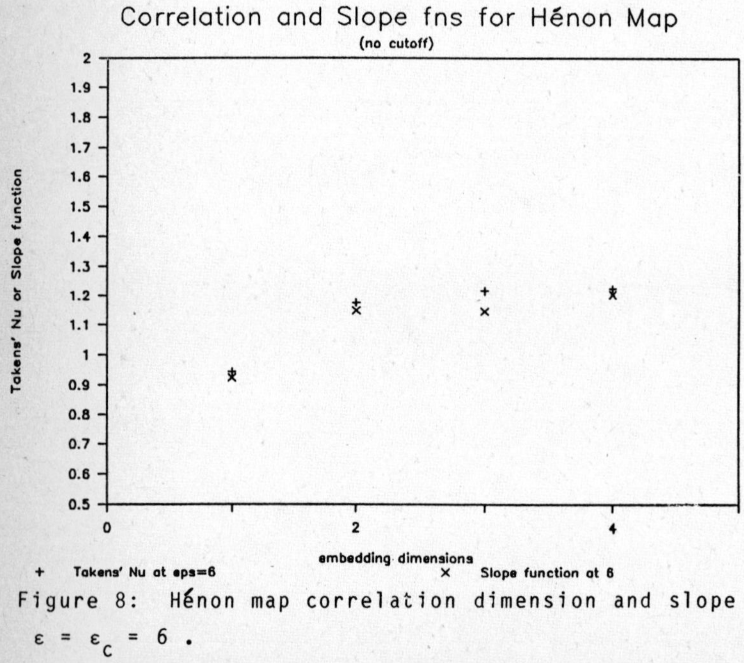

Figure 8: Hénon map correlation dimension and slope functions at $\varepsilon = \varepsilon_c = 6$.

At the point ε_c, the truncated and untruncated lattice slopes are within 10%. This suggests that in an untruncated data set with max $\varepsilon \approx 50$, $\varepsilon_c \simeq 6$ is about the largest ε that can be taken without incurring serious boundary effect errors. In Figure 8, we have therefore, plotted $\nu(\varepsilon_c)$, $S(\varepsilon_c)$ for the Hénon map vs. embedding dimension d. The curves do approach the expected value of 1.22.

Figure 9 shows a plot of $\nu(15)$, $S(15)$ for the truncated lattice data set. The curves reach plateaus at values 2.2, 2.1 respectively. The plots of $\nu(\varepsilon_c)$ and $S(\varepsilon_c)$ are not significantly different. The dimensions 2.2 and 2.1 are somewhat higher than the expected 2.0. This is indicative of the presence of randomness and noise in the lattice data set.

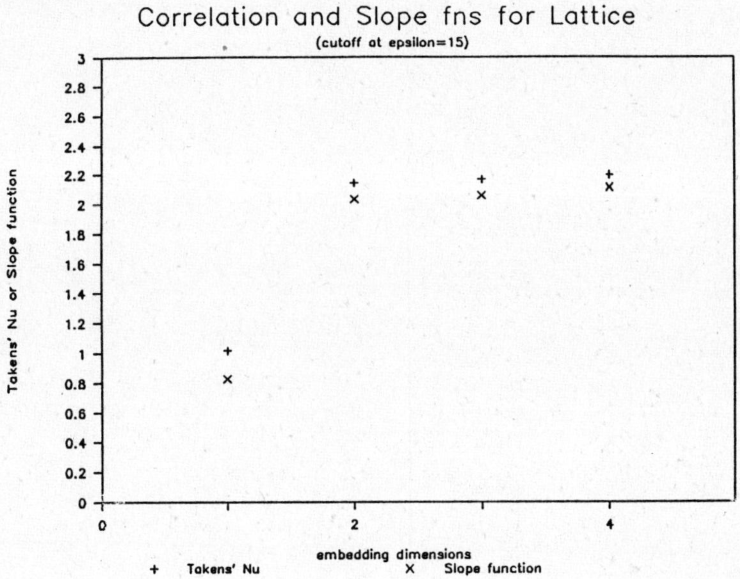

Figure 9: Lattice correlation dimension and slope functions at $\varepsilon = 15$.

7. CONCLUSION

In the data sets treated here, with N = 1200 points, Takens' method gives the correlation dimension to within 10%.

In an untruncated data set, the largest allowable ε seems to be about $\varepsilon_c \simeq .1 \times$ maximum ε. A somewhat larger $\varepsilon \simeq .3 \times \varepsilon_{max}$ can be used in a truncated data set.

ACKNOWLEDGEMENTS

This work was supported by the NSWC Independent Research Program, the Office of Naval Research and DARPA, under Order No. 5830, Program Code No: 6E20E.

REFERENCES

1. P. Grassberger and I. Procaccia, Phys. Rev. Lett. 50, 346 (1983).

2. F. Takens, "Invariants related to dimension and entropy", in Atas do 13° Coloqkio Brasiliero de Matematica, Rio de Janeiro, 1983, in "Dynamical Systems and Bifurcations", B. L. J. Braaksma, H. W. Broer and F. Takens, eds., L. N. Mathematics, No. 1125, Springer-Verlag, Berlin, 1985).

3. See, for instance, J. Doyne Farmer, Edward Ott and James A. Yorke, Physica 7D, 153 (1983) for discussion of a variety of notions of dimension.

4. P. Grassberger, Phys. Letters 97A, 227 (1983).

5. H. G. E. Hentschel and I. Procaccia, Physica 8D, 435 (1983); P. Grassberger and I. Procaccia, Physica 13D, 34 (1984).

6. H. S. Greenside, A. Wolf, J. Swift and T. Pignataro, Phys. Rev. A25, 3453 (1982).

7. L. P. Kadanoff, in "Perspectives in Nonlinear Dynamics", M. Shlesinger, R. Cawley, A. W. Saenz and W. W. Zachary, eds., World Scientific, Singapore, 1986; and I. Procaccia, in "Dimensions and Entropies in Chaotic Systems", G. Mayer-Kress, ed., Synergetics Series, Springer-Verlag, Berlin, 1986.

8. J. Holzfuss and G. Mayer-Kress, in "Dimensions and Entropies in Chaotic Systems", G. Mayer-Kress, ed., Synergetics Series, Springer-Verlag, Berlin, 1986.

9. F. Takens, in "Dynamical Systems and Bifurcations", B. L. J. Braaksma and H. W. Boer and F. Takens, eds., L. N. Mathematics, No. 1125, Springer-Verlag, Berlin, 1985.

10. H. D. Brunk, "An Introduction to Mathematical Statistics", Ginn and Company, Boston, 1960.

11. M. Henon, Commum. Math. Phys. 50, 69 (1976).

12. N. H. Packard, J. P. Crutchfield, J. D. Farmer and R. S. Shaw, Phys. Rev. Lett. 45, 712 (1980).

13. F. Takens, in "Dynamical Systems and Turbulence, Warwick, 1980", D. A. Rand and L.-S. Young, eds., L. N. Mathematics, No. 898, Springer-Verlag, Berlin, 1981.

14. John Guckenheimer in "Dynamical Systems and Chaos", L. Garrido, ed., L. N. Physics, No. 179, Springer-Verlag, Berlin, 1983. Guckenheimer attributes the delay-coordinate embedding procedure as apparently due to Ruelle, see also Ref. 12. Formulation of the procedure in the context of dynamical systems theory is given by Takens in Ref. 13. Also, the first physical demonstration of the representation of a system in a higher dimensional space from a single time-series, as well as the intuitive idea of the embedding procedure, was given in Ref. [12].

15. W. E. Caswell and J. A. Yorke, "Invisible Errors in Dimension Calculations: Geometric and Systematic Effects", in "Dimensions and Entropies in Chaotic Systems", G. Mayer-Kress, ed., Synergetics Series, Springer-Verlag, Berlin, 1986.

B. QUANTUM CHAOS

MICROWAVE IONIZATION OF HIGHLY EXCITED HYDROGEN ATOMS: EXPERIMENT AND THEORY

P.M. Koch and K.A.H. van Leeuwen
Physics Department
State University of New York at Stony Brook
Stony Brook, NY 11794-3800

O. Rath and D. Richards
Faculty of Mathematics
The Open University
Milton Keynes MK7 6AA, United Kingdom

R.V. Jensen
Applied Physics
Yale University
New Haven, CT 06520

Abstract. This article elaborates on a talk delivered by the first author at the First International Conference on the Physics of Phase Space (University of Maryland, 20-23 May 1986). It reviews briefly our still limited, but rapidly growing understanding of a dynamical process, the ionization of highly-excited hydrogen atoms by a microwave electric field. Classical dynamics explains surprisingly well many recent experimental results from Stony Brook, on which the article focusses. Some experimental results not well explained, however, appear to be essentially quantal in origin. These are just now beginning to be understood. New, detailed questions continue to arise as older questions are answered.

INTRODUCTION TO THE MICROWAVE IONIZATION PROBLEM

Chaotic classical dynamics[1], of either dissipative or non-dissipative non-linear systems, has certainly been one of the unifying themes of this rather broad conference. A quantal system, however, is governed by the linear Schroedinger equation, which seems to be the wrong kind of equation to exhibit the non-linear, local instability of classical chaos. Though a driven, quantal system with a discrete spectrum is said to be doomed to quasi-periodic behavior[2], we focus here on a real, time-dependent quantal process involving a continuous spectrum, the ionization of highly-excited hydrogen atoms by a linearly polarized microwave electric field. Since its experimental discovery[3] in the 1970's, extensive theoretical work[4] in one-[5,6], two-[7], and three[8]-spatial-dimensions has shown that ionization may be described classically via a transition to chaos. [Elsewhere in this volume Professor Jensen treats these classical dynamics, and Professor Bayfield reviews the status of recent experimental and theoretical work at Pittsburg.] We shall see that purely classical theories have gone a long way toward explaining recent experimental data on microwave ionization.

But classical theories do not explain all features of our data; some will only be explained by a quantal treatment of the dynamics. Indeed, the microwave ionization (and excitation) of hydrogen has emerged as one of the two testing grounds for the study of so-called "quantum chaos," a field[9] whose only satisfactory definition, thus far, is "the study of quantal systems whose classical counterparts exhibit chaotic dynamics." [The other involves the hydrogen atom in an intense magnetic field[10], a complementary problem involving a time-independent Hamiltonian.]

Below we shall be using (reduced mass) atomic units (au) in which $e=\mu=\hbar=1$. For atomic hydrogen ($\mu/m_e=0.999455$) the au of electric field is 5.13665×10^9 V/cm; the au of angular frequency is 4.13189×10^{16} rad/s. The non-relativistic Hamiltonian for the present problem is[4,8,11]

$$H(\vec{r},t) = p^2/2 - r^{-1} + zA(t)F_0\cos\omega t \qquad [1]$$

where $\vec{p}$ and $\vec{r}$ are, respectively, the momentum and coordinate operators for the (reduced mass, spinless) electron, and A(t) is the envelope function which describes the slow turnon and turnoff of the microwave electric field $F_0 \cos\omega t$. When $F_0=0$, the atom is usefully described by the parabolic quantum numbers $(n,n_1,|m|)$, its energy depends only on n via $E_n=-1/2n^2$ au, and the energy splitting between levels separated by Δn units is $\Delta E \simeq -\Delta n/n^3$ au. When $F_0 \neq 0$, the linear polarization of the field preserves the magnetic quantum number m. In the classical limit the Kepler frequency of the orbiting electron approaches the $\Delta n=1$ frequency splitting. The natural scaled variables in the classical dynamics are $n^3\omega$ au, the ratio of the applied frequency to the Kepler frequency, and n^4F_0 au, the ratio of the applied field to the Coulomb binding field. When $n^3\omega=1$, the classical Kepler frequency equals the applied frequency, whereas the quantal photon energy $\hbar\omega$ is very near the $\Delta n=\pm 1$ energy splittings. The Coulomb n to $(n+\Delta n)$ and n to $(n-\Delta n)$ energy splittings are unequal, but differ only by about $3(\Delta n)^2/n^4$ au. This small difference for $\Delta n \ll n$ is probably ignorable if n and n^4F_0 are both large, since strongly driven quantal transitions (stimulated emission and absorption) leading to ionization are significantly power-broadened and couple together a large number of field-free states.

EXPERIMENTAL TECHNIQUES AND DATA

All published hydrogen experiments have been in the range 5–12 GHz. Table 1 shows some useful numbers for 10 GHz. From left to right the columns show (1) the principal quantum number; (2) the electric field strength near which static field ionization sets in; (3) the ratio of the binding energy to the microwave photon energy, or the number of photons energetically required to get to the $F_0=0$ ionization limit; (4) the ratio of the n to (n+1) energy splitting to the photon energy, or the number of microwave photons needed to drive a $\Delta n=1$ transition (this ratio is also approximately the inverse of the scaled frequency $n^3\omega$); (5)[6] the approximate power, which drops as n^{-8}, that is required to produce in a typical microwave cavity [rectangular waveguide] the F_0-values shown in column 2.

For $\sim 10^1$ W of continuous microwave power, ionization experiments are limited to n>45 in a nonresonant waveguide, which allows ω to be varied over an entire microwave band, and to n>25 in a typical cavity, which resonates only at certain frequencies. Varying n for fixed ω, however, produces a stepwise variation of the scaled frequency $n^3\omega$. We used this strategy[12] in our recent experiments at Stony Brook, taking data at 9.9233(4) GHz throughout the range n=32–90. Throughout the resulting wide range of scaled frequencies, $n^3\omega$=0.05–1.1, ionization always requires a large number of photons, rising to several hundred at the lower n-values. Inter-n splittings correspond to the energy of many fewer photons. Classically, when $n^3\omega$ is of order unity, one would expect the most interesting dynamics: The oscillatory motion due to the microwave field could couple strongly with the Kepler orbital motion.

Fig. 1 shows some features of our experimental apparatus. An $\simeq$14-keV proton beam was partially neutralized in a Xe gas cell (not shown), producing a fast beam of H(n) atoms weighted in n approximately as n^{-3}. A double-resonance method[13] employing two CO_2 lasers excited those in the $(n,n_1,|m|)=(7,0,0)$ extremal parabolic state, via the (10,0,0) state, to a selected (n,0,0) state. Neither laser beam

n	F_0 for $n^4F_0=0.1$	$E/\hbar\omega$	$\Delta E_1/\hbar\omega$	P_{cav} (W)	P_{wg} (W)
10	51.4 kV/cm	3288.1	570.7	10 kW	3.1 MW
20	3.2 kV/cm	822.0	76.4	40 W	12 kW
30	0.6 kV/cm	365.3	23.2	1.5 W	0.4 kW
40	0.2 kV/cm	205.5	9.9	0.2 W	0.05 kW
50	0.08 kV/cm	131.5	5.1	0.03 W	7 W
60	0.04 kV/cm	91.3	3.0	6 mW	2 W
70	0.02 kV/cm	67.1	1.9	2 mW	0.5 W
80	13 V/cm	51.4	1.3	0.4 mW	0.1 W
90	8 V/cm	40.6	0.9	0.3 mW	0.07 W

Table 1. The assumed microwave frequency is 10 GHz.

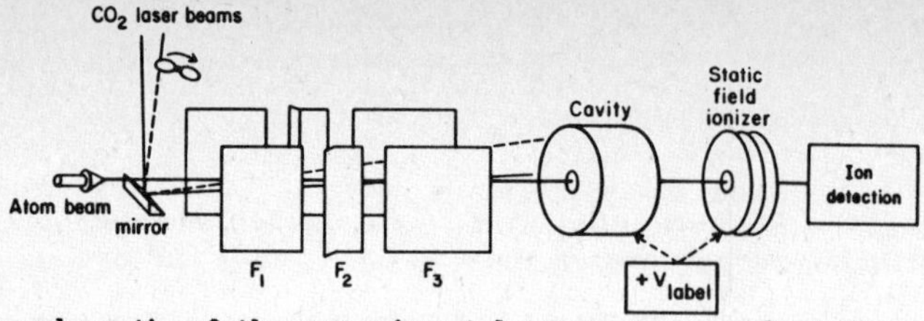

Fig. 1. A schematic of the experimental apparatus used at Stony Brook.

entered the microwave cavity. The unique substate distribution produced by the laser excitation was altered by stray fields before the atoms entered the cavity; auxiliary tests showed that it could be characterized by equally populated substates and unchanged n. This corresponds to a microcanonical distribution of classical electron trajectories filling all three spatial dimensions. Thus, a complete theoretical modeling of these experiments would need to be three dimensional. [It is possible experimentally[14] to begin with a very narrow substate distribution in the microwave region, but a superimposed static electric field F seems to be required[15] to preserve the "quasi-one-dimensionality" of the atoms in the oscillatory field. Correct theoretical modeling of this situation would necessitate adding a term +zF to Eq. 1 and evaluating its influence on the dynamics.]

We used two different methods, hereafter called "ionization" and "quenching," to cover the range n=32-90. For both methods, each atom experienced in its rest frame about 300 microwave oscillations inside the cavity with constant amplitude F_0 (determined to an estimated accuracy of $\pm$ 5%) between a slow rise and fall of the field [A(t) in Eq. 1] over about 40-80 oscillations in the spatially-varying microwave "fringe fields" near the holes in the cavity endcaps. In the "ionization" method, a static voltage V_{label}~200 V applied to the cavity body enabled "energy-labeled" detection[13] of protons produced inside the cavity. This greatly enhanced the signal-to-background ratio of the data.

Fig. 2 shows an "ionization" probability (P_{ion}) curve for n=64 at 9.923 GHz. Below $F_0 \simeq 18$ V/cm, P_{ion} is near zero and below the "noise" level in the data; above $F_0 \simeq 38$ V/cm, P_{ion} "saturates" at 100%. The smooth, monotonic [but see below, Fig. 6]

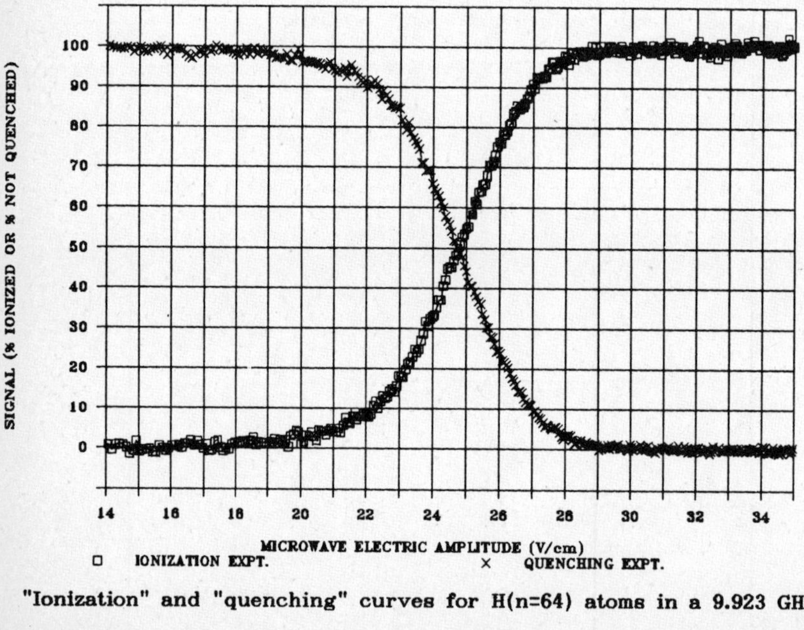

Fig. 2. "Ionization" and "quenching" curves for H(n=64) atoms in a 9.923 GHz field.

curve appears reasonable: P_{ion} increases as F_0 increases. Though the "ionization" method looks directly at ionization, the longitudinal static electric field F_{label} ~10^1 V/cm produced outside the cavity by V_{label} is its main disadvantage. This F_{label} can ionize[16] atoms with n>mid-70's.

The "quenching" method, for which the cavity V_{label}=0, avoids this problem. Referring to Fig. 1, the highly-excited atoms were ionized and detected well downstream of the cavity in a longitudinal, static electric field, and one recorded how this signal was "quenched" by the microwaves. That it does not directly measure ionization is the principal disadvantage of this method. Fig. 2, however, compares "quenching" and "ionization" curves taken under the same microwave system conditions. Notice the close match between P_{ion} and $P_{quenched}=1-P_{not\ quenched}$. We have determined from many such comparisons that for n=32 to about the lower 70's, even if the curves exhibit "structure" such as in Fig. 6, the two methods always give very nearly the same information. This is no longer the case for n>mid-70's, because of F_{label} in the "ionization" method. For n≃74 to n=90 we recorded only "quenching" data, but we interpret such data in terms of ionization.

Using a preliminary analysis of our data, Fig. 3 shows on a semi-logarithmic scale the F_0-values required to produce for each n-value P_{ion}=0.1 and P_{ion}=0.9, respectively. [The 10%-curve extends the set of data presented previously in Ref. 12, where is also discussed the effect of superimposing a static electric field with the microwave field. The 90%-curve is entirely new.] Both curves show a distinct "staircase"-like structure, with the "length" (in Δn) of the steps increasing with n and varying somewhat with P_{ion}. Both curves are non-monotonic in n for several ranges of n, especially, for P_{ion}=0.1, just below n=70 and from 83 to ~88. This is rather remarkable: As n increases the Coulomb binding force decreases as n^{-4}, yet F_0 actually increases.

COMPARISONS OF EXPERIMENT WITH THEORY: SOME ANSWERS, MANY NEW QUESTIONS

Fig. 4 presents the same data in classcially scaled variables, n^4F_0 versus $n^3\omega$. Now the staircases of Fig. 3 become series of bumps. At local maxima, the atoms are relatively more stable against microwave ionization; at local minima, they are relatively less stable. The stability may be associated[12] closely with rational

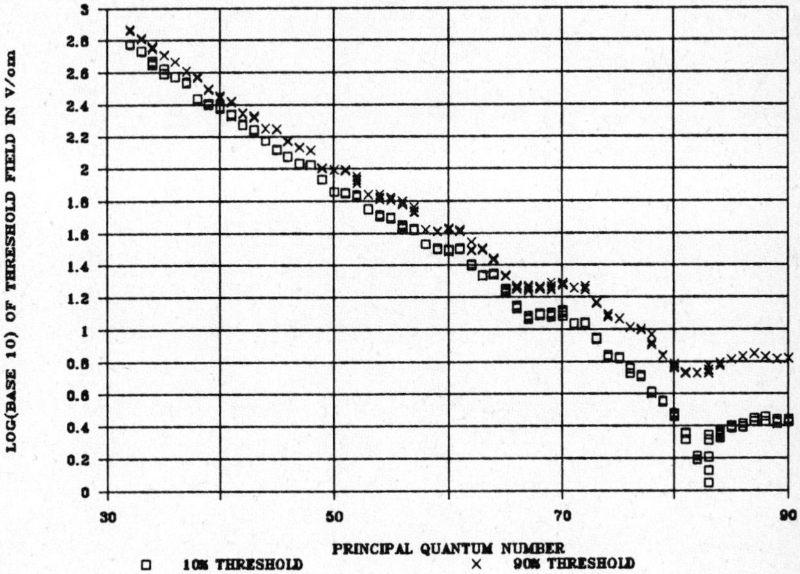

Fig. 3. 10% and 90% ionization threshold fields (at 9.923 GHz) in V/cm for H(n=32-90).

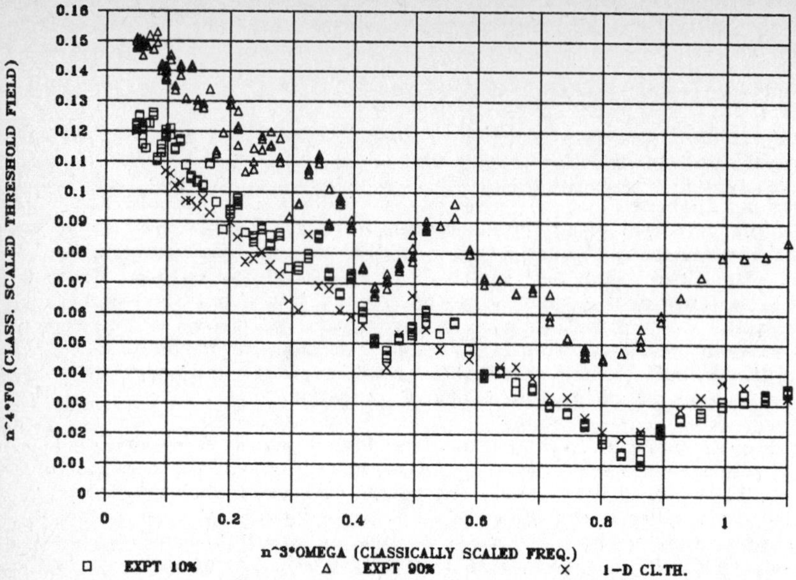

Fig. 4. Classically scaled ionization threshold fields n^4F_0 versus classically scaled frequency $n^3\omega$: P_{ion}=0.1 (3-D experiment and 1-D classical estimates); P_{ion}=0.9 (3-D experiment).

frequency ratios such as 1/1, 2/3, 1/2, 2/5, 1/3, 1/4, 1/5, and 1/6. Classical stability may be correlated[6-7] with non-linear resonance trapping regions, or "islands" in the classical phase space. These may be visualized in "phase-space portraits" obtained from 1-D treatments[5,6] of the dynamics. Fig. 1 in Professor Jensen's contribution elsewhere in this volume (which describes the stabilizing role of the slow turnon [A(t) in Eq. 1] of the field in the experiments) clearly shows

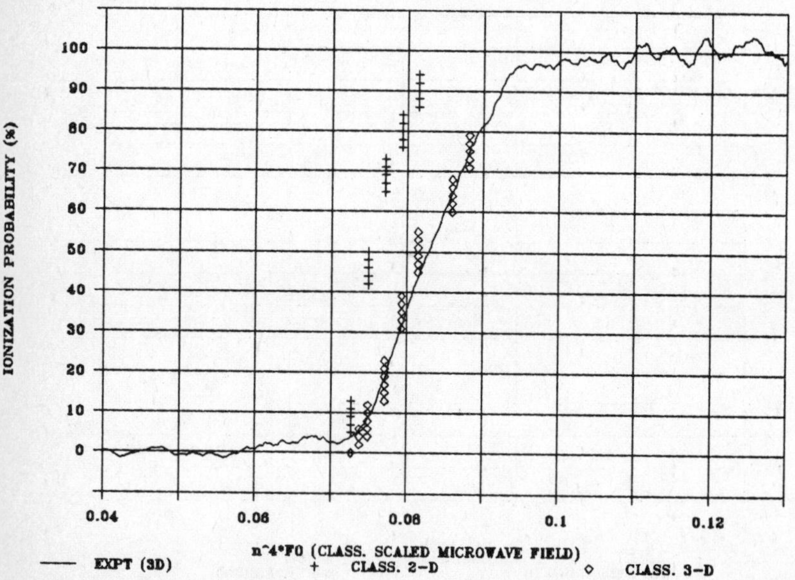

Fig. 5. Ionization of H(n=58) at 9.923 GHz: experiment (3-D) and classical Monte Carlo calculations (2-D and 3-D).

for n^4F_0=0.03 and $n^3\omega$ =1 (for n=88) the island associated with the frequency ratio 1/1 embedded in a "chaotic sea." Fig. 4 also shows 1-D, classical estimates (for a slow turnon of the field) of ionization thresholds[5], which follow rather closely the n-dependence of the experimental 10% thresholds.

Recalling that it is a 3-D ensemble of atoms in the Stony Brook experiments, this agreement with 1-D thresholds may be somewhat surprising. Though the 1-D, classical dynamics produces reasonable thresholds, it does not accurately reproduce experimental ionization curves. This requires calculations in higher dimensions. Fig. 5 compares a (3-D) experimental n=58 curve with 2-D and 3-D classical Monte Carlo simulations[8] of the experiment, including 300 oscillations of the field and its slow turnon and turnoff. With no adjustable parameters, the classical 3-D calculation excellently reproduces the experimental curve. Though the 2-D curve begins its rise at about the same F_0-value, its slope is too large. A 1-D curve (not shown) would rise even more abruptly.

Detailed comparisons[17] between experimental[18] and classical 3-D[8] curves are in progress. Agreement appears to be excellent for many n-values, but there are exceptions that seem to fall into two classes. One class involves n-values on or very near the "bumps" in Fig. 4. For example (not shown), for n=69, where $n^3\omega$=0.496 is closest to the "1/2 resonance," the experimental ionization curve lies significantly above the classical 3-D curve, although for n=68 and n=70, agreement is quite good. Conversely, for n=61 (not shown), where $n^3\omega$=0.343 is close to the "1/3 resonance," the experimental ionization curve lies significantly below the classical 3-D curve. Agreement is excellent for n=60, however, and very good for n=62 when P_{ion}<0.2. Though the experiments involve large quantum numbers, this does not necessarily mean that the "classical limit" has been reached. The resonance islands in the 1-D classical "phase space portraits" have a width in action of only a few units of $\hbar$, thus supporting only a few quantal states. One suspects the importance of quantal effects in such cases.

The other class of exceptions includes about a dozen ionization curves between n=32-61 that have "structure." [11,18] In some cases this may involve abrupt changes of slope or steps; in a few exceptional cases, a non-monotonic "bump." Fig. 6 shows two such cases. The n=34 experimental curve has a step near threshold. The n=38 curve has a distinct bump; as F_0 increases, P_{ion} rises, falls, then rises again. Classical 3-D Monte Carlo calculations[8] for n=38 accurately reproduce the second rise of the experimental curve but miss the bump entirely. It appears here, too, that quantal effects are playing a role.

This suspicion is strengthened by very recent 1-D quantal calculations[19] carried out for a sudden turnon of the field. The projection operator method couples a

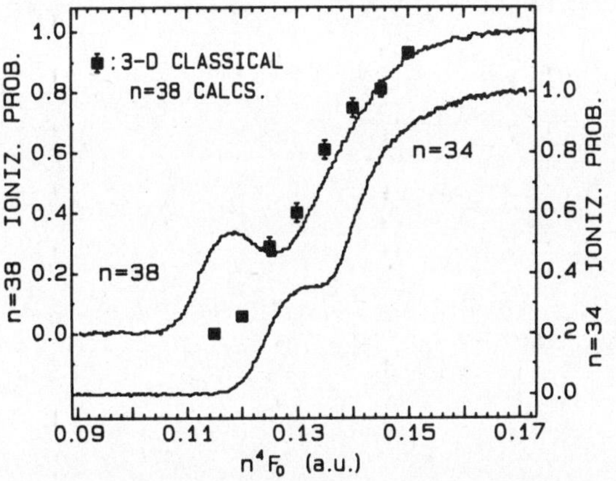

Fig. 6. Solid lines: 3-D experimental ionization curves at 9.923 GHz for H(n=38) (left vertical axis) and for H(n=34) (right vertical axis); squares: 3-D classical Monte Carlo calculations for H(n=38).

large number of bound states with themselves and with the continuum, while neglecting continuum-continuum coupling, whose influence is said to be less important. The proposed ionization mechanism involves those states excited into a relatively narrow "ionization window" being strongly coupled to the continuum. The method predicts "subthreshold" (i.e., sub-classical-threshold) non-monotonic structure. This is correlated with anti-crossings between "quasi-energy states," the eigenstates of the quantal, one-cycle, "Floquet" operator.[30] It appears from preliminary comparisons[21] that such calculations produce at least some of the experimental structure, particularly the large n=38 bump shown in Fig. 6. One remarks again that 1-D dynamics, here quantal, seems to capture much of the relevant physics of the 3-D experiments.

Notice that this "subthreshold" ionization mechanism makes the quantal system less stable than the corresponding classical system. This is in contrast to another predicted[22] quantal effect that would make the quantal 1-D system be more stable than the classical system against microwave ionization. This quantal effect is somewhat related[23] to Anderson localization in disordered solids. "Delocalization" of the quasi-energy states (their spreading throughout the bound basis and into the continuum) is a quantal route to ionization. For lower n-values or for a range of $n^3\omega>1$ somewhat outside the range of present experiments, "localization" is predicted[22] to raise the quantal ionization threshold above the classical one. This "localization" effect, however, is predicted[22] to play a less important role in dimensions higher than one, a potentially crucial point for experiments designed to search for it. Recent 1-D quantal calculations[24] for a sudden turnon of the field found no evidence for this effect between $n^3\omega=0.4-1.05$. Over this range the 1-D quantal[24] and 1-D classical[6] 10% ionization thresholds (but not the 90% thresholds) are in good agreement. Because of the sudden turnon, there is only semi-quantitative agreement with the 10% thresholds shown in Fig. 4.

Others, including, independently, the last two authors of this contribution, are actively working on the quantal dynamics. Many questions are now open. Space limitations preclude summarizing here the points raised throughout this contribution, but one thing is clear: The problem of the microwave ionization (and excitation) of highly-excited hydrogen atoms has emerged as a fundamental experimental and theoretical testing ground for the study of quantal dynamics in a system whose classical counterpart is known to be chaotic. This provides an important route to understanding what is-- and what is not-- "quantum chaos."

ACKNOWLEDGMENTS

P.M.K., K.A.H.v.L., and R.V.J. appreciate research support from the National Science Foundation. R.V.J. is grateful for receipt of an Alfred P. Sloan Foundation Fellowship and O.R. for an Open University Studentship. D.R. acknowledges support from the S.E.R.C. Many people, especially K. Altman, C. Baars, T. Bergeman, J. Bowlin, S. Hink, L. Lilly, A. Mortazawi-M., G. v. Oppen, D. Ramos, S. Renwick, B. Sauer, P. Scharf, and K.-U. Starke contributed to the experiments at Stony Brook. All the authors appreciate the many communications with the growing number of researchers in this field.

REFERENCES

1. From a vast literature we suggest A.J. Lichtenberg and M.A. Lieberman, "Regular and Stochastic Motion," Springer-Verlag, New York (1983); articles by L.P. Kadanoff and by M.C. Gutzwiller in Physica Scripta, Vol. T9 (1985), a special volume on "The Physics of Chaos and Related Problems"; J.R. Ackerhalt, P.W. Milonni, and M.-L. Shih, Phys. Rep. 128:205(1985).

2. T. Hogg and B.A. Huberman, Phys. Rev. Lett. 48:711(1982); Phys. Rev. A28:22(1983).

3. J.E. Bayfield and P.M. Koch, Phys. Rev. Lett. 33:258 (1974); J.E. Bayfield, L.D. Gardner, and P.M. Koch, Phys. Rev. Lett. 39:76 (1977); reviewed in P.M. Koch, J. Phys. (Paris), Colloq. 43:C2-187 (1982).

4. J.G. Leopold and I.C. Percival, Phys. Rev. Lett. 41:944(1978) and J. Phys. B12:709 (1979); D.A. Jones, J.G. Leopold, and I.C. Percival, J. Phys. B13:31 (1980); B.I. Meerson, E.A. Oks, and P.V. Sasorov, J. Phys. B15:3599 (1982); Soviet and other literature on this problem is reviewed in N.B. Delone, V.P. Krainov, and D.L. Shepelyansky, Usp. Fiz. Nauk 140:355 (1983) [Sov. Phys. Usp. 26:551 (1983)].

5. R.V. Jensen, Phys. Rev. A30:386 (1984).

6. J.G. Leopold and D. Richards, J. Phys. B18:3369 (1985).

7. J.G. Leopold and D. Richards, J. Phys. B19:1125(1986).

8. O. Rath and D. Richards, in preparation.

9. From a growing literature, we suggest G.M. Zaslavsky, Phys. Rep. 80:157(1981); see, also, many of the articles in "Chaotic Behavior in Quantum Systems: Theory and Applications," edited by G. Casati, Plenum, New York (1985).

10. D. Delande and J.C. Gay, Phys. Rev. Lett. 57:2006(1986).

11. P.M. Koch, in Ref. 3. D.R. Mariani and W. van de Water contributed importantly to the microwave ionization results presented in this reference.

12. K.A.H. van Leeuwen, G. v. Oppen, S. Renwick, J.B. Bowlin, P.M. Koch, R.V. Jensen, O. Rath, D. Richards, and J.G. Leopold, Phys. Rev. Lett. 55:2231 (1985); P.M. Koch, in "Fundamental Aspects of Quantum Theory," A. Frigerio and V. Gorini, editors, Plenum, New York (1986).

13. P.M. Koch, in: "Rydberg States of Atoms and Molecules," R.F. Stebbings and F.B. Dunning, editors, Cambridge University Press, New York (1983).

14. J.E. Bayfield and L.A. Pinnaduwage, Phys. Rev. Lett. 54:313 (1985) and J. Phys. B18:L49 (1985); J.N. Bardsley, B. Sundaram, L.A. Pinnaduwage, and J.E. Bayfield, Phys. Rev. Lett. 56:1007(1986); J.E. Bayfield, elsewhere in this volume.

15. J.G. Leopold and D. Richards, J. Phys. B (to be published, 1987).

16. P.M. Koch and D.R. Mariani, Phys. Rev. Lett. 46:1275 (1981).

17. K.A.H. van Leeuwen, P.M. Koch, O. Rath, D. Richards, J.G. Leopold, and R.V. Jensen, in preparation.

18. K.A.H. van Leeuwen and P.M. Koch, in preparation.

19. R. Bluemel and U. Smilansky, Proceedings of Adriatico Research Conference on Quantum Chaos, Trieste, June 1986, to be published in Physica Scripta, 1987.

20. Structure in microwave experiments with helium Rydberg atoms has also been linked to avoided crossings of its Floquet eigenstates; see W. van de Water, K. van Leeuwen, P. Koch, and T. Bergeman, Bull. Am. Phys. Soc. 31:942(1986).

21. R. Bluemel, U. Smilansky, P.M. Koch, K.A.H. van Leeuwen, O. Rath, and D. Richards, in preparation.

22. G. Casati, B.V. Chirikov, and D.L. Shepelyansky, Phys. Rev. Lett. 53:2525(1984); G. Casati, B.V. Chirikov, D.L. Shepelyansky, and I. Guarneri, Phys. Rev. Lett. 57:823(1986).

23. S. Fishman, D.R. Grempel, and R.E. Prange, Phys. Rev. Lett. 49:509(1982); Phys. Rev. A29:1639(1984).

24. J.N. Bardsley and M.J. Comella, J. Phys. B19:L565(1986).

HIGHLY EXCITED HYDROGEN IN MICROWAVES: MEASUREMENTS ON THE EXTERNALLY DRIVEN BOUND ELECTRON AT THE CLASSICAL THRESHOLD FOR CHAOS

J. E. Bayfield and L. A. Pinnaduwage
Department of Physics and Astronomy
University of Pittsburgh
Pittsburgh, PA 15260

Many-photon n-changing transitions in electrically polarized highly excited hydrogen atoms have been studied experimentally [1-4] as well as theoretically using both classical theory [5-6] and quantum theory [4-7]. The experiments involve stretching the atom along the direction of an applied external static electric field, and then driving its electron by a linearly polarized microwave field directed along the stretching direction. This configuration results in close to one-dimensional behavior, making possible the many-state numerical quantum calculations needed to investigate possible quantum correspondences to various features of the system's classical nonlinear dynamics.

The experiments expose the stretched highly excited atoms to a pulse of about 3000 microwave field oscillations. At low microwave fields, resonant multiphoton n-changing transitions are observed. At somewhat higher fields, a second type of n-changing that is independent of microwave frequency also occurs. At still higher fields, relatively large changes in n become observable along with finite probabilities for ionization.

There appears to be a correspondence between the above observed types of quantum transitions and the types of classical electron motion conveniently displayed in Poincare phase space plots. It is suggested by the sum total of correspondence principles, including the one between electric dipole matrix elements and the Fourier components of the classical time-development of the position of the electron [8]. The resonant multiphoton transitions correspond to the classical resonance structures called islands and island chains; if the classical trajectory winding number for periodic trajectories is defined as the ratio of the number of free-atom orbital periods to the number of microwave periods, then it is a rational number equal to the ratio of the change in quantum number to the number of photons absorbed in the quantum transition. Thus, as expected by the correspondence principle, the sets of classical and quantum resonances of the system are essentially the same at large quantum numbers. The correspondence principle does not appear rigorously established for the nonperturbative regimes corresponding to nonresonant n-changing and ionization. Yet it seems that the nonresonant transitions correspond to motion on non-island KAM surfaces; at low fields this would be expected to be primarily adiabatic, being limited to just one surface. Breakup of the KAM surfaces at high fields characterizes the transition to the classically stochastic regime; this would correspond to the nonadiabatic n-changing to high n-values, along with ionization.

Some idea of the nature of the nonresonant n-changing can be obtained by considering the value of the single largest two-state Rabi-flopping or population-changing probability amplitude for each field-free state of the atom. These probability amplitudes are of order unity for states with quantum number down close to the initial state, as well as for those states higher. Then the Rabi-flopping frequencies become comparable to both the free-atom electron orbital frequencies and the microwave frequency; classically this is the situation where the electron orbits are noticeably changed every free-atom orbit period. Many one-photon n-changing resonances become broadened so that they highly overlap. Beginning then with atoms in a state of a given quantum number, as time develops a larger and larger number of states become populated and participate in determining the net amounts of n-changing. Thus we have the already suggested idea of a "diffusion" of microwave energy into the atom.

The quantum calculations must close-couple many states above some lowest value to obtain final-state probability distributions. The quantum calculations have identified a transition region of quantum number values, below which the quasienergy states of the atom-field system primarily are localized and above which they primarily are delocalized; for many choices of microwave field and frequency in the region of our experiments, this quantum transition region corresponds to the classical transition region near the "top KAM" surface in the Poincare plots [4-9]. Totally delocalized quasienergy states correspond to classically stochastic electron trajectories. If one starts a quantum calculation out at a quantum number value in the region of delocalized quasienergy states, then a roughly exponential final-state quantum number distribution results [9].

The measured final-state population distributions show a narrow peak about the initial quantum number, identified with the adiabatic n-changing, and an exponential tail at large n to be identified with the nonadiabatic n-changing [4] . It has been suggested that such tails in the state-distribution of the quasienergy states of the atom-field system are related to the minimal positive Lyapunov exponent for the classical motion [10]. There may be frequencies where the measured final-state distributions are dominated by one quasienergy state that has a noticeably delocalized character; this could account for the observed 3000-cycle exponential tails.

Thus, many of the outward classical features of this nonlinear dynamical system that classically exhibits the period-doubling bifurcation route to deterministic chaos have been seen both in experiments and in numerical quantum calculations.

1. J. E. Bayfield and L. A. Pinnaduwage, Phys. Rev. Lett. 54, (1985) 313.
2. J. E. Bayfield and L. A. Pinnaduwage, J. Phys. B 18, (1985) L49.
3. J. E. Bayfield, invited paper presented at the conference on Fundamental Aspects of Quantum Theory, Como, Italy, Septermber, 1985, to be published by Plenum Press.
4. J. N. Bardsley, B. Sundaram, L. A. Pinnaduwage and J. E. Bayfield, Phys. Rev. Lett. 56, (1986) 1007.
5. R. V. Jensen, Phys. Rev. Lett. 54, (1985) 2057.
6. J. G. Leopold and D. Richards, J. Phys. B 18, (1985) 3369.
7. G. Casati, B. V. Chirikov and D. L. Shepelyansky, Phys. Rev. Lett. 53, (1984) 2525.
8. I. C. Percival and D. Richards, Adv. Atomic and Molecular Phys. 11, (1975) 1-82, Section 3.
9. B. Sundaram, Ph. D. dissertation, University of Pittsburgh, 1986.
10. D. L. Shepelyansky, Phys. Rev. Lett. 56, (1986) 677.

TRANSITION STRENGTH FLUCTUATIONS AND THE ONSET OF CHAOS

Y. Alhassid

A. W. Wright Nuclear Structure Laboratory
Yale University, New Haven, Connecticut 06511

Abstract
The maximum entropy formalism is used to examine the signature of classical chaos in the fluctuations of the transition strength in the corresponding quantum mechanical system.

1. Introduction

The onset of chaotic motion in Hamiltonian classical systems of few degrees of freedom is by now well understood [1]. Quantum mechanics on the other hand is a linear theory so that the concept of "quantum chaos" is a vague one. However, one does expect to observe signatures of the onset of classical chaos in the corresponding quantum systems[2]. Recently, considerable effort in this direction has been concentrated on statistical properties of the energy spectrum[3,4]. Although counter-examples exist, it seems that in a generic situation the nearest neighbor level spacing distribution **changes from a Poisson** to a Wigner distribution as the corresponding classical motion changes from regular to chaotic. It was further suggested[3,4] that the chaotic regime has the statistics of a random matrix model, namely that of the Gaussian orthogonal ensemble (GOE). The GOE is known to provide a reasonable description of the statistical properties of complex heavy nuclei where the number of degrees of freedom is large[5]. However, more incisive tests beyond spectral statistics are necessary before a similar description is adopted in systems with few degrees of freedom in their chaotic regime. We thus suggest a study of transition strength distributions as well, to find a signature of the associated classical chaos[6]. The transition strength of a particular probe (operator) T is defined by

$$y = |\langle f|T|i\rangle|^2, \tag{1}$$

where $|i\rangle$ is a fixed initial eigenstate and $|f\rangle$ is any final eigenstate if the system is Hamiltonian. We can construct a density function P(y) such that P(y)dy measures the probability of locating the transition strength in the interval dy around y. We shall argue that in the extreme chaotic limit we expect to obtain a Porter-Thomas[7] distribution and that for non-chaotic systems the distribution is narrower. Our conclusions are shown to follow from random matrix theory or alternatively from maximal entropy considerations. The latter are better suited for a system whose Hamiltonian is well defined.

2. Random-matrix models

The collection of final states $|f\rangle$ of a particular Hamiltonian is represented by the n'th eigenstates (n fixed) of an ensemble of Hamiltonians (GOE). If the dimension N of the Hilbert space is large, we obtain for the strength (1) a Porter-Thomas distribution[5]

$$P(y) = (2\pi\langle y\rangle)^{-1/2}\, y^{-1/2} \exp(-y/2\langle y\rangle). \tag{2}$$

A simple ansatz used[4] to describe intermediate situations (between regular and chaotic motion) is an ensemble where each GOE matrix element is modified by an $\exp(-(i-j)^2/\sigma^2)$ cut-off factor. The width σ interpolates between the Poisson (σ=0) and GOE ($\sigma\to\infty$) limits. The strength distribution predicted by this random matrix ensemble has a width decreases from that of a Thomas-Porter distribution as $\sigma\to 0$.

3. Maximal entropy approach

The strength function is constrained by a sum rule $\sum_f |\langle f|T|i\rangle|^2 = \langle i|T^\dagger T|i\rangle$ satisfied by T. When the amplitude $x = \langle f|T|i\rangle$ is real this sum rule can be written as

$$\int_{-\infty}^{\infty} x^2 p(x)dx = \langle i|T^\dagger T|i\rangle/N \quad . \tag{3}$$

We assume that when the system is fully chaotic, the final states are devoid of any individual characteristics so that no other constraints except the sum-rule (3) are imposed. Maximizing the distribution entropy $S[P] = -\int dx\, P(x)\ell nP(x)$ under that constraint we find the Porter-Thomas distribution (2). In statistical terms it is just a χ^2 distribution with one degree of freedom.

When the system is more regular we expect to have more accurate semi-classical estimates so that propensity rules become important and additional constraints have to be imposed. This will necessarily make the distribution narrower when compared with a Porter-Thomas one with the same average strength $\langle y\rangle$. A simple ansatz for such an additional constraint is the averaged deviance of the strength y from it average value (given by an average "surprisal") $I = -\int_0^\infty dy\, P(y)\, \ell n(y/\langle y\rangle)$. The maximal entropy distribution is now

$$P(y) = \left[(\nu/2^{\frac{\nu}{2}}/\Gamma(\tfrac{\nu}{2})\right](y/\langle y\rangle)^{\frac{\nu}{2}-1} \exp(-\nu y/2\langle y\rangle) \quad . \tag{4}$$

It is a χ^2 distribution in ν degrees of freedom and its width $\sqrt{2/\nu}$ decreases with increasing ν as shown in Figure 1. It should be emphasized that we do not necessarily expect (4) to describe the strength distribution in intermediate situations. Rather, it is used to illustrate the deviation from the limiting Porter-Thomas distribution and has a parameter ν that allows for simple control of the width. We expect ν to decrease towards 1 as the system becomes more chaotic.

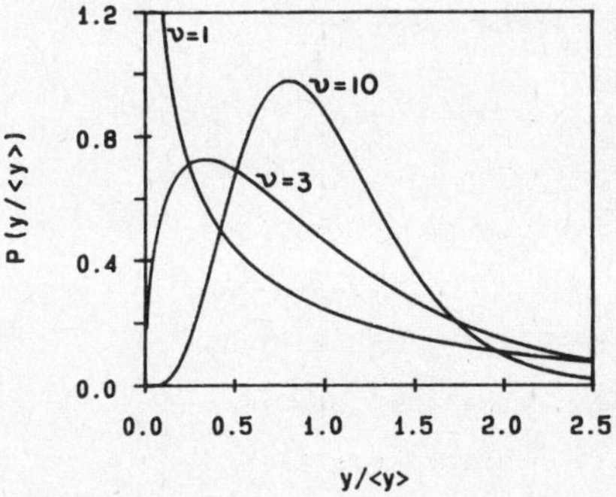

Fig. 1. χ^2 distributions with ν=1,3 and 10.

4. Henon-Heiles type potential

To illustrate our ideas we use a system with two degrees of freedom - an Henon-Heiles potential with a stabilizing term $V(X,Y) = \frac{1}{2}(X^2+Y^2) + \varepsilon(X^2Y-Y^3/3) + C(X^2+Y^2)^2$.

As a generic probe we have chosen the operator which displaces the potential surface by given amounts α and β along the X and Y directions, respectively. It is important to note that in constructing the strength distribution, strengths with different energies can be grouped together so that it is necessary to factor out their secular variation with energy[6].

The onset of classical chaos is at an energy E_c, about 2/3 of the dissociation energy of the C=0 potential. For every initial state we have divided the final states into two groups - below and above E_c - and calculated the strength distribution separately for each group. We have then fitted a χ^2 distribution by optimizing ν. Typical results are shown in Figure 2. Since the matrix elements are complex (while the theory considers real matrix elements) the limiting chaotic case corresponds to $\nu \simeq 2$. Our results confirm these expectations. When both the

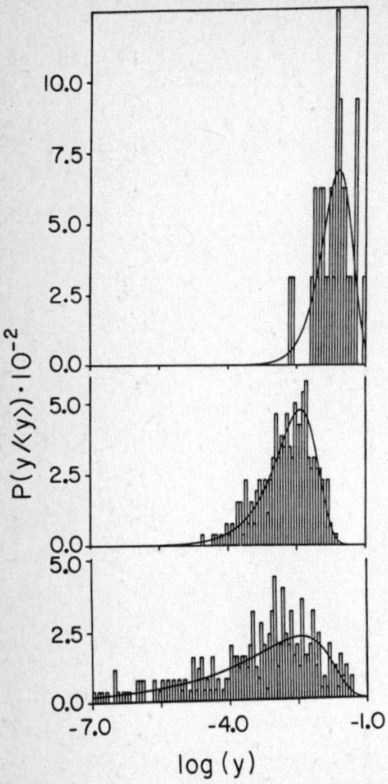

initial and final states are above E_c, $\nu \simeq 2.5$ (top of Fig.2). When the initial state is around E_c and the final states above, $\nu \simeq 3.8$(middle); and when both initial and final states are below E_c, $\nu \simeq 5.4$ and the fit is not quite acceptable. The width effect is best seen when initial and final states are of comparable energies, since for for states widely separated in energy the effect is hindered by the smallness of the matrix elements. To get enough statistics we can consider all initial states in a given energy band.

Fig. 2. Histograms of computed transition strengths for the shift operator with $\alpha=\beta=0.5$ and their fit by (4). Note the increased width as the corresponding classical system becomes more chaotic (see text).

To conclude, it seems that although the distributions depend on the probe, the general trend of ν decreasing towards the limiting Porter-Thomas value when the system becomes more chaotic is a universal feature.

I thank my collaborator R.D. Levine. I also thank Y.M. Engel and J.M. Brickman for their contribution in the computational study. This work was supported in part by the Department of Energy contract DE-AC02-76ER 03074. Y.A. is an Alfred P. Sloan Fellow.

References

1. For review see A.J. Lichtenberg and M.A. Libermann, Regular and Stochastic Motion (Springer-Verlag, Berlin, 1983).
2. For review see E.B. Stechel and E.J. Heller, Ann. Rev. Phys. Chem. 35, 563 (1984).
3. O. Bohigas, M.J. Giannoni, and C.Schmidt, Phys. Rev. Lett. 52, 1 (1984).
4. T.H. Seligmann, J.J.M. Verbaarschot, and M.R. Zirnbauer, Phys. Rev. Lett. 53, 215 (1984).
5. T.A. Brody, J. Flores, J.B. French, P.A. Mello, A. Pandey, and S.S.M. Wong, Rev. Mod. Phys. 53, 385 (1981).
6. Y.Alhassid and R.D. Levine, Yale preprint YNT86-11, to be published.
7. C.E. Porter and R.G. Thomas, Phys. Rev. 104, 483 (1956).

APPLICATION OF PHASE SPACE TO QUANTUM STATICS AND CLASSICAL ADIABATICS

J H Hannay
HH Wills Physics Laboratory
University of Bristol
Tyndall Avenue
Bristol BS8 1TL
U.K.

This contribution is divided into two entirely separate parts with nothing in common except that they are both firmly rooted in phase space. The second is purely classical - to do with a slowly changing Hamiltonian, while the first involves a static Hamiltonian, and, ultimately, some quantum mechanics. First though, I want to argue for the purely classical principle on which it is based - the 'principle of uniformity'[1].

I'll sacrifice generality and describe only the simplest circumstance - a dynamics governed by an area preserving (Poincaré) map of a rectangle to itself. (Instances being the 'bounce map' of free motion in a 2D enclosure - 'billiards', or the abstract linear map of a torus to itself - the 'cat map' of Arnold[2]). Moreover, I'll take the map to be an ergodic map so that a typical point $\underline{r}_o$ jumps about gradually filling the whole rectangle evenly. If a δ-function spike is sited on each iterate of the point we get a perfect 'lawn of δ-grass'. The function that the lawn tends to if it is suitably normalized by dividing by the number of spikes is a box function of unit volume over the lawn

$$\lim_{N\to\infty} \frac{1}{N}\sum_{n=1}^{N} \delta(\underline{r}-\underline{r}_n) = \text{Rectangle area}^{-1},$$ <u>a constant</u> independent of r and the initial position r_o.

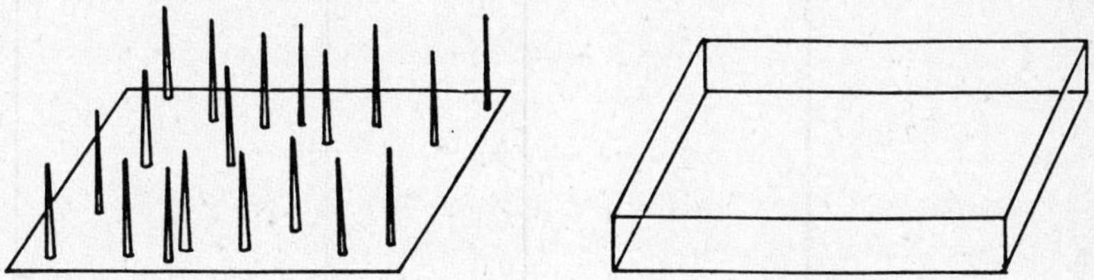

Now we let the formula bite its own tail by taking $\underline{r}=\underline{r}_o$ and interpreting what it then says. As a function of $\underline{r}_o$, $\delta(\underline{r}_o-\underline{r}_n)$ picks out with a spike those points $\underline{r}_o$ which return to themselves after n mappings - the 'periodic points' with period n. So this function itself consists of not one, but a (finite) number of spikes which, however, need not have the same 'strength' as each other because it is a δ-function of a function ($\underline{r}_n=r_n(\underline{r}_o)$), so there is an implicit Jacobian derivative giving the weighting strengths. Taken all together though, the superpositon of these functions for all n, yields, according to the formula, a uniform distribution. Periodic orbits taken with their natural weightings are uniformly distributed in phase space. And in a definite sense this holds, too, for the opposite extreme of integrable systems (for example, the circular billiard). The argument just given certainly does not claim to be completely watertight (it implicitly involves the interchange of two limits). Nonetheless it invites pursuit. The application to quantum energy eigenvalues in the next paragraph requires not the full principle of uniformity but rather the 'sum rule' for periodic orbit weights obtained by integrating the uniform function over the area of the rectangle.

The basis for the application is the semiclassical ($\hbar\to 0$)'periodic orbit sum' formula for the density of states due to M.C.Gutzwiller[3]. It identifies correction terms to the simple minded 'one state per Planck cell in phase space' rule for the average density of states. Each correction is associated with a periodic orbit: it is an oscillatory exponential whose 'period' is the action around the orbit divided by $\hbar$, and whose amplitude is the square root of the natural weighting just described. The oscillatory corrections superimpose on the average density of states curve to give, as contributions from longer and longer orbits are included, a function approximating a δ-function on each energy level.

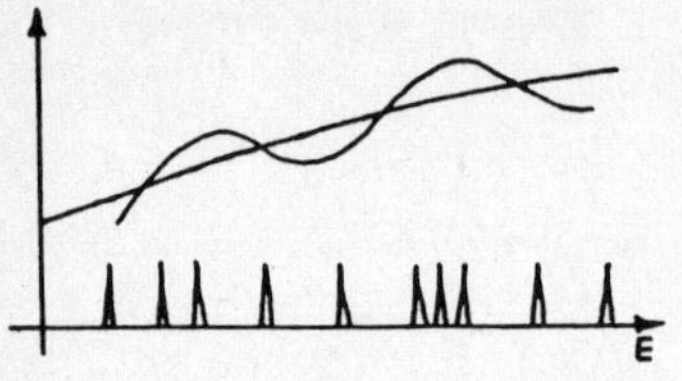

To make a connection with uniformity, we would like to square up the amplitudes to get the orbit weights. This can be done by considering not the density of states itself, but its correlation function. Appropriately defined, the double summation this produces reduces to a single one containing the required squares. The sum rule on the weights then supplies not the entire correlation function, but its behaviour for pairs of states separated by many others. The implied behaviour for ergodic systems mimics the equivalent limit of the correlation function of eigenvalues of random matrices, while that for integrable systems mimics a Poisson random process in accordance with expectations based on other arguments.

The other topic concerns the adiabatic change of a one freedom classical Hamiltonian. The standard illustration of this is the 'shortening pendulum' where a weight swings to and fro frictionlessly on the end of a string which is pulled up slowly through a little hole. A question is: how does the amplitude of the swing change in this process? Work is done on the pendulum so its energy is not conserved, but there is a well known substitute, the action, whch is conserved if the change is slow enough, to an accuracy, in fact, which is exponential in the slowness for a smooth process. In this example the action is the swing energy divided by the frequency - in general it is the area of the phase space contour around which the particle is moving. The Hamiltonian changes (long pendulum becomes short pendulum), but the particle keeps on that contour of the instantaneous Hamiltonian which has the same area as the original.

Now I want to consider the limitations of this rule[4]. It relies on the 'frequency' of the oscillation being much faster than the time scale of change of the Hamiltonian, so it may break down if the frequency is made to go to zero. The way to make this happen is to consider a double potential well in which the particle is oscillating initially above the barrier. One gradually raises the barrier until at a crucial moment the particle has to decide to 'fall into' one side or the other. Near the crucial moment it spends a long time either just making it over the barrier, or just being reflected from it so its oscillation frequency is near to zero, as required.

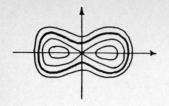

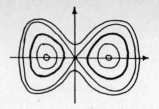

Phase spatially, the contours of the Hamiltonian function are loops inside and outside a figure of eight shape - the separatrix contour whose energy is that of the barrier top. Initially the separatrix is small and the particle traverses a dog's bone-shaped orbit outside it. But the separatrix grows as the barrier is raised and the dog's bone narrows in the middle trying to maintain its area until at the crucial moment the separatrix passes through it and it is split into two loops, around just one of which the particle moves. Evidently there is a dramatic reduction in the particles action - it is cut in half, at least approximately. But it is the approximately that we are interested in, the halving is in a sense trivial. How accurately is the action halved?

To find out, one must concentrate on the immediate neighbourhood of the crucial moment. The particle's contour hugs the separatrix (or one half of it) very closely except just near the hyperbolic point at the barrier top. The motion divides into two alternating phases, 'sweeps' and 'creeps'. During sweeps the particle moves very fast around one of the lobes of the figure eight, while during the creeps it moves very slowly on a hyperbolic trajectory near the barrier top. Its action (the area of its instantaneous contour) changes in both phases of motion increasing in creeps and decreasing in sweeps.

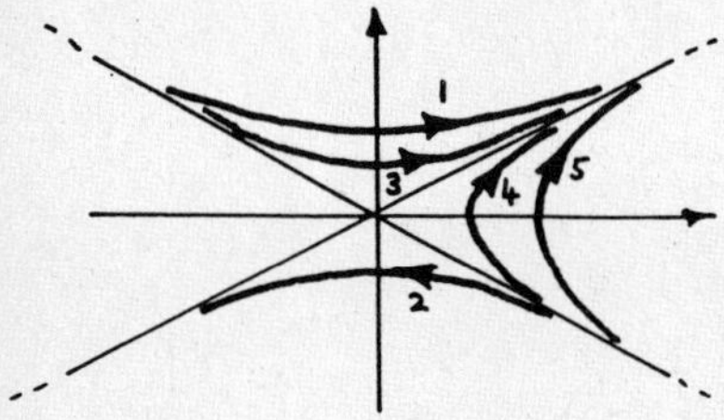

In creeps the particle is hardly moving but its contour is hugging the separatrix which is growing uniformly, so its area too grows uniformly. The net action change in this phase is therefore simply this uniform rate $\dot{S}$ times the creep duration. The duration is the characteristic time τ of the hyperbolic point times the ln of the creep energy measured with respect to the barrier top. (It is convenient to think of the barrier top energy as fixed and the wells falling instead.) In sweeps the action changes for a different reason - they are too swift for the separatrix action to have grown noticeably. Work is done on the particle however in its sweep (an amount equal to the uniform rate $\dot{S}$ mentioned earlier) so that it returns to the hyperbolic neighbourhood with a different energy, i.e., on a shifted contour with, therefore, a different area. The shift produces an area change equal to the change in energy (ln energy -1).

Summing up the action changes from sweeps and creeps from the infinite past to the infinite future (it does not matter that the description just given is invalid then) gives a net action change. For the symmetric double well just described, it is

$$S_{final} = 1/2\, S_{initial} - \dot{S}\tau \ln\,(2 \mathrm{Sin}\, \pi x)$$

where $\dot{S}$ and τ are the constants defined earlier and x is a 'random' variable, depending ultimately on where the particle was at the crucial moment. More directly, the energy ladder up which the particle climbs from creep to creep has fixed energy spacing $\dot{S}$, but variable shift defined by its lowest positive creep energy $x\dot{S}$ (where $0 \leqslant x \leqslant 1$). If the particle was caught for a very long time on the barrier at the crucial moment, x is near zero or near unity and the final action is indefinitely large. On the other hand, if $x \sim \frac{1}{2}$ the final action is a little less than half the initial one. In the limit of slow change, x is uniformly distributed on 0 to 1 so that $\Delta S = S_f - 1/2 S_i + \frac{\pi}{2}\dot{S}\tau$ has the distribution

$$P(S_f) = \frac{2}{\pi \dot{S}\tau} \exp\,(-\Delta S/\dot{S}\tau)/(1-\exp(-2\Delta S/\dot{S}\tau))^{1/2}$$

with mean value zero as one expects.

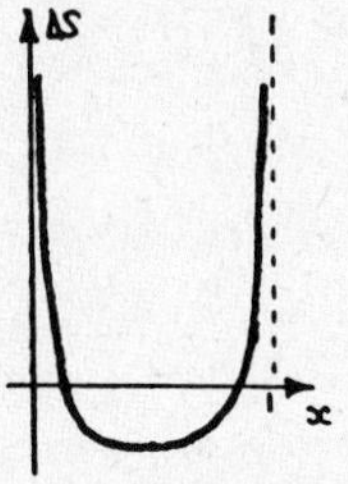

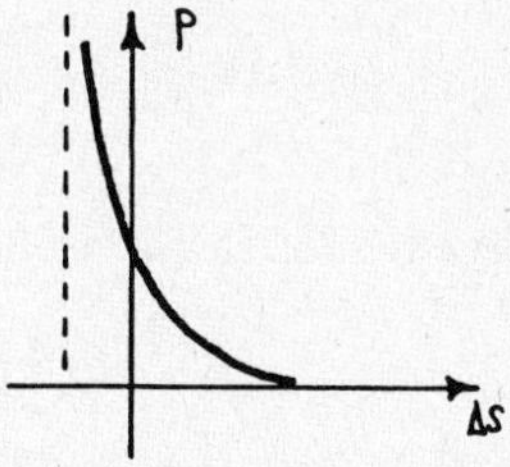

The most important feature of the distribution is its width, which is directly proportional to the (slow) rate of change $\dot{S}$ contrasting with the exp(-const/$\dot{S}$) accuracy of the ordinary adiabatic change. So the accuracy has indeed been reduced by the process described. Moreover, in the case of an asymmetric well it can be shown that the accuracy is even slightly worse, going like $\dot{S}$ ln $\dot{S}$ instead. Finally, I should mention that shortly before the presentation of this contribution a preprint covering the same material, with happily the same conclusions, was received[5].

REFERENCES

1 J.H.Hannay and A.M.Ozorio de Almeida (1984) J.Phys.A.17 3429

2 V.I.Arnold, 'Mathematical mehods of Classical Mechanics (1978)(Berlin: Springer)

3 M.C.Gutzwiller (1971) J.Math.Phys. 12 343

4 J.H.Hannay, to be submitted to J.Phys.A.

5 J.L.Tennyson, J.R.Cary, and D.F.Escande (1986) Phys.Rev.Lett. 56 2117

EVOLUTION AND EXACT EIGENSTATES OF A RESONANCE QUANTUM SYSTEM

Shau-Jin Chang and Kang-Jie Shi
Department of Physics
University of Illinois at Urbana-Champaign
1110 West Green Street
Urbana, IL 61801

The model that we shall describe is the quantum Chirikov map defined by

$$p_{n+1} = p_n + \frac{k}{2\pi} \sin 2\pi q_n \tag{1}$$

$$q_{n+1} = q_n + p_{n+1} \qquad (\text{mod } 1) \tag{2}$$

and

$$[p_n, q_n] = -i\hbar \ . \tag{3}$$

This map may be obtained from a periodically kicked free rotor with the amplitude of the kick to be a periodic function of q. The classical system has been studied thoroughly by Chirikov, Greene, and others. At k=0, p is constant and the system is integrable. At large k, the system appears to be ergodic. For an intermediate k such as k=1, the system can have all kinds of orbits depending on the initial values of (p,q). Fig. 1 describes some of the typical orbits for classical Chirikov map.

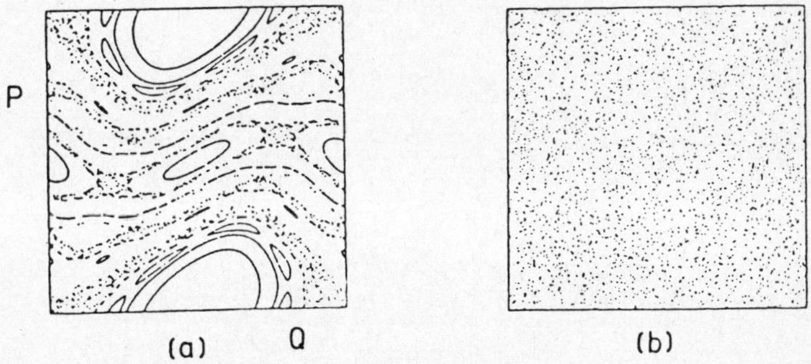

Fig. 1. Iterations of classical Chirikov maps. (a) k=1 and (b) k=10.

The quantum iteration is produced by a unitary evolution operator U,

$$p_{n+1} = U^{-1} p_n U, \ q_{n+1} = U^{-1} q_n U, \tag{4}$$

$$U = \exp\left(-\frac{i}{2\hbar} p_n^2\right) \exp\left(-\frac{ik}{(2\pi)^2 \hbar} \cos 2\pi q_n\right). \tag{5}$$

The eigenstate of U obeys

$$U|\psi\rangle = e^{-i\omega}|\psi\rangle, \tag{6}$$

where ω is known as the pseudoenergy.

Since q is periodic, the eigenvalues of the momentum p are discrete,

$$\langle m|p = 2\pi\hbar m\langle m|, \qquad m = \text{integer}. \tag{7}$$

In the momentum space, Eq. (6) becomes

$$\sum_{m'} U_{mm'}\psi_{m'} = e^{-i\omega}\,\psi_m, \tag{8}$$

where the $\infty \times \infty$ matrix elements are

$$U_{mm'} = e^{-i2\pi^2 m^2\hbar}\,(-i)^{m'-m}\,J_{m'-m}(z) \tag{9}$$

with

$$z = k/(4\pi^2\hbar). \tag{10}$$

It is difficult to study $U_{mm'}$ in its present form.

Under the quantum resonance condition,

$$2\pi\hbar = M/N, \quad M,N \text{ integer}, \tag{11}$$

the matrix elements $U_{mm'}$ become periodic in m, m',

$$U_{m+N,\ m'+N} = (-1)^{MN}\,U_{mm'}. \tag{12}$$

For simplicity, we choose MN=even. We now decompose the momentum variable m into

$$m = \ell N+s, \tag{13}$$

where

$$\ell = \text{Int}\,(m/N),\ s = m \bmod N. \tag{14}$$

In other words, we partition m into periodic cells of size N. The integer ℓ labels the ℓ-th cell, and the integer s describes an internal variable within the cell. In terms of variable ℓ and s, we can decompose $U_{mm'}$ as

$$U_{s+N\ell,s'+N\ell'} = \int_0^{2\pi}\frac{da}{2\pi}\,e^{-ia(\ell-\ell')}\,U(a)_{ss'}, \tag{15}$$

where $\underline{a}$ ($0 \leq a < 2\pi$) is the analog of the wave vector in Bloch wave, and $U(a)_{ss'}$ is an N×N unitary matrix in the internal space. We can compute U(a) straightforwardly as

$$U(a)_{ss'} = \frac{1}{N} \exp\left(-i2\pi^2 s^2 \hbar + i(s'-s)a/N\right)$$

$$\times \sum_{j=1}^{N} \exp\left(\frac{i2\pi(s'-s)j}{N} - iz \cos\left(\frac{a+2\pi j}{N}\right)\right). \qquad (16)$$

Each eigenstate of U(a) corresponds to an eigenstate of $U_{mm'}$ with the same pseudo-energy. Their eigenfunctions are related by

$$\psi_m = e^{-ia\ell}\, \psi_s(a). \qquad (17)$$

It is much simpler to diagonalize the finite matrix U(a).

We have studied numerically the pseudoenergies and the eigenfunctions for different values of $2\pi\hbar$, k, and a. In Fig. 2, we plot the $|\psi|^2$ of some eigenstates in the coherent state representation. As we can see, in the coherent state representation, the eigenstates follow quite closely to the classical orbits. By comparing Fig. 2 with the classical map Fig. 1, we can identify clearly each of these quantum states with a classical analog. In particular, the classical cycles are represented in the quantum map by isolated peaks of width $\sqrt{\hbar/2}$ and the KAM curves are represented in the quantum map by walls with width $\sqrt{\hbar/2}$. The chaotic regions are represented in the quantum map by ripples which no longer have well-defined wall structure.

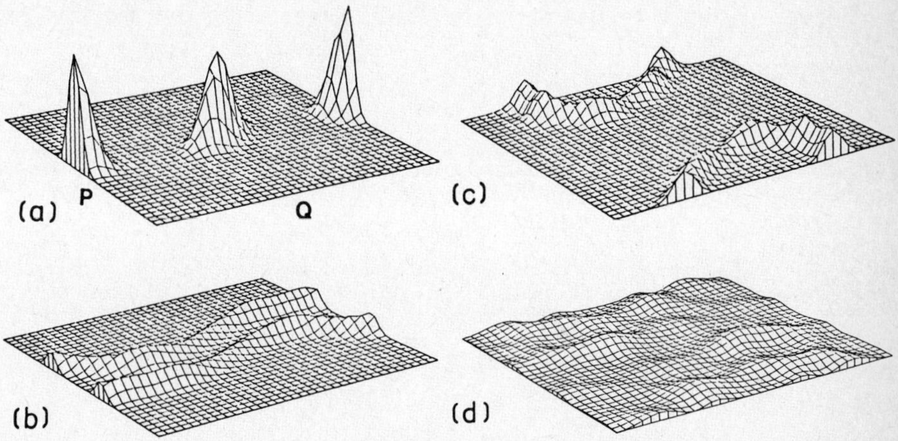

Fig. 2. The quantum eigenstate $|\psi|^2$ in the coherent-state representation. Parameters used are $2\pi\hbar = 1/102$, a=0, and k=1 in (a)-(c) and k=10 in (d).

This work was suppported in part by the US National Science Foundation under Grant No. PHY-82-01948.

QUANTUM KAM THEOREM

Kang-Jie Shi and Shau-Jin Chang
Department of Physics
University of Illinois at Urbana-Champaign
1110 West Green Street
Urbana, IL 61801

We study an n-dimensional Hamiltonian system, in which the q's are angle variables and the momenta are discrete. Because the Hamiltonian is periodic in q's, we can expand it as a Fourier series

$$H = A(\vec{p}) + B(\vec{p},\vec{q}) \equiv A(\vec{p}) + \sum_{k\neq 0} B_k(\vec{p})e^{ik.q} \tag{1}$$

where A is diagonal in p space. We assume that $A(\vec{p})$ and $B_k(\vec{p})$ are analytic in a "real band" $[\gamma]$ defined by $|\mathrm{Im}\, p_i| \leq \gamma$.

In analog with the classical KAM theorem, we require that

$$|B_k(\vec{p})| \leq \mu_B e^{-\beta k}, \quad k \equiv |\vec{k}|, \tag{2}$$

in the region (γ) defined by $|\vec{p}-\vec{p}_o| \leq \gamma$, where $\vec{p}_o$ satisfies

$$\nabla A(\vec{p}_o) = \vec{\omega}, \quad |\vec{\omega}.\vec{k}| \geq \kappa k^{-\nu} \text{ for each integer } \vec{k} \neq 0. \tag{3}$$

Furthermore we assume that

$$|A(\vec{p})| \leq M_A(1+|\vec{p}|)^J, \quad |B_k(\vec{p})| \leq M_B e^{-\beta k} \tag{4}$$

in the "real band" $[\gamma]$ to facilitate the derivation.

We consider a unitary transformation of the form

$$U = e^{\frac{i}{\hbar} C(\vec{p},\vec{q})}, \quad C(\vec{p},\vec{q}) = \sum_{k<k_c} C_k(\vec{p})e^{ik.q},$$

for a finite k_c and it has a definite classical limit. The purpose of U is to reduce the off-diagonal part of H to the order of B^2 in a subregion (γ') containing $\vec{p}_0$. We calculate the new Hamiltonian

$$H' = A + \{\frac{i}{\hbar}[A,C] + B\} + \{\frac{i}{\hbar}[B,C] + \frac{1}{2}(\frac{i}{\hbar})^2 [[A+B,C],C] + \ldots\}. \tag{5}$$

The first term $A(\vec{p})$ contains only the diagonal elements. The major part of the off-diagonal elements is in the second term. We construct $C_k(\vec{p})$ such that

$$\frac{i}{\hbar}[A,C] + B = 0. \tag{6}$$

Moving $e^{ik.q}$ to the left and equating the coefficients, we have

$$C_k(\vec{p}) = i\hbar B_k(\vec{p})/(A(\vec{p}) - A(\vec{p}-\hbar\vec{k})). \tag{7}$$

If $\hbar k$ is small and $|\vec{k}|<k_c$ (k_c depends on γ), we have

$$|C_k(\vec{p})| \leqslant |B_k(\vec{p})|/(\tfrac{1}{2}\kappa k^{-\nu}) \leqslant 2\mu_B\, e^{-\beta k}k^{\nu}/\kappa. \tag{8}$$

Note that $C = \sum C_k(\vec{p}) \exp(ik.q)$ is a finite summation. When $k \geqslant k_c$, $B_k(\vec{p})$ becomes neglibible because of the rapid damping of $e^{-\beta|\vec{k}|}$ in eq. (2). Now we treat the remaining part. Putting

$$B = \sum_{k\neq 0} B_k(\vec{p})e^{ik.q}, \quad C = \sum_{k<k_c} C_k(\vec{p})e^{ik.q} \tag{9}$$

into eq. (5), we obtain a series of an infinite summation,

$$\frac{i}{\hbar}[B,C] + \frac{1}{2}\left(\frac{i}{\hbar}\right)^2[[A+B,C], C] + \ldots = \Sigma\, D_k(\vec{p})e^{ik.q} \tag{10}$$

where $D_k(\vec{p})$ has a finite limit as $\hbar\to 0$. We can estimate the magnitude of $D_k(\vec{p})$ by the Cauchy integration formula. We can see that the argument of B_k and C_k must be shifted along the direction of real axis. We can modify $C_k(\vec{p})$ slightly to $\tilde{C}_k(p)$ to compensate the above shifting, obtaining

$$|\tilde{C}_k(\vec{p})| \leqslant M_c e^{-\beta_1 k}(1+|p|)^{-J}, \quad M_c\sim \mu_B, \quad \beta_1\sim\beta, \tag{11}$$

in a "real band" $[\gamma_1]\subset[\gamma]$ to ensure that $D_k(\vec{p})$ is $O(B^2)$ in the shrunk region (γ'), and that the off-diagonal part of the new Hamiltonian is bounded in $[\gamma']$. After the unitary transformation $U = \exp\left(\frac{i}{\hbar}\tilde{C}\right)$ we can define new A' and B_k' such that

$$H' = A'(\vec{p}) + \Sigma\, B_k'(\vec{p})e^{ik.q} \tag{12}$$

where

$$|B'_k(\vec{p})| \leqslant \mu_B'\, e^{-\beta' k} \text{ in } (\gamma'), \quad \mu_B'\sim \mu_B^2, \quad \beta'\sim\beta \tag{13}$$

and

$$|A'(\vec{p})| \leqslant M_A'(1+|\vec{p}|)^J, \quad |B_k'(\vec{p})| \leqslant M_B' e^{-\beta' k} \text{ in } [\gamma'] \tag{14}$$

with $\gamma' < \gamma_1 < \gamma$. That is, $B'\sim(B^2)$ in region (γ') defined by $|\vec{p}-\vec{p}_0| \leqslant \gamma'$. Since $A(\vec{p})$ is changed after the unitary transformation, we must find another $\vec{p}_0'$ such that $\nabla A'(\vec{p}_0') = \vec{\omega}$. Under similar conditions as used in establishing classical KAM theorem, we can treat this shifting of $\vec{p}_0$ as well.

We repeat this procedure many times. But unlike the classical mechanics, the procedure must terminate. This is because we need $\gamma \gg \hbar$ for further reduction. However, we can always obtain a final B^* much smaller than any given power of $\hbar$. After a sequence of unitary transformations the Hamiltonian becomes almost diagonal in a certain part corresponding to the region (γ^*). In p space we have

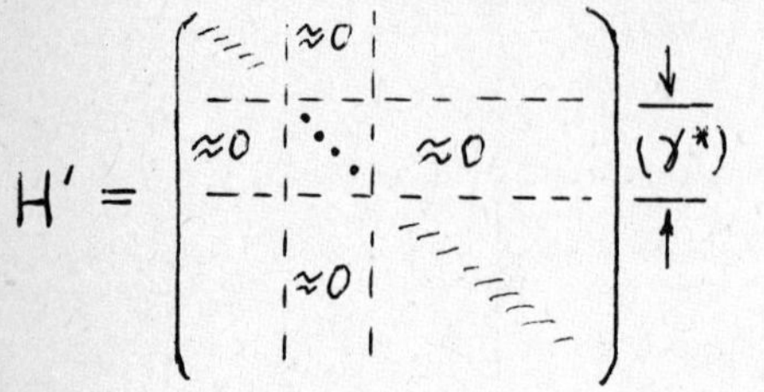

$$
H' = \begin{pmatrix} \ddots & \approx 0 & \\ \approx 0 & \ddots & \approx 0 \\ & \approx 0 & \end{pmatrix} \begin{array}{c} \downarrow \\ \overline{(\gamma^*)} \\ \uparrow \end{array}
\tag{15}
$$

The above derivation is for a Hamiltonian system. If we consider an $H(\vec{p},\vec{q},t)$ which is periodic in t, we can reach the same conclusion for discrete area-preserving map system. Quantum Chirikov map belongs to this category.

We next discuss the eigenstates of H'. We can use perturbation theory in region (γ^*). As we know, perturbation theory requires that the differences of the diagonal elements are much greater than the magnitudes of the off-diagonal elements. We set $\vec{p}_k = \hbar \vec{k} + \vec{\alpha}$ with $\vec{\alpha}$ to be arbitrary in a system. In the part of H' corresponding to region (r^*) whose size is much larger than $\sqrt{\hbar}$, and for most $\vec{\alpha}$ we can show

$$
|A(\vec{p}_1) - A(\vec{p}_2)| \gg \mu_B^*.
\tag{16}
$$

From eq. (16) we can prove that any eigenstate can have at most one big component in (γ^*). Other components in (γ^*) must be very small. Consequently, in the coherent-state representation, the absolute square of the wave function can either form a wall structure or have a very small amplitude in the region,

$$
|\vec{p}-\vec{p}_0^*| \le \gamma^*, \quad \text{all q.}
\tag{17}
$$

What we have studied above are properties of a Hamiltonian and its eigenstates after unitary transformations

$$
U = \prod_j \exp \left(\frac{i}{\hbar} C^j(\vec{p},\vec{q}) \right).
\tag{18}
$$

We can prove that as long as B is sufficiently small in the original Hamiltonian (the requirement of the magnitude is independent of $\hbar$), a coherent state can remain localized in both the p and the q space after these unitary transformations. Thus we can conclude that similar phenomena also appear in the original system. This work is supported in part by the National Science Foundation under contract No. NSF PHY-82-01948.

ATOMS IN STRONG FIELDS: CANDIDATES FOR LABORATORY STUDIES OF QUANTUM CHAOS

J. B. Delos, S. K. Knudson, R. L. Waterland, M. L. Du
College of William and Mary
Williamsburg, VA 23185

Classical chaos is certainly not well-understood, but at least it is a well-defined subject. In contrast, even the meaning of the term "quantum chaos" is still ill-defined. The words are used to describe those special or unique properties of quantum systems that emerge in the classical limit ($\hbar \to 0$, or $m \to \infty$, etc.) when the corresponding classical trajectories show irregular, chaotic behavior. The subject was born in theoretical speculations about the behavior of eigenvalues, eigenfunctions and transition amplitudes under such circumstances. There have now been many calculations, some of which demonstrate quite convincingly that these observable quantities can show very interesting behavior. Nevertheless, in contrast to the classical case, in quantum mechanics there are no simple, unambiguous criteria for distinguishing between regular and chaotic behavior. Therefore, it is not absolutely clear in any instance whether we are examining a phenomenon for which the word chaos is appropriate, or whether we are seeing regular behavior in which the simple pattern is eluding us.

Most important, there have as yet been very few laboratory experiments that address the fundamental issues. It is proposed that atomic systems in strong external electric and magnetic fields may be ideal candidates for the study of quantum chaos.

1. These systems have a small number of degrees of freedom. A one-electron atom has three spatial degrees of freedom if electric and magnetic fields are parallel, then L_z is conserved, the azimuthal angle is an ignorable coordinate, and the number of significant degrees of freedom is reduced to two. (Spin-effects are normally irrelevant in highly excited states.)

2. The Hamiltonian is known to a high degree of accuracy. This may be contrasted with the situation that occurs in nuclei or in molecules, in which the relevant forces are rarely known accurately.

3. The density of states is high. Even if classical trajectories show chaotic behavior, it is unlikely that any unusual quantum phenomenon will occur unless there is a large number of quantum states in the relevant region of phase-space. Near the ionization threshold, the density of states becomes infinite.

4. The Hamiltonian contains parameters that can be varied in the experiment. The strength of E and B are easily varied, and the nuclear charge can be selected.

5. The Hamiltonian is time-independent. Classical chaos can arise in time-independent or time-dependent systems. The former have been studied more extensively, and our knowledge of them can help to interpret corresponding quantum phenomena. Furthermore, quantum chaos might manifest itself very differently in wave-packets than it does in stationary states.

6. Trajectories of these systems show every type of motion, from nearly complete order to apparently complete chaos. The approach to chaos is as interesting as is chaos itself, and it should be accessible in these systems.

7. Atoms can be isolated from other perturbing influences in beam or ion traps, and the relevant states are experimentally accessible.

We have studied very extensively the case of a one-electron atom in a strong external magnetic field, and studies of atoms in combined electric and magnetic fields are in progress.

For the pure magnetic-field case, the important part of the Hamiltonian can be written in suitably scaled variables ($\rho z p_\rho p_z$) as

$$H = \frac{1}{2}(p_\rho^2 + p_z^2) + \hat{L}^2/2\rho^2 - (\rho^2 + z^2)^{-1/2} + \frac{1}{8}\rho^2 . \qquad (1)$$

The last term is called the "diamagnetic" term, and interesting effects arise when it is comparable to the other terms. The quantity $\hat{L}$ is the z-component of angular momentum, scaled by the field-strength according to

$$\hat{L} = L_z B^{1/3} / m^{2/3} Z^{2/3} e c^{1/3} = L_z B^{1/3} / 61.7 \tag{2}$$

where in the last equation, L_z is measured in units of $\hbar$ and B in Tesla.

Since the Hamiltonian (1) depends on just one parameter, the structure of the fields of trajectories in phase space depends only upon L and upon the energy, and a detailed picture of the classical behavior can be attained.

An overview of this behavior is shown in Fig. 1. For small $\hat{L}$ and for energies not too close to ionization, a trajectory can be regarded as a Kepler ellipse that rocks, tilts and flips in space as its orbital parameters vary slowly with time. For large $\hat{L}$ the trajectories have a helical structure: the electron circles around a magnetic field line and bounces slowly back and forth in the z direction. Between these limiting cases there is a transition regime where a 2:1 resonance occurs, and an irregular regime.

How much of this regime is experimentally accessible? In Fig. 1 we see that the irregular regime extends in a very narrow strip just below the ionization energy to $\hat{L}\sim 0$. In fact, current experiments on near-threshold ionization of atoms in magnetic fields are sampling this irregular regime. Presently, however, the resolution of these experiments has been inadequate for measurement of individual states.

More information can be obtained if L is increased. We see from Fig. 1 that for $\hat{L} = 0.5$ the irregular regime constitutes about 1/5 of the total domain of bound energies. Increasing the magnetic field is not practical (some 10^4 Tesla would be required), but increasing L is possible. In a 5T field $L_z\sim 18$ corresponds to $\hat{L}\sim 0.5$ and $L_z\sim 54$ takes us to the very center of the transition regime.

Hulet and Kleppner[2] have already shown that they can populate the "circular" states having $n\sim 20$, $L_z\sim 20$ in a field-free region. It would now be interesting to make a beam of atoms (such as Hydrogen) in a state such as n = 21, ℓ = 20, m = 20, pass the beam through a field of about 5T, and measure the spectrum of energy levels, which ranges from about 25 to 300 cm^{-1} above the circular state. Such an experiment would give invaluable information about the transition to chaos and the behavior of atoms in classically chaotic regions.

References

1. J. B. Delos, S. K. Knudson, and D. W. Noid, Phys. Rev. Lett. 50, 579 (1983); Phys. Rev. 28, 7 (1983); 30, 1208 (1984); D. W. Noid, S. K. Knudson, and J. B. Delos, Chem. Phys. Lett. 100, 367 (1983); S. K. Knudson and D. W. Noid, Chem. Phys. 89, 353 (1984); R. L. Waterland, M. L. Du, and J. B. Delos (papers in preparation).
2. R. G. Hulet and D. Kleppner, Phys. Rev. Lett. 51, 1430 (1983).

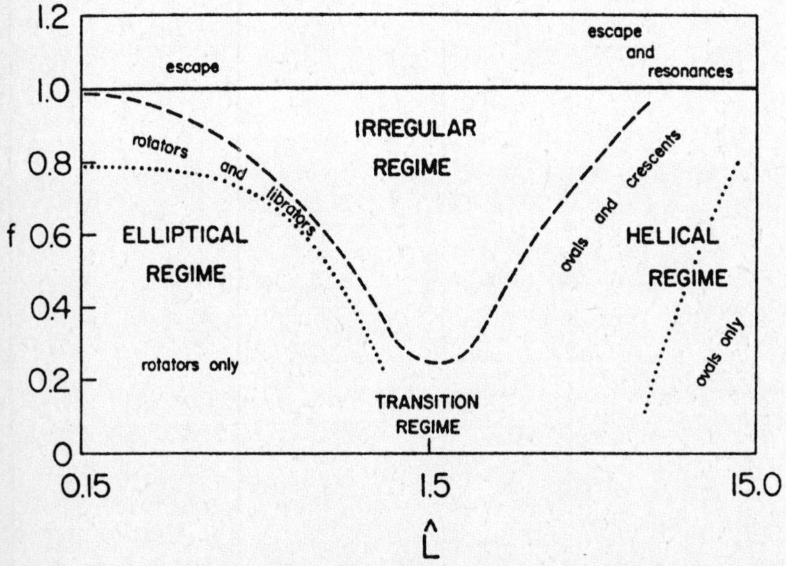

FIG. 1. Types of trajectories that are most common at various values of dimensionless energy f, and scaled angular momentum $\hat{L}$. The dimensionless energy is $(\hat{E}-\hat{E}_{min})/(\hat{E}_{escape}-\hat{E}_{min})$, and the scaled angular momentum is given in Eq. (2.6). One may equivalently regard the horizontal axis as being proportional to the $\frac{1}{3}$ power of the magnetic field. [See also Eq. (8.1).]

Quantum Analysis of States near a Separatrix*

John R. Cary, Department of Astrophysical, Planetary, and Atmospheric Sciences and Department of Physics, Petre Rusu, Department of Physics, and Rex T. Skodje, Department of Chemistry, University of Colorado, Boulder, Colorado 80309.

Quantum modifications of classical chaos have been of great interest in recent years. This interest is due in part to a basic contradiction. In certain classical systems the existence of chaos, as defined, for example, by the lack of existence of isolating integrals, can be proven. In contrast, such chaos does not exist in bounded quantum systems, in which quasiperiodicity is easily shown. This may not mean that quantum systems are not chaotic, but only that the correct definition of quantum chaos is not yet known.

Indeed, the search for the correct definition of quantum chaos is the thrust of much of recent research. As examples we mention the work on the relationship of the sensitivity[1,2] and distribution of eigenvalues to nonseparability[3-5]. Along these lines, the idea is that a quantum analysis of classically chaotic systems should reveal the nature of quantum chaos. It is furthermore expected that such a connection will most likely be revealed in the study of states of large quantum number, i.e., in the semiclassical limit.

The semiclassical limit by itself has also been an area of much recent study. Examples of such work include that of Berry concerning the relation of the quantum adiabatic phase[6] to the previously known classical adiabatic phase[7] and the consequence of tunneling[8] on the quantum adiabatic theorem[9,10]. We note also the burgeoning literature on wavepacket evolution and coherent states[11-16]. Finally, we mention the use of adiabatic switching coupled with classical calculations to obtain quantum spectra[17,18].

In classical systems, chaos first appears near separatrices of integrable systems as they are perturbed. In part this is due to the exponential separation of near-separatrix orbits. Given the previous introduction, we are, therefore, motivated to study the quantum mechanics of states near a separatrix in the semiclassical limit.

Our result is that quantum mechanics drastically modifies the system even in the limit of very large quantum number, $N \approx 10^7$. The reason is that the classical excitation frequency, the orbit frequency, vanishes for orbits with energy E equal to the separatrix energy E_x. In contrast, the quantum excitation frequency, the frequency separation between neighboring states $\Delta\omega_{n,n+1} \equiv \omega_{n+1} - \omega_n$, cannot vanish, because there are no degeneracies in one-dimension in a quantum system. Moreover, the separation of quantum eigenvalues near a separatrix vanishes very slowly with the number of states. Hence, quantum effects can be observed by determining the excitation frequency of states having energy near the separatrix energy.

The system of interest is the motion of a nonrelativistic particle in one dimension in a double-well potential. The classical Hamiltonian for this system is $H = p^2/2m + V(x)$. A double-well potential is one with a single local maximum. The symbol E_X is used to denote the value of $V(x)$ at the local maximum, which is taken to occur at x_0. At the local maximum the potential is assumed to have nonvanishing second derivative, so that near the local maximum, the potential has the form, $V(x) = E_X - m\omega^2(x-x_0)^2/2 + O((x-x_0)^3)$. The symbol ω has been introduced to parameterize the second derivative of the potential at the maximum. It equals the rate of exponential divergence of orbits near the unstable fixed point.

An equation for the energy eigenvalues in the semiclassical limit for E close to E_X, the separatrix energy, was obtained previously by Connor[19]. His result can be put into the form,

$$\cos(\pi I_c/h - \varphi) = -\cos(\pi I_a/h - \pi I_b/h) / (1 + e^{2\pi\gamma})^{1/2} . \qquad (1)$$

The symbols in this expression are defined as follows. The action I_a is given by the usual loop integral for $E < E_X$. For $E > E_X$ it is simply twice the integral, $\int p\,dq$, on a constant-energy curve from the left turning point to the location of the maximum of V. The action I_b is defined similarly, and $I_c = I_a + I_b$. Both actions are functions of the energy. The symbol $\gamma \equiv (E - E_X)/\hbar\omega$, is a normalized energy. In the region of interest γ is of order unity, since we are considering states within a few quanta of the separatrix. Finally, the phase $\varphi = \gamma - \gamma \ln|\gamma| + \arg(\Gamma(1/2 + i\gamma))$.

Of interest here is the case in which γ is less than or of order unity. For this analysis, it is necessary to know the behavior of the classical action near the separatrix. This has recently been discussed in detail in Refs. 20 and 21. According to Refs. 20 and 21, the classical action for a particle with $E < E_X$ is given by

$$I_\alpha = Y_\alpha + [(E-E_X)/\omega][1 + \ln|E_\alpha/(E-E_X)|], \qquad (2)$$

in which α (=a or b) denotes the well in which the particle is trapped. The symbol Y_α denotes the classical separatrix action, *i.e.*, the value of I_α for $E = E_X$. The constants E_α are of the order of typical energies in the system.

Inserting Eq. (2) into Eq. (1) yields the following equation,

$$\cos[\pi N_c + \gamma\ln|\gamma_c| + \arg(\Gamma(1/2 - i\gamma))] = -(1 + e^{2\pi\gamma})^{-1/2}$$

$$\times \cos[\pi(N_a - N_b) + (\gamma/2)\ln|\gamma_a/\gamma_b|], \qquad (3)$$

In this equation, $N_\alpha = Y_\alpha/h$ is the phase-space area in units of h contained by lobe-α of the separatrix. The sum is defined to be $N_c \equiv N_a + N_b$. In addition, we have defined $\gamma_\alpha \equiv E_\alpha/\hbar\omega$ and $\gamma_c \equiv (\gamma_a \gamma_b)^{1/2}$. The quantities γ_α are large in the semiclassical limit, since they are the ratio of a macroscopic energy to the energy of one quantum. Indeed, they are of the order of N_c, the number of "trapped" quantum states.

The separation of energy states with γ less than or of order unity can be found from Eq.(3). The typical value for an asymmetric well is $\Delta\gamma = \pi/\, [\ell n|\gamma_c| + 1.96\,]$. A typical classical frequency is ω, which corresponds to $\Delta\gamma$ being order unity. We, therefore, see that quantum effects greatly regularize classical theory, in that the near separatrix frequencies do not approach zero, but are smaller than typical frequencies only by the logarithm of the number states. Thus, it is likely that these effects are observable in a macroscopic system such as a Penning trap[22]. In addition, this analysis has implications for semiclassical theory, in particular for adiabaticity, in which the minimum frequency is very important.

The authors acknowledge useful discussions with Profs. J. D. Hanson and R. G. Littlejohn.

References:

1. N. Pomphrey, J. Phys. B **7**, 1909 (1974). 2. I. C. Percival, Adv. Chem. Phys. **36**, 1 (1977). 3. M. V. Berry, Philos. Trans. Roy. Soc. (London) A **287**, 237 (1977). 4. M. V. Berry, J. Phys. A **10**, 2083 (1977). 5. G. M. Zaslavskii, Zh. Eksp. Teor. Phys. **73**, 2089 (1979). [Sov. Phys. JETP **46**, 1094 (1977).] 6. M. V. Berry, J. Phys. A **18**, 15 (1985). 7. J. H. Hannay, J. Phys. A **18**, (1985) in press. 8. M. V. Berry, J. Phys. A **17**, 1225 (1984). 9. T. Kato, Prog. Theor. Phys. **6**, 485 (1950). 10. M. Born and V. Fock, Zs. f. Phys. **51**, 165 (1928). 11. J. R. Klauder, J. Math. Phys. **4**, 1058 (1963). 12. E. J. Heller, J. Chem. Phys. **65**, 4979 (1976). 13. M. J. Davis. and E. J. Heller, J. Chem. Phys. **75**, 3916 (1981). 14. N. DeLeon and E. J. Heller, J. Chem. Phys. **78**, 4005 (1983). 15. R. G. Littlejohn, Phys. Rev. Lett. **56**, 2001 (1986). 16. M. M. Nieto, L. M. Simmons, jr., and V. P. Gutschick, Phys. Rev. D **23**, 927 (1981) and references therein. 17. R. T. Skodje, F. Borondo, and W. P. Reinhardt, J. Chem. Phys., (1986) to be published. 18. R. T. Skodje and F. Borondo, Chem. Phys. Lett. **118**, 409 (1985). 19. J. N. L. Connor, Chem. Phys. Lett. **4**, 419 (1969). 20. J. L. Tennyson, J. R. Cary, and D. F. Escande, Phys. Rev. Lett. **56**, 2117 (1986). 21. J. R. Cary, D. F. Escande, and J. L. Tennyson, submitted to Phys. Rev. A (1986). 22. L. S. Brown and G. Gabrielse, Revs. Mod. Phys. **58**, 233 (1986) and references therein.

ADIABATIC INVARIANTS, RESONANCES, AND MULTIDIMENSIONAL SEMICLASSICAL QUANTIZATION

Frank R. Johnston and Philip Pechukas
Department of Chemistry
Columbia University
New York, New York 10027

In 1978 Solov'ev published an interesting suggestion for how to do semiclassical bound state calculations on systems with many degrees of freedom.[1] The problem with the straightforward way of doing such calculations is not integrating an individual trajectory; that's a problem that grows only linearly with the number of degrees of freedom. The problem is finding those particular trajectories that satisfy the semiclassical quantization conditions; that's a problem that grows exponentially with the number of degrees of freedom, when attacked in the straightforward way by laying a grid of initial conditions in phase space, integrating many trajectories, and then rejecting almost all of them as bad guesses. Solov'ev suggested starting with a simple Hamiltonian for which semiclassical quantization is trivial; selecting a trajectory of this simple Hamiltonian that satisfies semiclassical quantization conditions; and following the trajectory in time as the Hamiltonian is slowly changed into the one of interest. This "adiabatic switching" method will work if the classical actions of the multidimensional system are adiabatic invariants, for then one ends with a trajectory of the final Hamiltonian that still satisfies semiclassical quantization conditions; and to calculate the semiclassical energy eigenvalue one just reads off the energy of this final trajectory according to the final Hamiltonian. In principle, the method of "adiabatic switching" requires calculation of only a single--albeit long--classical trajectory. The practice of the method has been explored in a number of recent papers.[2-4]

Solov'ev deserves credit not so much for the originality of his suggestion--the idea goes back to a 1916 paper by Ehrenfest[5]--as for the courage to put it in print, because it seems on the face of it that the method cannot work. Resonances will ruin it. Imagine a resonant invariant torus of the "simple" Hamiltonian--that is, a torus on which some of the frequencies of motion are rationally related. This torus, as it evolves under the slowly changing Hamiltonian, should always remain close to the corresponding invariant torus of the instantaneous Hamiltonian; in particular, all points on the torus should have essentially the same instantaneous energy, and therefore the average rate of change of the energy--determined by the time-average of $\partial H/\partial t$--should be independent of position on the torus. But on a resonant torus this time-average will generally vary from trajectory to trajectory: motion on a resonant torus is not ergodic.

That resonances are trouble was recognized already in the flurry of literature on adiabatic invariants that followed Ehrenfest's 1916 paper. The most substantive contribution to this literature was a 1925 paper by Dirac[6] which showed that the action integrals survive as adiabatic invariants provided a certain, rather complicated, inequality is satisfied at each resonance encountered during the adiabatic switch. Solov'ev's response to the resonance problem was a sentence to the effect that the frequencies on a torus of given action typically change during an adiabatic switch of Hamiltonians and therefore the fraction of time spent in resonance is typically zero, so typically there is no resonance problem. This is not quite right, but the basic attitude--that resonances can do no harm if the system is almost always out of resonance--is sound, as we shall see below.

The purpose of this paper is to set forth a rather simple sufficient condition for adiabatic invariance of the actions in two-dimensional systems; to remark on the implications for adiabatic switching as a practical method of semiclassical calculation; and to show, in an elementary model problem, the interesting way in which conservation of action fails when the condition is violated. We shall assume that all the Hamiltonians encountered--the initial Hamiltonian, the final Hamiltonian, and all the Hamiltonians run through in the adiabatic switch--are integrable, because that is the natural setting for this discussion. Dana and Reinhardt[7] have just completed a fascinating study of what happens when one switches adiabatically from an integrable to a nonintegrable system.

Let $H(p,q,\lambda)$, $0\leq\lambda\leq 1$, be a family of n-dimensional integrable Hamiltonians; let $S(q,I,\lambda)$ be a generating function of the canonical transformation to action-angle variables (I,Θ) that sends $H(p,q,\lambda)$ to $H(I,\lambda)$; and let λ vary with time, as $\lambda = \varepsilon t$ where ε is a "slowness" parameter. The adiabatic limit, of course, is $\varepsilon\to 0$.

The rules of the game say that if we want to integrate Hamilton's equations in the new variables (I,Θ), we have to add to the Hamiltonian H an extra piece, the time derivative of S:

$$\tilde{H}(t) = H + \varepsilon(\partial S/\partial\lambda)_{q,I} \equiv H(I,\lambda) + \varepsilon h(I,\Theta,\lambda) \qquad (1)$$

where the "nonadiabatic coupling" h is $\partial S/\partial\lambda$ expressed in the variables (I,Θ). It is perhaps helpful to think of h as the Hamiltonian that generates the canonical transformation $(p,q) \to (I,\Theta)$, in the following way: the equations $dI/d\lambda = -\partial h/\partial\Theta$, $d\Theta/d\lambda = \partial h/\partial I$ generate a trajectory

$\mathbf{I}(\lambda)$, $\Theta(\lambda)$ through $(\mathbf{I},\Theta)$ space, but in the original $(\mathbf{p},\mathbf{q})$ space this trajectory goes nowhere--$\mathbf{I}(\lambda)$, $\Theta(\lambda)$ are the action-angle coordinates, at λ, of the phase point labeled $\mathbf{I}(0)$, $\Theta(0)$ at $\lambda = 0$. The stronger the "nonadiabatic coupling" h, the faster the change in action-angle coordinates of a given phase point.

The equations of motion under $\tilde{H}(t)$ are

$$\dot{\mathbf{I}} = -\partial\tilde{H}/\partial\Theta = -\varepsilon\,\partial h/\partial\Theta, \quad \dot{\Theta} = \partial\tilde{H}/\partial\mathbf{I} = \boldsymbol{\omega}(\mathbf{I},\lambda) + \varepsilon\,\partial h/\partial\mathbf{I} \tag{2}$$

where $\boldsymbol{\omega}(\mathbf{I},\lambda)$ is the frequency vector, at λ, of the invariant torus of H with action vector $\mathbf{I}$. The "nonadiabatic coupling" h is periodic in Θ, $h(\mathbf{I},\Theta+2\mathbf{n}\pi,\lambda) = h(\mathbf{I},\Theta,\lambda)$, so the phase average of $\partial h/\partial\Theta$, over any invariant torus of H, is equal to zero: on average, over a torus, the rate of change of $\mathbf{I}$ is zero. The question is whether the time average of $\dot{\mathbf{I}}$ agrees with this phase average, over the long interval from $t = 0$ to $t = 1/\varepsilon$ during which λ varies from 0 to 1.

To see what might happen in various circumstances, consider a simple two-dimensional model problem in which $h(I_1,I_2,\Theta_1,\Theta_2,\lambda) = \alpha\cos(\Theta_1-\Theta_2)$, where α is a constant independent of $\mathbf{I}$ and λ. Then $\dot{I}_1 = \varepsilon\,\alpha\sin(\Theta_1-\Theta_2) = -\dot{I}_2$, $\dot{\Theta}_1 = \omega_1(\mathbf{I},\lambda)$, $\dot{\Theta}_2 = \omega_2(\mathbf{I},\lambda)$. Watch what happens to a resonant torus ($\omega_1 = \omega_2$ at $t = 0$), during adiabatic passage from $\lambda = 0$ to $\lambda = 1$, under various assumptions on the frequency function $\boldsymbol{\omega}(\mathbf{I},\lambda)$:

1. $\boldsymbol{\omega}$ independent of $\mathbf{I}$ and λ

Then $\omega_1 = \omega_2$ at all times; the resonance cannot be broken. From the equations of motion, $\Theta_1 - \Theta_2$ = constant and the total change in I_1 is $\Delta I_1 = \alpha\sin(\Theta_1-\Theta_2) = -\Delta I_2$. The actions are not adiabatic invariants.

2. $\boldsymbol{\omega}$ depends on λ but not $\mathbf{I}$

Now the resonance is broken as λ varies. Make a linear approximation, $\omega_1-\omega_2 = \beta\lambda = \beta\varepsilon t$. Then $\Theta_1-\Theta_2 = c + \beta\varepsilon t^2/2$, $\Delta I_1 = \varepsilon\alpha\int_0^{1/\varepsilon}\sin(c+\beta\varepsilon t^2/2)dt = O(\varepsilon^{\frac{1}{2}})$. The actions are adiabatic invariants; the fluctuations in $\mathbf{I}$ induced by the resonance are $O(\varepsilon^{\frac{1}{2}})$.

3. $\boldsymbol{\omega}$ depends on $\mathbf{I}$ but not λ

A more subtle case. <u>If</u> $\mathbf{I}$ = constant, then the resonance is not broken as λ varies, and so one expects $\mathbf{I}$ to be driven off, $\mathbf{I} \neq$ constant; but then the resonance is broken, since $\boldsymbol{\omega}$ varies with $\mathbf{I}$. What happens? Make a linear approximation; since $I_1(t) + I_2(t)$ = constant (from the equations of motion), we can write $\omega_1(t) - \omega_2(t) = \gamma\{(I_1(t)-I_2(t)) - (I_1(0)-I_2(0))\} = \gamma\{\Delta I_1(t)-\Delta I_2(t)\}$. Let $x = \Theta_1 - \Theta_2$; then $\dot{x} = \omega_1-\omega_2$, $\ddot{x} = \gamma(\dot{I}_1-\dot{I}_2) = 2\varepsilon\alpha\gamma\sin x$. Therefore $\dot{x}^2/2 + 2\varepsilon\alpha\gamma\cos x$ = constant, and $x(t)$ oscillates in a well of depth $O(\varepsilon)$. At any time, $\Delta I_1 = -\Delta I_2 = (\omega_1-\omega_2)/2\gamma = \dot{x}/2\gamma = O(\varepsilon^{\frac{1}{2}})$. As in case (2), the actions are adiabatic invariants, and the fluctuations induced by the resonance are $O(\varepsilon^{\frac{1}{2}})$.

4. $\boldsymbol{\omega}$ depends on λ and **I**

In light of (2) and (3), what can possibly go wrong in the general case, when $\boldsymbol{\omega}$ varies with both λ and **I**? Watch. Again let $x = \Theta_1 - \Theta_2$; $\dot{x} = \omega_1 - \omega_2 = \beta\varepsilon t + \gamma(\Delta I_1 - \Delta I_2)$; $\ddot{x} = \varepsilon(\beta + 2\alpha\gamma \sin x)$. If $|2\alpha\gamma| > |\beta|$, there are stationary points ($\ddot{x} = 0$) and therefore solutions $x(t)$ = constant. Then at the end ($t = 1/\varepsilon$) we have $\dot{x} = 0 = \beta + \gamma(\Delta I_1 - \Delta I_2)$, $\Delta I_1 = -\Delta I_2 = -\beta/2\gamma$. What's conserved on this trajectory is not action, but trouble--the resonance: $\omega_1 - \omega_2 = \dot{x} = 0$ for all time.

For adiabatic invariance of the actions in this model problem we want $|2\alpha\gamma| < |\beta|$; i.e., weak nonadiabatic coupling and weak dependence of $\boldsymbol{\omega}$ on **I**, compared to the variation of $\boldsymbol{\omega}$ with λ. The Solov'ev argument--that for fixed **I** the frequencies typically vary with λ, so resonances are instantly broken if **I** is invariant and therefore are not a concern--is not quite right. Resonances must be broken by the full nonadiabatic dynamics of the system. In the model problem, the condition $|2\alpha\gamma| < |\beta|$ implies that $d(\omega_1 - \omega_2)/dt \neq 0$ no matter where one starts on the resonant torus $\omega_1 = \omega_2$. In general, a two-dimensional resonance will be broken by the dynamics if $d\boldsymbol{\omega}/dt$ does not lie along $\boldsymbol{\omega}$, and this condition turns out to be sufficient for adiabatic invariance of the actions:

2-d adiabatic theorem. Suppose that in some neighborhood of **I**, and for all Θ and λ, $\boldsymbol{\omega}$ and $d\boldsymbol{\omega}/dt$ are nonzero and linearly independent; then **I** is an adiabatic invariant.

Note that $d\boldsymbol{\omega}/dt = \varepsilon\{(\partial\boldsymbol{\omega}/\partial\mathbf{I})\cdot(-\partial h/\partial\boldsymbol{\Theta}) + \partial\boldsymbol{\omega}/\partial\lambda\}$ and what is inside the brackets depends on **I**, $\boldsymbol{\Theta}$, and λ but not on ε; the condition that $d\boldsymbol{\omega}/dt$ be nonzero and not along $\boldsymbol{\omega}$ is a condition that is independent of the rate of adiabatic switching.

Proof of this 2-d result will be published elsewhere. It involves breaking the integration range from $t = 0$ to $t = 1/\varepsilon$ into intervals that grow in length as an appropriate fractional power of $1/\varepsilon$; approximating the equations of motion, in each interval, by $\dot{\boldsymbol{\Theta}} = \boldsymbol{\omega}$, $\dot{\mathbf{I}} = \varepsilon \sum_{\mathbf{n}\neq 0} \mathbf{f}_{\mathbf{n}} e^{i\mathbf{n}\cdot\boldsymbol{\Theta}}$ where the $\mathbf{f}_{\mathbf{n}}$ are constant vectors; and recognizing that, in a term like $e^{i\mathbf{n}\cdot\boldsymbol{\Theta}} \propto e^{i\mathbf{n}\cdot\int\boldsymbol{\omega}dt}$, if $\mathbf{n}\cdot\boldsymbol{\omega}$ happens to be zero at some instant--a resonance--$\mathbf{n}\cdot d\boldsymbol{\omega}/dt$ is certain to be nonzero. The worst that can happen, in calculating the change in **I** due to a particular resonance, is an integral like $\varepsilon\int e^{i\varepsilon t^2}\, dt = O(\varepsilon^{\frac{1}{2}})$.

A similar result, for time-independent perturbations of integrable 2-d Hamiltonians, was published many years ago by Arnol'd.[9]

In one dimension the condition for adiabatic invariance of the action is $\omega \neq 0$, and the fluctuations in I are $O(\varepsilon)$; in two dimensions, the conditions are $\boldsymbol{\omega}$ and $\dot{\boldsymbol{\omega}} \neq 0$ and linearly independent, and the fluctuations in **I** are $O(\varepsilon^{\frac{1}{2}})$. It is tempting to extrapolate to n dimensions; if $\boldsymbol{\omega}$ and its first n-1 time derivatives are linearly independent, the worst resonance one can meet has $\mathbf{n}\cdot\boldsymbol{\omega} = 0$ and, somewhere on the resonant torus, **n** also perpendicular to the first n-2 time derivatives of $\boldsymbol{\omega}$. The adiabatic theorem is saved by that last linearly independent time derivative, and one expects the fluctuations in **I** from such a resonance to be $O(\varepsilon^{1/n})$. If this is correct, it implies that the advantage of adiabatic switching for semiclassical quantization in many dimensions--the advantage of being a single-trajectory calculation--is illusory, because to achieve a given level of accuracy in the face of fluctuations in **I** that are $O(\varepsilon^{1/n})$ requires running the single trajectory for a time $t = 1/\varepsilon$ that grows exponentially with n.

What happens to tori, in an adiabatic switch, when the conditions for adiabatic invariance of the actions are violated? Return to the model problem, case (4), which is simple enough to be analyzed completely. The equation of motion for $x = \Theta_1 - \Theta_2$ is $\ddot{x} = \varepsilon\{\beta + 2\alpha\gamma \sin x\}$; i.e., the motion is that of a one-dimensional particle in the time-independent potential $\varepsilon\{2\alpha\gamma \cos x - \beta x\}$. This is a linear potential modulated by ripples; if $|2\alpha\gamma| > |\beta|$ the ripples are large enough to produce a well within each 2π-wide interval of x. Start on a resonant torus, $\omega_1 = \omega_2$ at $t = 0$; since $\dot{x} = \omega_1 - \omega_2$, the points on the torus lie at turning points of the classical x motion, and the set of initial conditions is the set of turning points of the x motion between, say, $x = 0$ and $x = 2\pi$. All points at or below the top of the well are trapped forever; this is a band on the torus which is independent of the "slowness" parameter ε. The kinetic energy of the motion in the well is $O(\varepsilon)$, so at all times $\dot{x} = O(\varepsilon^{\frac{1}{2}})$; but $\dot{x} = \omega_1 - \omega_2$, so <u>all points from this band on the resonant torus stay in resonance for all time during the adiabatic switch</u>, to within $O(\varepsilon^{\frac{1}{2}})$. We have "resonant locking" between the two vibrations of the system, enforced by strong nonadiabatic coupling. "Resonant locking" turns out to be a general phenomenon, not just an artifact of this simple model.[8]

Outside the band and not too near its edges, "particles" escape to large $|x|$ over the top of the well, driven by the overall linear decline of the potential. It is easy to verify directly from the equations that action is conserved, in the adiabatic limit, along these trajectories.

Finally, right outside the band is a thin transition region, whose width shrinks rapidly to zero with ε, within which "particles" spend a significant fraction of the total time $t = 1/\varepsilon$ crawling over the top of the well. Their final velocity is then less than it must be for conservation of the actions--because they haven't got far enough out in the linearly decreasing potential--but greater than it would be for "resonant locking".

In the adiabatic limit the resonant torus develops schizophrenia: a piece of it conserves the actions, the rest preserves the resonance. It is an interesting way for the adiabatic theorem to fail.

What happens to a nonresonant torus, $\omega_1 \neq \omega_2$ at $t = 0$, that would pass through resonance if the adiabatic theorem held? The initial conditions correspond to a set of x motions all starting with the same velocity and heading for collision with the potential. Most of these trajectories bounce off the potential without spending much time crossing the top of the last well before collision, and the actions are conserved along these trajectories. Only a narrow band of trajectories, whose width goes rapidly to zero with ε, spends significant time near the top of the well; these are trajectories with "energy" very close to the potential energy at the top, and these trajectories are temporarily trapped by the resonance. On passing through resonance the torus leaves behind a little tendril of phase points temporarily locked in resonance; the adiabatic theorem fails, because no matter how slow the adiabatic switch, a portion of the torus suffers large changes in action; but it fails in an essentially unobservable way, because this portion shrinks to zero so rapidly with ε.

Acknowledgments: This work was supported by a grant from the National Science Foundation. Frank R. Johnston is an NSF Graduate Fellow.

(1) E.A. Solov'ev, Sov. Phys. JETP 48, 635 (1978).

(2) R.T. Skodje, F. Borondo, and W.P. Reinhardt, J. Chem. Phys. 82, 4611 (1985).

(3) B.R. Johnson, J. Chem. Phys. 83, 1204 (1985).

(4) T.P. Grozdanov, S. Saini, and H.S. Taylor, Phys. Rev. A 33, 55 (1986).

(5) P. Ehrenfest, Versl. Kon. Akad. Amsterdam 25, 412 (1916); abridged English translation in Sources of Quantum Mechanics, ed. B.L. van der Waerden, North Holland, Amsterdam (1967).

(6) P.A.M. Dirac, Proc. Roy. Soc. (London) A107, 725 (1925).

(7) I. Dana and W.P. Reinhardt, preprint.

(8) F.R. Johnston and P. Pechukas, manuscript in preparation.

(9) V.I. Arnol'd, Sov. Math. Dokl. 6, 331 (1965).

INTRINSIC NONADIABATICITIES ON THE FAREY TREE

I. Dana and W. P. Reinhardt

Department of Chemistry and the Laboratory
for Research on the Structure of Matter,

University of Pennsylvania,
Philadelphia, Pennsylvania 19104, U.S.A.

This note reports new results[1] in the problem of adiabatic invariance in nearly-integrable systems. The relevance of the problem emerges naturally, for example, in the context of the semiclassical quantization of multidimensional systems by the recently-developed method of "adiabatic switching".[2] For multidimensional systems which remain integrable during the slow changes of parameters, adiabatic invariance is known[3] to depend crucially on the passage through an infinity of resonances (see also the recent work[4] on separatrix crossing in one-dimensional systems). These fill densely the phase space, but usually have zero width and occupy regions of zero measure. On the other hand, resonances in nearly-integrable systems generally correspond to island chains having finite widths and a chaotic separatrix. It is shown here that this fact leads to intrinsic nonadiabatic effects, which cannot be eliminated in the limit of infinitely slow change. The main island chains contributing to these effects are determined from the Farey tree in an interval containing the range of variation of the winding number.

As a typical system we have considered the "standard", or Taylor-Chirikov map,[5,6] under slow changes of the stochasticity parameter K,

$$I_{n+1} = I_n + K\,\hat{\lambda}(n/2N)\,\sin\theta_n \quad , \tag{1a}$$

$$\theta_{n+1} = \theta_n + I_{n+1} \quad , \tag{1b}$$

where N is a large integer, and

$$\hat{\lambda}(x) = \begin{cases} \lambda(2x), & 0 \leqslant x \leqslant 1/2 \quad , \\ \lambda(2-2x), & 1/2 \leqslant x \leqslant 1 \quad . \end{cases} \tag{2}$$

Here $\lambda(x)$ is the "switching function", monotonously increasing from 0 to 1 for $0 \leqslant x \leqslant 1$. Thus, one switches from $K = 0$ (the "integrable" case) to some $K > 0$, and then back to $K = 0$ following the same "path". We iterate with (1) all points θ on a given unperturbed torus ($K = 0$) of action J:

$$I = J, \qquad 0 \leqslant \theta < 2\pi \quad . \tag{3}$$

This gives, after n iterations ($n \leqslant 2N$), the "adiabatic curve"

$$I_n = F(\theta;\, J,\, n) \quad , \tag{4a}$$

$$\theta_n = G(\theta;\, J,\, n) \quad . \tag{4b}$$

Since the map (1) is area preserving, the curve (4) always "oscillates" around the position of the initial torus (3). Defining the nonadiabaticity $\Delta J(N)$ as

$$\Delta J(N) = \left\{\frac{1}{2\pi}\int_0^{2\pi} [F(\theta;\, J,\, 2N) - J]^2 d\theta\right\}^{1/2} , \tag{5}$$

one has then adiabatic invariance if $\Delta J(N) \longrightarrow 0$ as $N \longrightarrow \infty$. This possibility seems, however, to be excluded by the following phenomenological approach to the problem.

During the switching process (1) the instantaneous frequency or winding number "changes", in the sense that tori of action J at different values of K are generally associated with different irrational winding numbers $w(K;\, J)$. Thus, the curve (4) necessarily "crosses" regions of island chains, associated with rational winding numbers. To see how this crossing can introduce nonadiabatic effects as $N \longrightarrow \infty$, consider first the case of a single island chain, labeled by the index i, and associated with the winding number p_i/q_i. We thus assume that for $n \approx n_i$ (the "crossing time") the adiabatic curve is mostly located in the immediate vicinity of this island chain. The crossing time n_i, and the value K_i of K at the crossing are approximate solutions of

$$\hat{\lambda}(n_i/2N) = K_i/K \quad , \tag{6a}$$

$$w(K_i;\, J) = p_i/q_i \quad . \tag{6b}$$

We assume in (6b) that the chaotic separatrix is thin enough so that the winding--number function $w(K;\, J)$ is still well defined near $K = K_i$. This function can be calculated accurately by perturbation theory, or by the adiabatic-switching method itself.[1] For each K_i, Eq. (6a) admits two solutions for n_i, corresponding to the crossing of the same island chain twice, when switching on and off in (2). At the moment of the crossing, one can associate with the island chain two main quantities. The first one is the width δJ_i, defined as the difference in action between the two closest tori bounding the lower and the upper branch of the separatrix from below and from above, respectively (boundary circles). For K_i small enough one has, from perturbation theory,

$$\delta J_i \propto K_i^{q_i/2} \quad . \tag{7}$$

The second quantity is a typical time scale T_i. This can be interpreted as the average rotation period around an elliptic point, or as the number of iterations required for a stretching by a given factor to take place in the direction of the unstable manifold. In both cases we find, from residue theory,[6]

$$T_i \approx q_i\, K_i^{-q_i/2} \quad . \tag{8}$$

Because of the width δJ_i there is an uncertainty δK_i in the determination of K_i, roughly proportional to the width. This leads to a corresponding uncertainty δn_i in (6a), which can be interpreted as the time duration of the crossing. Assuming that

$\delta K_i/K_i \ll 1$, so that the width δJ_i does not change much during the crossing, we obtain

$$\delta n_i \approx N\, \delta K_i/(K\, \lambda_i'), \tag{9}$$

where λ_i' is the derivative of $\lambda(x)$ at $x = n_i/N$. Let us now assume that just before the crossing ($n \lesssim n_i$) the curve (4) can be represented by a single-valued function, for example that it approximates one boundary circle. Then, if the duration of the crossing (9) is of the same order of magnitude as the time scale (8), the curve will start developing multivaluedness in the form of "whorls" and "tendrils",[7] near the elliptic and the hyperbolic points, respectively. As $N \longrightarrow \infty$, it becomes convoluted to such an extent as to exclude the possibility of adiabatic invariance. In fact, our numerical data[1] indicates that the nonadiabaticity (5) tends to a nonvanishing value as $N \longrightarrow \infty$. This value, the "intrinsic nonadiabaticity", is found to be nearly proportional to the width δJ_i, provided the other island chains crossed have much smaller widths.

The width δJ_i generally decreases with increasing q_i at fixed K_i, and increases with K_i at fixed p_i/q_i (see, e.g., (7)). Thus, to determine the main island chains crossed (having the largest widths), we first arrange the rational numbers p/q, within the range of variation R_w of the function $w(K; J)$, in order of increasing denominators q. This can be done by using the idea of the Farey tree.[8] We start from a pair of rational numbers $w = p/q$ and $w' = p'/q'$ ($w > w'$), defining a small interval containing R_w and having the property of being "neighbours", namely

$$p\,q' - p'\,q = 1 \quad .$$

Such a pair can be easily found, for example, within the sequence of successive truncations of the continued-fraction expansion of some irrational number in R_w. We then construct the mediant $\bar{w} = (p + p')/(q + q')$. It can be shown that among all rational numbers in the interval (w', w), $\bar{w}$ has the smallest denominator. Moreover, $\bar{w}$ is neighbour to both w and w', so that from the pairs $(w', \bar{w})$ and $(\bar{w}, w)$ one can construct, by the mediant operation, two rational numbers having the smallest denominators in the respective intervals. This procedure generates at the m'th iteration 2^{m-1} new numbers forming with the previous ones 2^m neighboring pairs. As $m \longrightarrow \infty$, the set of all numbers generated in this way is the Farey tree associated with the interval (w', w). Any rational number in this interval appears exactly once in the tree. It is easy to see from the construction how the tree can be used to arrange all rationals in R_w in order of increasing denominators.

Next, we calculate the widths δJ_i on the Farey tree. It can well happen that for $q_j > q_i$, $\delta J_j > \delta J_i$, if, for example, $K_j > K_i$. Usually, however, these cases are relevant only on the first few generations of the tree, and can be easily identified by inspection once the behaviour of the function $w(K; J)$ is known. As an example, consider the case of $J/2\pi = (3 - \sqrt{5})/2$ (equivalent to the golden mean). Here, the function $w(K; J)$ turns out to be monotonic decreasing in the domain $0 \leqslant K \leqslant 0.97$. The range of variation R_w is only about 2% of $J/2\pi$. A neighboring pair covering this range is (1/3, 2/5), from which the Farey tree is constructed. Using the fact that K_i increases with decreasing p_i/q_i in (6b), the main island chains crossed can be easily determined by inspection of the first few generations of the tree. Thus, when switching to maximal values of K = 0.8, 0.9, 0.97, the main island chains crossed are, respectively, those corresponding to p/q = 11/29, 14/37, 3/8. Using the switching function $\lambda(x) = x - \sin(2\pi x)/(2\pi)$ in all these three cases, the adiabatic curve (4) (n = 2N) becomes quite convoluted already for $10^3 < N < 10^4$, reflecting typical structure of the dynamics in the corresponding island chain. Fig. 1 shows such a curve in the case of the switching to K = 0.8 for N = 10946 (the coordinates I and θ in the figure stand for I_{2N} and θ_{2N} in (4), and I is defined relative to the position of the initial torus (3)). The curve exhibits clearly strong Fourier components of orders multiples of 29, reflecting structure of the 11/29-island chain which is crossed near $K_i = 0.78$. We have observed[1] that for $10^3 < N < 10^4$ the nonadiabaticity (5) starts saturating, in all three cases, around asymptotic values, the intrinsic nonadiabaticities. These are due to the convoluted adiabatic curves, and are nearly proportional to the widths of the corresponding main island chains crossed.

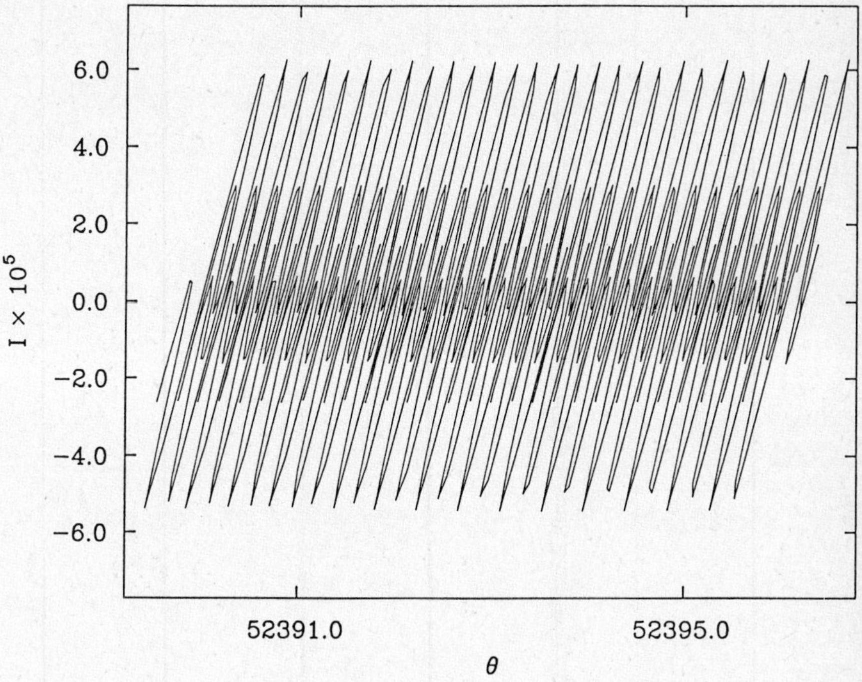

FIGURE 1

Acknowledgements

We would like to thank Prof. J. R. Cary and Mr. R. Gillilan for illuminating discussions, and Prof. I. C. Percival for sending us a copy of the manuscript in Ref. 8. The support of the National Science Foundation through grants DMR85-19059 and CHE84-16459 is gratefully acknowledged.

References

1. I. Dana and W. P. Reinhardt, "Adiabatic Invariance in the Standard Map", University of Pennsylvania preprint, to be published.
2. E. A. Solovev, Sov. Phys. - JETP 48, 635 (1978); T. P. Grozdanov and E. A. Solovev, J. Phys. B 15, 1195 (1982); B. R. Johnson, J. Chem. Phys. 83, 1204 (1985); R. T. Skodje, F. Borondo, and W. P. Reinhardt, J. Chem. Phys. 82, 4611 (1985); R. T. Skodje and F. Borondo, Chem. Phys. Lett. 118, 409 (1985); C. W. Patterson, J. Chem. Phys. 83, 4618 (1985); T. P. Grozdanov, S. Saini, and H. S. Taylor, Phys. Rev. A 33, 55 (1986); J. W. Zwanziger, E. R. Grant, and G. S. Ezra, to be published.
3. P. A. M. Dirac, Proc. R. Soc. 107, 725 (1925).
4. J. L. Tennyson, J. R. Cary, and D. F. Escande, Phys. Rev. Lett. 56, 2117 (1986), and to be published.
5. B. V. Chirikov, Phys. Rep. 52, 263 (1979), and references therein.
6. J. M. Greene, J. Math. Phys. 20, 1183 (1979).
7. M. V. Berry, N. L. Balasz, M. Tabor, and A. Voros, Ann. Phys. (N.Y.) 122, 26 (1979).
8. See, e.g., R. S. Mackay, J. D. Meiss, and I. C. Percival, "Resonances in Area-Preserving Maps", preprint QMC DYN 86.4, submitted to Physica D. Some of the terms related to the Farey tree in this work have been used by us.

CHAOTIC IONIZATION OF HIGHLY EXCITED HYDROGEN ATOMS

Roderick V. Jensen

Mason Laboratory, Yale University

New Haven, CT 06520, USA

Recent experimental measurements of the microwave ionization of highly excited hydrogen atoms with principal quantum numbers ranging from $n = 30$ to 90 are well described by a classical treatment of the nonlinear electron dynamics. In particular, the predictions of the threshold field for the onset of significant ionization is found to coincide with the onset of ***classical chaos*** in a one-dimesional model of the experiment. In this brief note I emphasize that this excellent agreement between the theoretical and experimental ionization thresholds requires a proper theoretical treatment of the slow, ***adiabatic*** turn-on of the microwave perturbation in which the persistence on nonlinear resonances in the chaotic phase space plays a crucial role.

Experimental studies of the interaction of highly excited hydrogen atoms with strong microwave fields provide a unique opportunity to examine the quantum behavior of a Hamiltonian system which can exhibit classical chaos.[1–4] A hydrogen atom in an oscillating electric field corresponds, classically, to a nonlinear oscillator subject to a periodic perturbation. When the perturbation exceeds a critical threshold, the classical oscillator exhibits a transition from regular behavior to global chaos.[5] Since the regular electron orbits remain bound, while the chaotic trajectories can wander away from the nucleus, the onset of classical chaos results in the ionization of the atom[6–10].

An important issue in the so-called problem of *quantum chaos* is whether and to what extent the effects of classical chaos persist in corresponding quantum systems. Since the Schrödinger equation is a linear equation that appears to be incapable of exhibiting the nonlinear, local instability which leads to classical chaos, one would expect the classical chaos to be suppressed. Nevertheless, detailed comparisons of the threshold fields for the onset of chaotic ionization with the microwave fields for significant (10 %) ionization in the experiments are in excellent agreement for initial quantum states with principal quantum numbers ranging from $n = 30$ to 90.[3,4] These results indicate that the effects of classical chaos can persist in these strongly perturbed hydrogen atoms and that a classical treatment of the electron dynamics provides a remarkably good description of true dynamics.

In particular, the classical theory successfully predicts the measured dependence of the threshold fields, n^4F, on the scaled frequency, $n^3\Omega$, which exhibits curious peaks at rational values of $n^3\Omega = 1, 2/3, 1/2, 2/5, 1/3$.[3,4] The purpose of this brief paper is to show that these peaks are associated with the presence of nonlinear resonances (island structures) in the classical phase space and that the excellent agreement between the classical theory and experiment requires a detailed treatment of the classical dynamics near these resonances as the perturbation is slowly (adiabatically) turned on.

The experiments were performed with carefully prepared beams of highly excited hydrogen atoms which were exposed to $\sim$ 300 periods of a $\sim$10 GHz microwave field as they passed through a microwave cavity. In addition, due to fringe fields, the atoms experienced a slowly increasing and decreasing field for $\sim 40 - 80$ periods as they entered and exited the cavity. For low microwave fields no ionization was observed. However, as the field was increased beyond a threshold field the entire beam was rapidly ionized. The experimental threshold was chosen to be the field at which 10 % of the beam was ionized during the transit time through the cavity.

The classical calculations were performed by integrating the nonlinear equations for a microcanonical ensemble of initial conditions with energies appropriate for the given quantum state. The fractional ionization was then computed by counting the number of classical trajectories which became chaotic and wandered away from the nucleus. Although the classical calculations were performed in one, two, and three spatial dimensions, the one-dimensional model appeared to be adequate for predicting the threshold fields for the onset of ionization since the nearly one-dimensional orbits, elongated along the field direction, are likely to be the easiest to ionize.

Although the classical thresholds were relatively insensitive to dimension, they were found to depend strongly on whether the perturbation was turned on suddenly or adiabatically, especially for values of n near rational values of the scaled frequency, $n^3\Omega$. As a consequence, great care was taken in modeling the experiment to include a slow turn-on of the perturbing field over 20-100 microwave periods. If the classical calculations are performed with a sudden turn-on, the distinctive peaks at resonant $n^3\Omega$, where the microwave frequency is a rational multiple of the classical Kepler frequency $1/n^3$, are suppressed.

The physical explanation for this dependence on the scaled frequency is clearly evident in Poincaré sections or phase-space portraits of the classical dynamics. These are generated by integrating the classical equations of motion for a number of different initial conditions for one-dimensional hydrogen atom and plotting the location of the trajectory once every period of the perturbation in the two-dimensional action-angle space of the unperturbed Hamiltonian. In the absence of any perturbation, the orbits lie on straight, horizontal lines in action-angle phase space. For weak fields, below the threshold for chaos, the straight line orbits are distorted but remain bounded by smooth invariant curves in phase space. The strongest distortion occurs for orbits which are nearly resonant with rational multiples of the microwave frequency. In this case the orbits are trapped in nonlinear resonances which form characteristic island structures in the phase space. These islands grow wider in action as the perturbation is increased, until they begin to interact or overlap, at which point the orbits near the edges (separatrices) of the islands become chaotic and begin to wander over large regions of phase space.[7–9]

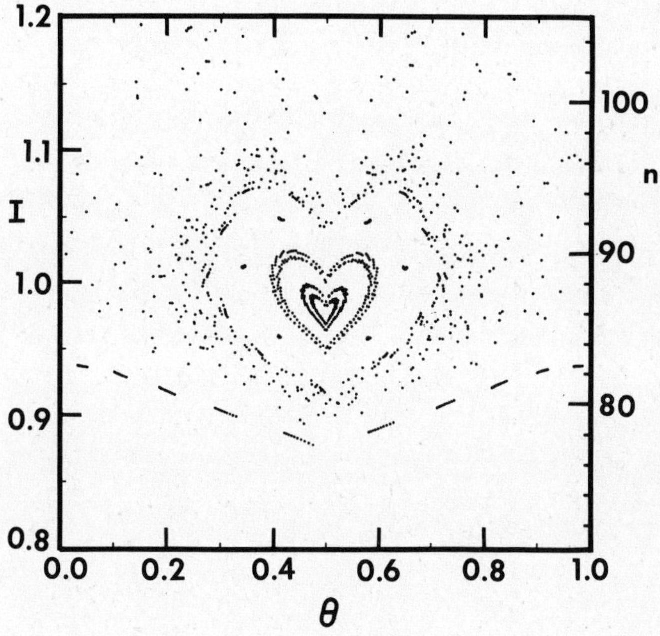

Fig. 1: The Poincaré section of action-angle space for a one-dimensional Hydrogen atom in an oscillating electric field with amplitude, $n^3F = .03$ and frequency $n^3\Omega = 1$ (with $n = 88$). The right vertical axis shows the corresponding values of the principal quantum numbers for the experimental parameters of Koch et al.[4]

Figure 1 shows a typical phase-space portrait for initial conditions which are near the $n^3\Omega = 1$ resonance for a microwave field strength of $n^4F = .03$ which is above the threshold for the onset of global chaos. Although many orbits are clearly chaotic due to interaction with the $n^3\Omega = 3/2$ and 2 islands (not shown in the figure), the phase space is dominated by a large island in the chaotic sea. In terms of the experimental parameters[3,4], this island spans a range of quantum states from $n = 82 - 94$, which corresponds to the measured bump in the ionization threshold for $n^3\Omega \sim 1$.

Similar island structures dominate the phase space at the critical threshold for $n^3\Omega = 1/2, 1/3, \ldots$. Classically, these nonlinear resonances mix different actions so that a whole range of initial actions, corresponding to a band of n's, responds to the perturbing fields in a similar way. In particular, one of the signatures of the classical ionization mechanism is that the unscaled values of the microwave fields (in V/cm) for the n's spanned by the resonances should be nearly equal. These plateaus are clearly visible in experimental plots of F(V/cm) versus n.[3,4]

If the classical perturbation were turned on suddenly, Fig. 1 indicates that approximately 50% of the initial conditions corresponding to $n = 88$ and 89, which are uniformly distributed on lines of constant action or n, would lie in the chaotic region of phase space and ionize. However, both the measured value for the threshold field for 10% ionization and the classical predictions for the onset of chaotic ionization, based on calculations which included the slow turn-on of the perturbing field, are higher than $n^4F = .03$.[4] The excellent agreement between the numerical simulations and the experiment is a consequence of the fact that the classical dynamics is more stable near resonance when the perturbation is turned on slowly.

The reason for this stability is the existence of an adiabatic invariant associated with the phase-space area enclosed by the invariant tori within the resonance island. As the perturbation is increased, these islands grow wider in action. In order for the area to be preserved, these islands must also contract in angle, moving away from the x-points and separatrices where the chaos sets in. This stabilization mechanism can be clearly seen by plotting a series of phase-space portraits for a slowly increasing perturbation.

An analysis of this adiabatic stabilization also indicates that initial states above and below the resonance are likely to be easier to ionize than the resonant n's. In fact, experimental measurements[4] show that across the $n^3\Omega = 1$ resonance the threshold fields in V/cm actually increase with increasing n. For example, stronger fields are required to ionize the more weakly bound $n = 90$ state than the $n = 82$ state. This remarkable feature of the experimental measurements is also reproduced by the classical theory.

In conclusion, the experimental measurements of the ionization of highly excited hydrogen atoms not only reflect the onset of chaos in the classical theory, but also the stabilizing effects of the resonance islands for a slowly increasing perturbation. The challenge now remains to understand the quantum description of these classical effects. Since the microwave frequencies are low ($\sim$ 100 or more photons must be absorbed for ionization) and the fields are so strong ($\sim$ 5-10% of the Coulomb binding field) conventional quantum treatments based on time-dependent perturbation theory are inadequate. However, a nonperturbative theory based on the quasi-energy states (QES) of the periodically perturbed Hamiltonian and how these QES are populated as a perturbation is slowly turned provides a useful description of the quantum dynamics with close analogies to the classical effects of nonlinear resonances and adiabatic stabilization.[11] This classical adiabatic stabilization mechanism and its quantum analog are also likely to have important applications to other dynamical systems where strong perturbation are slowly applied.

This work was supported by the National Science Foundation and the Alfred P. Sloan Foundation.

1. J.E. Bayfield and P.M. Koch, Phys. Rev. Lett. *33*, 258 (1974).
2. J.E. Bayfield and L.A. Pinnaduwage, Phys. Rev. Lett. *54*, 313 (1985).
3. K.A.H. van Leeuen, G.v. Oppen, S. Renwick, J.B. Bowlin, P.M. Koch, R.V. Jensen, O. Rath, D. Richards, and J.G. Leopold, Phys. Rev. Lett. *55*, 2231(1985).
4. P.M. Koch et al., "Chaotic Ionization of Highly Excited H Atoms", in this volume.
5. A.J. Lichtenberg and M.A. Lieberman, ***Regular and Stochastic Motion***, (Springer, New York, 1983). See also R.V. Jensen, "Classical Chaos", (to appear in American Scientist, January 1987) and references therein.
6. J.G. Leopold and I.C. Percival, J. Phys. B *12*, 709 (1979).
7. R.V. Jensen, Phys. Rev. Lett. *49*, 1365 (1982).
8. R.V. Jensen, Phys. Rev. *30A*, 386 (1984).
9. R.V. Jensen, in ***Chaotic Behavior in Quantum Systems***, edited by G. Casati, (Plenum, New York, 1985) p. 171.
10. N.B. Delone, V.P. Krainov, D.L. Shepelyansky, Usp. Fiz. Nauk *140*, 355 (1983) (Sov. Phys. Usp. *26*, 551 (1983).
11. R.V. Jensen, "The Effects of Classical Resonances on the Chaotic Microwave Ionization of Highly Excited Hydrogen Atoms", to appear in the Proceedings of the Adriatico Research Conference on Quantum Chaos, Trieste, 1986, published by Physica Scripta.

THE ROLE OF KAM-TORI FOR THE DYNAMICS OF NONLINEAR QUANTUM SYSTEMS

G. Radons, T. Geisel, and J. Rubner

Institut für Theoretische Physik
Universität Regensburg
D-8400 Regensburg, W. Germany

In nonintegrable classical Hamiltonian systems KAM-tori can confine the stochastic motion to certain regions of phase space.[1] In quantum systems the situation is quite different: Strictly speaking, KAM-tori do not exist, wave packets spread in time, the equation of motion is linear, etc. Therefore, we have asked the questions to which extent a quantized system has 'knowledge' of its classical counterpart, whether the classical KAM-tori still act as dynamical barriers and how large is the transition probability into classically inaccessible regions of phase space.

Similar questions have been adressed in the context of intramolecular energy transfer[2] for conservative Hamiltonian systems of 2 degrees of freedom. For a quantitative study, however, nonautonomous Hamiltonians of the form $H = K(p) + V(\theta)\tilde{\delta}(t)$ are more appropriate ($\tilde{\delta}(t)$ is the one-periodic δ-function). The classical dynamics of these periodically kicked systems is described by area-preserving mappings

$$(\theta_{t+1}, p_{t+1}) = (\theta_t + K'(p_t), p_t + V'(\theta_{t+1})), \tag{1}$$

and quantum mechanically by[3-5]

$$|\Psi_{t+1}\rangle = U|\Psi_t\rangle, \text{ with } U = \exp[-\tfrac{i}{\hbar} V(\hat{\theta})] \exp[-\tfrac{i}{\hbar} K(\hat{p})], \ [\hat{\theta},\hat{p}] = i\hbar, \tag{2}$$

θ_t, p_t, and $|\Psi_t\rangle$ denote angle, momentum, and wave function immediately after the kick at time t.

The kicked rotator Hamiltonian[4,5] is especially suited for our problem. Here $K(p) = p^2/2$ and $V(\theta) = k\cos\theta$. Classically its dynamics is described by the well-known standard mapping.[1] Above a critical nonlinearity parameter $k = k_c = 0.9716 \ldots$ KAM-tori cease to confine the diffusive motion in momentum space. At $k = k_c$ the last KAM-tori divide the phase space into cells around the integer resonances at $p = 2n\pi$, and around the half-integer resonances at $p = (2n+1)\pi$. This means that iterating a distribution $\rho_0(\theta,p) = \frac{1}{2\pi}\delta(p-p_0)$, e.g., with $p_0 = 3.2$, results in an asymptotic distribution

$$\bar{\rho}(\theta, p) = \lim_{T\to\infty} \frac{1}{T} \sum_{t=1}^{T-1} \rho_t(\theta,p) \tag{3}$$

which is confined to an interval (p_c^-, p_c^+) in momentum space with $p_c^- = 2.015$ and $p_c^+ = 4.269$.

We have performed an analogous quantum mechanical calculation, i.e., iterated a momentum eigenfunction $|\Psi_0\rangle = |p_0\rangle$ with eigenvalue $p_0 = 3.2$ and determined

$$P(p|p_0) = \lim_{T\to\infty} \frac{1}{T} \sum_{t=0}^{T-1} |\langle p|U^t|p_0\rangle|^2 \tag{4}$$

for $k = k_c$ (Fig. 1). This is the analog of the classical distribution $\bar{\rho}(\theta,p)$ projected onto the momentum axis.

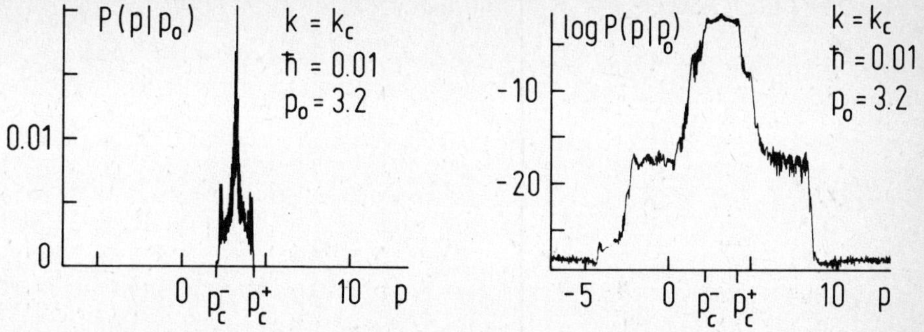

Fig. 1: The asymptotic distribution $P(p|p_0)$ on linear and logarithmic scale. Classically the distribution is confined to the interval (p_c^-,p_c^+) by KAM-tori.

From Fig. 1a one can see that most of the time-averaged probability remains within the classically accessible interval (p_c^-,p_c^+). The logarithmic display of $P(p|p_0)$, however, clearly shows that there is also a probability to find the system in classically inaccessible regions of phase space. This probability decays exponentially near certain values p_e that correspond to tori (near $p_c^{\pm}$) and broken tori (cantori). The fact that exponential decays may be related to cantori (and not only tori) can be observed for $k = 1.1$ (Fig. 2a), where classically the tori near $p_c^{\pm}$ have turned into cantori. For even larger values of k (e.g., $k = 2$) the cantori become irrelevant for the classical dynamics and correspondingly $P(p|p_0)$ only exhibits an overall exponential decay associated with Anderson localization.[5]

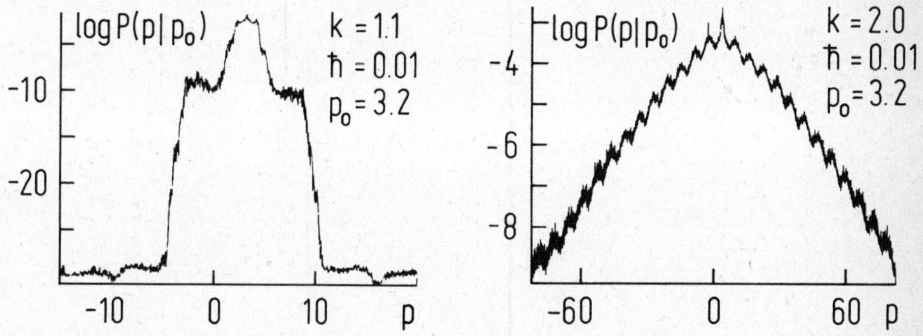

Fig. 2: Asymptotic distributions above the critical value k_c.

The exponential decay $\exp(-\lambda|p-p_e|)$ near tori or cantori for values $k \approx 1$ can be understood by means of Wigner functions corresponding to quasienergy eigenstates $|j\rangle$ (eigenfunctions of U) that are located on tori[3] or broken tori[6] in the semiclassical limit: $P(p|p_0)$ can be expressed as $P(p|p_0) = \Sigma_j |\langle p_0|j\rangle|^2 |\langle p|j\rangle|^2$, where $|\langle p|j\rangle|^2$ is the projected Wigner function $W_j(\theta,p)$ corresponding to $|j\rangle$, which is known to decay exponentially.[7] Although from this argument one would expect the $\hbar$-dependence to be $\lambda \propto \hbar^{-1}$ for $\hbar \to 0$, we find $\lambda \propto \hbar^{-2/3}$ as is shown in Fig. 3a.

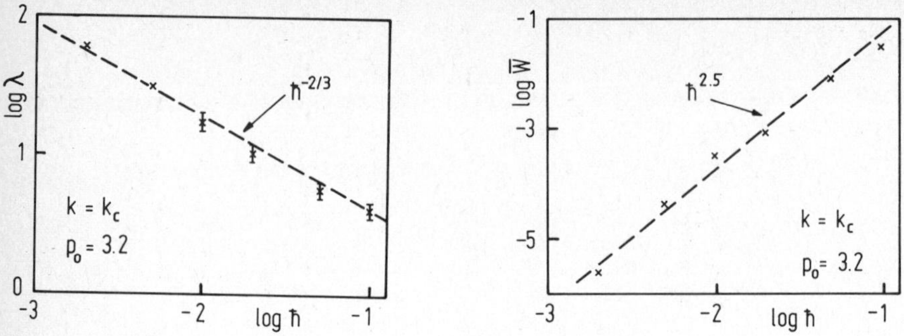

Fig. 3: $\hbar$-dependence of the exponent λ and of the transition probability into classically forbidden regions.

Fig. 3b shows the $\hbar$-dependence of the total probability $\bar{W}$ to find states outside the interval (p_c^-, p_c^+), $\bar{W} = \sum_{p \lessgtr p_c^\pm} P(p|p_0)$. We find that $\bar{W}$ vanishes as $\hbar^\gamma$ with $\gamma \approx 2.5$. If we choose an initial wave function $|p_0\rangle$ located in the much larger cells around integer resonances and record the probability in classically forbidden regions, we find different exponents which also depend on p_0. The reason for this probably is a partial localization of the wave function within the cell before the KAM-like barriers are reached.

We have also investigated the k-dependence of the quantity $\bar{W}$. As is clear from Figs. 1 and 2, this quantity strongly increases as k is increased beyond k_c, the increase getting stronger for smaller h. The main increase occurs above $k \approx 1.1 > k_c = 0.9716$ similar to observations for the Henon-Heiles potential.[2] The reason seems to be the importance of broken tori below k = 1.1. Details of this behaviour and of time-dependent quantities like the time-dependent probability W_t of finding the system in classically forbidden regions will be published elsewhere. Here we only mention that W_t increases algebraically at short times before turning into quasi-periodic behaviour at long times. The algebraic time dependence disappears when the dynamical barriers become irrelevant.

Finally we note that our results imply the possibility of KAM-like localization in Anderson models with pseudorandom diagonal disorder,[5] i.e., with short range correlations in the site energies. This is different from the usual localization mechanism which is found for large nonlinearities $k \gg 1$.

1 A.J.Lichtenberg, M.A.Lieberman, Regular and Stochastic Motion (Springer N.Y. 1983)
2 J.S.Hutchinson, R.E.Wyatt, Phys. Rev. A23, 1567 (1981)
3 M.V.Berry et al., Ann. Phys. N.Y. 122, 26 (1979)
4 G.Casati et al., Lecture Notes in Physics 93, 334 (Springer Berlin 1979)
5 D.R.Grempel, R.E.Prange, S.Fishman, Phys. Rev. A29, 1639 (1984)
6 R.B.Shirts, W.P.Reinhardt, J. Chem. Phys. 77, 5204 (1982)
7 M.V.Berry, Phil. Trans. Roy. Soc. Lond. 287, 237 (1977)

QUANTUM CHAOS, IS THERE ANY?

Joseph Ford
Georgia Institute of Technology
Atlanta, Gerogia 30332

The contribution by Professor Ford is not reproduced in these proceedings, because the material has already been published in the following references.

1) Chaotic Dynamics and Fractals, edited by M. F. Barnsley and S. G. Demko (Academic Press, New York, 1986).
2) The New Physics, edited by S. Kaplan (Cambridge Univ. Press, London, 1987).
3) Directions in Chaos, edited by Hao-Bai Lin (World Scientific Pub., Singapore, 1987).

C. WIGNER DISTRIBUTIONS

THE GENERAL PROPERTIES OF THE DISTRIBUTION FUNCTION AND REMARKS ON ITS WEAKNESS

E. P. Wigner
Joseph Henry Laboratories
Princeton University
Princeton, New Jersey 08544

INTRODUCTION

It was an unexpected pleasure to hear about the conference on the quantum mechanics of phase space, and I very much appreciate the pleasure to be invited thereto. I will be able to contribute very little to it that is not contained in the Physics Report article [Phys. Rep. 106, 121 (1984)] by M. Hillery, R. F. O'Connell, M. O. Scully, and myself - an article to which I have actually contributed, in contrast to Dr. Scully, very little. But I will admit that the underlying reformulation of the Schrödinger equatin was started by me, in 1932 (Phys. Rev. 40, page 749). I was interested in the thermodynamic behavior of macroscopic objects which is given with high accuracy, at ordinary or high temperatures, by classical statistical mechanics. At low temperatures quantum effects can become important and this manifested itself also in the "equation of state" (temperature and density dependence of the pressure) of the He gas. It was natural, therefore, to develop a substitute for the classical expression for the density in phase space (to be described below) which forms the basis for the calculation of the thermodynamic behavior in the temperature region in which classical mechanics can be assumed to be valid, and which easily provides a good approximation in the temperature region not too far away from the validity of classical physics. This means a probability function of the position and momentum variables q and p, defined in terms of the wave function ψ or the density matrix M, a probability function which is a hermitean expression of the wave function, hence linear in the density matrix. It is not too difficult to calculate and does give accurate results for the equation of state, and I hope also for other quantitites.

It must be admitted, of course, that the interpretation of the phase space density funciton is much less direct in the situation in which quantum effects play an important role than it is in the area of classical physics. The variables of the phase space are the 3N position and 3N momentum coordinates of the N-particle system to which the density function of the phase space refers. We'll write n for 3N. The classical phase space function's value at a point of phase space is the probability that the position and momentum coordinates of the N particles have the values given by the coordinates of corresponding points in phase space. If the phase space function has to be so closely defined that quantum effects play a role, this interpretation is not possible because there is no state in which both position and momentum coordinates have definite values. In fact the states of the system are not specified in terms of these coordinates. It follows that the interpretation of the density function of phase space is much less straightforward in the region in which quantum effects play a significant role than it is in classical theory - that is if the probabilities do not change significantly within distances in which the products of the p and q are not far from h. All this shows that the definition - and hence also the meaning - of the phase space functions is not as unique in the quantum region as it is in the region of the classical theory. The next section will therefore discuss the meaning and the properties of the quantum distribution function as defined in 1932. This definition does not take care of the existence of the spin and the extension of the theory to the description of the spin state will be discussed afterwards.

PROPERTIES OF THE PROPOSED QUANTUM DISTRIBUTION FUNCTION

As is apparent from the preceding discussion, the quantum distribution function to be discussed does not have such a simple meaning as the classical phase space function. It may be useful, therefore, to describe its basic properties before discussing its applications.

Let us first define the distribution function P to be considered. It will be defined as a function of n position and n momentum variables q and p, the n being three times the number of particles (n = 3N). If the state of the system is given by a position-dependent wave function $\psi(x_1,x_2,\ldots,x_n)$, the distributuion function is

$$P(q_1,\ldots,q_n;p_1,\ldots,p_n) = \frac{1}{(\pi\hbar)^n} \int \ldots \int dy_1\ldots dy_n\, \psi^*(q_1+y_1,\ldots,q_n+y_n)$$
$$\times\, \psi(q_1-y_1,\ \ldots,q_n-y_n)\, e^{2i(p_1y_1+\ldots+p_ny_n)/\hbar} \quad . \qquad (1)$$

This is, clearly, a nonrelativistic definition - as is fundamentally also that of the classical distribution function - but has, similar to that, some useful properties. These will remain also after the introduction of the spin variables. Before enumerating its useful properties, it may be good to give the P for a density matrix $M(q_1,\ldots;q_n;q_1',\ldots,q_n')$. This can be decomposed into orthogonal and normalized wave functions $\psi_1,\psi_2,\ldots$ which appear with probabilities $w_1,w_2,\ldots$. If the distribution function is assumed to be an additive function of these:

$$M(q_1,\ldots,q_n;q_1',\ldots,q_n') = \sum w_k\psi_k(q_1,\ldots,q_n)\psi_k^*(q_1',\ldots,q_n') \ , \qquad (2)$$

it is nautral to define the corresponding P as

$$P(q,p) = \frac{1}{(\pi\hbar)^n} \int dy\ M(q-y,q+y)e^{2i(p\cdot y)/h} \ . \qquad (2a)$$

In this equation, as in many later ones, the symbols q, y and p represent n variables each, $\int dy$ means integration over the n variables y, and $(p \cdot y)$ is the scalar product $\Sigma\, p_k y_k$. The notation used in these equations renders several future equations much simpler. Similar to the meaning of dy, the dq will mean integration over the n variables q and dp means integration over the n variables p. These notations simplify several of the following equations.

Let me now come to the properties of the distribution functions (1) and (2a) which I consider to be of significance. The proofs will be given for (1) but, because of the definition (2) of M, it will be evident that they apply also for the more general form (2a).

1. $P(q,p)$, if integrated over p (that is over $p_1, p_2, \ldots, p_n$), gives the probability of the configuration q, that is the probability that the position coordinates are $q_1, q_2, \ldots, q_n$. This is easily verified.

2. Similarly, if $P(q,p)$ is integrated over the q, it gives the probability that the momentum coordinates have the values $p_1, p_2, \ldots, p_n$.

These two properties can be easily verified, and it is clear also that they are less significant than the basic property of the classical P which represents the probability for both the positions to be given by $q_1, \ldots, q_n$, that is by q, and the momenta by p. But they do show that the average value of the classical energy, being the sum of two functions, one of the momentum, the other of positions, can be easily obtained.

3. These two observations suggest that q and p play similar roles in the definition of the distribution function P. Indeed, if the $\psi(q)$ is expressed in terms of its Fourier transform, the wave function $\chi(p)$ of the momentum coordinates:

$$\psi(q) = \int \chi(p) e^{ip \cdot q/\hbar} dp \,, \tag{3}$$

where we neglect constant factors temporarily, we obtain for $P(q,p)$

$$P(q,p) = \iiint \chi^*(p')\, e^{-ip' \cdot (q+y)/\hbar}\, \chi(p'')\, e^{ip'' \cdot (q-y)/\hbar}\, e^{2ip \cdot y/\hbar}\, dp' dp'' dy \,. \tag{3a}$$

The factors involving y give a delta function $\delta(2p-p'-p'')$ so that, again disregarding a constant factor, we can set $p' = p+z$, $p'' = p-z$ and the integration over z will replace the integration over p' and p''. Hence (3a) becomes

$$P(q,p) = (\pi\hbar)^{-n} \int dz\, \chi^*(p+z)\, \chi(p-z)\, e^{-2iz \cdot q/\hbar} \,. \tag{3b}$$

The numerical constant before the integral sign follows from the fact that the integral of P remains 1 and that the χ are also normalized. Eq.(3b) is a close analogue of (1), except for the fact that i is replaced by -i - which is natural - and shows that position and momentum coordinates play essentially the same role in the definition of our distribution function - just as they do in classical theory.

4. The transformation properties of P are the classical ones with respect to any of the classical transformations. The substitution of q+a for q clearly gives P(q+a,p) from P(q,p) - and this remains true even if a is not the same vector for all particles. If ψ is replaced - we use (1) in this discussion - by $e^{i\kappa\cdot q/\hbar}\psi$ the distribution function so obtained assumes the values of the original distribution function if p + κ is substituted for p - actually κ can be an arbitrary n dimensional vector, but naturally independent of the q.

The past three points are natural demands and are easily verified.

5. The so-called transition probability between two states, ϕ and ψ for instance, is, as a rule, not really observable. If the system is in the state ψ and an observation is made as a result of which the system's state vector becomes ϕ, the probability of this result of the observation is (if both ϕ and ψ are normalized) $|(\phi,\psi)|^2$, the absolute square of the scalar product of the two state vectors. The observation in question is not possible for every state vector ϕ, but the existence of the scalar product, or at least the measurability of its absolute value, is often assumed for all ϕ. If the original state of the system is best given by a density matrix M, the probability that the measurement transfers it into the state ϕ is given by the scalar product of ϕ and Mϕ, that is (ϕ,Mϕ). It is worth noting therefore that if P_ψ and P_ϕ correspond, by (1), to ψ and ϕ, the so-called transition probability between them becomes

$$|(\psi,\phi)|^2 = (2\pi h)^n \iint dpdq\ P_\psi(q,p)P_\phi(q,p) \quad . \tag{4}$$

It follows from (4) also that if P_M and P_N are the distribution functions which correspond to the density matrices M and N, then

$$\mathrm{Trace}(MN) = (2\pi\hbar)^n \iint dpdq\ P_M(q,p)P_N(q,p) \quad . \tag{4a}$$

All the preceding observations are easily verified and are also contained in the aforementioned article of Hillery, O'Connell, Scully and myself - most are in fact also contained in the aforementoined 1932 article. Apparently, there is a great deal of arbitrariness in the definition (1) of the distribution function but R. F. O'Connell has shown that some of the preceding properties already fully determine it. This was not known when (1) was originally proposed but is well worth remembering.

The last observation, eq.(4), also shows that most distribution functions, though real, are not everywhere positive. For two orthogonal wave functions, $(\psi,\phi) = 0$, the integral over $P_\psi P_\phi$ must vanish. They can not be both positive everywhere - for most ψ and ϕ neither is. But, as Heisenberg pointed out, there is no state for which both p and q have definite values. Transition probabilities are observable, at least many of them, and it is satisfactory that, according to (4), the expressions for these can not be negative. The fact that most functions of p and q do not represent possible states renders the quantum distribution function to be a less simple quantity than is the classical distribution function, since, in classical theory, all everywhere non-negative distribution functions are conceivable. This point will be mentioned again later, together with the fact that the condition which an arbitrary function of p and q must obey in order to be a possible distribution function is not simple. Clearly, it must be possible to write it in the form (2a) with a positive definite (or non-negative) self-adjoint matrix M, but this is not a simple condition.

6. The preceding observations on the properties of our quantum mechanical distribution functions gave properties which the classical distribution functions also had - in fact the properties of the latter were more general. We now come to an equation which shows the quantum mechanical nature of our distribution function - the equation of its time dependence. Essentially the same equation will be used afterwards to determine the distribution funciton for the thermodynamic equilibrium.

The equation for $\partial P/\partial t$ has two types of terms. The first type originates from the kinetic energy terms $-(\hbar^2/2m)\partial^2/\partial q^2$ of the expression for $i\hbar\partial\psi/\partial t$, the second one from the potential energy terms. Both are easily determiend and were long ago. Here only the first one will be reproduced in full detail. It gives for $(\pi\hbar)^n(\partial P/\partial t)_k$

$$\frac{i}{\hbar}\frac{\hbar^2}{2m}\int\Big[-\frac{\partial^2\psi^*(q+y)}{\partial q^2}\psi(q-y)+\psi^*(q+y)\frac{\partial^2\psi(q-y)}{\partial q^2}\Big]\,e^{2ipy/\hbar}\,dy\ . \tag{5}$$

The second derivatives with respect to q can be replaced by second derivatives with respect to y and a partial integration then be carried out. The two terms in which the products of both first derivatives appear then cancel and the terms in which the exponential is differentiated gives

$$\frac{i\hbar}{2m}\frac{2ip}{\hbar}\int\Big[\frac{\partial\psi^*(q+y)}{\partial y}\psi(q-y)-\psi^*(q+y)\frac{\partial\psi(q-y)}{\partial y}\Big]\,e^{2ipy/\hbar}dy\ . \tag{5a}$$

The differentiations with respect to y can be replaced by differentiations with respect to q - changing the sign of the second term. The result then is the same expression which appears in classical theory for the kinetic energy part $(\partial P/\partial t)_k$ of $\partial P/\partial t$ - if written in detail it is

$$\Big(\frac{\partial P}{\partial t}\Big)_k = -\sum_\kappa \frac{1}{m}\,p_\kappa\,\frac{\partial P}{\partial q_\kappa}\ . \tag{5b}$$

The potential part of $\partial P/\partial t$ can be expressed in two ways. One can expand the potential energy expression in

$$(\pi\hbar)^n\left(\frac{\partial P}{\partial t}\right)_p = \frac{i}{\hbar}\int \psi^*(q+y)[-V(q+y)+V(q-y)]\psi(q-y)e^{2ipy/\hbar}dy \qquad (6)$$

either into a power series of y, or represent it as a Fourier transform. The second possibility shows again that p and q play similar roles in the theory of distributions. But the epxansion of V(q-y)-V(q+y) as a power series of y gives

$$-V(q+y) + V(q-y) = -\sum_{\lambda_1\ldots\lambda_n} 2\left(\frac{\partial^{\lambda_1+\ldots+\lambda_n}}{\partial q_1^{\lambda_1}\ldots\partial q_n^{\lambda_n}}V(q)\right)\left(\frac{y_1^{\lambda_1}\,y_2^{\lambda_2}\ldots y_n^{\lambda_n}}{\lambda_1!\ldots\lambda_n!}\right), \qquad (6a)$$

the summation to be extended to all non-negative (integer) λ the sum of which is odd. This gives for the potential caused part of the time derivative

$$\left(\frac{\partial P}{\partial t}\right)_p = \sum_{\lambda_1\ldots\lambda_n}\left(\frac{(\hbar/2i)^{\lambda_1+\ldots+\lambda_n-1}}{\lambda_1!\lambda_2!\ldots\lambda_n!}\right)\left(\frac{\partial^{\lambda_1+\ldots+\lambda_n}}{\partial q_1^{\lambda_1}\ldots\partial q_n^{\lambda_n}}V\right)\left(\frac{\partial^{\lambda_1+\ldots+\lambda_n}}{\partial p_1^{\lambda_1}\ldots\partial p_n^{\lambda_n}}P\right) \qquad (6b)$$

in which, however, all the λ are non-negative and their sum odd. The first term of the series, in which one λ is 1, all others 0, gives the classical expression for $(\partial P/\partial t)_p$. The lowest order corrections contain the second power of $\hbar$. And they constitute for

$$\frac{\partial P}{\partial t} = \left(\frac{\partial P}{\partial t}\right)_k + \left(\frac{\partial P}{\partial t}\right)_p \qquad (6c)$$

the lowest order corrections. None of the preceding considerations is new, neither is the last point of this section.

7. The oldest use of the quantum mechanical distribution function was based on the calculation of the quantum effects on the equations of states of gases. If Bose or Fermi statistics of these is disregarded, (this was treated later, in 1984 by O'Connell and Wigner) the distribution function of these is the normalized form of $e^{-H/kT}$. Setting $1/kT = \beta$, this can be written as $e^{-\beta H}$ and the equation which replaces the equation for $\partial P/\partial t$ becomes

$$\frac{\partial P}{\partial \beta} = -HP\ . \qquad (7)$$

The expansion of P in terms of β has been discussed when the expression (1) or (2a) was first proposed and is reviewed also in the article by Hillery, O'Connell, Scully and Wigner mentioned several times before. There is no point repeating the calculation which replaces the calculation of the distribution function for $e^{tH/i\hbar}$ for the calculation of the distribution function for $e^{-\beta H}$. Perhaps I mention that the first application of the quantum mechanical distribution function concerned the equation of state of the He gas. At very low temperatures the experimental results deviated considerably from that given by classical theory, that is by the classical distribution function. The correction introduced by the quantum corrections to this discussed here were in the right direction but accounted only for about 2/3 of the deviations from the experimental measurements. It is possible that the reason for this was that the potential energy function was not known well enough. It would therefore be worthwhile to repeat that calculation. Its desirability was actually the stimulant for the introduction of our P.

THE SPIN VARIABLE

The preceding discussion largely disregards the spin variable - which is natural in the case of the He gas, since the He atoms have no spin. A possible way to add the description of the spin state to that of the other variables was discussed before (1983) for systems with spin 1/2 but that is easily generalized for larger spin.

For every particule of spin s the density matrix has $(2s+1)^2$ components. The problem is only to find such linear combinations of these components which have relatively simple properties. We can specify the $(2s+1)^2$ components with two index symbols: μ and μ' - the first giving the row index of the density matrix, and μ' the column index. Both run for each particle of spin s from -s to s in integer steps. We can then form, for each particle, another description of the spin state by combining the row and column components to have simple transformation properties. They will have transformation properties which correspond to the direct product of two representations $D^{(s)}$. It is possible then to produce linear combinations of the components characterized by μ and μ' which transform under rotations by the representations $D^{(0)}, D^{(1)}, \ldots, D^{(2s)}$. In the case of $s = 1/2$, which was considered before, there is a scalar and a vector component - the former giving the total probability, the others being formed by the components of $D^{(1)}$.

Let us denote the density matrix by $M(\xi,\mu;\xi',\mu')$, μ giving the row index of the spin variable of the particle in question, ξ denoting all other variables of the row, μ' and ξ' the same interpretation for the columns. The distribution function proposed would replace the μ and μ' by the indices S and m:

$$M'(\xi,\xi';S,m) = \sum_{\mu\mu'} (S,m;s,\mu,s',\mu')M(\xi,\mu;\xi',\mu') , \tag{8}$$

the first factor after the summation sign being the coefficient which transforms the representation of the direct product $D^{(s)*} \times D^{(s)}$ into $D^{(S)}$, and m, μ and μ' are the row indices of the representations S, and $D^{(s)*}$ and $D^{(s)}$. It would not be reasonable to produce here these coefficients in general but it may be worth noting that

$$M'(\xi,\xi;0,0) = \frac{1}{2s+1} \sum_{\mu} M(\xi,\mu;\xi,\mu) . \tag{8a}$$

For the case of s = 1/2, the S assumes only two values: 0 and 1. The coefficients for 0 are given in (8a), those for S = 1 the transferred M, that is the M', were given as the expectation values of the x, y, and z components of the spin operator, that is of s_x, s_y, s_z. In many cases the effects of the higher S components of M' are insignificant and in those cases the same transformation of the spin coordinates can be recommended.

The total transformation to the quantum mechanical distribution function P obeys then the same equation as in the absence of spin (4) and the μ and μ' for every particle are replaced, in terms of (8), by S and m.

This is a somewhat superficial description of the transformation of the spin variables for what I call the quantum mechanical distribution function, but I hope that it gives the proposed transformation clearly enough.

PROBLEMS OF THE PROPOSED QUANTUM MECHANICAL DISTRIBUTION FUNCTION

The quantum mechanical distribution theory here described has two weaknesses. One of these was mentioned before: given an arbitrary real function P(p,q), it is not clear whether it is a possible distribution function. If it is, it can be written in the form (2a) in terms of an acceptable density matrix M but the acceptability of a density matrix is also not easily verified. In particular, it must be positive definite, or semidefinite - that is no expectation value of the transition to any state, that is no $(\phi, M\phi)$ can be negative. This applies also to our P: no integral of the product of two quantum mechanical distribution functions can be negative.

Just as in the usual theory, it is sufficient to demonstrate that the product is non-negative with any distribution function representing a single state, that is having the form (1), but even this is an infinite task - just as it is in ordinary quantum mehcanics dealing with density matrices.

The other difficulty well worth mentioning is one also shared, at least to some degree, with the usual formulation of quantum mechanics: the postulate of the coherence with relativity theory. This causes difficulties also in the usual theory - it is necessary to introduce a field, that is an infinitely more complex definition of the state than is used in Schrödinger's old fashioned theory. In addition, the equations often lead to infinities and these must be eliminated by "renormalization". In summary, even the usual theory has weaknesses - I would say that its beauty is not absolute.

But the weakness of the theory here discussed is much more fundamental - at least it is so at present. It assumes that the interaction of the particles is instantaneous - that it depends only on their same-time positions. This is acceptable, and in fact generally accepted, in non-relativistic theory but is in conflict with the theory of relativity in which simultaneity is not independent of the state of motion of the coordinate system describing it. This renders, quite generally, the description of the states of systems by phase space functions unattractive - in phase space the interaction is assumed to depend on the simultaneous position of the particles and is, therefore, not relativistically invariant. It is possible to make it invariant, for instance by postulating that it depends on the distance in the coordinate system at rest with the temporary center of mass of the particles, or to depend on the two positions at the time when their relativistic distance is zero - when one is on the light cone of the other or conversely. It is even possible to assume a "force" depending on the integral of the distances between the light cones. But these possibilities have not been explored to my knowledge and the present theories assume ineractions of fields - i.e. only interactions at points of the same positions and times. These gave many apparently correct results but needed the introduction of "fields", in particular electromagnetic potentials, and are not in harmony with the phase-space theories. Perhaps this could be amended, but I do not know of serious attempts in that direction - not even by myself.

WIGNER DISTRIBUTION FUNCTION APPROACH TO THE CALCULATION OF QUANTUM EFFECTS IN CONDENSED MATTER PHYSICS

R. F. O'Connell
Department of Physics and Astronomy
Louisiana State University
Baton Rouge, LA 70803

In condensed matter physics, the most common technique used in the calculation of quantum effects is that involving Green's functions, supplemented to a lesser extent by path-integral methods. Here we point out the potential value of the Wigner distribution function approach and we amplify our remarks by considering specific examples. In particular, we discuss our recent work on the extension of the range of applicability of phase-space techniques for the study of quantum systems; this is achieved by developing an expansion for phase-space functions in powers of the interaction potential.

I. Introduction

The study of quantum effects in many-particle physics has relevance to investigations in many branches of physics, particularly condensed-matter physics[1)] and quantum optics.[2)] Whereas path-integral methods have found application for certain types of problems (particularly dissipative tunnelling calculations[3)]), by far the most popular method in use is that of Green's functions.[4)] However, a further alternative approach makes use of the Wigner distribution function (WDF)[5,6)] The essence of the latter method is to use the sophisticated and well-developed phase-space approach to classical mechanics to do quantum-mechanical calculations with the help of a quantum distribution function. Recently, we have shown that still another technique -- involving use of a generalized Langevin equation -- can be very useful for the study of quantum effects.[7)] However, our emphasis here will be on the use of the WDF, with particular attention being paid to a discussion of a recent method which we have developed for extending the range of usefulness of the WDF.[8)]

In essence, Green's functions are correlation functions of quantum field operators at different times. Such functions provide a wealth of information concerning the equilibrium and dynamical properties of a system. For example, an important approach to the study of non-equilibrium properties is the linear response theory of Kubo;[9)] a calculation of the linear response leads to a determination of the generalized susceptibility which, in turn, is related (via the fluctuation-dissipation theorem) to a correlation function describing the properties of the system in equilibrium. Thus, for instance, the starting point of many calculations of the conductivity of a system is the current-current correlation function appearing in the Kubo- Green[10] formula. By contrast, the goal of the generalized Langevin equation method[7] is to by-pass an explicit evaluation of the correlation function and calculate the generalized susceptibility directly, from which the conductivity and other transport properties immediately follow. Then, if one needs the correlation function for other purposes (such as an evaluation of the decay time of the correlations), one simply reverses the Kubo approach and uses the fluctuation-dissipation theorem to obtain the correlation function from the generalized susceptibility.

Traditionally, the WDF has been used to obtain quantum corrections in the near-classical limit. The implementation of this approach has been via the Wigner-Kirkwood (WK) expansion,[5,6] which involves an expansion in powers of $\hbar$. In Section II, we will present the salient points of the WDF formalism. Then, in Section III, we will briefly review some applications involving the WK expansion. We will also discuss some problems where exact results have been obtained either analytically or numerically. In Section IV, we discuss our own recent investigations which were aimed at extending the range of applicability of phase-space techniques by developing an expansion for the WDF and correlation functions in powers of an interaction potential.

II. The Wigner Distribution Function (WDF)

In quantum mechanics, the average of a function of the position and momentum operators, $\hat{A}(\hat{q},\hat{p})$ say, is given by[5,6]

$$< \hat{A} > = \mathrm{Tr}\,(\hat{A}\,\hat{\rho}) \quad , \tag{1}$$

where $\hat{\rho}$ is the density matrix (and we will designate all operators by a ^). The essence of the insight achieved by Wigner was to show that one could write this result in a form involving integrations over phase-space:

$$< \hat{A} > = \iint A(q,p)\; P(q,p)\; dq\; dp \quad , \tag{2}$$

where $P(q,p)$ is the WDF and $A(q,p)$ is the classical quantity corresponding to $\hat{A}(\hat{q},\hat{p})$. All integrations are from $-\infty$ to $+\infty$. The result is written in a one-dimensional form but can be easily extended in an obvious manner.[5,6]

In the case of a canonical distribution at temperature T, we have

$$\hat{\rho} = e^{-\beta\hat{H}}/Z(\beta) \equiv \hat{\Omega}/Z(\beta) \quad , \tag{3}$$

where $\hat{\Omega}$ is the unrenormalized density matrix and $Z(\beta) = \mathrm{Tr}(e^{-\beta\hat{H}})$. It immediately follows that $\hat{\Omega}$ satisfies the Bloch equation, from which one obtains the Wigner classical correspondence of $\hat{\Omega}$, $\Omega(q,p)$ say, in the form of a series solution (the Wigner-Kirkwood expansion[5,6)]) in powers of $\hbar$:

$$\Omega(q,p) = \exp\{-\beta H(q,p)\} \{1+\hbar^2 a_2 + O(\hbar^4)\} , \tag{4}$$

where a_2 involves derivatives of the potential.

In the case of a non-equilibrium situation, an exact result for the time dependence is given by

$$\frac{\partial P}{\partial t} = \frac{\partial_k P}{\partial t} + \frac{\partial_v P}{\partial t} , \tag{5}$$

where

$$\frac{\partial_k P}{\partial t} = - \frac{p}{m} \frac{\partial P(q,p)}{\partial q} , \tag{6}$$

and

$$\frac{\partial_v P}{\partial t} = \sum_\lambda \frac{1}{\lambda!} \left(\frac{\hbar}{2i}\right)^{\lambda-1} \frac{\partial^\lambda V(q)}{\partial q^\lambda} \frac{\partial^\lambda P(q,p)}{\partial p^\lambda} , \tag{7}$$

λ being restricted to odd integers. An alternative form for $\partial_v P/\partial t$ is given by

$$\frac{\partial_v P}{\partial t} = \int dj\, P(q,p \mp j)\, J(q,j) , \tag{8}$$

where

$$\begin{aligned} J(q,j) &= \frac{i}{\pi\hbar^2} \int dy\, [V(q+y) - V(q-y)]\, e^{-2ijy/h} \\ &= \frac{1}{\pi\hbar^2} \int dy\, [V(q+y) - V(q-y)] \sin(2jy/\hbar) \end{aligned} \tag{9}$$

is the probability of a jump in the momentum by an amount j if the positional coordinate is q.

III. Some Applications of the WDF Method in Condensed Matter Physics

Our purpose here is not to give a detailed review, but simply to delineate the usefulness--and also the limitations--of the WDF in the condensed matter area.

First of all, we will consider the case of thermodynamic equilibrium. In Wigner's original paper on the subject,[5] he used an equation of the form of (4) to calculate quantum corrections to the second virial coefficient of He. The same Wigner-Kirkwood (WK) expansion has been used recently in the calculation, to order $\hbar^6$, of the pair distribution function of liquid neon,[11] very good agreement with experiment being achieved. The success of this calculation arises from the fact that the behaviour of neon is almost classical. The WK method has also been used to calculate quantum corrections to simple molecular fluids such as D_2 and H_2.[12]

Quantum corrections to the thermodynamic properties of a classical one-component plasma have been considered by various authors, both for three-dimensional[13,14] and two-dimensional systems.[14] However, again the use of a WK expansion implies near-classical (high-temperature) conditions; more explicitly the WK dimensionless expansion parameter is the ratio of the deBroglie wavelength λ to some typical length appropriate to the system (such as the average inter-particle distance). Another feature of the WK expansion is that it involves derivatives of the potential, and thus the existence of a convergent expansion depends on the potential exhibiting a relatively smooth behaviour. For example, as noted by Jancovici,[15] in the case of a hard-sphere gas (an oft-times studied model in statistical mechanics) the terms of the WK expansion diverge and thus such an expansion is no longer useful.

Turning next to non-equilibrium problems, it is clear already from (5) to (9) that the use of the WDF provides one with quantum corrections to the Boltzmann and other transport equations, a fact exploited by various authors.[16-28] Many of these papers use either the small $\hbar$ expansion or else make other approximations. A notable exception is the work of Ferry and collaborators[27] who used (8) and (9) which, of course, are exact results, to describe ballistic transport through resonant-tunneling quantum wells. Another innovative use of the WDF is the work of Barker[28] who treats tunneling phenomena from the point of view of non-equilibrium electron transport.

It is difficult to summarize all the various papers which use the WDF for transport studies, not only because of space restrictions but also because of my personal belief that the subject is not in an entirely satisfactory state. For example, it is difficult to ascertain from many papers the magnitude of the quantum effects obtained by use of the WDF, particularly vis-à-vis other effects (such as memory effects[7]) which could be potentially as large. As a result, incisive conclusions are hard to come by.

Are there any cases for which exact solutions exist? The answer is that there are at least two cases which are of physical interest viz. the harmonic oscillator problem[6] and the case of a particle in a magnetic field B.[14] The latter problem is particularly relevant to the analysis of a metal-oxide-semiconductor (MOS) inversion layer since, as pointed out by Alastuey and Jancovici,[14] this is a system which is certainly not classical for B=0 but which may become so for strong B (since the Landau gyration radius ℓ replaces λ as a characteristic quantum length scale when $\ell<\lambda$).

The next question which arises relates to what is the best method of treating problems which can be described by a Hamiltonian

$$\hat{H} = \hat{H}_0 + \lambda \hat{H}' , \tag{10}$$

where the exact solution is known in the case where λ is zero. This is the problem which we will address in the following section.

IV. Extension of the Range of Applicability of the WDF

Solving (10) by the traditional WK method would, in the case $\lambda=0$, give a result correct to a certain order in $\hbar^2$ whereas in point of fact an exact solution is known. Thus, the question arises as to how to retain all the information contained in the exact solution to the H_0 problem. The answer is that if the $\hat{H}'$ term is a perturbation, one can carry out an expansion in powers of λ. This is a well-known procedure in the case of the Green's function formalism, and it was adapted recently by Dickman and the present author, for problems using the WDF method.

Our starting-point is the expression[4]

$$\hat{\Omega} \equiv e^{-\beta\hat{H}} = e^{-\beta\hat{H}_0} U(\beta,0) , \tag{11}$$

where

$$U(\beta,0) = 1 - \lambda \int_0^\beta \tilde{H}'(\sigma)d\sigma + \frac{\lambda^2}{2!} \int_0^\beta d\sigma_1 \int_0^\beta d\sigma_2 T\,[\tilde{H}'(\sigma_1)\tilde{H}'(\sigma_2)] + \dots \tag{12}$$

and where

$$\tilde{H}(\sigma) = e^{\sigma\hat{H}_0}\, \hat{H}'\, e^{-\sigma\hat{H}_0} \tag{13}$$

is the perturbation in the "interaction picture," T being the time-ordering operator.

We may now use Eqs. (11) and (12) and the well-known rule for the Wigner translation of an operator product to write

$$\Omega(q,p;\beta) = \Omega^{(o)}(q,p;\beta)e^{\hbar\Lambda/2i}[U(\beta,0)]_w$$

$$= \Omega^{(o)}[1 - \lambda e^{\hbar\Lambda/2i} \int_0^\beta d\sigma\, [\tilde{H}'(\sigma)]_w + \ldots], \qquad (14)$$

where the subscript "w" denotes the classical Wigner equivalent of an operator[6] and where

$$\Lambda \equiv \frac{\overleftarrow{\partial}}{\partial p}\frac{\overrightarrow{\partial}}{\partial q} - \frac{\overleftarrow{\partial}}{\partial q}\frac{\overrightarrow{\partial}}{\partial p} , \qquad (15)$$

the arrows indicating the direction of operation.

Thus, in principle we now have Ω to any order in λ. The next step is to consider a quantity whose calculation underlies the determination of a host of physical quantities. This is a correlation function and, in particular, we consider the correlation

$$C(t) = \frac{1}{2} \langle\hat{q}(t)\hat{q}(0) + \hat{q}(0)\hat{q}(t)\rangle \quad . \qquad (16)$$

The Wigner phase-space equivalent of C(t) is then evaluated. However, for the purpose of carrying out an expansion in powers of λ, it is best to turn attention to the Laplace transform

$$J(s) = \int_0^\infty dt\; e^{-st}C(t) \quad . \qquad (17)$$

Then, using $P_n(q,p;\beta)$ to denote the WDF correct to nth order in the coupling λ, and $\langle\ \rangle_n$ to denote a phase-space average with respect to P_n, we find that an expansion for J(s) may be written in the following form:

$$J(s) = \langle A\rangle_n \{1 + \lambda \langle B^{(1)}\rangle_{n-1}/\langle D\rangle_{n-1} + \text{---}$$

$$+ \lambda^n \langle B^{(n)}\rangle_0 / \langle D\rangle_0 \} + O(\lambda^{n+1}) , \qquad (18)$$

where A, $B^{(1)}$, ---, $B^{(n)}$, and D are various phase-space quantities, which are given explicitly in Refs. 8 and 29.

As an example of the above formalism, if we consider the one-dimensional anharmonic oscillator

$$\hat{H} = \hat{p}^2/2m + \frac{1}{2} m\, \omega_o^2\, \hat{q}^2 + \lambda\, \hat{q}^4 , \qquad (19)$$

we find that

$$J(s) = \langle q^2 \rangle_1 \frac{s}{s^2 + \omega_R^2} + O(\lambda^2) \, , \tag{20}$$

where

$$\omega_R^2 = \omega_o^2 + 12\lambda/m \, (m\omega_o^2 A) + O(\lambda^2) \, , \tag{21}$$

and

$$A = (2/\hbar\omega_o)\tanh(\beta\hbar\omega_o/2) \, . \tag{22}$$

Thus, we see, without having to evaluate $\langle q^2 \rangle$, that the anharmonicity results in a first-order frequency shift given by

$$\Delta = \omega_R - \omega_o = \frac{6\lambda}{m\omega_o} \left[\frac{\hbar}{2m\omega_o} \coth \frac{\beta\hbar\omega_o}{2}\right] . \tag{23}$$

The significance of this result is that it is correct to lowest order in λ and to all orders in $\hbar$ in contrast to the result obtained from the WK expansion[30]. As we have shown,[30] it is also the result obtained using the Green's function approach. Similar agreement is obtained in the case of the more complicated three-dimensional anharmonic lattice problem.[31]

In conclusion, we have extended the range of applicability of the WDF and we have pointed out that the WDF can provide an alternative method for tackling quantum problems in the condensed matter area. We also found that the WDF method is conceptually and computationally simpler for the problem we tackled. However, we must admit that more investigation is required before one can make a general claim of the same nature. As a final remark we note that it should be possible to extend the above equilibrium calculation to the non-equilibrium domain since response functions may be obtained from correlation functions via the fluctuation-dissipation theorem. Such an extension is presently under study.

Acknowledgments

This research was partially supported by the U. S. Office of Naval Research, Contract No. N00014-86-K-0002.

References

1. G. D. Mahan, Many-Particle Physics, Plenum (1981).

2. W. H. Louisell, Quantum Statistical Properties of Radiation, Wiley (1975).
3. A. O. Caldeira and A. J. Leggett, Ann. Phys. (N.Y.) 149, 374 (1983) and 153, 445 (E) (1984).
4. S. Donaich and E. H. Sondheimer, Green's Functions for Solid State Physicists, Benjamin (1974); G. Rickayzen, Green's Functions and Condensed Matter, Academic (1980).
5. E. P. Wigner, Phys. Rev. 40, 749 (1932).
6. M. Hillery, R. F. O'Connell, M. O. Scully, and E. P. Wigner, Phys. Rep. 106, 121 (1984).
7. G. W. Ford, J. T. Lewis, and R. F. O'Connell, Phys. Rev. Lett. 55, 2273 (1985); ibid., J. Phys. B 19, 41 (1986); ibid. Phys. Rev. A (to be published).
8. R. Dickman and R. F. O'Connell, Phys. Rev. Lett. 55, 1703 (1985).
9. R. Kubo, J. Phys. Soc, Japan 12, 570 (1957); ibid., Rep. Prog. Phys. 29, 255 (1966).
10. M. S. Green, J. Chem. Phys. 19, 1036 (1951); R. Balescu, Equilibrium and Nonequilibrium Statistical Mechanics, Wiley (1975), pps. 463 and 660.
11. F. Barocchi, M. Neumann, and M. Zoppi, Phys. Rev. A 31,4015 (1985).
12. A. K. Singh and S. K. Sinha, Phys. Rev. A 30, 1078 (1984).
13. J. P. Hansen and P. Vieillefosse, Phys. Lett. 53A, 187 (1975).
14. A. Alastuey and B. Jancovici, Physica 97A, 349 (1979); ibid. 102A, 327 (1980).
15. B. Jancovici, Phys. Rev. 178, 295 (1969).
16. H. Mori, R. Oppenheim, and J. Ross, in Studies in Statistical Mechanics, edited by J. De Boer and G. E. Uhlenbeck, Wiley (1962), Vol. I.
17. D. C. Kelly, Phys. Rev. 134, A641 (1964).
18. G. Benford and N. Rostoker, Phys. Rev. 181, 729 (1969).
19. F. Brosens, L. F. Lemmens, and J. T. Devreese, Phys. Stat. Sol. (b) 74, 45 (1976); ibid. 81, 551 (1977).
20. T. Hasegawa and M. Shimizu, J. Phys. Soc. Japan 38, 965 (1975).
21. J. R. Barker, in Physics of Non-Linear Transport in Semiconductors, edited by D. K. Ferry, J. R. Barker and N. C. Jacaboni, Plenum (1980); J. R. Barker and S. Murray, Phys. Lett. 93A, 271 (1983).
22. J. R. Barker and D. K. Ferry, Solid-State Electronics 23, 519 (1980).
23. G. J. Iafrate, H. L. Grubin, and D. K. Ferry, J. de Physique Colloq. 42, C7 (1981).
24. H. L. Grubin, D. K. Ferry, G. J. Iafrate and J. R. Barker, in VLSI Electronics: Microstructure Science, Vol. 3, Academic (1982).
25. R. K. Reich and D. K. Ferry, Phys. Lett. 91A, 31 (1982).
26. A. P. Jauho, Phys. Rev. B 32, 2248 (1985).
27. U. Ravaioli, M. A. Osman, W. Potz, N. Kluksdahl and D. K. Ferry, Superlattices and Microstructures, in press; N. Kluksdahl, W. Potz, U. Ravaioli, and D. K. Ferry, Bull. Am. Phys. Soc. 31, (3), 395 (1986).
28. J. R. Barker, Superlattices and Microstructures, in press.
29. R. Dickman and R. F. O'Connell, Superlattices and Microstructures, in press.
30. R. Dickman and R. F. O'Connell, Phys. Rev. B 32, 471 (1985).
31. R. Dickman and R. F. O'Connell, to be published.

SIGNAL PROCESSING USING BILINEAR AND NONLINEAR TIME-FREQUENCY-JOINT-REPRESENTATIONS

HAROLD SZU

Naval Research Laboratory, Code 5709
Washington, DC 20375-5000

Abstract

Bilinear and nonlinear signal processing are described based upon the following observations:

(a) A phase space for signal processing is identified with a time-frequency joint representation (TFJR) that appears almost everywhere naturally, for example in bats, in music, etc.

(b) A sudden slow-down mechanism is responsible for the transition from a phase coherent-to-incoherent wavefront and provides us the sharpest tone transduction from a Békésy traveling wave in a model of the inner ear. The cause of the slowdown is physically identified to be due to three forces. This has been used to derive a cubic deceleration polynomial responsible for a cusp bifurcation phenomenon which occurs for every tone transducted along the nonuniform elastic membrane. The liquid-filled inner ear cochlea channel is divided by the membrane into an upper duct that has hair cells for the forward sound-generated flow and the lower duct for the backward balance-return flow.

(c) Both cross Wigner distribution (cross-WD) W_{21} (t_0, ν_0) and cross Woodward ambiguity function (cross-AF) $A_{21}(\tau,\mu)$ are bilinear TFJR's in the central (t_0, ν_0) and difference (τ,μ) coordinates for two independent signals $s_1(t_1)$ and $s_2(t_2)$. A neurogram is a nonlinear TFJR.

(d) Active probing uses (Doppler μ, delay τ)-weighted correlation, cross-AF, while passive listening uses (mean ν_0, central t_0)- selected convolution, cross-WD. Both are useful for post processing in a marginal probability sense. A neurogram is useful for reverberation and noise robust detection pre-processing.

Such an algorithm of neurogram is exemplified by a chirp signal in noise and reverberation.

1. Signal Processing in a Bat

On one hand, the notion of time frequency joint representations (TFJR) has evolved [1-23] for the purpose of two-dimensional (2-D) optical processing [2,13,17,23] of 1-D signals. On the other, knowledge in the physiology of the ear and bats' signal processing has matured [24-26] [30-32]. Putting together this work we can design a device that performs like the ear. The special kind of nonlinear dynamics required for sharp tone selectivity in our model of the inner ear is described in Sect. 2. Neurogram algorithms for TFJR output are described in Sect. 3. An example of a chirp signal in noise and reverberation will be given. Then TFJR of two signals is given in Sect. 4.

In the present section we review bat signal processing and begin with a remark that Mother Nature has niches for every surviving species. A flying bat at night can see things using ultrasound, echoed back to its acoustic fovea, shown in the surveillance mode in Fig. 1a. The sound is generated from the bats mouth and nose, and the echo is received by its sensitive ears tuned to the ultrasound.

In the tracking mode, because the echoes are upward Doppler shifted from an approaching moth, the bat downward shifts the emitted chirp in order to keep echoes within the sensitive acoustic fovea. This is similar to the human visual system (HVS) which can point and track an interesting object within the visual fovea. In order to achieve this, a correlation matching in terms of a TFJR and a negative feedback control in post processing must be developed as an integral part of the bats' signal processing capability.

In the attack mode, to achieve a temporal resolution of about 60 μsec, which corresponds to a spatial resolution of about 1 cm, a broad band sound is squeezed in a short time interval with a fast repetition rate. A decade ago, J. Simmons, et al. [25] gave an optimal filter analysis for bat signal processing and concluded that a broad-band FM sweep can give better time/distance resolution, while a pure tone can give better velocity/Doppler shift resolution. However, the hypothesis that bats are able to evaluate phase information as in a fully coherent cross-correlation receiver is recently in doubt [39].

Due to the significant overlapping time (60 m sec between the emitted sound $s_1(t_1)$ and the returned sound $s_2(t_2)$), and due to the fact that the inner ear performs like a Fourier mode analyser (in terms of a simple and robust method of zero crossing for frequency analyses at neuron level), we conjecture a correlation matching between two signals s_1 and s_2. The Doppler shift μ and the delay τ, measured in the central neural system, is equivalent in form to a Woodward ambiguity function.

$$\chi_{21}(\tau,\mu) = \int dt\, s_1(t)\, s_2(t-\tau) \exp(-i2\pi\,\mu\, t)\,. \tag{1}$$

It is a bilinear TFJR. A Kalman-like feedback is used to shift downward the emitted sound in the next time step

$$s_1(t_{n+1}) = s_1(t_n) + [s_2(t_n) - s_1(t_n)]\, |\chi_{21}(\tau_n,\mu_n)|_{\text{threshold}}\,, \tag{2}$$

where the nonlinear threshold of a good match (Eq. (1)) warrants a down shift for the next emission as shown in the left-hand side of Eq. (2). For no detection $s_2 = s_1$, then $s_1(t_{n+1}) = s_1(t_n)$ repeats itself. Likewise, some moths have also developed countermeasures against bats. Moths have learned to detect the sound and dodge the bat's chase in zigzag paths and, in a desperate emergency close up their wings and drop straight to the ground to minimize the horizontal Doppler shift. Some even learn to shout back to jam and confuse bats' echoes with irrelevant information. Historically, echolocations of horse-shoe bats have been discovered by D. Griffin in 1938. The carrier frequency (CF) of the emitted ultrasound is 83 kHz upon which the frequency modulation (FM) carries the pertinent information for seeing and tracking insects. The bat can fly up to the speed of 12 meters per second, which can generate a Doppler shift about 4 kHz ~ 6 kHz from an approaching moth. The sequence of time-frequency spectra is reprinted from Neuweiler [32], and slightly re-plotted for clarity, together with a flying bat preying on a moth at the instant labelled by number four, shown in the bottom panel of Fig. 1.

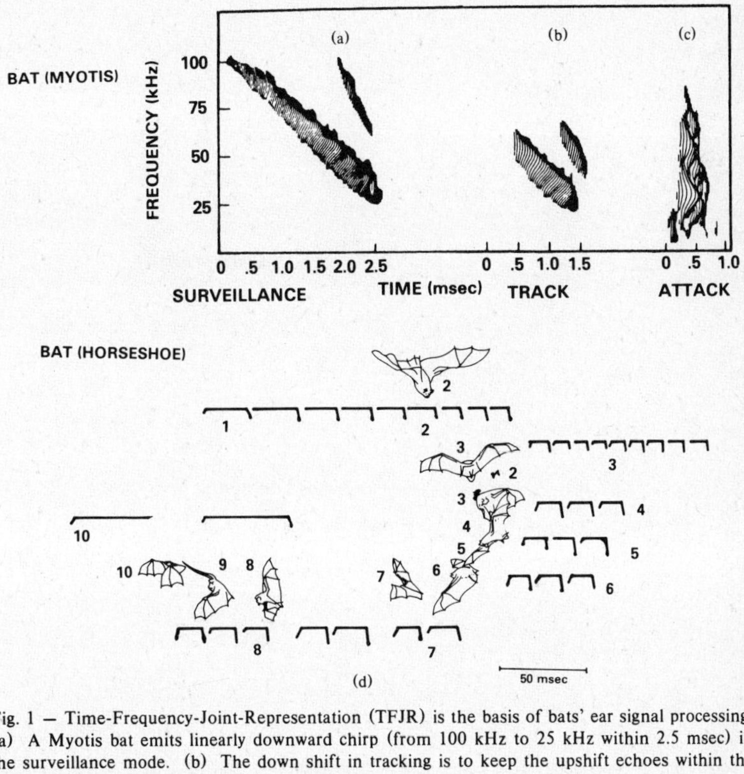

Fig. 1 — Time-Frequency-Joint-Representation (TFJR) is the basis of bats' ear signal processing. (a) A Myotis bat emits linearly downward chirp (from 100 kHz to 25 kHz within 2.5 msec) in the surveillance mode. (b) The down shift in tracking is to keep the upshift echoes within the most sensitive acoustic fovea. (c) A broad band at the moment of attack is to achieve a sharp spatial resolution (about 1 cm). (d) A horseshoe bat is preying on a moth at the instance labelled by number four and TFJR's are plotted along each snapshot. Three letters of number two indicate the bat, TFJR and the moth at the same instance of time.

A diagram of the snail-shaped and liquid-filled inner ear reprinted from [32] is shown in Fig. 2a, where the high frequency component of the ultrasound is resonant near the input end of the cochlea while the relatively low frequency sound penetrates inwards. A nonuniform wedge-shaped basilar membrane (BM), shown in Fig. 2b, bisects the fluid channel and the BM has hair-bundles for picking up the sharply attenuated downward pressure of a nonuniform traveling wave, discovered by 1961-Nobel laureate G. Von Bekésy [35] in 1928. The location of a peak between the trailing wake and the wave breaking corresponds to a sound frequency analyzed by the inner ear. Overall, the inner ear behaves like a low pass filter for low frequency sound penetrating toward the end of the cochlea, and it has been thus modeled as a linear dispersive transmission line.

The neural background material for a bat is briefly reviewed in order to support the neurogram signal processing described in Sect. 3. The number of neurons per octave is about 4×10^4 in the center of the acoustic fovea located at about 83 kHz. The number distribution exponentially drops below the peak number density (Fig. 2c). The single neuron firing rate is plotted in response to the actual time-frequency spectral density, obtained when a bat is approaching head on against a moth, and is compared with those when approaching from below the moth, shown respectively in the echo, the FM component, and the AM component in three columns of Fig. 2d.

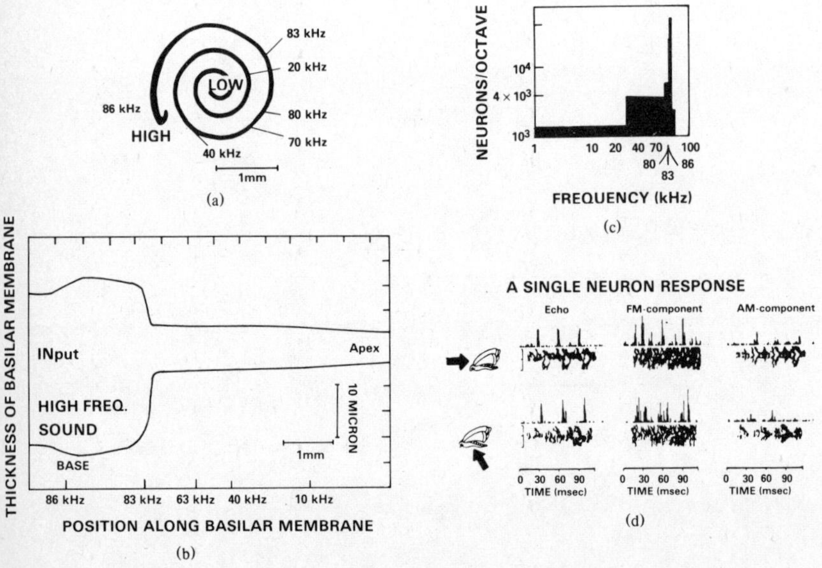

Fig. 2 — The inner ear of a horseshoe bat. (a) A snail-shaped inner ear is filled with a fluid. (b) A nonuniform wedge shape basilar membrane (BM) is, shown in the left below, bisecting the snail channel, so that an equal and opposite amounts of fluid has been displaced above and below BM. Note that the high frequency sound (~86 kHz) penetrated less along the viscoelastic BM, while the low frequency sound (~20 kHz) filters down to the apex. (c) Neurons per octave is plotted against sound frequency. (d) A single neuron response is plotted with respect to echo, FM and AM as well as TFJR components plotted below, when a bat is approaching a moth head on, and when a bat is approaching a moth from below (indicated by the arrow).

The neuron distribution seems to drop off rapidly similar to HVS peripheral vision [29]. Thus, we speculate that a $\Delta\,\nu$-scale invariance exists in the periphery of the acoustic fovea. Absolute tone perception is not needed outside the fovea because the nonuniform (neuron/and the associated sound pickup hair at the input location) x along the cochlea corresponding to ν

$$x/\Delta\nu \equiv \exp(\log x/\Delta\nu) = \exp(x') \tag{3}$$

is one-to-one mapped onto a uniformly packed neural fiber bundle at the output location

$$x' \equiv \log x - \log \Delta\nu \tag{4}$$

which achieves in parallel the logarithmic tone/coordinate transformation without actual computations. Note that the absolute location x for the tone detection is shifted by log $\Delta\,\nu$ in the output x′ in the brain in a scale-graceful degradation fashion. Concluding the brief review, we summarize the active sonar of a horseshoe bat in Fig. 3.

- SOUND DETECTION FOVEA 82 kHz - 86 kHz
  1. SEND MONOTONE 83 kHz (AT SPEED 12 m/sec) DOPPLER ~ 4 kHz
  2. SEND LOWER TONE VIA A NEGATIVE FEEDBACK IN ORDER TO MAINTAIN THE ECHO IN THE FOVEA
  3. LOWER THE FREQUENCY OF THE EMITTED SOUND $s_1(t_1)$ AND SEND IT AGAIN TO CORRELATE IT WITH THE RETURNING SOUND $s_2(t_2)$ WITH AN OVERLAP ABOUT 60 msec
- TIME RESOLUTION AT HOMING $\Delta\tau = |t_1 - t_2| \sim 60\ \mu$sec ~ 1 cm RESOLUTION
- TEXTURE AND TARGET DISCRIMINATION ~ 1 kHz – 80 kHz
  1. CHIRP DOWN, BROAD BAND FM SWEEP
  2. SEEING WITH SOUND IN THE ACOUSTIC FOVEA

Fig. 3 — Active sonar of horseshoe bats

2. Nonlinear Dynamics Bifurcation Model for Sharp Hearing

We present a new theory for sharp tone selectivity based on experimental evidence. Schroeder [27] adopted the Békésy observation of traveling waves in a cochlea to explain a two-tone suppression experiment in the limit of a long sound wavelength compared with the width of the sound duct. We will apply Schroeder's model [27] together with the hydrodynamics to obtain a new result. G. Von Békésy traveling waves are represented by their envelopes, shown in Fig. 4 schematically for a straightened cochlea. Nonuniform traveling waves in a viscoelastic fluid channel exist in the cochlea where a high frequency sound pressure penetrates less inward than a low frequency sound. The envelopes of these traveling waves with their wakes and sharp dropoffs are peaked at different depths, as shown in Fig. 4. This phenomenon is referred to as a placement theory of hearing. The cochlea has two liquid-filled ducts separated by the elastic BM (like a nonuniform drum membrane separating the air), and the two ducts are connected by the flow at the inner most end of the cochlea. However, the sound pressure enters only through the upper stapes window into the upper duct where the traveling wave on the BM is generated, and the return flow in the lower duct balances the forward flow in the upper duct by erasing the ripple on the BM and the lowest mode eventually survives and stops at the round window beneath the stapes window. Due to the nonuniform thickness,

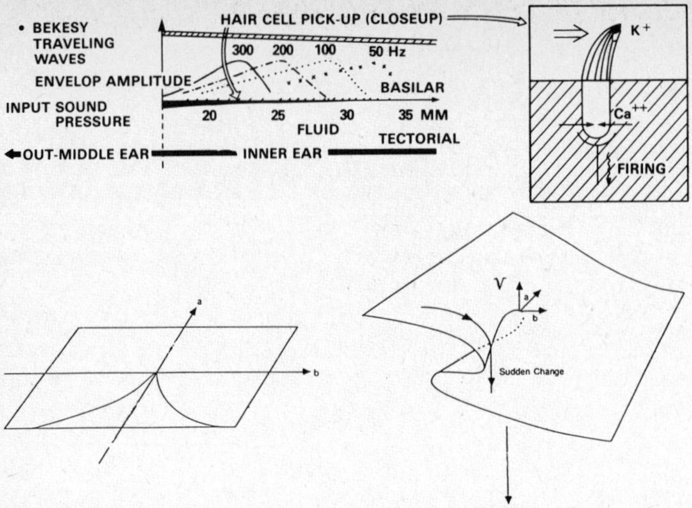

Fig. 4 — Sharp hearing has a cusp-bifurcation. A straightened cochlea and a blowup hair bundle cell (pickup by 20 neurons) on the basilar membrane is shown on the top (for a human inner ear). Then, the combination of the nonlinear impedance model and the hydrodynamics leads to a cusp-bifurcation at each resonant frequence position denoted by the wavebreaking slopes $(\partial V/\partial X)_o <> 0$ in the position parameter a, and the acceleration $(\partial V/\partial t)$ parameter b.

the BM has different elasticity. Moreover, a hair has grown over the upper surface of the BM and flips under the traveling wave. Then the ter Kile (1900) pressure-electrical effect, studied by H. Davis (1957) [36], is due to the fact that those hair bundles on the BM under the downward pressure of wave breaking pick up potassium ions and open up calcium ion currents shown [28] in the blowup in Fig 4. At the neurodynamic level (nut shell), our theory is consistent with Shamma's theory [40] in that the sudden slowdown of a particular mode at a specific position creates the discontinuity in neighborhood neuron firing rates. This produces an "acoustic tone" detection, similar to visual edge detection, and is referred to as a sound transduction theory.

The overall tone selectivity is due to such a sorting by sieving phenomenon among various modes of the traveling wave that broadens by losing high frequency modes along its way inward. The sharpness in tone conceivably due to the sharpness in wave breaking at a specific tone-placement on the BM and due to the fact that the wave components corresponding to the tone vanishes, thereafter this happens to every mode on its way inward. We would like to propose that such a sudden slow-down of a specific mode must correspond to a dynamic cooperation existing in a phase transition. This can not be solely attributed to the wave breaking, nor to the backflow beneath the BM, nor to the nonuniform rigidity of the BM that shrinks the rippling toward the thicker side. We attribute this effect to the synergism of three independent components of the velocity due to three independent forces (1) the hydrodynamic wave breaking inertial force (2) the onset of the (resonant tunnelling) return flow at the trough-backflow and (3) the nonuniform elastic membrane shrinking force. The stronger the return flow becomes; and the greater the transition discontinuity occurs, the better the frequency resolution becomes. Clearly, this must involve some special kind of nonlinear dynamics for such a bootstrapping effect in the frequency selectivity.

The precise nature of the nonlinear dynamics is an issue that we address and is our contribution to hearing theory. A full treatment requires both nonlinear membrane dynamics and nonlinear hydrodynamics. An interested reader may find some useful material in the references [30, 37, 38]. We begin with the two-tone suppression experiments. In the long wavelength limit, the Lesser and Berkley hydrodynamic model of a nonuniform transmission line for Békésy waves can be simplified at a resonance depth for a pure tone, using the following analogy

$$\text{Current} = \text{Voltage/Impedance} \tag{5}$$

and

$$\text{Velocity} = \text{Pressure gradient / Kinetic Inertia}. \tag{6}$$

Since the kinetic inertia must be nonlinear and scalar, Schroeder assumes the following form:

$$\text{Kinetic Inertia} = R_0 + \rho_0 \,|\,\text{Velocity}\,|^2; \tag{7}$$

where R_0 and ρ_0 are real positive constants respectively related to the fluid and the membrane inertias. Consequently, one solves from Eqs. (6) (7) that

$$\text{Pressure gradient} = R_0 \,\text{Velocity} + \rho_0 \,\text{Velocity}^3 \tag{8a}$$

$$\text{Velocity} \approx (\text{pressure gradient})^{1/3} \tag{8b}$$

which implies, for two pressure gradients ∇P_1 and ∇P_2 with $\nabla P_2 >> \nabla P_1$, a Taylor expansion

$$\text{Velocity} \approx \nabla P_2^{1/3} + \nabla P_1 \nabla P_2^{-2/3} \tag{9}$$

such that the first tone associated with ∇P_1 can be greatly suppressed by the second tone associated with ∇P_2. Both have shared a common portion of the BM when two modes of the traveling wave pass through. Consequently, suppression phenomenon seems to be occurring in the ear, rather than in the brain.

Adopting Eqs. (8a), we are now ready to derive the nonlinear dynamics at each resonant position along the cochlea BM. We consider Stokes hydrodynamics equation

$$\frac{\partial \vec{V}}{\partial t} + \vec{V} \cdot \left(\frac{\partial \vec{V}}{\partial \vec{X}} \right) = -\frac{1}{\rho} \frac{\partial P}{\partial \vec{X}} - \frac{\eta}{\rho} \nabla^2 \vec{V} \tag{10a}$$

where the viscous fluid has the velocity $\vec{V}$, and the kinematic viscosity η/ρ, and it is assumed to be incompressible with the density ρ. According to the hydrodynamic Eq. (10a) a sinusoidal wave in the liquid becomes naturally steepening and eventually breaks because of the nonlinear convective derivative. A wave breaking must involve the change of the slope near the position x_o where the wave breaks. Since an effective position parameter a $(a>0,\ a=o,\ a<o)$ will be related to the change of slope in our theory, we shall briefly review the hydrodynamic wave breaking phenomenon as follows. A one-dimensional fluid differential is defined for a specific mode λ

$$\frac{dV_\lambda}{dt} \equiv \frac{\partial V_\lambda}{\partial t} + \frac{\partial V_\lambda}{\partial x_\lambda}\frac{dx_\lambda}{dt}\ , \tag{10b}$$

where the first term is a partial derivative at a fixed spatial position, $dx_\lambda/dt \equiv\ V_\lambda$, and x_λ $= \cos(\omega_\lambda t + x_o)$ for a given sound dispersion relationship between the frequency and the mode denoted by ω_λ. A differential acceleration exists after the position of the crest (denoted as C_+) and before the position of the backflow trough (denoted T_-), because both the forward flow $((V_\lambda)_{C_+} > 0)$ in the upper duct and the backflow $((V_\lambda)_{T_-} < 0)$ in the lower duct have the negative slope $((\partial V_\lambda/\partial x)_{C_+} < 0$ and $(\partial V_\lambda/\partial x)_{T_-} < 0$. Consequently, at a fixed position between the crest

$$\left(\frac{\partial V_\lambda}{\partial t}\right)_{C_+} = -\,(V_\lambda)_{C_+}\left(\frac{\partial V_\lambda}{\partial x}\right)_{C_+} > 0\,, \tag{10c}$$

$$\left(\frac{\partial V_\lambda}{\partial t}\right)_{T_-} = -\,(V_\lambda)_{T_-}\left(\frac{\partial V_\lambda}{\partial x}\right)_{T-} < 0\,, \tag{10d}$$

so that the inertial force will move the crest and the trough of the wave close to each other leading to the wave breaking (as indicated by the pair of arrows in Fig. 5). We know that the acceleration and the fluid component V_λ associated with a particular mode λ stop when V_λ and $\dfrac{\partial V_\lambda}{\partial x}$ vanish at a specific placement x_o within the clamped ends of BM. Moreover, the resonant onset of the return flow happens to be coincident with the trough in the lower duct, and thus the return flow enhances the back flow of the trough. This furthermore follows the elastic pull toward the input side, the thicker part of the BM. To derive it mathematically, the transversal component V_λ of the righthand side of Eq. (10a) may be replaced in the long wavelength limit (λ> duct width) by the equivalent Schroeder's pressure gradient Eq. (8a).

$$\frac{\partial V_\lambda}{\partial t} + \left(\frac{\partial V_\lambda}{\partial \bar{x}}\right)_{0\pm} \cong -\,(R_0/\rho)\ V_\lambda - (\rho_0/\rho)\ V_\lambda^3\ . \tag{11a}$$

Because the acceleration near the onset of the return flow at $x_o\pm$ is a function of only V_λ, we can thus introduce the Lyapunov potential $W(V_\lambda)$ in V_λ defined by the resulting differential Eq. (11a). We can introduce to (11a) an arbitrary deceleration parameter, $-b$, scaled by the inertia ratio ρ_0/ρ, (namely $\rho\left(\dfrac{\partial V_\lambda}{\partial t}\right) = -\rho_0\, b$)

$$\frac{\partial V_\lambda}{\partial t} = -\,\frac{\partial W}{\partial V_\lambda} = -\,(\rho_0/\rho)\ V_\lambda^3 - (R_0/\rho)\ V_\lambda - V_\lambda\,(\partial V_\lambda/\partial x)_0\pm - (\rho_0/\rho) b\,. \tag{11b}$$

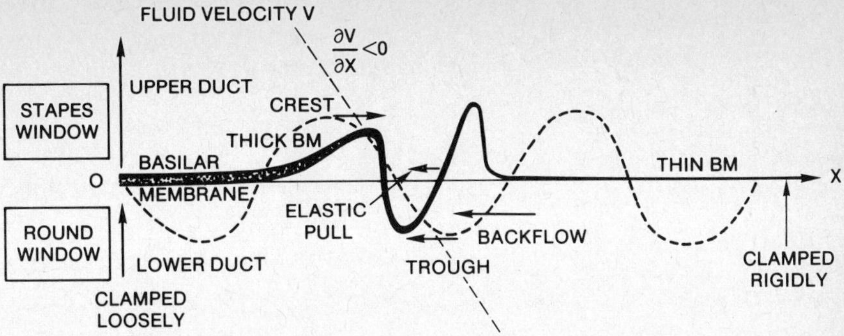

Fig. 5 — A physical mechanism of the sudden slow-down cusp catastrophe: A synergism exists among three independent effects requiring a third order deceleration polynomial for the nonlinear steepening, and the onset of the return flow at the backflow of the trough (in the lower duct below the basilar membrane (BM)), as well as the nonuniformly elastic pull of the BM back to the input front end (thicker side). When they work together to suddenly slow down the traveling of wavelets of a particular mode at a specific position, the orderly wavelets with a definite phase relationship (coherent phase) pile up and becomes disordely (random) in the phase relationship among wavelets. It creates a pronouncing non-inhibitory effect by such sharp discontinuity, similar to the visual edge detection among neighborhood neuron firing rates in the neurogram.

We can integrate Eq. (11b) to give the Lyapunov function

$$W = (\rho_0/4\rho)\ V_\lambda^4 + (R_0/2\rho)\ V_\lambda^2 + \frac{1}{2}\ V_\lambda^2\ (\partial V_\lambda/\partial x)_{0\pm} + (\rho_0/\rho)\ b\ V_\lambda \tag{12}$$

which has the canonical form of a cusp

$$W(V_\lambda,\ a,\ b) = (\rho_0/\rho)\left[\frac{1}{4}\ V_\lambda^4 + \frac{a}{2}\ V_\lambda^2 + bV_\lambda\right]. \tag{13}$$

The breaking position along the BM may be expressed in an effective position parameter

$$a = (R_0/\rho_0) + (\rho/\rho_0)\ (\partial V_\lambda/\partial x)_{0\pm} \tag{14}$$

which, for a thick membrane $\rho_0 >> R_0$, can have $a > 0$ for positive slope $(\partial V_\lambda/\partial x)_{0\_} > 0$, $a = 0$ for zero slope $(\partial V_\lambda/\partial x)_0 = 0$, and $a < 0$ for negative slope $(\partial V_\lambda/\partial x)_{0+} < 0$. We can cast the resonant phase transition phenomenon from multiple states (three real solutions) to single states (one real solution) in terms of Thom's 1975 theory of Catastrophe (i.e., the Greek word for a sudden "turn down" in the change of state among multiple equilibrium). This is depicted in Fig. 4 for a cusp bifurcation depending on the algebraic value of the parameters a and an arbitrary constant acceleration b. By eliminating the variable V_λ from the first and the second derivatives of Eq. (11b) one finds $(4a^3 + 27b^2 = 0)$ which shows 3 real solutions when $a < 0$, and one real solution for one unique combination of three independent causes when $a > 0$. We may describe these causes in terms of three forces: (1) the hydrodynamic inertial force, (2) the elastic restoring force of the nonuniform thickness membrane, and (3) the onset of the resonant return flow field. Therefore, a cubic polynomial, at a fixed spatial position shown by Eq (11), will generally be required for three independent solutions with three independent phases which are denoted as $V_1(x), V_2(x), V_3(x)$. The cubic acceleration polynomial can vanish

$$\frac{\partial V_\lambda}{\partial t} = (V_\lambda - V_1)(V_\lambda - V_2)(V_\lambda - V_3) = 0, \tag{15}$$

at three specific points near x_o, in general, for three solutions. A degenerate and interesting case of "sudden slow down" (catastrophe) occurs if $V_\lambda = V_1(x_o) = V_2(x_o) = V_3(x_o)$ all have a unique phase. In this case all three forces work together to stop the forward acceleration flow of the component V_λ after $x_{o\,+}$. Thereafter, it becomes physically unlikely to produce any sound confusion in the particular mode λ transduction because lower frequency modes are still traveling forward coherently and higher modes of V_λ have stopped either before x_o or at $x_{o\,+}$. In fact, the distance between the position x_o, where a coherent phase exists, and the stoppage position x_{o_+}, where the traveling waves of V_λ have three relative phase shifts [40], could be used as a measure of the sharpness of the λ-mode.

The catastrophe model conveys a bootstrap characteristic that a small deviation from a dynamic equilibrium becomes the cause of a further deviation. The basic hydrodynamic equations have the following perturbation in density, velocity, and energy fluxes denoted collectively as a column vector

$$\underline{\psi}\,(x,t) = (\Delta\rho\,, \Delta\underline{V}\,, \Delta E)^T \tag{16a}$$

where $\Delta\rho$, $\Delta\underline{V}$, ΔE symbols denote the density flux, the velocity flux, and the energy flux respectively, and the superscript T stands for the matrix transpose operation. A standard perturbation expansion can be used to derive the following general form,

$$\frac{\partial}{\partial t}\underline{\psi} = L\,\underline{\psi} + B\,(\underline{\psi}\,,\underline{\psi}) + C\,(\underline{\psi}\,,\underline{\psi})\underline{\psi}\,, \tag{16b}$$

where the bracket denotes an inner product which is invariant under the change of algebraic sign. In the case that $B >> C$ we can omit the cubic term in $\underline{\psi}$ and describe the nonlinear saturation of the growth rate.

$$\frac{\partial}{\partial t}\underline{\psi} = L\,\underline{\psi} + B\,(\underline{\psi}\,,\underline{\psi})\,. \tag{17}$$

However, due to the round trip nature of cochlea fluid perturbations above and below the membrane, a constant pure tone produces the flux acceleration with respect to the flux deceleration that must be related through the time reversal at the resonant point. When $t \rightarrow -t$ for $\underline{\psi} = -\underline{\psi}$ we demand the acceleration to be an odd function rather than the even quadratic, thus $B = 0$ and consequently the necessary form of a cusp catastrophe follows:

$$\frac{\partial}{\partial t}\underline{\psi} = L\,\underline{\psi} + C\,(\underline{\psi}\,,\underline{\psi})\underline{\psi} = NL\,\underline{\psi}\,. \tag{18}$$

In effect, we have a nonlinear growth rate that changes with the intensity of the flux, $NL = L + C\,(\underline{\psi}\,,\underline{\psi})$, this general formula includes Eq. (11b) as a special case when the flux is only the fluid velocity of the λ-mode without the density and the energy variations.

3. Nonlinear Signal Processing Using a Neurogram

A novel design approach for replacing a conventional hydrophone with a fiber-optic cochlea hydrophone will be described in another paper [43]. The cochlea hydrophone can be either coated with pressure-to-light conversion chemicals in a straightened version of cochlea or in a spiral version by modulating a laser beam (by acoustic pressure) inside the fiber optics cochlea. When use is made of nonlinear thresholding in counting detections in pairwise products, both methods can produce a neurogram. Neurograms [43] have been hypothesized in the physiology for reverberation and noise robust signal detection. Excellent reviews of neurogram's modeling have appeared in the literature [31], [42], [40]. The incorporation of the neurogram into a device such as fiber optics is however suggested for the first time.

We recapitulate the essential knowledge that may be useful for the design of a bionic ear in Appendix A. Several general remarks are given below.

A 20 dB discrepancy in the sharpness of the frequency selectivity exists between the basilar membrane response models (1-D, 2-D, 3-D, active) and the neural response experiments (see App. B). This prompted us to propose a cusp nonlinear dynamics model which sharpens the resonance and increases the channels density without cross talk (see Sect. 2). In analogy to the human cochlea microphone, the bionics cochlea, when coated with the chemical for the active pressure-light conversion along the inner coaxial fiber optics, seems to be a promising concept for implementing noise-reverberation robust neurograms in a hydrophone array. The novel design addressed here is based on fiber optics for a relatively cheap and disposable array.

The understanding of the nonlinear cusp bifurcation in Sect. 2 makes the sharp tone analysis possible in the acoustic fovea of the bionics cochlea. Since the conversion from the acoustic pressure to the optical light is located at each resonant and attenuation position along the cochlea, the light intensity gives a direct Fourier transform magnitude read out along the position of the cochlea. The light outputs are fiber-optics coupled with other outputs from pressure-light transducers and could be operated in the simple mode of on-and-off at a selected acoustic frequency for a particular application, or it could be also operated for a hot spot detection mode by a space-time integration.

We shall now describe an algorithm which implements a neuogram applicable to the fiber device. A neurogram is a nonlinear pre-processing. The neurogram may be statistically justified due to 20 neurons per hair bundle cell that each has the hard clipping (switch-like) pickup property along the basilar membrane. A block diagram for producing a neurogram is given in Table 1 and briefly summarized as follows.

Essentially the algorithm of a neurogram has three steps as follows: Each *Filter Bank* takes a short time segment (4 msec apiece in time domain). Each channel Histogram counts hard clipping zero crossing in the frequency domain. The correlation is measured by summing over pairwise products between *adjacent channels.* And the procedure is repeated for a later time segment. Such a nonlinear time frequency point representation is called a neurogram.

Table 1

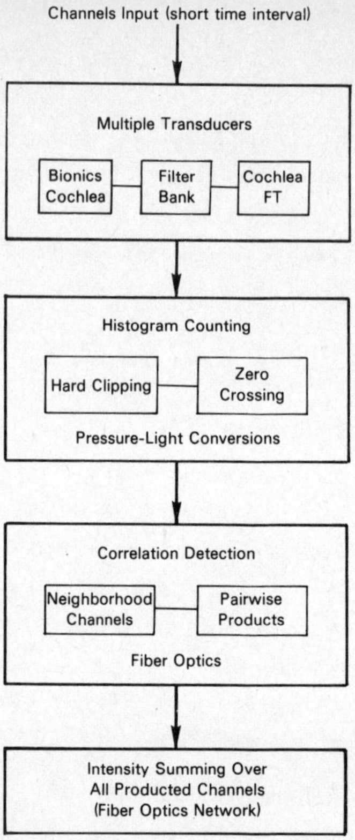

An example of an input chirp signal in noise and reverberation is demonstrated in Fig. 6, which shows: (a) multiple channels, (b) nonlinear clipping, (c) histogram of pulse lengths in a window Δt_1, etc. (d) multiply pairwise for the set of intersects, (e) sum the correlation products in Δt_1, and (f) accumulator, where all short-time-preprocessing segments become a time-frequency joint representation, from which all other TFJR follows in Sect. 4.

The principle of cusp catastrophe has also been associated with a Gabor elementary signal in hearing [41]. It may be applied to the fiber optic cochlear hydrophone device. A nonuniform membrane partitioning for each mode the forward flow, the backflow and the return flow produces as many efficient multiple channels as possible in the short time intervals without the cross talks and aliasing due to dense adjacent channels and short time sampling. In summary, our design concept is that a nonuniform membrane can divide the liquid filled hydrophone into two ducts and it wraps around the centeral fiber optic core in affecting the light propagation within the fiber optics and producing the neurogram in the output.

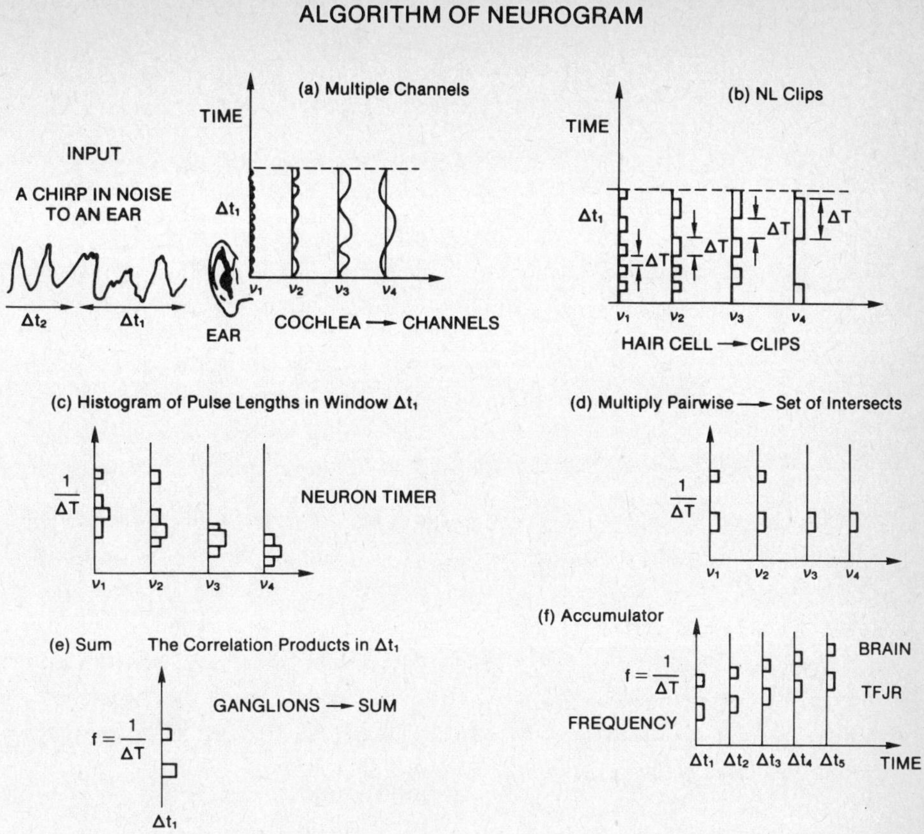

Fig. 6 — A Neurogram Algorithm for processing a chirp in noise and reverberation.

4. MATHEMATICS FOR JOINT REPRESENTATIONS OF TWO INDEPENDENT SIGNALS

We begin the mathematical foundation of post-processing with two questions. What is a phase space? What is a joint representation of one signal?

In physics, the canonical transform defines conjugate variables: coordinate q and momentum p, which form a phase space (q,p). In general, a phase space is a space of conjugate variables of a hetero respective transform, e.g., Fourier transform in conjugate variables time t and temporal frequency ν. It is useful when a bridge called joint representation (JR) is built as a compromise between two transform spaces. Such a "partial transform" changes a quadratic energy expression into bi-linear form. For example, ambiguity function (AF) $A(\tau,\mu)$ is defined [14], similar to the skew version Eq. (1), as follows,

$$A(\tau,\mu) = \int dt_0 \, s^* \left(t_0 - \frac{\tau}{2} \right) s \left(t_0 + \frac{\tau}{2} \right) \exp \left(- i \, 2\pi \, t_0 \, \mu \right). \tag{19}$$

Wigner distribution (WD) $W(t_0, \gamma_0)$ is real and defined as [10]

$$W(t_0, \nu_0) = \int d\tau \; s^* \left[t_0 - \frac{\tau}{2} \right] s \left[t_0 + \frac{\tau}{2} \right] \exp\left(- i \, 2\pi \, \tau \, \nu_0 \right), \tag{20}$$

and Kirkwood instantaneous power spectral (IPS) density is defined [11] [15]

$$\text{IPS}\,(t,\nu) = s(t) \; S^*(\nu) \; \exp\left(- i \, 2\pi \, \nu t \right) \,. \tag{21}$$

A simple prescription by integrating one variable can carry these JR's (A,W,IPS) from one domain to the other domain of the remaining variable. Then, such a bilinear JR is said to satisfy the *marginal property* of a quadratic energy expression

$$\int \text{JR}\,(t,\nu) \; dt = |S(\nu)|^2; \quad \int \text{JR}(t,\nu) d\nu = |s(t)|^2 \,. \tag{22}$$

All JR's defined by Eqs. (19, 20, 21) satisfy the marginal property, Eq. (22).

An essential difference between active and passive processing is due to the "ownership" of the signal in question. If one owns the signal, one can create a template for its echo correlation. A cross-ambiguity function (AF) becomes the Doppler shifted and time-delayed matched filter for detection. On the other hand, [3] if one does not own the signal, a convenient approach is testing the coherence among various receivers with respect to a possible set of incoming signal frequencies, ν_0. Thus, a cross-Wigner distribution (WD) between pairwise received signals may be utilized. This is particularly prevalent in passive acoustic undersea surveillance, as introduced [6]. As a result, optical signal processing based on Bragg cells has been reported [2].

For two independent signals $s_1(t_1)$ and $s_2(t_2)$, the phase space (Fig. 7) is four dimensional: t_1, ν_1 and t_2, ν_2. Both cross-AF and cross-WD follow naturally from a rotation in the phase space to the central coordinates denoted by the subscript zero

$$t_0 = (t_1 + t_2)/2; \quad \nu_0 = (\nu_1 + \nu_2)/2 , \tag{23}$$

$$\nu_1 t_1 - \nu_2 t_2 = \nu_0 \tau + \mu t_0$$

• CHANGE TO THE CENTER COORDINATE FOR THE SLOWLY-VARYING AND THE FAST VARYING

• DOUBLE PRODUCT

$$S_2^*(\nu_2)\, S_1(\nu_1) = FT_{t_2,\, \nu_2}{}^* \; FT_{t_1,\, \nu_1} \left\{ s_2^*(t_2)\; s_1(t_1) \right\}$$

WHERE $\qquad\qquad FT_{t_2,\, \nu_2}{}^* \; FT_{t_1,\, \nu_1} = FT_{t_0,\, \mu} \; FT_{\tau,\, \nu_0}$

• ALTERNATIVE ORDERS OF DOUBLE INTEGRALS YIELD BOTH THE CENTRAL FREQUENCY SELECTED CONVOLUTION (WD) AND THE DIFFERENCE FREQUENCY MATCHED CORRELATION (AF)

$$S_2^*(\nu_2)\, S_1(\nu_1) = FT_{t_0,\, \mu} \; W_{21}(t_0,\, \nu_0) \; ; \qquad W_{21}(t_0,\, \nu_0) \equiv FT_{\tau,\, \nu_0} \; s_2^* \left(t_0 - \frac{\nu}{2} \right) s_1 \left(t_0 + \frac{\nu}{2} \right)$$

$$= FT_{\tau,\, \nu_0} \; A_{21}(\mu,\, \tau); \qquad A_{21}(\mu,\, \tau) \equiv FT_{t_0 \mu} \; s_2^* \left(t_0 - \frac{\tau}{2} \right) s_1 \left(t_0 + \frac{\tau}{2} \right)$$

• SIMILARLY

$$s_2^*(t_2)\, s_1(t_1) = FT_{\nu_2,\, t_2}^{*\,-1} \; FT_{\nu_1,\, t_1}^{-1} \left\{ S_2^*(\nu_2)\, S_1(\nu_1) \right\} = FT_{\mu,\, t_0}^{-1} \; FT_{\nu_0,\, \tau}^{-1} \left\{ S_2^* \left(\nu_0 - \frac{\mu}{2} \right) S_1 \left(\nu_0 + \frac{\mu}{2} \right) \right\}$$

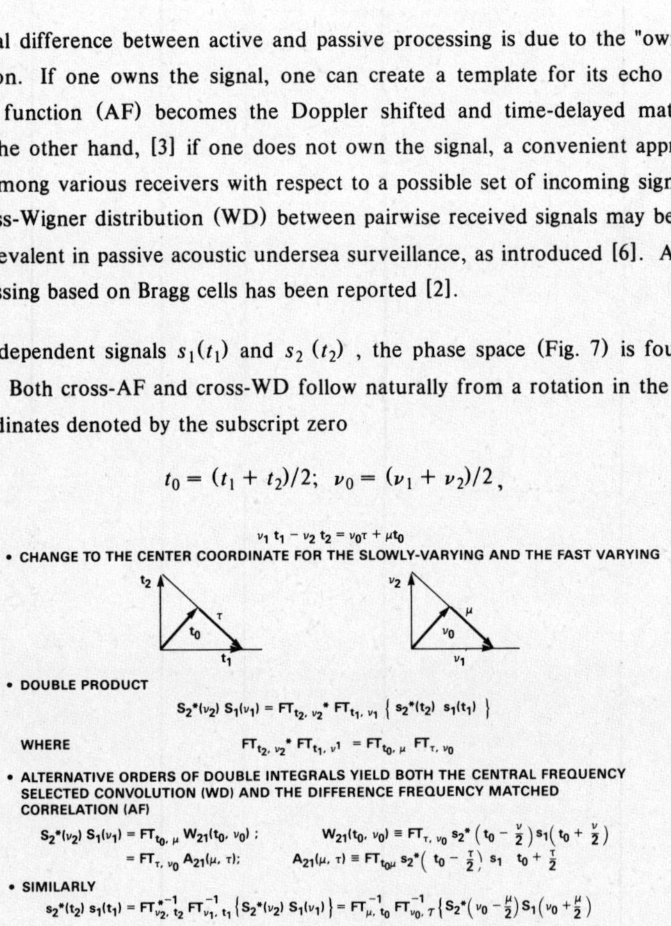

Fig. 7 — Derivations of cross-WD and cross-AF

and the difference coordinates

$$\tau = t_1 - t_2; \quad \mu = \nu_1 - \nu_2 . \tag{24}$$

It is straightforward to verify the invariant phase

$$\nu_1 t_1 - \nu_2 t_2 = \nu_0 \tau + \mu t_0 \tag{25}$$

and the Jacobian of the coordinate transform is invariant, as the physics does not change in the centered mass coordinates. The complex conjugate product of Fourier amplitudes of two signals is by definition given as follows

$$S_2^*(\nu_2)\, S_1(\nu_1) = FT^*_{t_2,\nu_2}\, FT_{t_1,\nu_1} \left\{ s_2^*(t_2)\, s_1(t_1) \right\} . \tag{26}$$

where

$$FT_{t,\nu}\{\cdot\} = \int dt \exp(-\, i\, 2\pi\, \nu t)\, (\cdot) . \tag{27}$$

Due to the phase invariance (25) of the central coordinate transformations (23, 24), we obtain the following identical pairs of Fourier operations

$$FT^*_{t_2,\nu_2}\, FT_{t_1,\nu_1} = F_{t_0,\mu}\, F_{\tau,\, \nu_0} . \tag{28}$$

Expressing the double integral (26) with (28) in the central coordinate, we can alternatively carry out any one Fourier transform in the new coordinate frame in obtaining either the cross-AF as the difference frequency μ-weighted correlation,

$$A_{21}(\tau,\mu) \equiv FT_{t_0,\mu} \left\{ s_2^* \left(t_0 - \frac{\tau}{2} \right) s_1 \left(t_0 + \frac{\tau}{2} \right) \right\}, \tag{29}$$

or the other cross-WD as the central frequency ν_0 - selected convolution,

$$W_{21}(t_0,\nu_0) \equiv FT_{\tau,\nu_0} \left\{ s_2^* \left(t_0 - \frac{\tau}{2} \right) s_1 \left(t_0 + \frac{\tau}{2} \right) \right\}, \tag{30}$$

where the correlation integral (29) has an identical sign in t_0 while the convolution integral (30) has the opposite signs in τ.

Consequently, we have derived from (26) both cross-AF (29) and cross-WD (30) as well as underlying relationships summarized as follows:

$$\begin{aligned} S_2^*(\nu_2)\, S_1(\nu_1) &= FT_{\tau,\nu_0}\, FT_{t_0,\mu} \left\{ s_2^* \left(t_0 - \frac{\tau}{2} \right) s_1 \left(t_0 + \frac{\tau}{2} \right) \right\} \\ &= FT_{\tau,\nu_0} \left\{ A_{21}(\tau,\mu) \right\} = FT_{t_0,\mu} \left\{ W_{21}(t_0,\nu_0) \right\} . \end{aligned} \tag{31}$$

Similarly, we could begin the derivation in the Fourier conjugate space, rather than in the temporal space (26),

$$s_2^*(t_2)\; s_1\,(t_1) = FT^{*-1}_{\nu_2,t_2}\; FT^{-1}_{\nu_1,t_1} \left\{ S_2^*(\nu_2) S_1(\nu_1) \right\}, \tag{32}$$

where the inverse Fourier transform is defined by

$$FT^{-1}_{\nu,t}\{\,\cdot\,\} = \int d\nu \,\exp\,(i\; 2\pi\; \nu t)\;(\,\cdot\,). \tag{33}$$

Identically due to the invariance of the coordinate transformation, we obtain

$$FT^{*-1}_{\nu_2,t_2}\; FT_{\nu_1^{-1},t_1} = FT^{-1}_{\mu,t_0}\; FT^{-1}_{\nu_0,\tau} \tag{34}$$

and

$$\begin{aligned} s_2^*(t_2)\; s_1\,(t_1) &= FT^{-1}_{\mu,t_0}\; FT_{\nu_0,\tau} \left\{ S_2^* \left(\nu_0 - \frac{\mu}{2}\right) S_1 \left(\nu_0 + \frac{\mu}{2}\right)\right\} \\ &= FT^{-1}_{\mu,t_0} \left\{ A_{21}(\tau,\mu)\right\} = FT^{-1}_{\nu_0,\tau} \left\{ W_{21}(t_0,\,\nu_0)\right\}, \end{aligned} \tag{35}$$

where

$$A_{21}(\tau,\mu) = FT^{-1}_{\nu_0,\tau} \left\{ S_2^* \left(\nu_0 - \frac{\mu}{2}\right) S_1\left(\nu_0 + \frac{\mu}{2}\right)\right\}, \tag{36}$$

$$W_{21}\,(t_o,\nu_0) = FT^{-1}_{\mu,t_0} \left\{ S_2^* \left(\nu_0 - \frac{\mu}{2}\right) S_1(\nu_0 + \mu_2)\right\}. \tag{37}$$

It is readily verified from (31) that the temporal definitions (29, 30) are identical to the temporal frequency definitions (36, 37). Moreover, the cross signal JR's can be reduced to the auto version JR's for a single signal (19)(20), and provide us the rationale for auto-Wigner distribution. The cross-WD of the mean coordinate (t_0,ν_0) describes a slowly varying and collective property, while the cross-AF of the difference coordinate (μ,τ) describes a rapidly decaying and correlation property. Thus, the slowly varying cross-WD finds its way into geometric ray propagation and the partial coherence, while the fast decaying cross-AF sits at the heart of sonar/radar correlation detection.

When both kinds of information are needed, a new joint representation which satisfies the marginal property (22) is constructed [1] (Fig. 8),

$$\mathrm{IPS}_{21}(t_0,\nu_0,\mu,\tau) \equiv W_{21}(t_0,\nu_0)\; A_{21}^*\;(\mu,\tau)\,\exp\left[-\,i\;2\pi\;(\mu t_0 - \nu_0\,\tau)\right], \tag{38}$$

such that from (20) (21) the ambiguity surface follow

$$\int\!\!\int dt_0 d\nu_0\; \mathrm{IPS}_{21}\,(t_0,\nu_0,\,\mu,\tau) = |A_{21}(\mu,\tau)|^2\,. \tag{39}$$

PROPERTY (1)

$$\int \text{TFJR}(\nu, t)\, d\nu = I(t)$$

$$\int \text{TFJR}(\nu, t)\, dt = \text{PSD}(\nu)$$

PROPERTY (2)

$$\text{TFJR}(\nu, t) > 0 \text{ EVERYWHERE}$$

WHEN ONLY (1) IS PREFERRED, SZU PROPOSED IN 1984

$$\text{IPS}_{21}(t_0, \nu_0, \tau, \mu) = W_{21}(t_0, \nu_0)\, A_{21}^*(\mu, \tau) \exp[-i2\pi(\mu t_0 - \nu_0 \tau)]$$

Fig. 8 — Marginal distribution of two signals

and the Wigner surface is obtained

$$\int\int d\mu d\tau\ \text{IPS}_{21}(t_0, \nu_0, \mu, \tau)\,| = |W_{21}(t_0, \nu_0)|^2. \tag{40}$$

Thus, we have systematically generalized Woodward, Wigner, and Kirkwood joint representations for two independent signals. The higher dimensional phase space for any number of signals can be statistically related to the set of pairwise mutual information, similar to the situation of a pair correlation function that is sufficient for describing the kinetic theory of gases. However, we shall not digress further for details [34] of the post processing (shown in Fig. 9), and make the remark that TFJR is natural (shown in Fig. 10). TFJR has better S/N because the reduced bandwidth in the center coordinate system admits less noise, as shown in Fig. 11.

$$W_{uv} \circledast\circledast W_{kz}^* = FT^{-1}_{\mu, t_0}\, FT_{\tau, \nu_0} \left\{ A_{uv} A_{kz}^* \right\} \quad (1)$$

$$A_{uv} \circledast\circledast A_{kz}^* = FT_{t_0, \mu}\, FT^{-1}_{\nu_0, \tau} \left\{ W_{uv} W_{kz}^* \right\} \quad (2)$$

EXAMPLE 1: WHY WD IS BETTER THAN (SHORT TIME) SPECTROGRAM OF A WINDOW FUNCTION g FOR LOUD SPEAKER QUALITY ANALYSIS?

$$W_{ss} \circledast\circledast W_{gg} = FT^{-1}_{\mu, t_0}\, FT_{\tau, \nu_0} \left\{ A_{ss} A_{gg}^* \right\} \underset{\uparrow}{=} |A_{sg}|^2 = \text{SPECTROGRAM}$$

WOODWARD CENTRAL PEAK THEOREM

EXAMPLE 2: WHY WD CORRELATION WITH WD IS BETTER THAN WD TEMPLATE MATCH FILTER?

$$W_{ss} \circledast\circledast W_{gg} = \sum_{n=0}^{\infty} \int\int \frac{1}{n!} [i2\pi(\mu t_0 - \nu_0 \tau)]^n A_{ss} A_{gg}^*\, d\tau d\mu$$

ALL MOMENTS OF AF TEMPLATE MATCHING WITHOUT THE NOISEY MOMENT DIVERGENCE

Fig. 9 — General smoothing formulas for loud speaker quality analysis and pattern recognition

- OCEAN WAVES
- MUSIC SCORES
- COCHLEA FT
- SPECTROGRAMS
- RADAR RANGE AND DOPPLER SHIFT
- SURVEILLANCE

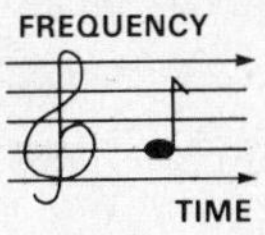

Fig. 10 — TFJR is natural

FT PAIR RECIPROCAL RELATIONSHIP $\Delta \nu \, \Delta T \gtrsim 1$

$$S(\nu) = FT_{t,\,\nu} \left\{ s(t) \right\}$$

$\updownarrow$ $\updownarrow$
$\Delta \nu$ ΔT

WIGNER DISTRIBUTION

WEIGHTED CORRELATION
COMPLEX PEAK

$$A(\tau,\mu) = FT_{t_0,\,\mu} \left\{ s^* \left(t_0 - \frac{\tau}{2} \right) s \left(t_0 + \frac{\tau}{2} \right) \right\}$$

$\updownarrow$ $\updownarrow$ $\updownarrow$
$\Delta \nu / \sqrt{2}$ ΔT ΔT

SELECTED CONVOLUTION
REAL SMOOTH

$$W(t_0,\nu_0) = FT_{\tau,\,\nu_0} \left\{ s^* \left(t_0 - \frac{\tau}{2} \right) s \left(t_0 + \frac{\tau}{2} \right) \right\}$$

WOODWARD
AMBIGUITY

THE <u>REDUCED BANDWIDTH ADMITS LESS NOISE</u> BY THE FACTOR LOG $\sqrt{2}$

MOREOVER, CONVOLUTION SMOOTHING REDUCES $\Delta \nu / 2$ WHICH IMPLIES
3 dB REDUCTION OF N/S

(RECT (t) + NOISE (t)) * (RECT (t) + NOISE (t)) = $\wedge$ (t) + . . .
ΔT ΔT $2\Delta T$

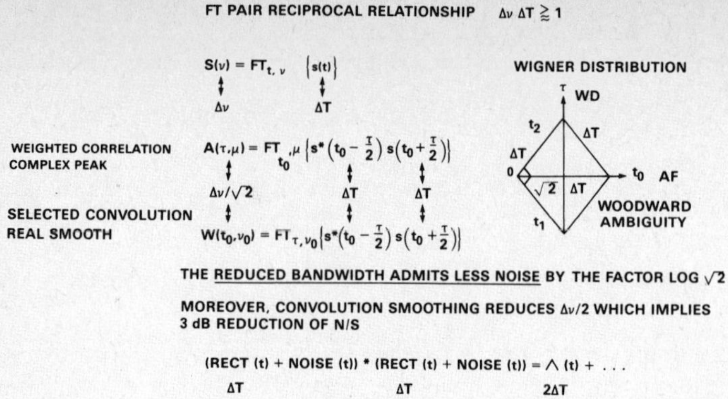

Fig. 11 — TFJR admits less noise

Acknowledgement

The author wishes to thank Professor Kim for his kind invitation to the plenary session, and to Dr. Zachary for waiting patiently for the final version. The design approach to a fiber optic cochlea hydrophone is obviously an outcome of the state of the art review of bi-linear TFJR and nonlinear TFJR signal processing. More follow-up work is needed in this area. Being a short review paper, only those early reviews and recent lead articles have been cited, and from which the reader should be able to find the contribution of many other researchers known to the author as well. The author also wishes to thank his colleagues, S. Gardner and R. Hartley at NRL, for their critical comments.

Appendix A

A REVIEW ABOUT HEARING

1. The outer ear admits only plane waves (below 27 KHz) because the small diameter of ear canal ($\sim 0.8 \times 2.3$ cm cylinder) eliminates all higher order modes. The zero at 7 KHz created by the shape of pinna transfer function is used for the elevation detection, while delay and head motions are used for lateralization.

2. The middle ear of three bones matches the impedance of an airborne-sound with the impedance of a fluid-borne-sound of the cochlea by the simple geometry namely the area of ear drum and the area of stapes footplate.

3. The inner ear is a linear dispersive transmission line built on viscoelastic-fluid membranes packing closely in a round trip snail shape. It can orderly display various frequency modes according to the order that higher frequency components are near the input end of the inner ear. This mode analysis is in effect a one-dimensional Fourier transform displayed along the basilar membrane. There exists a wake and a sharper cut off than those traveling waves discovered in a dead ear by Békésy in 1928. Thus, at least four models of the inner ear have been proposed to explain it, and here we propose another cusp model. The 1-D model by J. J. Zwislocki ('48) is summarized by T. J. Lynch et al ('82) in an analog electric circuit diagram. The 2-D model of fluid mechanics shows Békésy traveling wave using hydrodynamics by M. Leser and D. Berkeley ('72) and G. Zweig et al ('76). The 3-D model includes a tongue-like tectorial membrane with its own resonance frequency independently done by J. J. Zwislocki ('79) and J. Allen ('80). Some active spontaneous emission model producing a negative resistance has been studied by S. T. Neely in 1981.

REFERENCES

1. H.H. Szu and H.J. Caulfield, "The Mutual Time-Frequency Content of Two Signals," Proc. IEEE **72**, No. 7, 902-908, July 1984.
2. R.A. Athale, J.N. Lee, E.L. Robinson, and H.H. Szu, "Acousto-Optic Processors for Real-Time Generation of Time-Frequency Representations," Opt. Lett. **8**, No. 3, 166-168, Mar. 1983.
3. H.H. Szu, "Two-Dimensional Optical Processing of One-Dimensional Acoustic Data," Opt. Eng. **21**, No. 5, 804-813, Sept.-Oct. 1982.
4. H.H. Szu and J.A. Blodgett, "Wigner Distribution and Ambiguity Function," In: "Optics in Four Dimensions — 1980," edited by M.A. Machado and L.M. Narducci, Pub. by Am. Inst. Phys. Conf. Proc. **65**, No. 1, pp. 355-381, 1981.
5. H.H. Szu and J.A. Blodgett, "Image Processing for Acoustical Patterns," In: "Workshop on Image Processing for Ocean Acoustics," May 6-8, 1981, Woods Hole, Mass.
6. H.H. Szu, "Optical Data Processing Using Wigner Distribution," In: "Symposium on 2-D Signal Processing Techniques and Applications," March 3-5, 1981, Washington, D.C. (NRL Memo Report 4526, DTIC No. 1180069).
7. E. Wolf, "Coherence and Radiometry," J. Opt. Soc. Am. **68**, No. 1, 6-17, Jan. 1978.
8. A. Walther, "Radiometry and Coherence," J. Opt. Soc. Am. **58**, 1256-1259, 1968.
9. A. Walther, "Radiometry and Coherence," J. Opt. Soc. Am. **68**, 1622-1623, 1973.
10. E. Wigner, "On quantum correction for thermodynamics equilibrium," Phys. Rev. **40**, 749-759, June 1932.

11. J.G. Kirkwood, "Quantum statistics of almost classical assemblies," Phys. Rev. **44**, 31-37, July 1933.

12. M.J. Bastiaans, "Wigner distribution functions and their applications to first order Optics," Opt. Comm. **32**, 3238, Jan. 1980.

13. H.O. Bartelt, K.H. Brenner, and A.W. Lohmann, "The Wigner distribution function and its optical production," Opt. Comm. **32**, 32-38, Jan. 1980.

14. S.M. Sussman, "Least-square synthesis of radar ambiguity functions," IRE Trans. Info. Theo., **IT-8**, 246-254, Apr. 1962.

15. O.D. Grace, "Instantaneous power spectra," J. Acoust. Soc. Am., **69**, No. 1, 191-198, Jan. 1981.

16. T.A.C.M. Claasen and W.F.G. Mecklenbrauker, "The Wigner distribution — a tool for time-frequency signal analysis, Part I: continuous-time signals," *Philips J. Res.*, Vol. 35, pp. 217-250, 1980. "Part II: discrete signals," *Philips J. Res.*, Vol. 35, pp. 276-300, 1980. "Part III: relations with other time-frequency signal transformations," *Philips J. Res.*, Vol. 35, pp. 372-389, 1980.

17. R.L. Easton, Jr., A.J. Ticknor, and H.H. Barrett, "Application of the Radon Transform to Optical Production of the Wigner Distribution Function," Opt. Eng. Vol. 23, No. 6, 738-744, Nov.-Dec. 1984.

18. B.V.K. Vijaya Kumar and C.W. Carroll, "Performance of Wigner Distribution Function Based Detection Methods," Opt. Eng. Vol. 23, No. 6, 732-737, Nov.-Dec. 1984.

19. C.P. Janse and A.J.M. Kaizer, "The Wigner Distribution: A Valuable Tool for Investigating Transient Distortion," J. Audio Eng. Soc. Vol. 32, No. 11, 868-882, Nov. 1984.

20. D. Chester, F.J. Taylor, and M. Doyle, "The Wigner Distribution in Speech Processing Applications," J. Franklin Inst. (USA), Vol. 318, No. 6, 415-430, Dec. 1984.

21. B.V.K. Vijaya Kumar and C.W. Carroll, "Effects of Sampling on Signal Detection Using the Cross-Wigner Distribution Function," Appl. Opt. Vol. 23, No. 22, 4090-4094, Nov. 15, 1984.

22. N. Weidenhof and J.M. Waalwijk, "Wigner Distributions: A Refined Mathematical Tool for Appraising Loudspeakers," Funk-Tech. (GERMANY), Vol. 39, No. 9, 371-373, Sept. 1984.

23. N. Subotic and B.E.A. Saleh, "Generation of the Wigner Distribution Function of Two-Dimensional Signals by a Parallel Optical Processor," Opt. Lett. Vol. 9, No. 10, 471-473, Oct. 1984.

24. G. Neuweiler, In: "Animal Sonar Systems," R.G. Busnel and J.F. Fish, editors, Plenum, New York 1980.

25. J.A. Simmons, D.J. Howell, N. Suga, "Information Content of Bat Sonar Echoes," Amer. Scient. **63**, 204 (1975).

26. D.R. Griffin, "Listening in the Dark," Yale, New Haven 1958.

27. M.R. Schroeder, "Models of Hearing," Proc. IEEE, Vol. **63**, No. 9, pp. 1332-1350, Sept. 1975.

28. A.J. Hudspeth, "The Hair Cells of the Inner Ear," Scient. Amer., pp. 54-64, Jan. 1983.

29. H. Szu and R. Messner, "Adaptive Invariant Novelty Filters," Proc. IEEE, Vol. **74**, p. 518-519, Mar. 1986.

30. M.B. Lesser and D.A. Berkley, "Fluid Mechanics of the Cochlea, Part 1," J. Fluid Mech. Vol. **51** pt. 3, pp. 497-512, 1972.

31. J.B. Allen, "Cochlear Modelling," IEEE ASSP Magazine **2**, No. 1, pp. 3-29, Jan. 1985.

32. G. Neuweiler, "How bats detect flying insects," Physics Today, pp. 34-40, Aug. 1980.

33. R. Thom, "Stabilitè Structurelle et Morphogenèse," (New York, Benjamin, 1972).

34. H.H. Szu, "Applications of Wigner and Ambiguity Functions to Optics," Proc. IEEE Int. Symp. Circuits and Systems, San Jose, CA, May 5-7, 1986.

35. G. Von Békésy, "Experiments in Hearing," New York McGraw Hill, 1960.

36. H. Davis, "Biophysics physiology of the inner ear," Physio. Rev. **37**, pp 1-49 (1957).

37. H. Szu, "Brown motion of elastically deformable bodies," Physical Review A Vol. **11**, No. 1, pp. 350-359, Jan. 1975.

38. H. Szu, "Laser scattering from droplets: A theory of multiplicative and additive stochastic processes," Physics of Fluids, Vol. **21**, No. 8, pp. 1243-1246, Aug. 1978.

39. D. Menne and H. Hachbarth, "Accuracy of Distance Measurement in the Bat Eptesicus Fuscus: Theoretical Aspects and Computer Simulations," J. Acoust. Soc. Am. **79** (2), pp. 386-397, Feb. 1986.

40. S. Shamma, "Speech Processing in the Auditory System I: The representation of speech sounds in the response of the auditory nerve, vol. **78** (5), pp. 1612-1622, NOV. 1985; II. ibid. pp. 1622-1632.

41. T.W. Barrett, "Cochlear Fluid Mechanics Considered as Flows on a Cusp Catastrophe," Acustica Vol. **38**, pp. 118-123, 1977.

42. J. Caelen, "Space/time Data-Information in the ARIAL Project Ear Model," Speech Commu., Vol. **4**, pp. 163-179, 1985.

43. H. Szu, "Nonlinear Signal Computing Using Neurograms," to appear in a book: "Optical and Hybrid Computing," (Edited by H. Szu, Oct. 1986, published by SPIE).

INTERFERENCE IN PHASE SPACE

Wolfgang Schleich*,** and John A. Wheeler*

* Center for Theoretical Physics
Department of Physics
University of Texas at Austin
Austin TX 78712 U.S.A.

** Max-Planck-Institut
für Quantenoptik
D-8046 Garching b. München
West-Germany

It is amazing that the two central ideas of early quantum mechanics -- quantization in phase space and the correspondence principle of Bohr [1] -- combined with the concept of interference [2] can provide us with the most vivid insight available into such a problem as the distribution of photons in a squeezed state [3,4]. Whereas the probability W_m of finding m photons in a coherent state [5] is well-known to follow the formula of Poisson, the corresponding distribution in the case of a highly squeezed state exhibits oscillations [6,7]. In the semiclassical limit of large quantum numbers, the probability W_m is closely related [6] to the two areas of overlap in phase space between the band representing the m-th photon state and the long, thin cigar of a strongly squeezed state. It has been argued [6] that it is <u>interference</u> between the two contributions which gives rise to these modulations. Ref. 6 discusses these ideas quantitatively. The present article derives a simple analytic expression for the photon distribution of a highly squeezed state based on the area-of-overlap algorithm. In order to focus on the essential ideas we suppress the detailed calculations, to appear in Ref. 7. It is appropriate at the start to recall the simplest properties of a state of definite photon number.

A single mode of the electromagnetic field in a number state is equivalent [5] to a harmonic oscillator of frequency ω and mass μ with coordinate q and momentum $\tilde{p}$.

In terms of the dimensionless coordinate x and momentum p, defined by $q = \sqrt{\hbar/\mu\omega}\, x$ and $\tilde{p} = \sqrt{\hbar\mu\omega}p$, the energy in the m-th state reads

$$m + 1/2 = (p_m{}^2 + x^2)/2. \tag{1}$$

The trajectories in phase space are circles of radius $\sqrt{2(m+1/2)}$. Each state takes up an area 2π in phase space. Therefore, we associate with the m-th number state an occupied band of inner radius $\sqrt{2m}$ and outer radius $\sqrt{2\,(m+1)}$.

In a squeezed state the fluctuations in one of the quadrature components, x (or p), are reduced [3,4] at the expense of the other, p (or x), expressed by the distribution in phase space [7],

$$P_{sq}(x,p) = (1/\pi) \exp \{-(2/\epsilon)(x-\sqrt{2}\,\alpha)^2 - (\epsilon/2)p^2\} . \tag{2}$$

For the sake of simplicity, we have assumed strong squeezing in the x-variable; that is, $\epsilon << 1$. Thus, the contour lines have the shape of a long thin cigar, vertical, and centered at the point $x_0 = \sqrt{2}\,\alpha$ on the positive x-axis. According to the area-of-overlap concept [6], the probability W_m of finding m photons in the squeezed state is governed by the overlap in phase space between circular band representing the m-th number state and the tall cigar representing the squeezed state. The cigar and the ring overlap either not at all, or in two diamond-shaped areas, according as the excitation m is appropriately greater or less than α^2. The area of overlap associated with either *one* of these diamonds we obtain by weighting each point of phase space according to the distribution function P_{sq} of (2); thus,

$$A_m = (1/2) \int dx \int dp\, P_{sq}(x,p) \tag{3}$$

m-th band

$$= (\epsilon/4\pi)^{1/2} \ (m+1/2 - \alpha^2)^{-1/2} \exp [\epsilon(\alpha^2 - m - 1/2)]. \qquad (4)$$

The probability to be compared with experiment is not the sum, 2 A_m, of the areas of the two diamonds. Neither is the intensity on the photographic plate in the familiar double-slit experiment equal to the sum of the intensities that would arrive through the two slits separately!

Quantum mechanics instructs us, we know, to add, not probabilites, but probability amplitudes. Interference is as inescapable in the physics of the squeezed state as it is in the Young experiment. Here, however, the interference takes place in a more sophisticated arena: not in coordinate space, but in phase space.

No interference? Probability of excitation of the m-th number state calculated as

$$W_m = 2\ A_m\ ? \qquad (5)$$

Wrong! Wrong, first, because we don't get in this way the oscillations mentioned in the introduction; and second, because we should have been adding probability amplitudes <u>before</u> any evaluation of probability; thus,

$$W_m = \left| \sqrt{A_m} \exp (i\phi_m) + \sqrt{A_m} \exp (-i\phi_m) \right|^2 = 4\ A_m \cos^2\phi_m\ . \qquad (6)$$

The physics of this result is clear. In the one diamond the momentum of the field oscillator is positive, corresponding to motion to the "right"; in the other, negative, as appropriate for motion to the "left". The phase ϕ_m, when doubled, is governed by the area in phase space between the center of the cigar and the center of the ring in the limit of strong squeezing and large quantum numbers m (that is, $[2(m + 1/2)]^{1/2} >> \sqrt{2}\,\alpha$) and is then

$$\phi_m = (1/2)\int_{\sqrt{2}\alpha}^{\xi_m} dx \int_{-p_m}^{p_m} dp - (\pi/4) \ . \qquad (7)$$

Here $\xi_m = [2\,(m + 1/2)]^{1/2}$ denotes the right-hand turning point of the semiclassical path (1) with $p_m = [2\,(m + 1/2) - x^2]^{1/2}$. The phase ϕ_m increases rapidly with increase of the quantum number m. As a consequence the photon distribution W_m is a rapidly oscillating function of m.

This result, derived here by the simplest of semiclassical reasoning, agrees well with the expression derived in [4], by the full quantum-mechanical apparatus of Hermite polynomials, for the distribution in photon number of a squeezed state.

In conclusion, we recognize the distribution of the photon number of a squeezed state as the straightforward consequence of "*interference in phase space*".

The authors thank H. Carmichael, C.-S. Cha, R.Y. Chiao, L. Cohen, K. Dodson, K. Kraus, M. Hillery, R.F. O'Connell, M.O. Scully, D.F. Walls and H. Walther for useful and stimulating dicussions. Preparation of this article was assisted by the University of Texas Center for Theoretical Physics and by NSF Grant PHY 8503890.

References:

1. M. Born, Vorlesungen über Atommechanik, in: Struktur der Materie in Einzeldarstellungen (eds. M. Born and J. Franck) (Springer, Berlin, 1925).

2. J.A. Wheeler and W.H. Zurek, eds., Quantum Theory and Measurement (Princeton University Press, Princeton NJ, 1983); R.P. Feynman, R.B. Leighton and M. Sands, The Feynman Lectures on Physics (Addison-Wesley, Reading MA, 1964)

Vol. 3.

3. D.F. Walls, Nature 306, 141 (1983).

4. H.P. Yuen, Phys. Rev. A 13, 2226 (1976).

5. M. Sargent, M.O. Scully and W.E. Lamb Jr., Laser Physics (Addison-Wesley, Reading MA, 1974).

6. J.A. Wheeler, Lett. Math. Phys. 10, 201 (1985).

7. W. Schleich and J.A. Wheeler, to be published.

8. M. Hillery, R.F. O'Connell, M.O. Scully and E.P. Wigner, Phys. Rep. 106, 121 (1984).

A QUANTUM MECHANICAL MOMENT PROBLEM

Francis J. Narcowich
Department of Mathematics
Texas A & M University
College Station, TX 77843

I will talk about a moment problem that was first mentioned by Moyal in his seminal 1949 paper [1]: For a sequence of scalars $m_{0,0}, m_{0,1}, m_{1,0}, \cdots$, under what conditions will there be a Wigner distribution $P(q,p)$ for which

$$(1) \qquad m_{j,k} = \int p^j q^k \, P(q,p)\,dq\,dp \;?$$

In discussing this quantum mechanical moment problem, I will take an approach that is similar to the one used in the classical Hamburger moment problem, and I will derive conditions necessary for the problem to have a solution; these conditions include as a special case the position-momentum uncertainty relation. I will also discuss the possibility that these conditions are sufficient as well as necessary.

The notation that I'll adopt will be the same as that R. F. O'Connell and I used in [2], where one may find some of what I'll say here. Let $z = (q,p)$ denote a point in phase space, $a \equiv (u,v)$ denote a point in the "dual" of phase space, and $\sigma(a,z) \equiv qv - pu$ be the usual symplectic form. For a function $g(z)$ define

$$(2) \qquad \tilde{g}(a) \equiv \int g(z) e^{i\sigma(a,z)}\,dz \,, \quad dz \equiv dq\,dp \,;$$

$\tilde{g}(a)$ is called the symplectic Fourier transform of g. Given a Wigner distribution P, one can express its moments in terms of $\tilde{P}$:

$$(3) \qquad m_{j,k} = i^{j-k} \frac{\partial^{j+k}}{\partial u^j \partial v^k} \tilde{P}\Big|_{a=0} \,.$$

Characterizing symplectic Fourier transforms of Wigner distributions is similar to characterizing Fourier transforms of positive measures. I begin by defining functions of <u>$\hbar$-positive type</u> [3]: Let $\tilde{g}(a)$ be a continuous function. I will say that $\tilde{g}(a)$ is of <u>ħ-positive type</u> if, for every finite set of points $\{a_1,\ldots,a_n\}$ and for every set of n complex numbers $\{\lambda_1,\ldots,\lambda_n\}$,

$$(4) \qquad \sum_{j,k=1}^{n} e^{\frac{i\hbar}{2}\sigma(a_k,a_j)} \tilde{g}(a_j - a_k)\lambda_j^* \lambda_k \geq 0 .$$

Equivalently [2], $\tilde{g}(a)$ is of ħ-positive type if, for every continuous function $\lambda(a)$ having compact support,

$$(5) \qquad \int e^{\frac{i\hbar}{2}\sigma(a,a')} \tilde{g}(a' - a)\lambda(a')^* \lambda(a)da\,da' \geq 0 .$$

Long ago, Kastler [3] and Loupias and Miracle-Sole (see the refs. in [2]) showed that for $\tilde{P}(a)$ to be the symplectic Fourier transform of a Wigner distribution P it is necessary and sufficient that $\tilde{P}$ satisfy these: (i) $\tilde{P}(a)$ must be continuous and of ħ-positive type; (ii) $\tilde{P}(0) = 1$. These conditions are certainly known, but not well known. In [2], R. F. O'Connell and I discussed them in detail, and we named them the KLM conditions.

With these condition in hand, I can easily derive <u>necessary</u> conditions for the $m_{j,k}$ to be moments of a Wigner distribution. In (5), let $\tilde{g} = \tilde{P}$, and set

$$(6) \qquad \lambda(a) = \{ \sum_{i,k=0}^{N} c_{j,k}(-1)^{j+k} \frac{\partial^{j+k}}{\partial u^j \partial v^k} \}\xi(a) ,$$

where the $c_{j,k}$'s are complex constants, and $\xi(a)$ is a smooth function having compact support. In the resulting expression, integrate by parts and then let $\xi(a) \rightarrow \delta(a - b)$. Evaluating the integrals yields this inequality:

$$(7) \qquad [\sum_{j,k,\ell,m} c_{j,k}^* c_{\ell,m} \frac{\partial^{j+k+\ell+m}}{\partial u'^j \partial v'^k \partial u^\ell \partial v^m}][e^{\frac{i\hbar}{2}\sigma(a,a')} \tilde{P}(a' - a)]|_{a=a'=b} \geq 0 .$$

After carrying out the differentiations and evaluations, one can use (3) to replace derivatives of $\tilde{P}$ at 0 by the corresponding moments. The result is an inequality that involves $\hbar$, the m's, the c's, and nothing else.

I want to illustrate this inequality in a simple case. Suppose that the original Wigner distribution P is such that the expectation values for p and q are both 0; that is, $m_{0,1} = m_{1,0} = 0$. Take $b = 0$, $\lambda(a) = c_1 \frac{\partial \xi}{\partial u} + c_2 \frac{\partial \xi}{\partial v}$, and note that $m_{0,0} = \tilde{P}(0) = 1$; (7) becomes this:

$$(8) \qquad |c_1|^2 m_{2,0} + |c_2|^2 m_{0,2} - c_1 c_2^* (m_{1,1} + \frac{i\hbar}{2}) - c_1^* c_2 (m_{1,1} - \frac{i\hbar}{2}) \geq 0 ,$$

which holds for all possible complex numbers c_1 and c_2. The usual conditions for a quadratic form being non-negative then imply that $m_{2,0} m_{0,2} \geq m_{1,1}^2 + \frac{\hbar^2}{4}$, from which the position-momentum uncertainty relation follows immediately.

Whether the inequalities in (7), when the derivatives of $\tilde{P}$ at 0 are replaced by the corresponding m's, give a set of sufficient as well as necessary conditions for the $m_{j,k}$'s to be moments of a Wigner distribution is an open question. In [2], O'Connell and I conjectured that they would be. We did this because something similar is true in the one-dimensional Hamburger moment problem. However, it's possible that the situation is more like the two-dimensional moment problem, where a similar set is again necessary but not sufficient.

In closing, I want to thank my colleauges, Ingrid Daubechies, Peter Lax, Robert O'Connell, and Eugene Wigner, for stimulating conversations and helpful comments. I also want to thank the Courant Institute and the Mathematics Department of The Ohio State University for their hospitality, and Texas A & M's Association of Former Students for its support during the 1985-1986 academic year.

[1] J. E. Moyal, Proc. Cambridge Phil. Soc. 45 (1949), 99-124.

[2] F. J. Narowich and R. F. O'Connell, "Necessary and sufficient conditions for a phase-space function to be a Wigner distribution", Phys. Rev. A, to appear.

[3] D. Kastler, Comm. Math. Phys. 1 (1965), 14-48.

TOMOGRAPHIC PROCEDURE FOR CONSTRUCTING PHASE SPACE REPRESENTATIONS

J.Bertrand * and P.Bertrand **
* LPTM- University Paris VII
2, place Jussieu, F-75251 Paris
** ONERA, F-92320 Chatillon

1. General outline.

Consider a wave theory (quantum mechanics, signal theory,...) where states are represented by vectors ϕ in a Hilbert space H and where observations are given by the sesquilinear form:

$$\langle A \rangle = (\phi, A_{op}\phi) = \mathrm{Tr}\,(A_{op} P_\phi). \qquad (1)$$

In these expressions, A_{op} denotes the physical observable and P_ϕ the projector on the ϕ state. There exists generally an invariance group G (Galilei's, time and frequency translations,...) represented unitarily on H, which guarantees the independence of the theory by change of observer.

The phase space version of such a theory is obtained through a linear one-to-one correpondence between operators and functions on a phase space Γ satisfying the following constraints:

i) The invariance group G acts in Γ by point transformations.

ii) The expectation values of observables are given by

$$\langle A \rangle = \int_\Gamma A(\gamma)\, f(\gamma)\, d\mu(\gamma), \qquad (2)$$

where $A(\gamma)$ and $f(\gamma)$ correspond respectively to A_{op} and to the projector P_ϕ, and where $d\mu(\gamma)$ is a measure on Γ.

In spite of its form, condition ii) cannot usually be interpreted as a mathematical expectation. This is due to the impossibility of finding a linear one-to-one correspondence between P_ϕ and a function f everywhere positive on Γ. However, if restricted to classes of simultaneously diagonalizable observables, equation (1) has a probabilistic interpretation and can be written:

$$\langle A \rangle = \int \alpha(\beta)\;\; \rho(\beta)\, d\beta \qquad (3)$$

where $\rho(\beta)$ is the positive diagonal part of the projector P_ϕ. The classes we consider in the following are those of observables invariant by some subgroup of G.

In phase space, observables invariant by a subgroup of G are represented by functions which have to be constant on the subgroup orbits. It results that (2) can be reduced by integrating on the orbits, thus introducing a kind of marginalization of f. At this stage, a natural requirement is that the marginalized f be identified with the corresponding density $\rho(\beta)$ appearing in (3). This is what we will call the tomographic constraint. The exploitation of this constraint permits a determination of f founded on the inversion of a Radon transform.

For the sake of illustration, the tomographic construction is applied to Quantum Mechanics in the next section. It leads to the original Wigner function [1] . The same procedure has also been used for the time-frequency representation of signals in another paper [2] .

2. Application to Quantum Mechanics.

The invariance group is Galilei's for fixed time; it acts on ϕ and f as follows

$$\begin{array}{ccc} \phi(\vec{x},t) & \longrightarrow & e^{-i(m/\hbar)\,\vec{v}.\vec{x} + i\psi(\vec{v},t)}\,\phi(\vec{x}+\vec{v}t+\vec{a},t) \\ \downarrow & & \downarrow \\ f(\vec{x},\vec{p}) & \longrightarrow & f(\vec{x}+\vec{v}t+\vec{a},\vec{p}+m\vec{v}) \end{array}$$

where $\psi(\vec{v},t)$ is a phase we need not know explicitly.

The constraint of galilean covariance on the correspondence $P_\phi \longrightarrow f$ is expressed by the commutativity of the above diagram [3] . In fact, this condition is consistent with a whole family of distribution functions [4] and we shall apply the tomographic constraint to remove the ambiguity.

In a first step, we restrict to one space dimension. Then the subgroups G_α of G are characterized by $\alpha \in \mathbb{R}$ such that

$$a = \alpha v .$$

The improper orthonormal basis diagonalizing the observables invariant by G_α is found to be

$$Z^\alpha_\lambda = \left| \frac{1}{2\pi(t+\alpha)} \frac{\partial\lambda(\beta,\alpha)}{\partial\beta} \right|^{1/2} \exp\left\{ \frac{i}{t+\alpha} \left[m/2\hbar)x^2 - \lambda(\beta,\alpha)x \right] \right\},$$

where $\beta \in \mathbb{R}$ and λ is an arbitrary function.

The diagonal part of P_ϕ in this basis is

$$\rho(\lambda(\beta,\alpha),\alpha) = \left| \int dx\, Z^\alpha_\lambda\, (x,\beta)\, \phi^*(x) \right|^2 \tag{4}$$

so that (3) becomes

$$\langle A \rangle = \int a(\beta)\, \rho(\lambda(\beta,\alpha),\alpha)\, d\beta. \tag{5}$$

On the other hand, for G_α-invariant observables, the phase space expression (2) can be written as

$$\langle A \rangle = \int \tilde{a}(\beta)\, I(\beta,\alpha)\, d\beta \tag{6}$$

where

$$I(\beta,\alpha) = \int dx\, dp\, f(x,p) \quad \delta(\beta + (t+\alpha)(p/m) - x).$$

Now the tomographic constraint consists in identifying the density ρ with the function I. The arbitrary function λ which takes into account different parametrizations in (5) and (6), is determined through the requirement of covariance by galilean transformations and space inversion. The resulting constraint is

$$\int dx\, dp\, f(x,p) \quad \delta(\beta + (t+\alpha)(p/m) - x) =$$

$$= \frac{m}{2\pi\hbar(t+\alpha)} \int dx\, dx' \exp\left\{\frac{im}{\hbar(t+\alpha)}\left[\frac{1}{2}(x'^2 - x^2) - \beta(x'-x)\right]\right\} \phi(x)\, \phi^*(x').$$

Thus, f is given by its Radon transform [5] and inversion yields:

$$f(x,p) = \int du\, e^{-2i\pi up}\, \phi(x+\pi\hbar u)\, \phi^*(x-\pi\hbar u).$$

When going over to the 3-dimensional case, we have to consider 3-parameter subgroups of Galilei's for fixed t and the corresponding classes of invariant observables. In that case, only invariant observables belonging to special classes can be diagonalized, namely those corresponding to subgroups whose phase space orbits are lagrangian hyperplanes. The tomographic constraint then has the form of an overdetermined Radon transform [6]. However, adding the requirement of rotational covariance of the correspondence $P_\phi \longrightarrow f$ yields a meaningful equation having Wigner's distribution function as a unique solution.

References.

[1] E.P.Wigner, Phys. Rev. 40,749 (1932).
[2] J.Bertrand and P.Bertrand, Rech. Aérosp. 1985-5, p. 1-7. See also: "Time-frequency representations of broad-band signals" in these Proceedings.
[3] E.P.Wigner in:"Perspectives in Quantum Theory", eds. W.Yourgrau and A. van der Merwe (Dover, New York) 1979.
[4] L.Cohen, J. Math. Phys. 7, 781 (1966).
[5] I.M.Gelfand, M.I.Graev and Ya.Vilenkin, "Generalized functions", vol. 5.
[6] A.Debiard and B.Gaveau, C.R.Acad.Sc.Paris 296,423 (1983).

WIGNER DISTRIBUTION ON SU(2)*

R. Gilmore
Department of Physics & Atmospheric Science
Drexel University, Philadelphia, PA 19104

Algebraic models have been used with increasing frequency recently to describe the properties of physical systems including atoms, molecules, solids, and nuclei. These models have been based on various Lie groups, including: SU(2), SU(3), SU(4), SU(6), SO(8), U(4)$\otimes$U(4), SU(8), and ISl(4)$\otimes$ISp(2n+2). In many instances a geometric interpretation of these algebraic models has emerged from the use of suitable coherent states.

Coherent states [1,2] have played an important role in the description of nonrelativistic systems whose Lie algebra is the Weyl algebra h(3) with generators q, p, and I. The Wigner distribution function [3] has played an important role in the description of such systems. It is therefore expected that an analogous distribution would also play an important role in understanding the properties of systems described by algebraic models.

The difficulty with extending the definition of the Wigner distribution from the Weyl algebra h(3) to other Lie algebras is the following. The Wigner distribution on phase space is constructed from the wave functions defined over configuration space. For a general Lie group a configuration space is not naturally defined. However, there is a natural phase space associated with an algebraic model. This is the space on which the associated coherent states are defined. As a result, it is necessary to adopt a definition for the Wigner distribution which is (a) useful, and (b) produces the standard Wigner distribution for the Lie group H(3). A similar problem was encountered in the construction of generalized coherent states [4,5] for any Lie group. In that case the properties that H(3) coherent states possess could not all be preserved for arbitrary Lie groups. Some of the most cherished properties had to be relinquished [(a) eigenstate of the annihilation operator and (b) minimum uncertainty states] in order to develop systems with desirable and extremely useful properties.

The two most useful properties of the Wigner distribution appear to be (a) the association of a function on phase space with an operator on a Hilbert space and (b) the computation of Hilbert space averages by taking integrals over phase space. Therefore, we take as the defining requirements on Wigner distributions on SU(2) the following two conditions:

1) $\hat{A} \longleftrightarrow W_J(\hat{A},\Omega)$

2) $(\hat{A},\hat{B})$ ↙ ↘

$$Tr_J \hat{A}^\dagger \hat{B} = \frac{2J+1}{4\pi} \int W_J^*(\hat{A},\Omega)\, W_J(\tilde{\hat{B}},\Omega)\, d\Omega .$$

Here $\hat{A}$, $\hat{B}$ are operators in the Hilbert space of dimension 2J+1, Ω are the coherent state parameters which are identified with the sphere $S^2 = SU(2)/U(1)$, and $(2J+1)/4\pi\ d\Omega$ is the usual Haar measure.

Two phase space distributions already exist on SU(2)/U(1). These are the Q- and P- representations. The three distributions, P-, W-, and Q-, have very similar properties. In particular, the phase space image of a spherical tensor operator, $\mathcal{Y}^L_M(\hat{j})$, is proportional to the corresponding spherical harmonic [6]. The proportionality factor is

$$\left(\frac{4\pi}{2L+1}\right)^{1/2} [J(J+1)]^{L/2} \left|\left\langle \begin{matrix} J & L & J \\ J & 0 & J \end{matrix} \right\rangle\right|^n$$

where n = -1, 0, +1 for P-, W-, and Q.

It is possible to introduce a superoperator, $\hat{W}(\hat{J},\Omega)$, with 'one foot' in each of the spaces H_J and S^2. Using this operator, the Wigner representative of an operator $\hat{A}$ can be constructed by taking the trace of $\hat{W}$ against $\hat{A}$ in H_J. In the other direction, the operator associated with a phase space distribution $f(\Omega)$ can be constructed by integrating $\hat{W}$ against $f(\Omega)$ over S^2. Superoperators $\hat{Q}$, $\hat{P}$ also exist.

A convolution kernel has been defined

$$K^2(\Omega, \Omega') = \int <\Omega|g|\Omega> <\Omega'|g-1|\Omega'>\ d\mu(g) .$$

This kernel can be used to relate the P-, W-, and Q- representations of an operator $\hat{A}$ by convolution

$$P \underset{K^{-1*}}{\overset{*K}{\rightleftarrows}} W \underset{K^{-1*}}{\overset{*K}{\rightleftarrows}} Q .$$

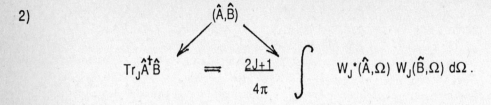

The superoperators $\hat{P}$, $\hat{W}$, and $\hat{Q}$ are related similarly. An entire class of phase space distributions, $D_J{}^n$, can be constructed by convolution: $D_J{}^n = W * K^n$, where n is not necessarily either possitive or integral. The Wigner distribution is unique among these (n=0) in that it is the only one which is coupled to itself in the computation of inner products.

Three important limits can be taken. These are the classical spin limit: $J \to \infty$, $N \to \infty$, $J/N \to$ constant; the contraction limit $U(2) \to H(4)$; and the classical phase space limit $\hbar \to 0$. All the standard properties of the usual coherent states and phase space distribution functions are recovered in the second limit. They are also recovered by application of this generalized construction to the Weyl group H(4).

This construction generalizes in a straightforward way to arbitrary Lie groups. The existence of distributions (n<0) and the invertibility of mappings may become delicate for noncompact groups.

* This work is supported in part of NSF Grant PHY 884-1891.

1. E. Schrödinger, Naturwiss. **14**, 644 (1927).
2. R. J. Glauber, Phys. Rev. **130**, 2529 (1963), ibid., **131**, 2766 (1963).
3. E. P. Wigner, Phys. Rev. **40**, 749 (1932).
4. F. T. Arecchi, E. Courtens, R. Gilmore, and H. Thomas, Phys. Rev. **A6,** 2211 (1972).
5. R. Gilmore, Ann. Phys. (NY), **74**, 391 (1972); Rev. Mex. de Fisica **23,** 143 (1974).
6. R. Gilmore, J. Phys. **A9**, L65 (1976).

PHASE SPACE CALCULATIONS OF COMPOSITE PARTICLE PRODUCTION

E.A.Remler

The College of William and Mary

Williamsburg, Virginia 23185 USA

ABSTRACT *In order to study complex reactions such as those encountered in nuclear heavy ion physics, reaction theory must be re-expressed in terms of the density operator. Exact expressions relating inclusive production cross sections to scattering solutions of the density operator have been derived. When these are applied to various dynamical approximations of the density operator, a simple, useful, and physically intuitive picture of production as seen in classical phase space emerges. These methods should also be useful in molecular dynamics.*

Nuclear heavy ion physics [1] studies collisions between heavy nuclei with laboratory bombarding energy of a few Mev/nucleon to a few Gev/nucleon having as its goal the understanding of statistical (esp. thermodynamic) properties of highly excited nuclear matter. The dynamics of systems of this size (typically a few hundred nucleons) must be statistical and hence is described using the density operator. Reaction theory however, was developed primarily to deal with pure states which are described by single wave functions. Thus, to treat such systems, the reaction theory of waves is in the process of being extended to a "reaction theory of densities".

One of the most basic formula in standard reaction theory [2] is the expression for the transition matrix ,

$$T_{fi} = \langle f|V_f|i^{(+)}\rangle \ , \qquad (1)$$

Here $|i^{(+)}\rangle$ is the outgoing wave solution for incident channel i, $\langle f|$ is the final state channel wave function, typically a product of plane waves, one for each final state fragment and, V_f is the inter (as opposed to intra) fragment interaction (i.e.$(H-V_f)|f\rangle=E|f\rangle$). The importance of Eq.(1) comes from the fact that for systems of more than a few particles, it is impossible to calculate $|i^{(+)}\rangle$ without making approximations which generally break down as the system separates. Thus, to find production amplitudes, one cannot simply calculate $|i^{(+)}\rangle$ out to very large separations and pick out amplitudes of interest from appropriate regions of configuration space. Approximations to $|i^{(+)}\rangle$ are at best valid only in the near wave zone where all particles are strongly interacting. Eq.(1) allows us, remarkably, to circumvent this problem. It can do this because the factor V_f confines configuration space integrals to near wave zone regions where practical approximations have hopes of being valid. In this talk, I wish to describe an extension of this formula to a similar exact formula (Eq.(2)) relating the density operator to cross sections which is important for the same reasons. Although examples are drawn from nuclear heavy ion physics, these results should also apply to systems of atoms, quarks, etc.

An exact analog of Eq.(1) which describes inclusive production of composite fragments C in terms of the density operator was introduced some years ago [3,4,5,6,7];

$$\Delta\sigma_{C'}(\vec{p}_C) = tr(|C,\vec{p}_C\rangle\langle C,\vec{p}_C| \; [-iV_C \; , \; \textstyle\int dt \; \rho] \;) \; . \tag{2}$$

The role of $|i^{(+)}\rangle$ in Eq.(1) is taken by the time integral of the density, $\int dt\rho$. Both are solutions to the time-independent scattering equation. The role of V_f is taken by $[-iV_C \; , \;]$ where V_C is the interaction between composite fragment C and all other constituents of the system. Again, this factor enables us to make calculations, now of cross sections, which require information only from configuration space regions where approximations tend to be most valid. The role of $\langle f|$ is taken by $|C,\vec{p}_C\rangle\langle C,\vec{p}_C|$ where $|C,\vec{p}_C\rangle$ is an eigenstate of C with momentum $\vec{p}_C$.

On the left hand side of Eq.(1) is the transition amplitude, which can be defined formally by

$$\langle f|\exp(iH_ft)\;\exp(-iHt)|i\rangle^{+\infty}_{-\infty} =$$

$$\textstyle\int^{+\infty}_{-\infty} dt \; (\partial/\partial t) \; \langle f|\exp(iH_ft)\;\exp(-iHt)|i\rangle \; = \; -2\pi i\delta(E_f-E_i) \; T_{fi} \; , \tag{3}$$

where $H_f = H - V_f$. This can be interpreted as saying that T_{fi} gives the net change in the amplitude of |f> over the course of the collision. Its place is taken by $\Delta\sigma_{C'}$ on the left hand side of Eq.(2), which is defined analogously by

$$tr[|C,\vec{p}_C><C,\vec{p}_C|\rho(t)]^{+\infty}_{-\infty} =$$

$$\int^{+\infty}_{-\infty} dt\ (\partial/\partial t)\ tr[|C,\vec{p}_C><C,\vec{p}_C|\rho(t)] = \Delta\sigma_{C'}(\vec{p}_C)\ . \qquad (4)$$

The first expression for $\Delta\sigma_{C'}$ gives it in terms of asymptotic values of $\rho(t)$ and thus defines it in terms of inital and final state observables. Defining the contribution from the final state as $\sigma_{C'}$, one finds [4]

$$\sigma_{C'}(\vec{p}_C) = \sigma_C(\vec{p}_C) + \textstyle\sum_D \int d\vec{p}_D\ \mathcal{N}_{C/D}(\vec{v}_{CD})\ \sigma_D(\vec{p}_D)\ ;$$

$$\vec{v}_{CD} = \vec{p}_C/M_C - \vec{p}_D/M_D\ . \qquad (5)$$

Here, the sum over D is over all composites larger than C, $\mathcal{N}_{C/D}(\vec{v}_{CD})$ are coefficients calculable from the bound state wave functions of C and D and, assuming ρ describes an incident beam of unit flux, σ_C $(=\partial\sigma_C/\partial\vec{p}_C)$ is the inclusive differential (taken with respect to its momentum) cross–section of C .Thus $\sigma_{C'}$, given in Eq.(2), is not the observed inclusive cross–section for C; rather, it equals σ_C plus contributions from cross–sections for composites which are larger than C. $\sigma_{C'}$ is called the inclusive cross–section for "primordial" C or, for "C–like correlations" (written with a prime). $\mathcal{N}_{C/D}(\vec{v}_{CD})\ d\vec{p}_C$ is the effective number of C' found in a larger composites D with momentum $\vec{p}_D$. Since Eq.(5) gives $\sigma_{C'}$ in terms of experimental cross sections, it serves to define $\sigma_{C'}$ as experimental observables which one compares to values calculated theoretically via Eq.(2).

When Eq.(2) is expressed in the Wigner representation using any one of a number of popular approximations for ρ, simple and physically intuitive formulas emerge. Consider first a Monte Carlo Intranuclear Cascade (INC) calculation of a collision of two nuclei. In this model [8], trajectories $\Phi(t)=\{\vec{X}_a(t),\vec{P}_a(t);\ a=1...A\}$ are generated by allowing nucleons to move freely until two of them reach their point of closest approach. If they are within an area equal to two-nucleon total cross–section, they collide and their momenta discontinuously change with probabilities determined by their differential cross–section. Each cascade γ generates a 6A dimensional phase space trajectory which is simply a sequence of straight line segments. Each Φ^γ is a member of an ensemble $\{\Phi^\gamma ; \gamma = 1...\Gamma\}$ where Γ is the total number of cascades. The Wigner representation of the density [3] is then given by the ensemble average

$$\rho(\varphi,t) = \int dy \exp(ipy) <x-y/2|\rho|x+y/2> \ ; \ x = \vec{x}_1...\vec{x}_A \ ; \quad \text{etc. p, y, dy.}$$

$$\sim (1/\Gamma) \sum_1^\Gamma \delta(\varphi - \Phi^\gamma) = < \delta(\varphi - \Phi^\gamma) >_\gamma \ ;$$

$$\delta(\varphi - \Phi^\gamma) = \prod_1^A h^3 \delta(\vec{x}_a - \vec{X}_a^\gamma(t)) \ \delta(\vec{p}_a - \vec{P}_a^\gamma(t)) \ ; \ h = 2\pi. \tag{6}$$

The physical picture underlying INC is that of a gas of nucleons, dilute enough so that any pair complete a two–body scattering while they remain far from third bodies. INC approximates the time required to complete the two–body scattering by zero. Although this picture is at best marginally correct for cases of interest in nuclear heavy-ion physics, it gives reasonably good results in practice and in any case serves well as a basic paradigm.

The general form given in Eq.(6) can now be inserted into Eq.(2) using any C one wishes. It is instructive to consider the trivial case first – that in which C is not even a composite, but is just a nucleon. When C= proton 'a' with momentum $\vec{p}$,

$$\Delta\sigma_{a'}(\vec{p}) = <\sum_i (\delta(\vec{p} - \vec{P}_{i+}) - \delta(\vec{p} - \vec{P}_{i-}))>_\gamma \ ; \ \vec{P}_{i\pm} = \vec{P}_1^\gamma(t_{1i}^\gamma \pm) \tag{7}$$

where $(t_{ai}^\gamma \pm)$ is the time immediately after/before the 'i'*th* collision of 'a' during the γ*th* cascade. Thus, due to collision 'i' with some other particle, 'a' jumps in momentum from $\vec{P}_{i-}$ to $\vec{P}_{i+}$. Since the momentum in INC is constant between collisions, $\vec{P}_{i-}=\vec{P}_{(i-1)+}$. As a consequence, the sum over 'i' in Eq.(7) reduces to $<\delta(\vec{p}-\vec{P}_{final})-\delta(\vec{p}-\vec{P}_{initial-})>_\gamma$. The term, $<\delta(\vec{p}-\vec{P}_{initial})>_\gamma$, describes the Fermi momentum distribution of 'a' in the initial state and comes from the $t=-\infty$ contribution to $\Delta\sigma_{a'}(\vec{p})$ in Eq.(4). It therefore does not contribute to the cross section. Summing over all 'a' which are protons, Eq.(7) leads to the usual expression for an INC prediction of the summed charge cross–section:

$$\sigma_{\text{summed charge}}(\vec{p}) = \sum_{\text{protons}} <\delta(\vec{p}-\vec{P}_{final})>_\gamma \ . \tag{8}$$

Note the form of Eq.(7) giving $\Delta\sigma_{a'}$ as a sum of changes due to collisions. This is characteristic of all the results obtained from Eq.(2) and is a reflection of the fact (see Eq.(4)) that an integral of a time rate of change is being calculated.

For this case (C=one proton) one sees that Eq.(5) is only correct if $\mathcal{N}_{C/D}(\vec{v}_{CD})d\vec{p}_C$ is indeed the probability for finding C in cell $d\vec{p}_C$ in a composite D with momentum $\vec{p}_D$. In fact, Eq.(5) takes into account the Fermi momentum of particles bound in D whereas the formula usually used for $\sigma_{\text{summed charge}}(\vec{p})$ ignores the Fermi momentum width in which case $\mathcal{N}_{C/D}(\vec{v}_{CD})$ would be proportional to a delta function in $\vec{v}_{CD}$ and

$$\sigma_{C'}(\vec{p}_C) \sim \sigma_C(\vec{p}_C) + \sum_D \int d\vec{p}_D\, \mathcal{N}_{C/D}(\vec{v}_{CD})\, \sigma_D((M_D/M_C)\vec{p}_C)$$

$$= \sigma_C(\vec{p}_C) + \sum_D \mathcal{N}_{C/D}\, (M_D/M_C)^3\, \sigma_D((M_D/M_C)\vec{p}_C)\ ;$$

$$\sigma_{C'}(\vec{p}_C)\, d\vec{p}_C \sim \sigma_C(\vec{p}_C)\, d\vec{p}_C + \sum_D \mathcal{N}_{C/D}\quad \sigma_D(\vec{p}_D)\, d\vec{p}_D\ ;\ \vec{p}_D = (M_D/M_C)\vec{p}_C$$

$$\mathcal{N}_{C/D} = \int d\vec{p}_C\, \mathcal{N}_{C/D}(\vec{v}_{CD}) = \int d\vec{p}_D\, (M_D/M_C)^{-3} \mathcal{N}_{C/D}(\vec{v}_{CD}).$$

When C is a nucleon, $\mathcal{N}_{C/D} = 1$ if it is inside D and 0 otherwise. Applying this to the single proton case and summing over all protons yields,

$$\sigma_{\text{summed charge}}(\vec{p}) = \sigma_{\text{protons}}(\vec{p}) + \sum_D \text{(charge of D)}\ (M_D/M_{\text{proton}})^3\, \sigma_D((M_D/M_{\text{proton}})\vec{p}) \quad (9)$$

which is the usual formula.

Non-trivial $\mathcal{N}_{C/D}$ have been calculated only for the following cases: $C=H^2$, $D=H^3$ or He^3, $\mathcal{N}_{C/D}=(3/2)(.88)=1.32$; $C=H^2$, $D=H^4$, $\mathcal{N}_{C/D}=(3)(.75)=2.25$; $C=H^3$, $D=H^4$, $\mathcal{N}_{C/D}=(3/4)(.97)$. The first factors[4,9] come from spin–isospin overlaps while the second [10]are corrections due to spatial overlaps integrals. These coefficients take into account all the different combinations of nucleons in D which can form a C–like correlation.

The generalization of Eq.(7) to C which are composite reads as follows,

$$\Delta\sigma_{C'}(\vec{p}) = [(2S+1)/2^N]\, \sum_{a1...aN} \langle \sum_i \left(\delta(\vec{p}-\vec{P}_{i+})\, W_C(\Phi_{i+}) - \delta(\vec{p}-\vec{P}_{i-})\, W_C(\Phi_{i-}) \right) \rangle_\gamma . \quad (10)$$

The first term on the right hand side is a statistical spin factor; S is the spin of C and there are N nucleons in it. This factor is the effective fraction of spin states of the N nucleons which contribute to the possible spin states of C. The $\sum_{a1...aN}$ means that one is to sum over all subsets of N nucleons which can possible form C. For example, if C were a Deuteron, the sum would extend over all possible neutron–proton pairs. We now pick a cascade γ and look at the sequence of collisions this N–tuple makes with the rest of the system. Any collision an N–tuple member makes with a nucleon not in the N–tuple counts as a collision 'i'. The total momentum of the N–tuple after/before the 'i'*th* collision is $\vec{P}_{i+/-}$. The function W_C is defined via

$$\delta(\vec{p} - \vec{p}_C)\ W_C(\varphi) = \int dy\ \exp(ipy)\ \langle x-y/2|\ C,\vec{p}_C\rangle\langle C,\vec{p}_C|x+y/2\rangle\ . \tag{11}$$

Thus, $W_C(\varphi)$ is the Wigner representation of the spatial bound state wave function of C and depends only on relative positions and momenta (assuming for simplicity that spin and space are decoupled). The term $W_C(\Phi_{i+})$ in Eq.(10) is therefore the value of W_C immediately after the 'i'*th* collision and differs from $W_C(\Phi_{i-})$ only because the momentum of one of the nucleons a1...aN has changed due to scattering by an external particle. Positions do not change from before to after. Positions do change between consecutive 'i' whereas momenta do not. Finally, the first negative term in the sum over 'i' must be deleted for the same reason as before. Note that because of the W_C factors, there is no longer exact cancellation by pairs in this oscillating sum. In certain cases however [5], there is approximate cancellation in which case only the last term survives as in Eq.(8). This is a significant computational simplification.

Eq.(10) may be interpreted as follows. Each interaction 'i' of the N–tuple results in an "annihilation" of (their probability of being) a C' (=primordial C or C–like correlation) with momentum $\vec{p}=\vec{P}_{i-}$ and a "creation" of a C' with momentum $\vec{p}=\vec{P}_{i+}$. The probability that before/after collision they could be considered as forming a C' is determined by the relative positions in phase space before/after via the value of W_C. The W_C factor ensures that nucleons which are far apart on the scale set by the bound state wave function of C do not contribute.

So far, Eq.(10) has been used [5] to calculate deuteron production in nuclear heavy ions collisions, but has not been used for heavier nuclei because of the large amount computer time needed to do the sum over all possible N–tuples in a collision. For such cases, further approximations are needed.

One way to improve on the INC model is to include various mean field effects. For example, between relatively short-range two-body collisions nucleons can feel the long-range part of the interaction with other nucleons. This has been taken into account by adding motion between collisions due to a classical force in the INC calculation [11]. Another model which has been extensively used in both nuclear and molecular physics [12] is pure classical dynamics (no stochastic forces). Eq.(10) can be generalized to take care of such models by simply noting that in numerical calculation, the effect of a classical force is just a succession of closely spaced deterministic momentum impulses. These can be therefore taken into account in the formula in exactly the same way as the stochastic impulses generated in an INC calculation. It is also possible to obtain an analytic expression Eq.(10) directly in terms of forces [4]. It is worth noting that although classical trajectory models can exhibit clustering in the final state and therefore directly predict composite production cross sections, the quantum mechanical formula derived here should give intrinsically more accurate results.

Eq.(2) can also be applied directly to single-particle distributions which may be obtained either from direct solutions to a Boltzmann equation or indirectly from fluid dynamics [6]. In this case one finds

$$\sigma_{C'}(\vec{p}) \sim [(2S+1)/2^N]\,[\mathrm{Comb}] \int d\varphi\, \delta(p-\textstyle\sum_1^N \vec{p}_n)\; W_C(\varphi)\times$$

$$[\partial/\partial t - \textstyle\sum_1^N \vec{v}_n\cdot\partial/\partial\vec{x}_n\,]\prod_1^N [a(\varphi_n,t)/A]\;;$$

$$\varphi_n = \vec{x}_n, \vec{p}_n\;;\; d\varphi = \textstyle\prod_1^N d\vec{x}_n d\vec{p}_n/h^3\,. \qquad (12)$$

Here, [Comb] stands for an isospin combinatorial factor giving the number of ways neutrons and protons can be picked out of the A nucleons in the system to form a composite C, and, $a(\varphi_n,t)$ is the time-dependent number distribution function in phase space. The term $[\partial/\partial t-\sum_1^N\vec{v}_n\cdot\partial/\partial\vec{x}_n\,]\prod_1^N[a(\varphi_n,t)/A]$, which is the rate of change of the distribution due to forces on the particles, is essentially the collision term in a Boltzmann equation. Thus, Eq.(12) says that changes in the primordial C number are due to the rate at which collisions change the occupancy of the portion of phase space defined by $\delta(p-\sum_1^N\vec{p}_n)\; W_C(\varphi)$. One can show [4] that Eq.(12) is equivalent to Eq.(10) in the absence of correlations between particles. Eq.(12) when applied to fluid dynamics can be shown to lead to the interesting conclusion that net composite production rate in a fluid is proportional to its rate of expansion.

In summary, we see a number of ways in which the phase space representation of basic quantum reaction theory gives rise to new views of complex production processes. These have proven useful in nuclear heavy-ion physics. Their application to molecular processes seems to be an interesting area for future development.

This work was supported in part by the National Science Foundation.

REFERENCES

[1] S.Nagamiya and M.Gyulassy, *High-Energy Nuclear Collisions* Advances in Nuclear Physics Vol.13 (1984) 201. (Plenum Pub. Co., Ed.J.W.Negele and E.Vogt)

[2] M.Goldberger and K.Watson, *Collision Theory*,J. Wiley & Sons, N.Y., 1964.

[3] E.A.Remler, *Ann.Phys.N.Y.***95**(1975)455.

[4] E.A.Remler, *Ann.Phys.N.Y.***136**(1981)293.

[5] M.Gyulassy, K.Frankel and E.A.Remler, *Nuc.Phys.***A296**(1983)596

[6] E.A.Remler, *Phys.Rev.***C25**(1982)2974.

[7] E.A.Remler, Phys.Lett.**B159**(1985)81.

[8] J.Cugnon, *Phys.Rev.***C22**(1980)1885.

[9] G.Bertsch and J.Cugnon, *Phys.Rev.***C25**(1982)2514.

[10]E.A.Remler,*Composite Particle Cross Sections from the Density Operator II*,(unpublished).

[11]J.Aichelin and G.Bertsch, *Phys.Rev.***C31**(1985)1730.
H.Kruse, B.V.Jacak, J.J.Molitoris, G.D.Westfall, and H.Stocker, *Phys.Rev.***C31**(1985)1770.

[12]A.R.Bodmer and C.N.Panos, *Nuc.Phys.***A356**(1981)517.

THE GEOMETRY OF WIGNER'S FUNCTION

N. L. Balazs
State University of New York at Stony Brook
Stony Brook, L.I., N.Y. 11733

I. INTRODUCTION

Consider a one-dimensional system described by a Hamiltonian operator $\hat{H}$. The state of the system can be characterized either by a density operator $\hat{\varrho}$, or by Wigner's function $f(p,q) = \mathrm{Trace}(\hat{\Delta}(p,q)\hat{\varrho})$, $\hat{\Delta}(p,q) = h^{-1}\int du\,dv \exp[iu(\hat{p}-p)+iv(\hat{q}-q)]$.

What are the geometrical properties of the (p,q) manifold in which the function f is displayed ?

II. THE DETERMINATION OF THE INVARIANCE GROUP

In order to answer this question we must study the transformations of the p,q labels, and find those transformations which transform physically realizable state into physically realizable states.

If Planck's constant is zero, this manifold becomes the classical phase-space, with p and q being the canonical momenta and coordinates. Thus, a point characterizes a physically realizable state, or equivalently, the phase-space density $\delta(p-p_1)\delta(q-q_1)$ describes a physically realizable state. The motion of a phase point is described by the Hamiltonian equations of motion, and the allowed transformations are the canonical transformations which preserve these equations of motion. The phase space is not a metric space, there is no meaning attached to the separation between two general phase points.

If Planck's constant ceases to be zero, the manifold is no longer a phase space; we will call it a mock phase space. A point can no longer specify a physically realizable state. In fact, $f=\delta(p-p_1)\delta(q-q_1)$ is not an admissible Wigner function, since the associated density operator has negative eigenvalues. On the other hand, the functions $\delta(p-p_1)$ or $\delta(q-q_1)$ are permitted Wigner functions. Their singularities are spread out on lines. On can show[1] that Wigner functions which are delta functions can only be singular on lines if they are to represent physically realizable states; i.e., they must be of the form $\delta(ap+bq-c)$ with a,b,c constant. These functions describe the "thinnest" states possible. We may erect Wigner functions based on any curve, in the sense that the function is large only on a strip around this curve, while the associated density operator is still positive. Thus the width of physically realizable Wigner functions can be used to distinguish straight lines from curves in the mock phase space. Consequently, only those transformations are permitted which preserve straightness, i.e., the linear inhomogeneous transformations. We deal with an affine geometry.

III. AFFINE GEOMETRY

A point (or vector) x has coordinates $x_1 = p$, $x_2 = q$. (Notice that the components carry different physical dimensions; the vector x has no definite physical dimension.) Let x(t) be a curve, and t a parameter. The affine separation ds between the points x(t + dt) and x(t) on the curve is defined as

$$ds = \|(\dot{x} \wedge \ddot{x})\|^{1/3} dt ,$$

where $(\dot{x} \wedge \ddot{x}) = \begin{vmatrix} \dot{x}_1 & \dot{x}_2 \\ \ddot{x}_1 & \ddot{x}_2 \end{vmatrix} = \dot{x}_1\ddot{x}_2 - \dot{x}_2\ddot{x}_1$. The quantity ds is additive and invariant under the required transformations, hence qualifies as a separation. If the curve is straight, ds is zero. Its dimension is $h^{1/3}$. Having obtained ds we can now find geodesics, intrinsic equations of a curve, etc. However, this ds contains second derivatives as well, thus the geometry is not Riemannian. (For example, there are an infinite number of geodesics connecting two points !)

Identify now t with s, and denote the derivatives with respect to s by a prime. Then $x' \wedge x'' = 1$, and redifferentiating, we find that $x''' + k(s)x' = 0$. Here $k(s) = x'' \wedge x'''$ is the affine curvature; x is the tangent vector, and x turns out to be the affine normal. If $ds \neq 0$. x', x'' are never proportional, thus form a natural set of basis vectors. (In the Appendix we show the geometrical meaning of ds and the affine normal.)

The separation along a curve is given by s; s is proportional to $h^{1/3}$. The distance from the point $x_A = x(s_A)$ on a curve to a point x_B on the affine normal issuing from x_A is given by $D_{AB}\, x\, (s_A) = x_B - x_A$. Thus D has the dimension $h^{2/3}$. A phase cell with edges $s = h^{1/3}$ and $D = h^{2/3}$ has area h.

IV. WIGNER'S FUNCTION REPARAMETRIZED

Consider a bound state of the Hamiltonian $\hat{H}$, with energy E. Classically, this would correspond to a phase space density $\delta(H - E)$, with H being the classical Hamiltonian. The associated Wigner function has lost its singularity and is large in a ribbon like region around the curve H = E. In terms of s and $\mathcal{D}$ the semiclassical approximation to Wigner's function is very simple. Transcribing Berry's expression[2] we find

$$f(p,q) \sim \frac{1}{\sqrt{\pi}\hbar} \frac{Ai(-2\mathcal{D}/\hbar^{2/3})}{(1/\omega)(ds/dt)(1/\hbar^{1/3})} .$$

Here ω is the circular frequency of the bound motion at energy E. Let p(s), q(s) be the point on the H = E curve , such that the affine normal $x''(s)$ issuing from this point continues in a line passing through the point p,q which is the argument of f(p,q), then $\mathcal{D}\cdot x''(s) = (p - p(s), q - q(s))$; $x''(s)$ is taken positive on the concave side of the curve. Thus the ribbon alluded to has a width $h^{2/3}$ measured along the affine normal. We also see that f is oscillatory on the concave side of H = E, and is decaying exponentially on the other side.

The complicated expression originally obtained by Berry had that appearance because the simple scalar argument with a simple geometrical meaning was written out long-hand in a particular coordinate system.

We see that the distance and separation between points within <u>one</u> stationary state can be satisfactorily interpreted in this way. May we use these notions to describe a distance, or separation between <u>two</u> stationary states ?

V. THE DISTANCE BETWEEN TWO STATIONARY STATES

Classically no invariant meaning can be attached to the distance between two points, or states in the phase space. (This is the ultimate reason why so many classical definitons of stability exist!)

In quantum mechanics the distance between two states ψ_A, ψ_B is $\|\psi_A - \psi_B\| = \{2[1 - \mathrm{Re}\,p(\psi_A \cdot \psi_B)]\}^{1/2}$. Thus, the scalar product of the states determine the distance. In terms of the associated Wigner functions f_A and f_B this scalar product is given as $\iint dp dq\, f_A f_B$.

Consider two particularly simple states. ψ_A is the non normalizable stationary state of $\hat{H}_A = (\hat{p} - a)^2/2m + F\hat{q}$; ψ_B is a **stationary** state of $\hat{H}_B = (\hat{p} - a)^2/2m - F\hat{q}$. The eigenvalues have the same value E. In the p representation the eigenfunctions are: $\psi_A \sim \exp(-iEp/F\hbar + (p-a)^3/(6mF\hbar))$; ψ_B is the same with the sign of F and a reversed. The associated Wigner functions are given by Eq. (1) with the ω factor omitted. The two curves $H_A = E$, $H_B = E$ are shown in Figure 1.

The scalar product of the two wavefunctions can be easily evaluated, and we find that it is proportional to $Ai(a^2-2mE)/(mF\hbar^{2/3})$. One can check that the argument of the Airy function is just $D/\hbar^{2/3}$, where D is the least separation along the common affine normal between the two curves $H_A = E$ and $H_B = E$. This is the separation along the q axis. (In this geometry a parabola is a geodesic. The axis of a parabola must be an affine normal direction by symmetry; but normals to a geodesic must be all parallel to each other; thus all affine geodesics must be lines parallel to the q axis). Again, by symmetry, the least separation must be on the q = 0 line. The separation between these two states can then be also interpreted in accordance with our geometrical picture.

VI. THE LOSS OF THE METRIC IN THE CLASSICAL LIMIT

As h goes to zero the width of the ribbon decreases. When h is zero the width disappears, and with it our ability to distinguish straightness . The metric introduced is thus no longer relevant, and we can no longer use it. The canonical transformations will now play their usual role. They do not preserve straightness. Indeed, the transformation which leads to action and angle variables is precisely that one which "straightens" the phase space trajectories. We see here the geometrical reason why action and angle variables play such a role in the passage toward quantum theory.

APPENDIX

The line element ds

Take a curve x(t) and two points on it, $x_a = x(t_a)$ and $x_b = x(t_b)$. Construct the triangle x_a, x_b, z from the tangents and the chord; (z is the intersection of the two tangents). See Figure 2. The area of this triangle is $a = (1/2)\ (z-x_a) \wedge (x_b - x_a)$. Since $z = x_a + \alpha \dot{x}_a = x_b - \beta \dot{x}_b$, we find $z - x_a = \alpha \dot{x}_a$, $x_B - x_A =$

$\alpha \dot{x}_a + \beta \dot{x}_b$, or $(x_b - x_a) \wedge \dot{x}_b = \alpha \dot{x}_a \wedge \dot{x}_b$. Eliminating $z - x_a$ and α we obtain

$$\text{area} = (1/2) \frac{\dot{x}_a \wedge (x_b - x_a)(x_b - x_a) \wedge \dot{x}_b}{\dot{x}_a \wedge \dot{x}_b}.$$

Expand x_a and x_b in powers of $\tau = t_b - t_a$ and substitute. We immediately get the area as $(\dot{x}_a \wedge \ddot{x}_a)\tau^3/8$. If we wish to use this area to produce an additive measure we must extract a cube root to make it proportional to τ. This way we find ds to be proportional to area$^{1/3}$. Since an area is invariant under linear inhomogeneous transformations, and ds is obviously additive, we found our line element.

The affine normal

x (s) is the affine normal. Geometrically we construct it as follows. Take the point x(s) on the curve, and draw the tangent. Then draw a parallel to the tangent and construct a chord . Bisect it.(This can be done by symmetry, so no distance measure is needed). Draw a line through the bisecting point and the point x(s). As the chord approaches the tangent, this line approaches the affine normal direction.

ACKNOWLEDGEMENT

I wish to thank the Organizing Committee for their kind invitation, and the National Science Foundation for their financial support.

REFERENCES

[1]N.L. Balazs, Physica 102A, (1980) 236

[2]M.V. Berry, Phil. Trans. Roy. Soc. 287,(1977) 237

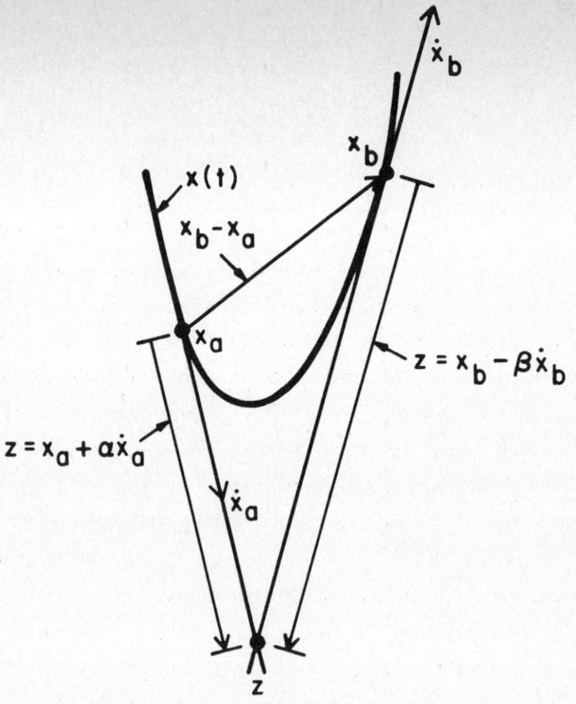

Fig.1. Area construction used to express separation between two points on a curve.

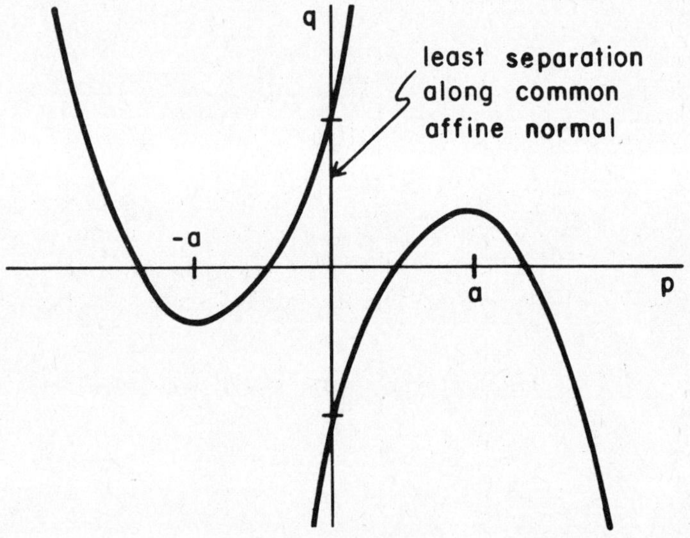

Fig.2. Least distance along affine normal between two states erected over parabolae.

DISTRIBUTION FUNCTIONS IN ELEMENTARY PARTICLE PHYSICS

F. Zachariasen
California Institute of Technology
Pasadena, CA 91107

This talk is a very brief overview of the prospects for the use of distribution functions and transport methods in the theory of elementary particles. There are (at least) two areas in which such methods show promise—multiparticle production in high-energy collisions, and the theory of quark-gluon plasmas. In both of these, ideas borrowed from classical hydrodynamics have been extensively used, but without any direct justification from the underlying quantum field theory.

To provide this justification has so far been impossible, not only because of the intrinsic difficulty of dealing in a practical way with field theory, but also because some of the conceptual ideas about how to formulate hydrodynamics in a field theory context are still missing. But there is hope that these problems can be overcome, and if they can, a transport approach to at least some problems in elementary particle physics should become a very valuable asset.

Much attention is paid in particle physics to inclusive reactions, in which some observed set of particles is produced together with any number of unobserved particles in a high energy proton-proton (or proton-antiproton) collision. Labelling the particles by their four momenta, one writes such reactions as

$$p + p' \to q_1 + q_2 + \cdots + q_n + X : \tag{1}$$

p and p' are the incident protons, $q_1 \ldots q_n$ are the observed particles (mostly pions) and X is anything.

The transition amplitude to any particular out-state $\langle X^{(-)}|$ for this reaction can be written

$$\begin{aligned} S_{fi} = (-i)^n \int d^4x_1 \cdots \int d^4x_n \frac{e^{iq_1x_1}}{\sqrt{2w_1}} \cdots \frac{e^{iq_nx_n}}{\sqrt{2w_n}} \\ (\Box_{x_1} + \mu^2) \ldots (\Box_{x_n} + \mu^2) \langle X^{(-)}| (\phi(x_1) \cdots \phi(x_n))_+ |pp'^{(+)}\rangle. \end{aligned} \tag{2}$$

Here $\phi(x)$ is the field operator for the produced particle (taken as spinless here for simplicity), $w = \sqrt{\vec{q}^2 + \mu^2}$ is its energy when it has three-momentum $\vec{q}$ and mass μ, so that its four-momentum $q = (w, \vec{q})$, and $\Box$ means $\partial^2/\partial t^2 - \vec{\nabla}^2$. The in-state $|pp'^{(+)}\rangle$ represents the two incident protons and $()_+$ is the time-ordering symbol. The probability for the reaction, when X is unobserved, is then

$$\begin{aligned} &\int d^4x_1 \cdots \int d^4x_n \int d^4y_1 \cdots d^4y_n \frac{e^{iq_1x_1}}{\sqrt{2w_1}} \cdots \frac{e^{iq_1x_n}}{\sqrt{2w_n}} \\ &\frac{e^{-iq_1y_1}}{\sqrt{2w_1}} \cdots \frac{e^{-iq_1y_n}}{\sqrt{2w_n}} (\Box_{x_1} + \mu^2) \cdots (\Box_{x_n} + \mu^2) \\ &(\Box_{y_1} + \mu^2) \cdots (\Box_{y_n} + \mu^2) \langle pp'^{(+)}| (\phi(x_1) \cdots \phi(x_n)) \\ &+ (\phi^*(y_1) \cdots \phi^*(y_n)) + pp'^{(+)}\rangle. \end{aligned} \tag{3}$$

Let us define $r_i = x_i - y_i$, $R_i = \frac{x_i+y_i}{2}$ and the functions

$$\begin{aligned}\tilde{F}_n(q_1, R_1, \ldots, q_n, R_n) = \int d^4r_1 \ldots \int d^4r_n e^{iq_1r_1} \ldots e^{iq_nr_n} \\ (\Box_{R_1-r_1/2} + \mu^2) \ldots (\Box_{R_n-r_n/2} + \mu^2)(\Box_{R_1+r_1/2} + \mu^2) \\ \ldots (\Box_{R_n+r_n/2} + \mu^2)\langle pp'^{(-)}|(\phi(R_1 - r_1/2) \ldots \phi(R_n - r_n/2)) \\ + (\phi^*(R_1 + r_1/2) \ldots \phi^*(R_n + r_n/2)) + |pp'^{(-)}\rangle,\end{aligned} \tag{4}$$

where the q_i are arbitrary four vectors, not on the mass shell. Then the transition probability is

$$\frac{1}{2w_1} \cdots \frac{1}{2w_n} \int d^4R_1 \ldots \int d^4R_4 \tilde{F}(q_1, R_1 \ldots q_n, R_n)|_{q_i^2=\mu^2}. \tag{5}$$

Evidently $\tilde{F}_n$ can be interpreted as a phase space distribution function. The probability to observe a particle of momentum $\vec{q}$ and energy w produced in a collision of two protons is

$$\frac{1}{2w} \int d^3\vec{R} \int dt \tilde{F}_1(\vec{q}, w; \vec{R}, t), \tag{6}$$

and therefore $\tilde{F}(q, R)$ is the probability to find one particle of four-momentum q (off shell) at position $\vec{R}$ at time t. Note that for this interpretation to be meaningful, the incident state $|pp'^{(+)}\rangle$ must not be a momentum eigenstate, it must be a wave packet. Otherwise translation invariance of the fields, $\phi(x) = e^{iPx}\phi(0)e^{-iPx}$ where P is the total energy momentum operator, would make $\tilde{F}$ independent of R.

The distribution function interpretation can be strengthened by defining a function without "the external legs amputated":

$$\tilde{F}_n(q_1, R_1, \ldots q_n, R_n) \equiv (q_1^2 - \mu^2) \ldots (q_n^2 - \mu^2) F_n(q_1, R_1, \ldots q_n, R_n). \tag{7}$$

Then

$$\begin{aligned}F_n(q_1, R_1, \ldots q_n, R_n) = \int d^4r_1 \ldots \int d^r r_n e^{iq_1r_1} \ldots e^{iq_nr_n} \\ \langle pp'^{(+)}|(\phi(R_1 - r_1/2) \ldots)_+ (\phi^*(R_1 + r_1/2) \ldots)_+ |pp'^{(+)}\rangle.\end{aligned} \tag{8}$$

This is just the n-particle relativistic off-shell Wigner function. (It has, as (7) shows, poles in each q^2 at its mass-shell value.) .

The F_n satisfy relativistic transport equations. For example,

$$q_\mu \partial/\partial R_\mu F_1(q, R) = \frac{1}{2i} \int d^4r e^{iqr} \langle pp'^{(+)}|j(R-r/2)\phi^*(R+r/2) - \phi(R-r/2)j^*(R+r/2)|pp'^{(+)}\rangle, \tag{9}$$

where $j \equiv (\Box + \mu^2)\phi$. The operator $q_\mu \partial/\partial R_\mu = q_0(\partial/\partial t + \vec{q}/q_0 \cdot \vec{\nabla})$ is just the generalization of the usual $\partial/\partial t + \vec{v} \cdot \nabla$; $\vec{q}/q_0 = \vec{q}/w$ is the relativistic velocity when q is on shell. The right-hand-side depends on the interactions among the particles, since $j = (\partial/\partial\phi)\mathcal{L}_{INT}$ where $\mathcal{L}_{INT}$ is the interaction Lagrangian. For example, in a ϕ^4 field theory, $j = \lambda(\phi^*\phi)\phi$, and in a γ_5 meson theory, $j = g\bar{\psi}\gamma_5\psi$. The RHS is thus the relativistic field theory analog of the Boltzmann collision term.

The interpretation of all this is quite attractive, but is it useful? Can one, using the intuition that analogies with conventional distribution functions and transport theory give, calculate anything via this approach? Evidently, as is nearly always true in field theory, exact calculations are impossible and one must resort to approximations.

One such is the field theory analog of the Hartree approximation. For example, in the ϕ^4 theory one replaces $\langle\phi^*(R-r/2)\phi(R-r/2)\phi^*(R+r/2)\phi(R+r/2)\rangle$ by $\langle\phi^*(R-r/2)\phi(R+r/2)\rangle$ $\langle\phi^*(R-r/2)\phi(R-r/2)\rangle$. Then (9) becomes

$$q_\mu \frac{\partial}{\partial_\mu} F(q,R) = \frac{\lambda}{2i} \int d^4r \int d^4q' F(q',R) e^{i(q-q')r} \int d^4q'' [F(q'',R-r/2) - F(q'',R+r/2)]. \quad (10)$$

This is now a closed Virasoro-like equation for the one-particle distribution.

Approximations such as this should be valid if correlations are weak, which is not far from being the case experimentally in high-energy collisions. They can exhibit collective modes, which are perhaps responsible for some of the observed clustering in particle production.

Another possible use of the distribution function and transport equation is to derive hydrodynamics. Relativistic hydrodynamic models of high-energy collisions have existed for years, but they have up to now largely been written down from intuition, rather than derived. As a result, the assumptions, such as local equilibrium, on which they are based remain unjustified.

As usual, we define a particle density by

$$\begin{aligned} n(R) &= \int \frac{d^4q}{(2\pi)^4} F(q,R) \\ &= \langle\psi|\phi^2(R)|\psi\rangle. \end{aligned} \quad (11)$$

(We have now generalized the definition of F by replacing the incident state $|pp'^{(+)}\rangle$ by any state $|\psi\rangle$, to allow ourselves the option of using the transport language for a wider class of problems.) We also define a current density

$$\begin{aligned} j_\mu(R) &= \int \frac{d^4q}{(2\pi)^4} q_\mu F(p,R) \\ &= i\langle\psi|\phi(R)\partial_\mu\phi^*(R) - \phi^*(R)\partial_\mu\phi(R)|\psi\rangle. \end{aligned} \quad (12)$$

For a neutral field (ϕ real) j_μ of course vanishes. Evidently we have current conservation

$$\partial_\mu j_\mu(R) = 0. \quad (13)$$

The choice of a stress tensor must depend on the dynamics. For example, for the $\lambda\phi^4$ theory, we define

$$T_{\mu\nu}(R) = \int \frac{d^4q}{(2\pi)^4} q_\mu q_\nu F(q,R) + \delta_{\mu\nu}\lambda \int \frac{d^4q}{(2\pi)^4} \int \frac{d^4q'}{(2\pi)^4} F_2(q,R,q',R) - \tfrac{1}{2}(\delta_{\mu\nu}\Box - \partial_\mu\partial_\nu)n(R). \quad (14)$$

This stress tensor satisfies the relativistic hydrodynamic equations of motion

$$\partial_\mu T_{\mu\nu}(R) = 0 \quad (15)$$

and furthermore its trace gives a kind of equation of state:

$$T_{\mu\mu}(R) = \mu^2 n(R), \quad (16)$$

where, we recall, μ is the mass of the field ϕ.

If we had a perfect fluid, then $T_{\mu\nu} = (\epsilon + p)u_\mu u_\nu - \delta_{\mu\nu}p$, where ϵ is the energy density, p is the pressure, and u_μ is the fluid velocity ($u_\mu u_\mu = 1$). In that case (16) reads

$$\epsilon - 3p = \mu^2 n. \tag{17}$$

In general, however, our field theoretic $T_{\mu\nu}$ does not describe a perfect fluid, and therefore has ten independent components. Equations (15) and (16) constitute only five equations; six more are therefore needed to close the system and determine both $T_{\mu\nu}$ and n. What these should be is unclear. In classical physics, for a perfect fluid the remaining dynamical equation needed to close the system is mass conservation. In a field theory, where particle production and annihilation takes place, there obviously is no such condition.

Ideally one eventually wants to apply this thinking to QCD. There the hydrodynamic equations will become a set of "color electro-magneto hydrodynamic" equations describing, for example, a quark-gluon plasma. There are, however, further difficulties with extending the formalism to gauge field theories. For example, in ordinary electrodynamics, the obvious definition of the distribution function is

$$F_{\mu\nu}(q, R) = \int d^4 r e^{iqr} \langle \psi | A_\mu(R - r/2) A_\nu(R + r/2) | \psi \rangle, \tag{18}$$

where A_μ is the vector potential, or photon field. But this F is not gauge invariant.

In QED one can construct a gauge invariant distribution function by using the field tensor instead of the vector potential:

$$F_{\mu\nu,\lambda\sigma}(q, R) \equiv \int d^r r e^{iqr} \langle G_{\mu\nu}(q, R - r/2) G_{\lambda\sigma}(q, R + r/2) \rangle \tag{19}$$

where

$$G_{\mu\nu} = \partial_\mu A_\nu - \partial_n u A_\mu. \tag{20}$$

In non-Abelian gauge theories, this fails since the field tensor is itself gauge dependent. The only recourse is to introduce complexities such as path-ordered exponentials:

$$F_{\mu\nu,\lambda\sigma}(q, R) = \int d^4 r e^{iqr} \langle F^a_{\mu\nu}(q, R - r/2) P \left(e^{\int_{R+r/2}^{R-r/2} A^b_\alpha(l) t^b dl_\alpha} \right) F^a_{\lambda\sigma}(q, R + r/2) \rangle. \tag{21}$$

This is now gauge invariant, but probably too complicated to be very useful. This whole question remains unresolved.

Obviously, to make this whole approach useful in the QCD context requires much work and much more understanding. But if it could succeed, it might give us a real dynamical formalism for dealing with quark-gluon plasmas, quark matter, etc., as well as with high energy collisions. So it is certainly worth pursuing.

SINGLE AND MULTIPARTICLE WIGNER DISTRIBUTIONS IN INHOMOGENEOUS FERMI SYSTEMS

M. Durand, P. Schuck and E. Suraud
ISN, 53 Avenue des Martyrs
F - 38026 Grenoble Cédex
and R. Hasse, GSI, Postfach 110541, D - 6100 Darmstadt 11

We here give some examples of quantities describing properties of nuclear matter or of heavy nuclei, which are calculated in a semi-classical way in the phase space. They are : the single particle Wigner distribution, the multiparticle-multihole level densities, and the in-medium two-particle Green's function.

I - The single particle Wigner distribution

It can be expanded by means of distributions :

$$\hat{\rho}_W = \left[\theta(\varepsilon_F - \hat{H})\right]_W$$

$$= \theta(\varepsilon_F - H_c) + \hbar^2 \left[A_1 \delta'(\varepsilon_F - H_c) + A_2\, \delta''(\varepsilon_F - H_c)\right] + \mathcal{O}(\hbar^4), \qquad (1)$$

where W stands for the Wigner transform
θ is the step function
$\hat{H}$ is the hamiltonian operator, whose classical counterpart H_c contains the potential V,
ε_F is the Fermi energy
A_1 and A_2 contain derivatives of the potential.

In the case of a ramp potential $V(\vec{r},\vec{p}) = a\,x$, the exact solution is known :

$$\hat{\rho}_W = \int_{x_o}^{\infty} Ai(x)\, dx \ ; \quad x_o = \left(\frac{8m}{\hbar^2 a^2}\right)^{1/3} (\varepsilon_F - H_c), \qquad (2)$$

and the semi-classical approximation reads

$$\hat{\rho}_W = \theta(\varepsilon_F - H_c) + \frac{\hbar^2 a^2}{24m}\, \delta''\,(\varepsilon_F - H_c) + \mathcal{O}(\hbar^4), \qquad (3)$$

an expansion which resembles the low temperature expansion :

$$\lim_{T\to 0} \left[1 + \exp\frac{E - \varepsilon_F}{T}\right]^{-1} = \theta(\varepsilon_F - E) + \frac{\pi^2 T^2}{6}\, \delta'\,(\varepsilon_F - E) + \mathcal{O}(T^4), \qquad (4)$$

known to be a good approximation if $T \ll \varepsilon_F$.

Therefore, (3) is an asymptotic expansion for the Airy integral Eq. (2) accounting for the fall-off width and partially for the oscillations, with $\hbar^2$ playing the role of T. More generally, (1) is an asymptotic expansion for the Wigner density which is no longer exactly known, but which can be approximately reconstructed by a partial $\hbar$ - resummation,[1] and which is valid if the diffuseness of the Wigner function is small as compared to the Fermi energy.

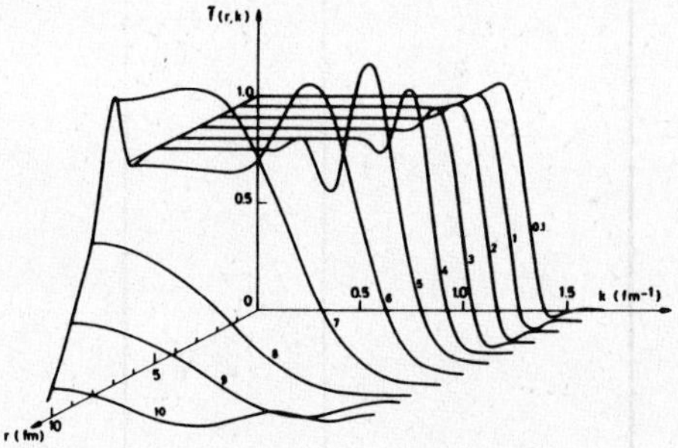

Fig. 1

Fig. 1 shows the semi-classical Wigner function[2], averaged over the directions of $\vec{p}$ for a heavy model nucleus (N=Z=112 in a spherical potential of Fermi shape). The nuclear shell-effects are washed out by the semi-classical approximation and only some of the Friedel oscillations survive.

II - The multiparticle-multihole level densities

Semi-classical np-nh level densities represent the average part of the corresponding exact quantities and they are meaningful because of the high density of states with high excitation energies. They are of importance in a number of physical problems.

By definitions of the single-particle, 1p-1h, and 2p-2h level densities:

$$g_{1p}(E) = \sum_{p} \delta(E-\varepsilon_p) \tag{5a}$$

$$g_{1p,1h}(E) = \sum_{p,h} \delta(E-\varepsilon_p+\varepsilon_h) \tag{5b}$$

$$g_{2p,2h}(E) = \sum_{p_1<p_2,h_1<h_2} \delta(E-\varepsilon_{p_1}-\varepsilon_{p_2}+\varepsilon_{h_1}+\varepsilon_{h_2}), \tag{5c}$$

where ε_i are the eigenenergies for the single-particle hamiltonian $\hat{H}$. (5a-b-c) may be written in a representation-independant way (5c is skipped because too lengthy):

$$g_{1p}(E) = \mathrm{Tr}\,\{\delta(E-\hat{H})\} \tag{6a}$$

$$g_{1p,1h}(E) = \mathrm{Tr}_1\,\mathrm{Tr}_2\,\{\delta(E-\hat{H}_1+\hat{H}_2)\,\theta(\varepsilon_F-\hat{H}_2)\,\theta(\hat{H}_1-\varepsilon_F)\}. \tag{6b}$$

In the Thomas-Fermi approximation, $\hat{H}$ is simply replaced by H_c and the traces are replaced by integrations over phase space.

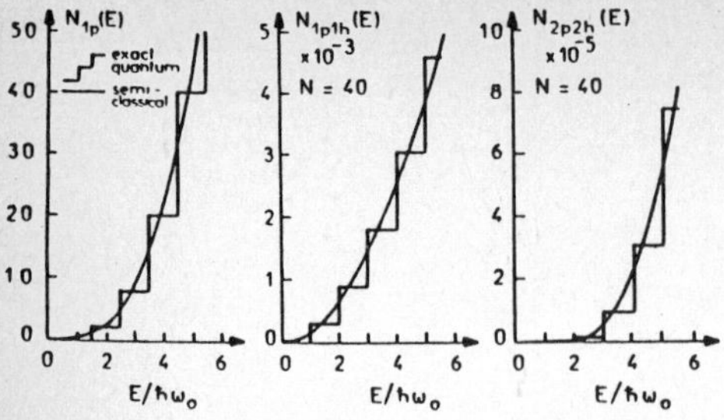

Fig. 2

Calculations have been performed for the spherical harmonic oscillator, for comparison with the exact solution[3]. Fig. 2 shows the number of states :

$$N_i = \int_0^E g_i(E')\, dE' \text{ for } N = 40$$

particles for the three cases (5a-b-c).

The semi-classical curves pass through the average of the exact step-like exact result.

III - The Green's function

As an example, the Brueckner T matrix uses the two-particle Green's function $\hat{G}_{12}$ and the two-nucleon interaction $\hat{v}_{12}$,

$$\hat{T}_{12} = \hat{v}_{12} + \hat{v}_{12}\, \hat{G}_{12}\, \hat{v}_{12} \tag{7}$$

with $\hat{G}_{12} = \hat{G}^o_{12} + \hat{G}^o_{12}\, \hat{v}_{12}\, \hat{G}_{12}$. (8)

The nucleons with total energy E scatter from states below the Fermi energy ($E < 2\,\varepsilon_F$) into states above the Fermi level, then :

$$G^o_{12}(\vec{p}_1,\vec{p}_2) = \Theta(p_1-p_F)\,\Theta(p_2-p_F)\,\left[E-(p_1^2+p_2^2)/2m\right]^{-1} \tag{9}$$

$\hat{G}_{12}$ is evaluated semi-classically in the phase-space of both nucleons, $\hat{G}_{12} \to G(\vec{r}_1,\vec{p}_1;\vec{r}_2,\vec{p}_2) = G_W$ to lowest order in $\hbar$:

$$G_W = G^o_{12}(\vec{p}_1,\vec{p}_2)\cdot\left[1-G^o_{12}(\vec{p}_1,\vec{p}_2)\,v(\vec{r}_1,\vec{p}_1;\vec{r}_2,\vec{p}_2)\right]^{-1} \tag{10}$$

From (7-9-10) we then can find T_{12}.

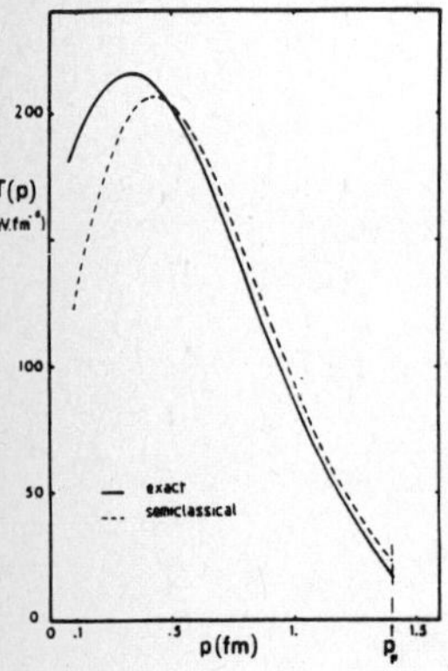

Fig.3 compares the T matrix evaluated in this way for a separable potential[4] with the exact solution (all momenta equal to p).

References

1) P.Ring and P. Schuck in "The nuclear many body problem",chap.13, Springer-Verlag, 1980

2) M.Durand and P. Schuck,Proceedings of "Topical Meeting on phase space approach to nuclear dynamics",1985,Trieste (Italy)

3) G. Ghosh, R. Hasse, P. Schuck and J. Winter, Phys.Rev.Lett.50, 1250 (1983)

4) P. Schuck and E. Suraud, Proceedings of "Topical meeting on phase space approach to nuclear dynamics", 1985, Trieste (Italy).

SQUEEZED STATES AND THEIR WIGNER FUNCTIONS

Antoine Royer
Centre de Recherches Mathématiques
Université de Montréal
Montréal, Québec H3C 3J7, Canada

A squeezed state may be defined as a quantum state whose Wigner function has contours akin to very flat ellipses: the uncertainty along one direction in phase space is "squeezed" to the detriment of that along the perpendicular direction. An infinitely squeezed state has its Wigner function sizable only on a straight line. Position and momentum eigenstates are examples of infinitely squeezed states. Squeezed states, especially Gaussian ones, have come to play a prominent role in quantum optics and in the theory of high precision measurements, in connection, notably, with the effort to detect gravitational waves[1].

We will here describe some ways by which squeezed states can be prepared, manipulated, and used to detect weak forces acting on a free particle. Our overall approach is very much inspired by a classic paper of Lamb[2]. The Wigner phase space representation plays an essential guiding role in this discussion. Reciprocally, insight is gained concerning the physical status of the Wigner phase space.

Let us start by considering a Gaussian wave packet $\psi(x)=e^{-\frac{1}{4}x^2/a^2}$, which has the Wigner function (we neglect normalizing constants)

$$W(x,p) = \exp[-\tfrac{1}{2}x^2/a^2 - 2a^2p^2/\hbar^2] \qquad (1)$$

We may represent this state by the contour $W(x,p)=(1/e)^{1/2}$, i.e., by the ellipse $x^2/a^2 + 4a^2p^2/\hbar^2 = 1$, as shown in Fig.1(a). When $a\to 0$ or $a\to\infty$, the ellipse gets squeezed to a vertical or horizontal straight line, i.e., $\psi(x)$ tends to a position or to a momentum eigenstate.

Since the state (1) is the ground state of a harmonic oscillator, it may be prepared[2] by turning on a suitable harmonic potential Ax^2, let the particle settle to the ground state, and then turn off the potential. A sharp position measurement on a particle leaves it in a highly squeezed state (not necessarily Gaussian however), which we may represent by a narrow vertical segment passing through the position observed, Fig.1(b).

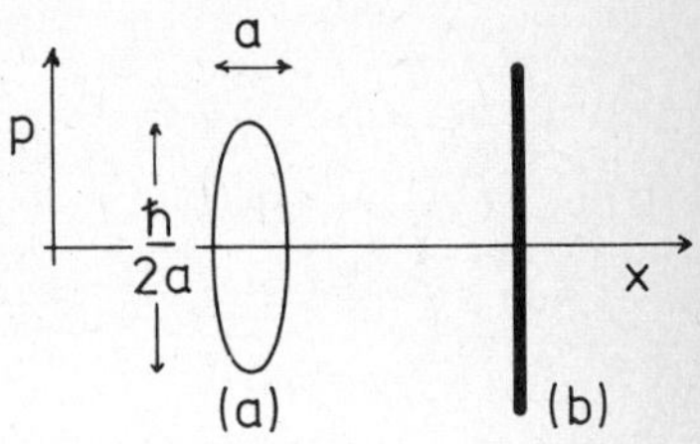

Fig.1

A fundamental property of the Wigner phase space representation is the fact that if the Hamiltonian governing the time evolution is a quadratic polynomial, then the evolution of the Wigner phase space is purely classical, i.e., each point (x,p) in that space follows a classical trajectory. Moreover, since classical evolution under a quadratic Hamiltonian induces a <u>linear</u> transformation of the phase space, straight lines remain straight lines, ellipses remain ellipses, etc., under time evolution (hence a Gaussian packet remains a Gaussian packet).

For instance, under free particle evolution (Hamiltonian $p^2/2m$), the (x,p) plane suffers a shearing motion parallel to the x axis, so that the state (b) in Fig.2 evolves into (c) after some time. The "spreading of the wave packet", i.e., of the position uncertainty Δx, here appears as a purely classical effect.

If we let the state (b) of Fig.2 evolve backwards in time, we get the state (a). When that state is let evolve freely, it goes back to (b), so that its position uncertainty initially contracts — such states have been called "contractive" by Yuen[3]. Lamb[2] has given a simple manner of preparing such a state: First prepare the dashed state (a'), in the manner indicated earlier (by turning on and off a harmonic potential); then apply an impulsive force $-Bx\delta(t)$, derived from the potential $\frac{1}{2}Bx^2\delta(t)$. Since this is quadratic, its effect is classical: each phase space point (x,p) gets displaced vertically by $-Bx$, $(x,p)\rightarrow(x,p-Bx)$, so that (a') goes to (a) for a suitable choice of B.

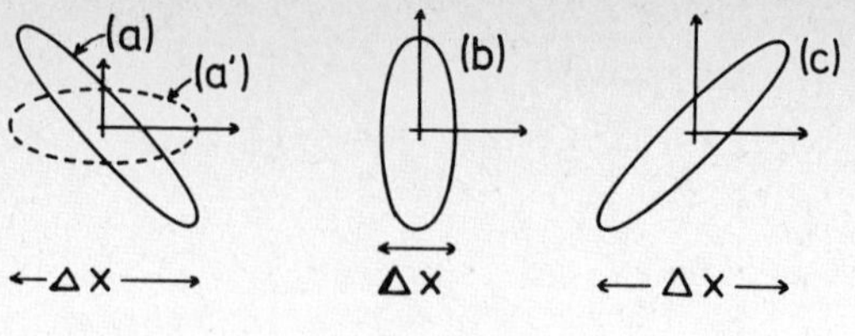

Fig.2

Let us now suppose that we wish to detect a weak classical force $F(t)$ acting on a particle [interaction Hamiltonian $-xF(t)$], by observing the displacement it induces. The laser interferometric method for detecting gravitational waves is an example of such a detection scheme based on position measurements[1]. Classically, the position of the particle can be monitored with arbitrary accuracy, so that an arbitrarily small force can be detected and measured. But quantally, a position measurement of precision Δx implies a momentum uncertainty $\Delta p \cong \hbar/\Delta x$, which "feeds back" into the position uncertainty at later times:

$$\Delta x(t) \cong \Delta x + (\Delta p/m)t \cong \Delta x + (\hbar/m\Delta x)t \gtrsim (\hbar t/m)^{\frac{1}{2}} \quad . \qquad (2)$$

Thus, for a (constant) force F to be detectable within a time interval t, the displacement $\frac{1}{2}(F/m)t^2$ it induces must be larger than $\Delta x(t)$, i.e.,

$$F \gtrsim (\hbar/mt^3)^{\frac{1}{2}} \quad . \qquad (3)$$

These lower bounds on $\Delta x(t)$ and F are referred to as the "standard quantum limit" (SQL)[1].

Yuen[3] proposed to beat the SQL by making use of contractive states. We here describe a modified and simplified (from an operational point of view) version of Yuen's scheme: By using Lamb's procedure mentioned above, prepare a highly squeezed contractive state, represented by the slanted segment (a) in Fig.3. After a time τ, this freely evolves into the vertical segment (b) through x_1: τ and x_1 are uniquely determined by the (known) characteristics of the state (a). A position measurement at time τ then yields a value in a small neighborhood of x_1.

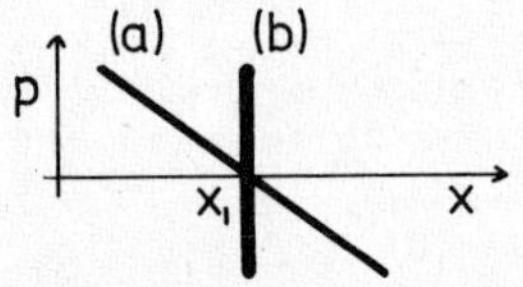

Fig.3

If an interaction $-xF(t)$ is acting, the (x,p) plane suffers during the interval $(0,\tau)$, in addition to its free evolution, an overall translation by the amount $(\delta x_{cl}(0,\tau),\delta p_{cl}(0,\tau))$, where

$$\delta x_{cl}(t_0,t_1) = \int_{t_0}^{t_1}dt \int_{t_0}^{t}dt'F(t')/m, \qquad \delta p_{cl}(t_0,t_1) = \int_{t_0}^{t_1}dtF(t) \quad . \qquad (4)$$

Thus, straight lines get displaced by the force $F(t)$, but not rotated, so that the state (a) in Fig.3 again becomes vertical at time τ, but through the point $x_2=x_1+\delta x_{cl}(0,\tau)$. It follows that a position measurement at time τ now yields a value in a small neighborhood of x_2, whence $\delta x_{cl}(0,\tau)$ can be deduced, since x_1 is known. One can thereby detect, in a finite time, an arbitrarily small force, by letting the initial contractive state (a) be sufficiently squeezed.

Let us now present another method which has certain advantages: Make a sharp position measurement at time t=0, yielding the value x_0 say; this leaves the particle in a squeezed state (not necessarily Gaussian) of very small position uncertainty – ideally a position eigenstate – represented by the vertical line (a) through x_0 in Fig.4. Let evolve freely during a time interval τ, so that the state becomes the slanted line (b). Apply the pulse potential $(m/\tau)(x-x_0)^2 \times\delta(t-\tau)$: this tilts (b) into the contractive state (c) (in effect, the pulse reverses the signs of the momenta). Let evolve freely for another interval τ, so that we get back the initial state (a), and make a position measurement. Because the particle is then in a position eigenstate (or nearly so), the measurement yields x_0, and essentially does not disturb the state (it is a "quantum non-demolition" measurement[1]). Note how this procedure is reminiscent of the spin echo technique in nuclear magnetic resonance.[4]

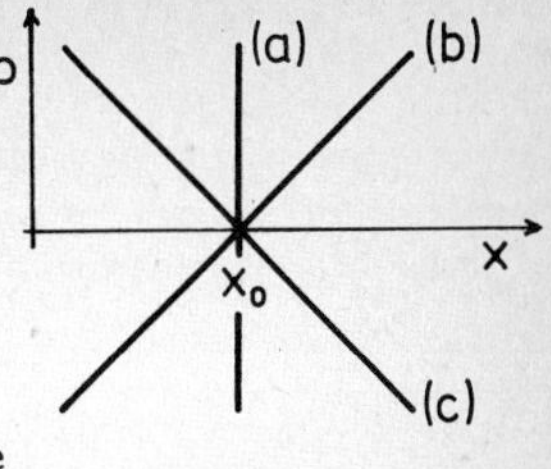

Fig.4

If a force F(t) is acting, the state again becomes vertical at time 2τ, but through the point

$$x_1 = x_0 + \delta x_{cl}(0,2\tau) - 2\delta x_{cl}(0,\tau) \tag{5}$$

[if F is constant during $(0,2\tau)$, then $x_1=x_0+(F/m)\tau^2$]. Thus, an arbitrarily small force can again be detected by this method. But an advantage over the previous method is that we can continue the procedure: let evolve freely from 2τ to 3τ, apply the pulse $(m/\tau)(x-x_1)^2\delta(t-3\tau)$, let evolve freely to 4τ, at which time a position measurement yields $x_2=x_1 +\delta x_{cl}(2\tau,4\tau)-2\delta x_{cl}(2\tau,3\tau)$, and so on. We can thereby monitor the force over a long period, each position measurement simultaneously serving as preparation for the next measurement [if F is constant, we get a cumulative displacement $n(F/m)\tau^2$ at the n-th position measurement].

In the case F=0, the succession of position measurements all yield x_0, so that we are effectively monitoring the classical trajectory $x_{cl}(t;x_0,0)$ of initial position x_0 and initial momentum 0. But one could in fact choose to rather monitor the trajectory $x_{cl}(t;x_0,p_0)$, for any p_0, by applying the pulses

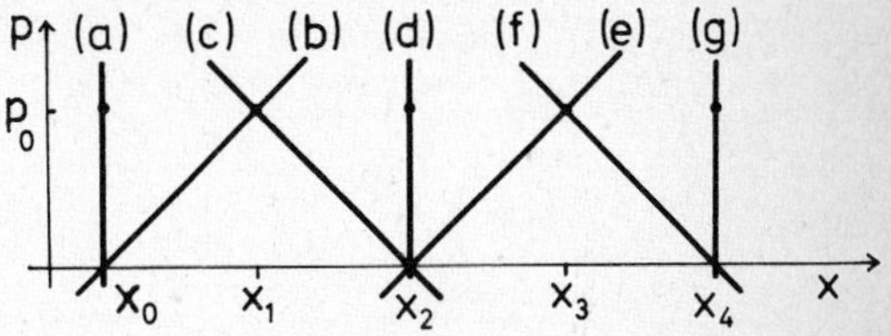

Fig.5: $x_n = x_0 + np_0\tau/m$

$$(m/\tau)(x-x_0-p_0\tau/m)^2\delta(t-\tau), \quad (m/\tau)(x-x_0-3p_0\tau/m)^2\delta(t-3\tau), \ldots,$$

the position measurements at times 2τ, 4τ,..., then yielding $x_0+2p_0\tau/m$, $x_0+4p_0\tau/m$,..., as shown in Fig.5. The possibility of monitoring the classical motion of a phase space point in the above manner (by manipulating a squeezed state by means of pulses which leave undisturbed the point being monitored) seems to confer to points in Wigner phase space a more than formal quality, despite the fact that a particle can never be localized at such points.

1. Quantum Optics, Experimental Gravitation and Measurement Theory, edited by P. Meystre and M.O. Scully, Plenum (New York 1983)

2. W.E. Lamb, Physics Today 22, 23 (1969)

3. H.P. Yuen, Phys. Rev. Lett. 51, 719 (1983)

4. C.P. Slichter, Principles of Magnetic Resonance, Springer (New York 1978)

WIGNER FUNCTION AND QUANTUM LIOUVILLE EQUATION IN CURVED SPACETIME

E. Calzetta and B. L. Hu
Department of Physics and Astronomy
University of Maryland
College Park, Maryland 20742

In this note we show how the state of a free scalar quantum field in a curved space can be described by means of a Wigner function obeying a quantum Liouville-Like equation.

The Wigner function[1] method has been extensively used in non-relativistic many-body problems[2] and in relativistic quantum field theory.[3] Generalizations of this formalism to a general relativistic context have recently been proposed.[4]

In relativistic quantum field theory (in flat space), one can introduce a Wigner function as a partial Fourier transform of one of the propagators of the theory. Consider for example a free, real scalar field $\phi(x)$, and the Wightman function

$$G(x,x') = \langle\phi(x')\ \phi(x)\rangle \tag{1}$$

$$(\ _x + m^2)G(x,x') = 0; \quad (\ _{x'} + m^2)G(x,x') = 0 \ , \tag{2}$$

where $\ = \eta^{\mu\nu}\ \partial_\mu \partial_\nu$, $\eta^{\mu\nu} = \mathrm{diag}(+,-,-,-)$. Make the ansatz

$$G(x,x') = i \int \frac{d^4k}{(2\pi)^4}\, e^{ik(x-x')}\, g(X,k) \tag{3}$$

where $X = \frac{1}{2}(x+x')$. Equations (2) imply

$$(k^2 - m^2 - \frac{1}{4}\ _X)\ g(X,k) = 0 \tag{3a}$$

$$\eta^{\mu\nu}\ k_\mu \frac{\partial}{\partial X^\nu}\ g(X,k) = 0 \ . \tag{3b}$$

Neglecting the derivatives with respect to X against the k^2-m^2 term in (3a), one finds that the most general solution for g is

$$g(X,k) = f(X,k)\ \delta(k^2-m^2) \tag{4}$$

where f is arbitrary. Using (4) in (3b) we find

$$(\eta^{\mu\nu}\ k_\mu \frac{\partial}{\partial X^\nu})\ f = 0 \quad (k^2 = m^2)\ , \tag{5}$$

which is a Liouville-like equation. One may thus identify f as a distribution function.

Now consider a free scalar quantum field propagating in curved spacetime. This situation arises in semiclassical theories of gravity, where the gravitational field is described by a c-number tensor field $g_{\mu\nu}$ while matter fields are quantized. An introduction to theory of quantum fields in curved spaces can be found in Ref. 5.

There are two obvious difficulties in the development of the

Wigner function method in curved space-time. The first is the well-known ambiguity in the concept of particles in a curved manifold. In order to define a distribution function we must first choose a set of modes, use those modes to define creation and destruction operators, and only then a meaningful concept of particle number can be introduced. The second problem is how to define the "average" and the "difference" of two points on a manifold.

We will solve these problems by use of Riemann normal coordinates.[6,7] This is a coordinate system in which the curved space "resembles" most closely the flat space in the neighborhood of a particular point z (the origin of coordinates). In this system there is a preferred set of modes, those which approach plane waves at the origin. This choice makes possible algebraic operations among the coordinates of different points. In this coordinate x^μ, the metric tensor $g_{\mu\nu}$ is expanded around flat space metric η_μ in increasing orders of the Rieman tensor $R_{\mu\rho\nu\sigma}$,

$$g_{\mu\nu}(x) = \eta_{\mu\nu} + \frac{1}{3} R_{\mu\rho\nu\sigma} x^\rho x^\sigma + \ldots . \tag{8}$$

Physical results should be independent of the choice of z. However, when approximations are made, e.g., by truncating the Taylor series Eq. (8), different choices of z will give approximations of different accuracy. In the fully covariant form, the Wightman function given by Eq. (1) obeys

$$\{g^{\mu\nu}(x) \nabla_\mu \partial_\nu + m^2 + \xi R(x)\} G(x,x') = 0 , \tag{9}$$

$$\{g^{\mu\nu}(x') \nabla_{\mu'} \partial_{\nu'} + m^2 + \xi (Rx')\} G(x,x') = 0 . \tag{9a}$$

Expanded in normal coordinates around z, Eq. (9a) becomes

$$\{[\eta^{\mu\nu} - \frac{1}{3} R^\mu{}_\rho{}^\nu{}_\sigma x^\rho x^\sigma] \frac{\partial^2}{\partial x^\mu \partial x^\nu} + \frac{1}{3} R^\nu_\rho x^\rho \frac{\partial}{\partial x^\nu} + m^2 + \xi R + \ldots\} \cdot G(X, k_\nu) = 0, \tag{10}$$

A similar equation is derived for (9b). Now we introduce the average $X = \frac{1}{2}(x+x')$ and make the ansatz

$$G(x,x') = i \int \frac{d^4k}{(2\pi)^4} e^{ik_\nu (x-x')^\nu} G(X, k_\nu) . \tag{11}$$

The four k_ν can be interpreted as the components of a vector at X. The function e^{ikx} is used formally here in the Fourier transform to enable the Wigner function to assume a form closest to that of flat space. This is acceptable with the use of Riemann coordinate as it addresses quasi-local effects.[8] The equations for G(x,k) are

$$\{- k^2 + m^2 + \xi R + \frac{\eta^{\mu\nu}}{4} \frac{\partial^2}{\partial X^\mu \partial X^\nu} + \frac{1}{3} R^\mu{}_\rho{}^\nu{}_\sigma (X^\rho X^\sigma - \frac{1}{4} \frac{\partial^2}{\partial k_\rho \partial k_\sigma}) k_\mu k_\nu - \frac{1}{3} R^\nu{}_\rho \frac{\partial}{\partial k_\rho} k_\nu + \ldots\} G(x,k) = 0 , \tag{12a}$$

$$\{\eta^{\mu\nu} k_\mu \frac{\partial}{\partial X^\nu} + \frac{1}{3} R^\mu{}_\rho{}_\nu{}_\sigma X^\rho \frac{\partial}{\partial k_\sigma} k_\mu k_\nu + \frac{2}{3} R^\nu{}_\rho X^\rho k_\nu + \ldots\} G(x,k) = 0. \tag{12b}$$

In Eqs. (12a) and (12b) we have retained only the first nontrivial corrections due to space-time curvature. The general solution of (12a) is given by

$$G(x,k) = (\quad -g(X) \;)^{-1/2} \quad \{ f(X,k)\delta(k^2-m^2)$$

$$-[(\xi-\tfrac{1}{3})Rf - \tfrac{1}{3} R^{\mu}{}_{\rho}{}^{\nu}{}_{\sigma} X^{\rho} X^{\sigma} k_{\mu} k_{\nu} f - \tfrac{1}{6} R^{\nu}{}_{\rho} k_{\nu} \frac{\partial f}{\partial k_{\rho}} + \tfrac{1}{4} (\quad _{X} f) \tag{13}$$

$$- \tfrac{1}{12} R^{\mu}{}_{\rho}{}^{\nu}{}_{\sigma} k_{\mu} k_{\nu} \frac{\partial^2 f_a}{\partial k_{\rho} \partial k_{\sigma}}]\delta'(k^2-m^2) + \tfrac{5}{24} R^{\nu\rho} k_{\nu} k_{\rho} f \,\delta''(k^2-m^2)\} + \ldots$$

where f is a function of X and k yet to be determined.

Using (13) in (12b), we get

$$\{\eta^{\mu\nu} k_{\nu} \frac{\partial}{\partial X^{\mu}} + \tfrac{1}{3} R^{\mu}{}_{\rho}{}^{\nu}{}_{\sigma} X^{\sigma} k_{\mu} k_{\nu} \frac{\partial}{\partial k_{\rho}} + \ldots\} f = 0 . \tag{14}$$

We recognize (14) as the leading terms in the normal coordinate expansion of the Liouville equation in curved space[9]

$$\{g^{\mu\nu}(X)\, k_{\nu} \frac{\partial}{\partial X^{\mu}} + \Gamma^{\rho}_{\mu\lambda}(X)\, k^{\lambda} k_{\rho} \frac{\partial}{\partial k_{\mu}} \} f = 0 . \tag{15}$$

This shows that results based on relativistic field theory in leading order conforms with that of classical kinetic theory. Details of this work will appear in Ref. 8

References

1. Wigner, E., Phys. Rev. 40, 749 (1932)

2. Kadanoff, L. P., and Baym, G., "Quantum Statistical Mechanics" (W. A. Benjamin, New York 1962).

3. Carruthers, P. and Zachariasen, F., Phys. Rev. D13, 950 (1976); Rev. Mod. Phys. 55, 245 (1983).

4. Winter, J., Phys. Rev. D32, 1871 (1985).

5. Birrell, N. D., and Davies, P.C.W., "Quantum Fields in Curved Space," Cambridge University Press, Cambridge (1982).

6. Bunch, T. S., and Parker, L., Phys. Rev.D20, 2499 (1979).

7. Our conventions are: Signature (+---), $R^{\mu}{}_{\nu\rho\sigma} = \partial_{\sigma} \Gamma^{\mu}{}_{\nu\rho} \ldots$, $R_{\mu\nu} = R^{\rho}{}_{\mu\rho\nu}$, $R = g^{\mu\nu} R_{\mu\nu}$

8. A similar method in field theory is the quasi-local expansion of the effective Lagrangian. See, Hu, B. L. and O'Connor, D. J., Phys. Rev. D30, 743 (1984).

9. E.g., J. M. Stewart, Non-Equilibrium Relativistic Kinetic Theory, lecture notes in Physics No. 10 (Springer-Verlag, Heidelberg, 1971).

10. Calzetta, E., Habib, S. and Hu, B. L., Phys. Rev. D (1986).

APPLICATION OF THE WIGNER FUNCTION IN THE THEORY OF ATOMIC AND MOLECULAR ELECTRONIC STRUCTURE

J. P. Dahl
Department of Chemical Physics
Technical University of Denmark
DTH 301, DK-2800 Lyngby, Denmark

The dynamical behaviour of electrons in atomic and molecular stationary states is governed by wave functions satisfying the time-independent Schrödinger equation. The modern computer has enabled us to construct good approximate solutions to this equation, and very accurate wave functions are now available for all atoms and a large class of molecules. Such wave functions have almost always been generated in position space, but there has been a substantial interest in their momentum space representatives as well.

The phase space representation of quantum mechanics allows us to include the position and momentum characteristics of a quantum state in a single picture, and hence it offers alternative ways of looking at the dynamical behaviour of electrons in atoms and molecules. A stationary state is now described by a pseudo-distribution function in phase space. This is the celebrated Wigner function[1]. Admittedly, there is a whole class of functions that may serve as phase space representatives of a given state[2], but there is a physically acceptable way of excluding all functions other than the Wigner function[3]. Hence we consider the Wigner function to be the canonical phase space function[4].

The phase space equivalent of the time-independent Schrödinger equation is a set of two coupled differential equations, denoted the dynamical equations elsewhere[5]. These equations determine the Wigner function uniquely[5,6], but they are in general of infinitely high order and quite intractable. Hence it is only in very special cases that it becomes possible to search for direct solutions. Considerable insight may be gained by studying these special cases[5,7], but in the general case it is more practical to construct the Wigner function from the wave function in position space, using Wigner's prescription[1]. This prescription gives the Wigner function as the Weyl transform of the density matrix[8].

In a previous paper,[9] we have performed a comprehensive study of the Wigner function $f_{1s}(\underline{r},\underline{p})$ for the ground state of the hydrogen atom and displayed its characteristic features by a series of contour maps. $f_{1s}(\underline{r},\underline{p})$ is a function in six-dimensional phase space. It is, however, fairly easy to see that it only depends on the three quantities r, p and u, where r and p are the lengths of $\underline{r}$ and $\underline{p}$ respectively, and u is the angle between $\underline{r}$ and $\underline{p}$. In other words, $f(\underline{r},\underline{p})$ is independent of the Euler angles which describe the orientation of the $\underline{r},\underline{p}$ plane. It is also found that the condition $u = \pi/2$ defines a dominant subspace in which the Wigner function finds its maximum support. Furthermore, the function is found to be everywhere non-negative in this subspace.

The dominant subspace is five-dimensional. It contains a three-dimensional subspace of particular interest, namely the subspace obtained by putting $r = a_0$ and $p = \hbar/a_0$, where a_0 is the radius of the first Bohr orbit. This is the subspace to which the ground state motion was restricted in early quantum mechanics. Hence we may call this subspace the classical subspace. The following statement may then be made: The Wigner function attains a large, positive and constant value in the classical subspace. It is also large and positive in a large region surrounding this subspace. In particular, it is everywhere non-negative in the dominant subspace. The regions in which the Wigner function become negative are well separated from the classical subspace.

In addition to the hydrogen atom, we have also studied a number of other atoms[10] and the LiH molecule[11]. In each case we constructed the Wigner function from the corresponding wave function, and the integrations involved were performed by expanding the atomic and molecular orbitals in a finite series of gaussians. For these several-electron systems we concentrated on the single-particle Wigner function which is the Weyl transform of the first order density matrix.

Our atomic and molecular studies have led to a number of interesting observations. Perhaps the most important one is the finding that the phase space formulation is a natural intermediate in the construction of local densities. The Wigner function allows us to define a canonical local density for any one-electron operator (momentum, kinetic energy, exchange energy, etc.)[12]. By means of these densities we have shown that a local virial theorem holds for the ground state of the hydrogen atom[9], and in the many-electron case we have made contact with Thomas-Fermi theory as well as other density functional theories[12,13].

Thus we believe that the phase space approach to atomic and molecular electronic structure theory is a useful and promising one, and that it shows us new and valuable ways of analysing both one- and many-electron states. The phase space approach is new in the sense that it has not been coherently studied before, but it has deep historical roots in the early works of Thomas, Fermi and Dirac, and phase space notions have always been more or less freely used by many workers in the field of atomic and molecular structure.

REFERENCES

1. E. Wigner, Phys. Rev. 40, 749 (1932).
2. L. Cohen, J. Math. Phys. 7, 781 (1966).
3. J. G. Krüger and A. Poffyn, Physica 85A, 84 (1976).
4. J. P. Dahl, Physica 114A, 439 (1982).
5. J. P. Dahl, Dynamical Equations for the Wigner Functions, in: Energy Storage and Redistribution in Molecules, ed. J. Hinze (Plenum Press, New York, 1983) pp. 557-571.
6. P. Carruthers and F. Zachariasen, Rev. Mod. Phys. 55, 245 (1983).
7. J. P. Dahl, The Phase Space Representation of Quantum Mechanics and the Bohr-Heisenberg Correspondence Principle, in: Semiclassical Descriptions of Atomic and Nuclear Collisions, eds. J. Bang and J. de Boer (Elsevier Science Publishers, Amsterdam, 1985) pp. 379-394.
8. H. J. Groenewold, Physica 12, 405 (1946).
9. J. P. Dahl and M. Springborg, Mol. Phys. 47, 1001 (1982).
10. M. Springborg and J. P. Dahl, to be published.
11. M. Springborg, Theoret. Chim. Acta 63, 349 (1983).
12. M. Springborg and J. P. Dahl, A Phase Space Approach to Energy Densities in Position Space, in: Local Density Approximations in Quantum Chemistry and Solid State Physics, eds. J. P. Dahl and J. Avery (Plenum Press, New York, 1984) pp. 381-412.
13. J. P. Dahl et al., to be published.

WIGNER PHASE-SPACE APPROACH IN THE MOLECULAR COLLISION THEORY - SEARCH FOR WIGNER TRAJECTORIES *

Hai-Woong Lee
Department of Physics
Oakland University
Rochester, Michigan 48063

The Wigner phase-space approach[1,2] currently in use in the molecular collision theory offers a semiclassical way of describing collisions: the collision dynamics is described using classical trajectories while initial and final collision states are represented quantum mechanically in terms of the Wigner distribution function. Over the last several years this approach--which we call the classical-trajectory (CT) Wigner approach--has been used with success to describe a variety of molecular collision and dissociation processes. Collision processes we have investigated in the past using the CT Wigner approach include collinear He-H_2 collision[3] and H-H_2 exchange reaction.[4] It now is well established that the CT Wigner approach yields more accurate transition probabilities than the widely used quasiclassical method. This is due to the quantum-mechanical assignment of initial and final states employed in the CT Wigner approach as opposed to the classical procedure of assigning initial and final states adopted in the quasiclassical method.

The accuracy of the CT Wigner approach, however, is not always as high as one desires. In particular, our recent calculation[4] indicates that there exist some noticeable quantitative disagreements between the Wigner probabilities and the exact quantum-mechanical probabilities for the H-H_2 exchange reaction. Any inaccuracy in the CT Wigner approach can be traced back to its use of classical trajectories. Although the Wigner approach owes much of its popularity to its ease of application due to the use of classical trajectories, it is this very aspect that limits its accuracy: quantum "dynamical" tunneling is totally missing in the CT Wigner approach.

As shown by Wigner,[5] the equation of motion that governs the time development of the Wigner distribution function W(q,p,t) is given by

$$\frac{\partial W(q,p,t)}{\partial t} = -\frac{p}{m}\frac{\partial W}{\partial q} + \int_{-\infty}^{\infty} dj\; W(q,p+j,t)J(q,j), \tag{1}$$

where

$$J(q,j) = \frac{i}{\pi\hbar^2} \int_{-\infty}^{\infty} dy \left[V(q+y)-V(q-y)\right] \exp(-2iyj/\hbar). \tag{2}$$

If we define V_{eff} as

$$\frac{\partial V_{eff}(q,p,t)}{\partial q} \frac{\partial W}{\partial p} = \int_{-\infty}^{\infty} dj\ W(q,p+j,t)J(q,j), \tag{3}$$

Eq. (1) can be solved to yield

$$\frac{dq}{dt} = \frac{p}{m}, \qquad \frac{dp}{dt} = -\frac{\partial V_{eff}(q,p,t)}{\partial q}. \tag{4}$$

Eq. (4) defines the Wigner trajectory[2] along which each phase-space point (q,p) of the Wigner distribution moves. The use of classical mechanics for the description of dynamics in the CT Wigner approach is equivalent to replacing Eq. (1) by the equation

$$\frac{\partial W(q,p,t)}{\partial t} = -\frac{p}{m}\frac{\partial W}{\partial q} + \frac{\partial V}{\partial q}\frac{\partial W}{\partial p}, \tag{5}$$

which immediately yields the classical Hamilton's equations,

$$\frac{dq}{dt} = \frac{p}{m}, \qquad \frac{dp}{dt} = -\frac{\partial V}{\partial q}. \tag{6}$$

The key point is that the phase-space points of the Wigner distribution should in principle be propagated along the Wigner trajectories defined by Eq. (4). In the CT Wigner approach, however, they are propagated along classical trajectories defined by Eq. (6). Therefore, it is apparent that development of a more accurate Wigner approach requires finding Wigner trajectories or at least semiclassical trajectories (classical trajectories with quantum corrections) for the system being considered.

For our study of Wigner or semiclassical trajectories, we consider simple potential scattering of a particle off a square well, assuming that at the initial time the particle can be represented by a Gaussian wave packet. Even for this simple system, Eq. (1) cannot be solved exactly. To obtain an approximate solution we replace the Wigner distribution function in the integral of the right-hand side of Eq. (1) by the free-field Wigner distribution function. The integration can then be performed, and the effective potential V_{eff} defined by Eq. (3)

can be obtained. The trajectories we obtain with this effective potential are not exactly Wigner trajectories because of the approximation we made, but they contain quantum corrections to classical trajectories, i.e., they are semiclassical trajectories and are expected to yield a more accurate description of the dynamics of the particle than the corresponding classical trajectories.

Inspection of these semiclassical trajectories reveal the following:

(A) The center of the packet (or phase-space points near the center of the packet) travels through the well to the opposite side. The momentum of the center of the packet continuously increases as the packet approaches the well, reaches a maximum value at the center of the well, and then decreases as the packet moves away from the well. This continuous change in momentum, which is a clear indication of the nonlocal nature of quantum mechanics, is in contrast to the purely classical behavior according to which the particle experiences a sudden increase or decrease in momentum only at the edges of the well.

(B) Phase-space points sufficiently far away from the packet center show three different behaviors depending on the initial coordinate and momentum. Some pass through the well, some bounce back from the well (thus yielding a nonzero value for the reflection coefficient), and a small fraction remain in the neighborhood of the well even after the main part of the packet moves far away from the well.

The example considered above shows clearly that the quantum-mechanical dynamics can be drastically different from the corresponding classical dynamics. Application of the CT Wigner approach to this example would yield the transmission coefficient $T \cong 1$ regardless of the well depth V_0 and width 2a, which obviously is wrong. If quantum dynamical tunneling plays an important role in the collision process one wants to describe, one may have to rely on the semiclassical-trajectory Wigner approach or the Wigner-trajectory Wigner approach rather than the classical-trajectory Wigner approach.

REFERENCES
1. E.J. Heller, J. Chem. Phys. 65, 1289 (1976).
2. H.W. Lee and M.O. Scully, Found. Phys. 13, 61 (1983).
3. H.W. Lee and M.O. Scully, J. Chem. Phys. 73, 2238 (1980).
4. H.W. Lee and T.F. George, J. Chem. Phys, in press.
5. E. Wigner, Phys. Rev. 40, 749 (1932).

*Research supported by the National Science Foundation under Grant No. CHE-8512406

OPTICAL EIGENMODES AND THE WIGNER DISTRIBUTION

Walter Schempp
Lehrstuhl fuer Mathematik I
University of Siegen
D-5900 Siegen
Federal Republic of Germany

1. The Wigner Distribution and the Ambiguity Functions

Let $\mathcal{S}(\mathbf{R})$ denote the complex Schwartz space. The Wigner distribution associated with the wave function $\psi \in \mathcal{S}(\mathbf{R})$ is given by the expression

$$P(\psi;q,p) = \int_{\mathbf{R}} \psi(q+\tfrac{1}{2}s)\,\overline{\psi}(q-\tfrac{1}{2}s)\,e^{-2\pi ips}\,ds$$

for all $(q,p) \in \mathbf{R}^2$. An application of the Fourier cotransform $\overline{\mathcal{F}}_{\mathbf{R}^2}$ to the sesquilinear form associated with P yields the cross-ambiguity function

$$H(\psi,\varphi;x,y) = \int_{\mathbf{R}} \psi(t+\tfrac{1}{2}x)\,\overline{\varphi}(t-\tfrac{1}{2}x)\,e^{2\pi iyt}\,dt$$

on the real plane $\mathbf{R}^2$. In particular,

$$H(\psi,\psi;.,.) = H(\psi;.,.) = \overline{\mathcal{F}}_{\mathbf{R}^2}\, P(\psi;.,.).$$

If $(U_\lambda)_{\lambda \in \mathbf{R}^\times}$ denotes "the" family of infinite dimensional, topologically irreducible, continuous, unitary, linear representations of the three-dimensional real Heisenberg nilpotent Lie group $\tilde{A}(\mathbf{R})$ acting on the complex Hilbert space $L^2(\mathbf{R})$, then we have

$$H(\psi,\varphi;x,y) = \langle U_1(x,y,0)\psi | \varphi\rangle$$

for all $(x,y) \in \mathbf{R}^2$, i.e., H is given by the coefficient function of the linear Schrödinger representation U_1 of $\tilde{A}(\mathbf{R})$ modulo the center $\tilde{C} = \{(0,0,z) \mid z \in \mathbf{R}\}$ of $\tilde{A}(\mathbf{R})$.

2. The Diamond Solvable Lie Group

Let $S_\lambda : (e^{2\pi i\theta}, z) \longmapsto e^{2\pi iH_\lambda}$ be the projective linear representation of $\mathbf{T} \times \tilde{C}$ which intertwines unitarily U_λ and $(e^{2\pi i\theta}, z).U_\lambda$. Here

$$H_\lambda = \frac{1}{\lambda}\frac{d^2}{dt^2} - 4\pi^2\lambda t^2$$

denotes the Schrödinger Hamiltonian of the harmonic oscillator, which admits a pure point spectrum formed by the simple eigenvalues $\{-2\pi(\mathrm{sign}\,\lambda)(2n+1) \mid n \in \mathbf{N}\}$. Form the characters

$$\chi_\lambda^n(e^{2\pi i\theta}, z) = e^{2\pi i(\lambda z+2\pi(2n+1))}\mathrm{id}_{L^2(\mathbf{R})} \qquad (n \in \mathbf{Z})$$

of $\mathbf{T} \times \widetilde{C}$. Then the diamond group

$$D(\mathbf{R}) = \mathbf{T} \ltimes \widetilde{A}(\mathbf{R})$$

(= semi-direct product of the compact torus group $\mathbf{T}$ with $\widetilde{A}(\mathbf{R})$), which is a non-exponential solvable Lie group, has "the" family $S_\lambda U_\lambda \otimes \chi_\lambda^n$ ($n \in \mathbf{Z}$) of topologically irreducible, continuous, unitary, linear representations which restrict to U_λ. The waveform $\psi \in \mathscr{S}(\mathbf{R})$, $\psi \neq 0$, is a transverse <u>eigenmode</u> of a <u>circular</u> optical waveguide <u>if and</u> only if there exists a number $n \in \mathbf{N}$ such that the restriction $\sqrt{\chi_1^n \circ S_1} \mid \mathbf{T}$ acts trivially on ψ modulo $\mathbf{T}$. It follows

$$\psi = \zeta_n h_n \quad (\zeta_n \in \mathbf{C})$$

for an integer $n \geq 0$ (h_n = Hermite function of degree $n \geq 0$) and

$$H(h_m, h_n; x, y) = \sqrt{\frac{n!}{m!}}\,(\sqrt{\pi}(x+iy))^{m-n} L_n^{(m-n)}(\pi(x^2+y^2)), \quad (m \geq n \geq 0),$$

for all pairs $(x,y) \in \mathbf{R}^2$ ($L_n^{(\rho)}$ = Laguerre function of degree $n \geq 0$ and order ρ). See the illustrations of Section 4 infra. The photograph visualizes the intensity pattern of a higher transverse eigenmode excited on a silica graded-index optical fiber about 85 microns in core diameter.

3. Coaxial Coupling of Transverse Eigenmodes

An application of the oscillator representation evaluated at the linear symplectomorphisms which transform the reference plane to the coupling plane transverse to the beam direction, allows us to compute the coupling coefficients of transverse eigenmodes in circular optical waveguides in terms of the beam parameters (radii of beam and curvature) and the Gaussian hypergeometric function ${}_2F_1$. The realization of the linear Schrödinger representation U_1 as an isotypic component of the left regular representation of $\widetilde{A}(\mathbf{R})$ yields a similar result in the case of rectangular optical waveguides.

4. Illustrations

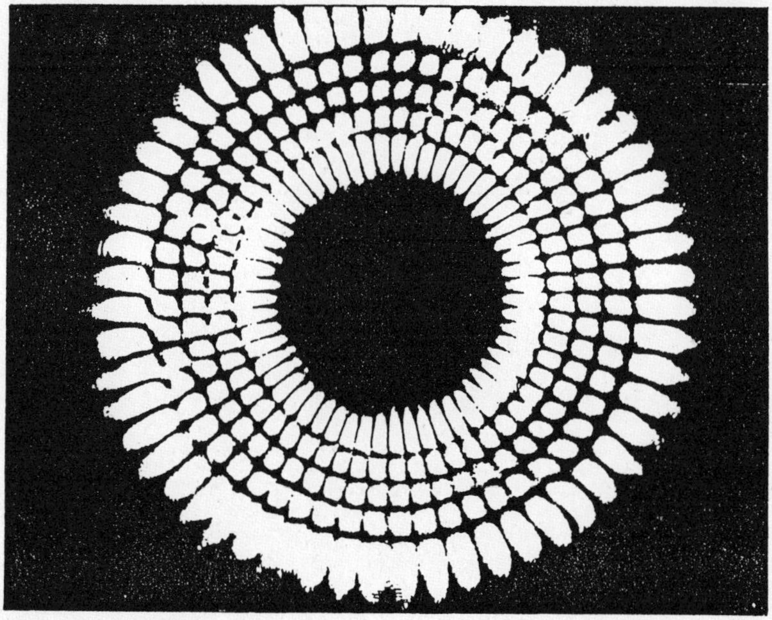

TIME FREQUENCY REPRESENTATION OF BROAD BAND SIGNALS

P. BERTRAND(*) and J. BERTRAND(**)

(*) ONERA - F- 92320 - Châtillon-sous-Bagneux
(**) LPTM - University Paris VII - F - 75251 - Paris

1 - General framework

The description of signals often requires an effective time-frequency (TF) representation that truly reflects their time-varying frequency content. Some examples of TF representations include written music and the short-time Fourier analysis by use of a sliding time window.

In practice, the signal $s(t)$ arises as a real function of time (acoustic pressure, electric voltage, ...) and has to be interpreted up to changes of phase and amplitude. Such transformations are expressed by :

$$s(t) \longrightarrow s'(t) = \lambda e^{-i\varphi} s_{(-)}(t) + \lambda e^{i\varphi} s_{(+)}(t) , \qquad (1)$$

where λ and φ are real constants and where $s_{(\pm)}$ refer to the positive and negative frequency parts of $s(t)$.

For convenience, the signal is usually characterized by its $s_{(+)}$ part (Gabor analytic signal [1]) and thus the state space of communication theory is a Hilbert space of positive frequency complex signals. Due to (1), the relevant information is represented by rays in this Hilbert space. In the following, we describe the signal by its Fourier transform :

$$S(f) = \int_{-\infty}^{\infty} e^{-2i\pi f t} \, s(t) \, dt \, . \qquad (2)$$

The fundamental invariance group of signal analysis is the affine group of clock changes :

$$t \longrightarrow a\,t + b \quad ; \quad a,b \in \mathbb{R} \, , \quad a > 0 \, .$$

In Hilbert space, this group is represented by :

$$S(f) \longrightarrow S'(f) = a^{\nu} e^{-2i\pi abf} S(af) \, , \qquad (3)$$

where for our purpose the constant ν is taken real (dimensional factor). This representation is unitary with respect to the scalar product :

$$(S_1 , S_2) = \int_0^{\infty} f^{2\nu-1} \, S_1^*(f) \, S_2(f) \, df \, . \qquad (4)$$

Introducing observables as hermitian operators, we will write observations as follows :

$$\langle A \rangle = \int_0^{\infty} df_1 \int_0^{\infty} df_2 \; A^*(f_1,f_2) \; S(f_1) \, S^*(f_2) \, (f_1 f_2)^{2\nu-1} \qquad (5)$$

with $A(f_1,f_2) = A^*(f_2,f_1)$.

In the TF version of the theory, we will represent operators by functions $\mathcal{A}(t,f)$ transformed by the affine group as :

$$\mathcal{A}(t,f) \longrightarrow \mathcal{A}'(t,f) = a^{\mu} \mathcal{A}(a^{-1}t-b, af) \quad , \mu \in \mathbb{R} \; . \tag{6}$$

The function corresponding to the projector on state S will be denoted by $P(t,f)$. The choice of μ in (6) depends on the required properties of $P(t,f)$. In this representation, the observations are given by invariant forms :

$$<A> = \int_{-\infty}^{\infty} dt \int_{0}^{\infty} df \quad \mathcal{A}(t,f)\, P(t,f)\, f^{2\mu} \; . \tag{7}$$

The expression of $P(t,f)$ will be derived in the next section, using a tomographic method [2] . The result is a new TF representation which reduces to the usual Wigner-Ville function [3] in the approximation of narrow band signals. Some properties are discussed.

2 - Construction of the time-frequency representation and applications

We first study observables invariant by subgroups of the affine group labelled by $\xi \in \mathbb{R}$ and defined by : $b = \xi(a^{-1}-1)$.

Forms (5), invariant by action of this subgroup, are diagonalized through a Mellin transform and can be written :

$$<A> = \int_{\mathbb{R}} A(\beta)\, |m_{\xi}^{\lambda}(\beta)|^{2}\, d\beta \quad , \tag{8}$$

where :

$$m_{\xi}^{\lambda}(\beta) = \int_{0}^{\infty} S(f)\, e^{2i\pi\xi f} f^{2i\pi\lambda(\beta,\xi)+\nu-1}\, |\lambda'_{\beta}|^{1/2}\, d\beta$$

and λ is a monotonous arbitrary function. Thus, for this class of observables, $<A>$ is a mathematical expectation.

In the TF representation, forms (7) invariant by action of the same subgroup on $P(t,f)$ can likewise be written in the form :

$$<A> = \int_{\mathbb{R}} \mathcal{A}(\beta)\, \rho_{\xi}(\beta)\, d\beta \tag{9}$$

where

$$\rho_{\xi}(\beta) = \int_{0}^{\infty} df \int_{-\infty}^{\infty} dt \quad P(t,f)\, \delta[(t-\xi)f-\beta]\, f^{\mu} \; .$$

Then, identifying $\rho_{\xi}(\beta)$ with the density $|m_{\xi}^{\lambda}(\beta)|^{2}$ appearing in (8), we obtain an equation for $P(t,f)$ depending on μ , ν and the function λ . The arbitrariness of λ is removed by invariance requirements leading to the constraint

$$\lambda(\beta,\xi) \equiv \beta \; .$$

Solving the corresponding equation yields the TF representation of the signal S [4] :

$$P(t,f) = f^{2\nu-\mu} \int_{-\infty}^{\infty} e^{-2i\pi uft} \left(\frac{u}{2\sinh\frac{u}{2}}\right)^{2\nu} S\left(\frac{ufe^{-\frac{u}{2}}}{2\sinh\frac{u}{2}}\right) S^{*}\left(\frac{ufe^{\frac{u}{2}}}{2\sinh\frac{u}{2}}\right) du \, . \tag{10}$$

Among properties of this expression, we note :

(i) If $S(f) = \delta(f-f_0)$, then the associated $P(t,f)$ is :

$$P(t,f) = f_0^{2\nu-\mu-1}\, \delta(f-f_0) \; .$$

(ii) The following diagrams are commutative :

$$\begin{array}{ccc} S(f) & \longrightarrow & S'(f) = a^{\nu} e^{-2i\pi abf} S(af) \\ \downarrow & & \downarrow \\ P(t,f) & \longrightarrow & P'(t,f) = a^{\mu} P(a^{-1}t-b, af) \end{array}$$

$$\begin{array}{ccc} s(t) & \longrightarrow & s'(t) = s(-t) \\ \downarrow & & \downarrow \\ P(t,f) & \longrightarrow & P'(t,f) = P(-t,f) \end{array}$$

(iii) $$\int_{-\infty}^{\infty} P(t,f)\, dt = f^{2\nu-\mu-1} |S(f)|^2$$

(iv)If P_1, P_2 are the representations of S_1, S_2 respectively, then we have :

$$\int_{-\infty}^{\infty} dt \int_0^{\infty} df \; P_1(t,f)\, P_2(t,f)\, f^{2\mu} = |(S_1,S_2)|^2 .$$

Property (iii) implies that for

$$\mu = 2\nu - 1 \qquad (11)$$

the TF representation (10) has the meaning of an instantaneous spectrum.

For narrow-band signals, the integration in (10) is restricted to a neighborhood of $u = 0$, and the class of $P(t,f)$ defined by (11) admits the unique Wigner-Ville function as an approximation. The latter could be obtained directly by replacing the affine group by the group of time and frequency translations [5].

The expression (10) is not everywhere positive but, like Wigner's function, it can be regularized by local integration. This is done by use of property (iv) and leads to the form [5] :

$$\tilde{P}(t,f) = f^{-\mu} \left| \int_0^{\infty} S(f')\, \Psi^*_{t,f}(f')\, f'^{\,2\nu-1}\, df' \right|^2 ,$$

where $\Psi_{t,f}$ is derived from a given signal $\varphi(f)$ by :

$$\Psi_{t,f}(f') = \varphi\left(\frac{f'}{f}\right) e^{-2i\pi f' t} f^{-\nu} .$$

This regularization has been applied in a radar imaging problem [5] , using coherent states relative to the affine group [6] . Alternative regularizations have been used in the literature [7] .

<u>Références</u>

1 - D. Gabor - J. Inst. Electr. Eng., III vol 93 (1946) p. 429-441

2 - J. Bertrand - P. Bertrand - Tomographic procedure for constructing phase space representations. In these proceedings.

3 - J. Ville - Câbles et transmissions n° 1 (1948), p. 60-74

4 - J. Bertrand - P. Bertrand - C.R. Acad. Sc. Paris, vol. 299, série 1 (1984) p. 635-638

5 - P. Bertrand - J. Bertrand - La Recherche Aérospatiale 1985-5, p.1-7, French and English Editions

6 - E.W. Aslaksen - J.R. Klauder - J. Math. Phys. vol 9 (1968) p. 206-211 and vol 10 (1969) p. 2267-2275

7 - A. Grossmann - J. Morlet - S.I.A.M. J. Math-Anal., vol 15 (1984) p. 723-736

QUASI-PROBABILITY DISTRIBUTIONS FOR ARBITRARY OPERATORS

Marlan O. Scully[+]

Max-Planck Institut für Quantenoptik
D-8046 Garching bei München, West Germany
and
Center for Advanced Studies
and Dept. of Physics and Astronomy
University of New Mexico
Albuquerque, New Mexico 87131

Leon Cohen[*]

Hunter College and Graduate Center
The City University of New York
New York, N Y 10021
and
Center for Advanced Studies
University of New Mexico
Albuquerque, New Mexico 87131

We consider the problem of writing joint quantum quasi-distributions for arbitrary non-commuting operators. A number of such expressions are derived and a method to construct an infinite number of them is given. These distributions satisfy the correct quantum individual probability distributions for each operator. We also show that, in general, the classical procedure for transforming random variables does not work for quantum quasi-distributions.

[+] Work supported by the Office of Naval Research.
[*] Work supported in part by the CUNY FRAP grant program.

1. INTRODUCTION

Quasi probability distributions for quantum mechanical operators have found many applications in numerous fields [1]. These distributions behave to a large extent like classical joint distributions since expectation values are obtained by phase space averaging rather than through the operator formalism of quantum mechanics. They are not positive definite and hence the name quasi-distributions. Although they can often be manipulated as a classical joint distribution in the phase space of position and momentum, there is one important aspect where the classical manipulation will give incorrect quantum mechanical results. This is the case when one transforms the probability distribution from one set of random variables to a new set of random variables corresponding to arbitrary operators. Since the usual transformations will not work, we derive here, from first principles, joint distributions for arbitrary operators which satisfy the quantum mechanical marginal distributions.

Joint distributions for operators other then position and momentum have been previously considered for special cases, for example, for the creation and annihilation operators. A significant contribution for the general case was made by Barut [2] who considered a particular class of distributions. As we will see, his class is obtained by a particular choice of the characteristic function. Further contributions to the general case have been made by Margenau and Hill [3] who used the symmetrization rule, and by Lax and Yuen [4] who developed a general operator algebra for particular orderings of the operators.

To contrast the classical and quantum situation we briefly review the classical procedure for obtaining new probability distributions from old. Suppose we have a *classical* joint distribution of position and momentum, $P(q,p)$, and wish to obtain the distribution for the random variables $a = a(q,p)$ and $b = b(q,p)$. In classical physics a simple way to do that is to form the characteristic function for the variables a and b (all integrals go from $-\infty$ to ∞).

$$M^{ab}(\theta,\tau) = < e^{i\theta a(q,p)+i\tau b(q,p)} > = \int\int P(q,p) e^{i\theta a(q,p)+i\tau b(q,p)}\, dq\, dp \,, \qquad (1.1)$$

and the probability distribution is then given by

$$P^{ab}(a,b) = \frac{1}{4\pi^2}\int\int M^{ab}(\theta,\tau) e^{-i\theta a - i\tau b} d\theta d\tau. \qquad (1.2)$$

By substituting Eq. (1.2) into Eq. (1.1) and doing the θ, τ integration we also have

$$P^{ab}(a,b) = \int\int \delta(a-a(q,p))\,\delta(b-b(q,p))\,P(q,p)\,dq\,dp. \qquad \textbf{(classical)}\,. \qquad (1.3)$$

Now, suppose we have two observables represented by the operators $A(Q,P)$ and $B(Q,P)$ which are functions of the position and momentum operators, Q and P, and where the classical functions corresponding to the two operators, A and B, are $a(q,p)$ and $b(q,p)$, respectively. Suppose further that we have the quantum quasi-distribution for position and momentum, $P(q,p)$, the Wigner distribution, for example. Can the same procedure as above be used to obtain quasi distributions for the operators A and B ? For the quantum case we must assure that the joint distribution satisfies the individual probability distributions as given by the Born rule, namely that

$$\int P(a,b)da = |\phi^A(a)|^2 \tag{1.4}$$

$$\int P(a,b)db = |\phi^B(b)|^2, \tag{1.5}$$

where $\phi^A(a)$ and $\phi^B(b)$ are the state functions in the A and B representations

$$\phi^A(a) = \int v_a^*(q)\psi(q)dq \tag{1.6}$$

$$\phi^B(b) = \int w_b^*(q)\psi(q)dq, \tag{1.7}$$

and $v_a(q)$ and $w_b(q)$ are the eigenfunctions of A and B with eigenvalues a and b respectively.

In general, if the procedure given by Eq. (1.3) is followed, then the resulting distribution will not satisfy the marginals Eqs.(1.4) and (1.5). There are certain situations where the classical procedure will work, and that will be discussed later. Therefore, to obtain joint distributions for arbitrary operators one must start at the beginning. We cannot use the joint quasi-distribution for position and momentum and transform it to obtain distributions for new operators.

2. DISTRIBUTIONS AND CHARACTERISTIC FUNCTIONS FOR ARBITRARY OPERATORS

For position and momentum, different ordering rules yield different distributions [5]. The most powerful approach to obtain these joint distributions is by the use of characteristic function method. For the case of position and momentum, a general theory relating distributions and ordering rules and characteristic functions has been developed [5]. We follow a similar approach here for arbitrary operators and start with the Weyl rule of ordering.

a) Weyl Correspondence

Moyal was the first to show that if we use the Weyl correspondence rule to form the quantum operators, then the characteristic function method will yield the Wigner distribution. For the case of position and momentum, the Weyl ordering for the characteristic function $M(\theta, \tau)$ gives (we take Planck's constant to be one),

$$M(\theta,\tau) = \langle e^{i\theta Q + i\tau P} \rangle = \int \psi^*(q) e^{i\theta Q + i\tau P}\psi(q)dq \tag{2.1}$$

$$= \int \psi^*(q) e^{i\theta\tau/2} e^{i\theta Q} e^{i\tau P}\psi(q)dq \tag{2.2}$$

$$= \int \psi^*(u-\tau/2) e^{i\theta u}\psi(u+\tau/2)du\,. \tag{2.3}$$

By taking the Fourier inverse the Wigner distribution is obtained,

$$P_W(q,p) = \frac{1}{4\pi^2}\int\int M(\theta,\tau)e^{-i\theta q - i\tau p}\,d\theta\,d\tau \tag{2.4}$$

$$= \frac{1}{2\pi}\int \psi^*(q-\tau/2)e^{-i\tau p}\psi(q+\tau/2)d\tau. \tag{2.5}$$

Similarly, for two arbitrary quantum operators we form their characteristic function by

$$M_1^{ab}(\theta,\tau) = < e^{i\theta A + i\tau B} > = \int \psi^*(q)e^{i\theta A + i\tau B}\psi(q)dq. \tag{2.6}$$

For arbitrary operators we cannot simplify the exponential as in the case for position and momentum since the Baker-Hausdorff theorem, as used to go from Eq. (2.1) to Eq. (2.2), cannot be applied with profit. An alternative procedure is as follows. We expand the wave function in terms of the complete set of functions $u_\gamma(q)$,

$$\psi(q) = \int \phi(\gamma)u_\gamma(q)d\gamma\ , \tag{2.7}$$

where the expansion functions are the eigenfunctions of the operator equation

$$(\theta A + \tau B)u_\gamma(q) = \gamma u_\gamma(q). \tag{2.8}$$

Inserting (2.7) into Eq. (2.6), the characteristic function is expressed

$$M_1^{ab}(\theta,\tau) = \int \psi^*(q)e^{i\theta A + i\tau B}\psi(q)dq \tag{2.9}$$

$$= \int\int\int \phi^*(\gamma')u_{\gamma'}^*(q)e^{i\theta A + i\tau B}\phi(\gamma)u_\gamma(q)d\gamma' d\gamma dq \tag{2.10}$$

$$= \int\int\int \phi^*(\gamma')u_{\gamma'}^*(q)e^{i\gamma}\phi(\gamma)u_\gamma(q)d\gamma' d\gamma dq \tag{2.11}$$

$$= \int\int \phi^*(\gamma')\delta(\gamma-\gamma')e^{i\gamma}\phi(\gamma)d\gamma' d\gamma \tag{2.12}$$

$$= \int |\phi^*(\gamma)|^2 e^{i\gamma}d\gamma\ . \tag{2.13}$$

Inverting Eq. (2.7) we have that

$$\phi(\gamma) = \int u_\gamma^*(q)\psi(q)dq \tag{2.14}$$

and substituting this into Eq. (2.13) we have another expression for the characteristic function,

$$M_1^{ab}(\theta,\tau) = \int\int \psi^*(q') < u_\gamma^*(q)\,|\,e^{i\gamma}\,|\,u_\gamma(q') > \psi(q)dq'dq \tag{2.15}$$

where

$$< u_\gamma^*(q) \,|\, e^{i\gamma} \,|\, u_\gamma(q') > \; = \int u_\gamma^*(q) e^{i\gamma} u_\gamma(q') d\gamma \; . \tag{2.16}$$

To obtain the distribution, we take the Fourier inverse and obtain

$$P_1^{ab}(a, b) = \frac{1}{4\pi^2} \int \int M_1^{ab}(\theta, \tau) e^{-i\theta a - i\tau b} \, d\theta \, d\tau \tag{2.17}$$

$$= \int \int \psi^*(q') G(a, b; q, q') \psi(q) dq \, dq' \; , \tag{2.18}$$

where

$$G(a, b; q, q') = \frac{1}{4\pi^2} \int \int \int u_\gamma^*(q) e^{i\gamma - i\theta a - i\tau b} u_\gamma(q') \, d\gamma \;\; d\theta \, d\tau \; . \tag{2.19}$$

Eqs. (2.17)-(2.19) show how one can write the analogue to the Wigner distribution for arbitrary operators by the use of the Weyl rule of association.

b) Symmetrization Rule.

Instead of taking Eq. (2.6) we could have taken the quantum characteristic function according to the product rule,

$$M_2^{ab}(\theta, \tau) = < e^{i\tau B} e^{i\theta A} > \; . \tag{2.20}$$

This corresponds to a different ordering rule for the operators than the Weyl rule. The symmetrical version of Eq. (2.20), namely Eq. (2.29), corresponds to the use of the so called symmetrization rule. To evaluate this characteristic function we define the transformation matrix, T_{ab} , from the A to the B representation, by

$$v_a(q) = \int T_{ab} w_b(q) db \tag{2.21}$$

where

$$T_{ab} = \int w_b^*(q) v_a(q) dq \; . \tag{2.22}$$

Hence,

$$e^{i\tau B} e^{i\theta A} \psi(q) = e^{i\tau B} e^{i\theta A} \int v_a(q) \phi^A(a) da = \int \int T_{ab} e^{i\theta a + i\tau b} \phi^A(a) w_b(q) \, da \, db \; . \tag{2.23}$$

Therefore

$$M_2^{ab}(\theta, \tau) = \int \psi^*(q) e^{i\tau B} e^{i\theta A} \psi(q) dq \tag{2.24}$$

$$= \int \int \int \phi^{*B}(b') w_{b'}^*(q) T_{ab} e^{i\tau b + i\theta a} \phi^A(a) w_b(q) dq \, da \, db \, db' \tag{2.25}$$

$$= \int \int \phi^{*B}(b) T_{ab} \phi^{A}(a) e^{i\tau b + i\theta a} \, da \, db \, . \tag{2.26}$$

The Fourier inverse of $M_2^{ab}(\theta, \tau)$ is

$$P_2^{ab}(q, p) = \frac{1}{4\pi^2} \int \int M_3^{ab}(\theta, \tau) e^{-i\theta a - i\tau b} \, d\theta \, d\tau \tag{2.27}$$

$$= \phi^{*B}(b) T_{ab} \phi^{A}(a). \tag{2.28}$$

This distribution and its generalization to N operators has been previously given by Barut [2] who derived it in a different way. In addition, Barut's paper gives a number of interesting consequences for this distribution. As pointed out by Barut, the real part is also a distribution that satisfies the marginals. The real part is derivable by taking the characteristic function to be

$$M_2^{ab}(\theta, \tau) = < \frac{e^{i\tau B} e^{i\theta A} + e^{i\tau A} e^{i\theta B}}{2} > , \tag{2.29}$$

which is equivalent to the use of the symmetrization rule.

c) General Correspondence

For the case of position and momentum, an infinite number of joint distributions which satisfy the marginals exist and can be readily generated. For each ordering rule or correspondence rule there will be a different distribution. A way to characterize explicitly all possible bilinear distributions is [5],

$$P(q, p) = \frac{1}{4\pi^2} \int \int \int e^{-i\theta q - i\tau p + i\theta u} f(\theta, \tau) \, \psi^*(u - \tfrac{1}{2}\tau) \, \psi(u + \tfrac{1}{2}\tau) \, du \, d\tau \, d\theta \ , \tag{2.30}$$

where $f(\theta, \tau)$ is the kernel which characterizes the distribution. For example, if we take $f = 1$ then we have the Wigner distribution, and if we take $f = e^{i\theta\tau/2}$ we have the symmetrical distribution

$$P(q, p) = \frac{1}{\sqrt{2\pi}} \psi(q) \phi^*(p) e^{iqp} \ . \tag{2.31}$$

By taking arbitrary kernels we can generate joint distributions at will. To assure that the marginals are satisfied one must take $f(\theta, \tau)$'s such that

$$f(\theta, 0) = f(0, \tau) = 1 \ . \tag{2.32}$$

To generalize to arbitrary operators we form a general characteristic function

$$M_g^{ab}(\theta, \tau) = f(\theta, \tau) \, M^{ab}(\theta, \tau) \ , \tag{2.33}$$

where $M^{ab}(\theta, \tau)$ is any characteristic function consistent with the marginals, for example, Eq. (2.6) or Eq. (2.20).

If we use Eq. (2.6) for M^{ab} the general joint distribution is then

$$P_g^{ab}(a, b) = \frac{1}{4\pi^2} \int \int M_g^{ab}(\theta, \tau) e^{-i\theta a - i\tau b} \, d\theta \, d\tau \tag{2.34}$$

$$= \int \int \psi^*(q') G_g(a, b; q, q') \psi(q) dq dq' \tag{2.35}$$

where now

$$G_g(a, b; q, q') = \frac{1}{4\pi^2} \int \int \int f(\theta, \tau) u_\gamma^*(q) e^{i\gamma - i\theta a - i\tau b} u_\gamma(q') \, d\gamma \, d\theta \, d\tau \, . \tag{2.36}$$

If we use Eq. (2.26) for $M^{ab}(\theta, \tau)$, the general distribution can be expressed as

$$P_g^{ab}(a, b) = \int \int K_g(a, b; a', b') \phi^{*B}(b') T_{a'b'} \phi^A(a') da' \, db' \tag{2.37}$$

$$= \int \int K_g(a, b; a', b') P_2(a', b') da' \, db' \tag{2.38}$$

where

$$K_g(a, b; a'b') = \frac{1}{4\pi^2} \int \int f(\theta, \tau) e^{i\theta(a-a') + i\tau(b-b')} \, . \tag{2.39}$$

The relation between G_g and K_g will be studied in another paper.

It is easy to show that the joint distributions derived above satisfy the proper marginal conditions Eq. (1.6) and (1.7). In terms of the characteristic function the marginal requirements are expressed by the conditions

$$M^{ab}(\theta, 0) = \langle e^{i\theta A} \rangle \tag{2.40}$$

$$M^{ab}(0, \tau) = \langle e^{i\tau B} \rangle \ , \tag{2.41}$$

and it is clear that M_g^{ab} satisfies Eq. (2.39) and (2.40) if $f(\theta, \tau)$ is chosen to have the properties given by Eq. (2.32).

3. CONCLUSION

As an illustration of the above, consider the case where $A = q$ and $B = -id/dq$, and we take for a and b the c-numbers q and p respectively. The eigenfunction equation given by Eq. (2.9) can be readily solved,

$$u_\gamma(q) = \frac{1}{\sqrt{2\pi\tau}} e^{i(q\gamma - \frac{1}{2}\theta q^2)/\tau} \, . \tag{3.1}$$

Calculating $G(q, p; x, x')$ by using Eq. (2.19) we have

$$G(q, p; x, x') = \frac{1}{2\pi} \delta(\frac{x' + x}{2} - q) e^{i(x' - x)p} \ , \tag{3.2}$$

and substituting into Eq. (2.18) straightforwardly yields the Wigner distribution,

$$P_1^{qp}(q,p) = \frac{1}{\pi}\int \psi^*(2q - x)e^{2i(q-x)p}\psi(x)dx\,. \tag{3.3}$$

For the symmetrization rule, the transformation matrix between position and momentum representation is of course

$$T^{ab} = \frac{1}{\sqrt{2\pi}}e^{iqp}\,, \tag{3.4}$$

which yields the distribution

$$P_2^{ab} = \frac{1}{\sqrt{2\pi}}\psi(q)\phi^*(q)e^{iqp}\,. \tag{3.5}$$

In conclusion, we discuss the circumstances when the classical transformation Eqs. (1.1)-(1.3) produce quantum distributions which satisfy the marginals. Consider the case where $a = a(q)$ and $b = b(p)$ and the quantum operators are given by $a(Q)$ and $b(P)$. In such a case the classical transformation gives

$$P_{ab}(a,b) = \int\!\!\int \delta(a - a(q))\,\delta(b - b(p))\,P(q,p)\,dq\,dp\,, \tag{3.6}$$

where $P(q,p)$ is any quasi distribution for position and momentum. For this case the marginals are satisfied. Consider for example the integration over the variable b,

$$\int P_{ab}(a,b)\,db = \int\!\!\int\!\!\int \delta(a - a(q))\,\delta(b - b(p))\,P(q,p)\,dq\,dp\,db. \tag{3.7}$$

$$= \int\!\!\int \delta(a - a(q))\;\; P(q,p)\,dq\,dp. \tag{3.8}$$

$$= \int \delta(a - a(q))\,b(p))\,|\psi(q)|^2\,dq \tag{3.9}$$

$$= |\psi(q(a))|^2\frac{dq}{da}\,, \tag{3.10}$$

which is the correct quantum result. However, we point out that that even though for this case the classical method yields a distribution that satisfies the marginals, it is not the same distribution obtained by Weyl ordering of the operators.

REFERENCES

1. For a general review see: N. I. Balazs and B. K. Jennings, Physics Reports 104, 347 (1984); M. Hillery, R. F. O'Connell, M.O. Scully, and E.P. Wigner, Physics Reports, 106, 121 (1984).
2. A. O. Barut, Phys. Rev. 108, 565 (1957).
3. H. Margenau and N. R. Hill, Prog. Theoret. Phys. 26, 722 (1961).
4. M. Lax and H. Yuen, Phys. Rev. 172, 362 (1968).
5. L. Cohen, J. Math. Phys. 7, 781 (1966); b) J. Math. Phys. 17, 1863 (1976); c) in: *Frontiers of Nonequilibrium Statistical Physics* , edited by G. T. Moore and M. O. Scully (Plenum Press, New York, 1986).
6. J.E. Moyal, Proc. Cambridge Phil. Soc. 45, 99 (1949).

OPERATOR RELATIONS, THE EIGENVALUE PROBLEM, AND REPRESENTABILITY FOR QUANTUM PHASE SPACE DISTRIBUTIONS

Leon Cohen
Hunter College of The City University
New York, N Y 10021

INTRODUCTION. There are an infinite number of quantum phase space distribution functions which can be used to calculate quantum mechanical expectation values in phase space. The main point of this paper is to give operator relations between distributions and phase functions and to show how one can use these relations to derive results in a straightforward way. Known or easily derived results for one distribution are readily transformed into equivalent equations for another distribution. This avoids rederivation of results whenever we are considering a new distribution. We will illustrate this by considering the eigenvalue problem. This approach also formulates so called quantum mechanical phase space in a distribution independent way in the same sense that ordinary quantum mechanics can be expressed in an arbitrary representation. We restrict ourselves to discussion of bilinear distributions.

We first give some mathematical results which allow manipulation of phase space functions in a particularly easy fashion. If g, h, and F are ordinary functions then

$$\frac{1}{4\pi^2}\int g(\theta,\tau)e^{+i\theta(q'-q)+i\tau(p'-p)}F(q',p')d\theta\, d\tau dq' dp' = g(i\frac{\partial}{\partial q}, i\frac{\partial}{\partial p})\ F(q,p)\ , \tag{1}$$

$$\frac{1}{4\pi^2}\int g(q',p')e^{-i\theta(q'-q)-i\tau(p'-p)}h(q-½\tau, p-½\theta)d\theta\, d\tau dq' dp' = g(q+½i\frac{\partial}{\partial p}, p+½i\frac{\partial}{\partial q})\ h(q,p)\ . \tag{2}$$

In addition, if we have two phase functions $A(q,p)$ and $B(q,p)$

$$g(i\frac{\partial}{\partial q}, i\frac{\partial}{\partial p})\ A(q,p)\ B(q,p) = g(i\frac{\partial}{\partial q_A}+i\frac{\partial}{\partial q_B}, i\frac{\partial}{\partial p_A}+i\frac{\partial}{\partial p_B})\ A(q,p)\ B(q,p)\ , \tag{3}$$

where the subscripts indicate on which functions the operations are performed. That is, in general we can use $\partial/\partial q = \partial/\partial q_A + \partial/\partial q_B$ and $\partial/\partial p = \partial/\partial p_A + \partial/\partial p_B$ if the operations are on a product of functions. Also, by expanding g in a Taylor series it is easy to show that

$$\int\int A(q,p)\ g(i\frac{\partial}{\partial q}, i\frac{\partial}{\partial p})\ B(q,p)dq\, dp = \int\int B(q,p)\ g(-i\frac{\partial}{\partial q}, -i\frac{\partial}{\partial p})\ A(q,p)\ dq\, dp\ . \tag{4}$$

TRANSFORMATION OF DISTRIBUTIONS AND PHASE FUNCTIONS. All bilinear distributions can be put in the form [1]

$$F(q,p) = \frac{1}{4\pi^2}\int e^{-i\theta q-i\tau p+i\theta u} f(\theta,\tau)\ \psi^*(u-½\tau)\ \psi(u+½\tau)\ du\, d\tau\, d\theta\ , \tag{5}$$

where $F(q,p)$ is the distribution and $f(\theta,\tau)$ is the kernel which determines the particular distribution being considered. For example, if we take $f = 1$ then we have the Wigner distribution and if we take $f = e^{i\theta\tau/2}$ we have the normally ordered distribution. The characteristic function is defined by

$$M(\theta,\tau) = \langle e^{i\theta q+i\tau p}\rangle = \int\int F(q,p)e^{i\theta q+i\tau p}dqdp = f(\theta,\tau)\int \psi^*(u-\tau/2)\, e^{i\theta u}\, \psi(u+\tau/2)\, du\ . \tag{6}$$

Now suppose we have two distributions F_1 and F_2 with kernels f_1 and f_2, then from (6) we see that their characteristic functions are related by

$$M_1(\theta,\tau) = \frac{f_1(\theta,\tau)}{f_2(\theta,\tau)}M_2(\theta,\tau)\ . \tag{7}$$

Using Eqs. (5), (6), and (7) we have that

$$F_1(q,p) = \frac{1}{4\pi^2}\int_{-\infty}^{\infty}\frac{f_1(\theta,\tau)}{f_2(\theta,\tau)}e^{i\theta(q'-q)+i\tau(p'-p)}F_2(q',p')\ d\theta\, d\tau\, dq'\, dp'\ , \tag{8}$$

and using (1) we have a general transformation from one distribution to another:

$$F_1(q,p) = \frac{f_1(i\frac{\partial}{\partial q}, i\frac{\partial}{\partial p})}{f_2(i\frac{\partial}{\partial q}, i\frac{\partial}{\partial p})} F_2(q,p) \ . \tag{9}$$

Now consider the random variables which are to be used to find expectation values via phase space integration. If $G(Q,P)$ is the quantum operator function of the position and momentum operators and $g(q,p)$ is the phase function, then for each distribution we want to assure that

$$< G > = \int \psi^*(q)\, G(Q,P)\, \psi(q)\, dq = \int\int g_1(q,p)\, F_1(q,p)\, dq\, dp = \int\int g_2(q,p)\, F_2(q,p)\, dq\, dp \ . \tag{10}$$

Substituting for F_2 as given by (9),

$$< G > = \int\int g_2(q,p) \frac{f_2(i\frac{\partial}{\partial q}, i\frac{\partial}{\partial p})}{f_1(i\frac{\partial}{\partial q}, i\frac{\partial}{\partial p})} F_1(q,p)\, dq\, dp = \int\int \left\{ \frac{f_2(-i\frac{\partial}{\partial q}, -i\frac{\partial}{\partial p})}{f_1(-i\frac{\partial}{\partial q}, -i\frac{\partial}{\partial p})} g_2(q,p) \right\} F_1(q,p)\, dq\, dp . \tag{11}$$

Therefore, we have

$$g_1(q,p) = \frac{f_2(-i\frac{\partial}{\partial q}, -i\frac{\partial}{\partial p})}{f_1(-i\frac{\partial}{\partial q}, -i\frac{\partial}{\partial p})} g_2(q,p) \tag{12}$$

for the transformation of phase functions. To obtain any particular phase function corresponding to the quantum operator we use [1]

$$g(q,p) = \frac{e^{\frac{i}{2}\frac{\partial^2}{\partial q \partial p}}}{f(-i\frac{\partial}{\partial q}, -i\frac{\partial}{\partial p})} G_Q(q,p) \ , \tag{13}$$

where $G_Q(q,p)$ is the quantum operator expressed in normal form, that is, the Q factors are made to precede the P factors and then one substitutes q and p for the operators. In this way one can generate so called correspondence rules at will. For each correspondence rule there is associated with it a distribution function. We emphasize that the classical-type function used in Eq. (10) as the random variable is not necessarily the classical function. It must be obtained from the quantum operator by Eq. (13). The quantum operator can be obtained from $g(q,p)$ by [1]

$$G(Q,P) = \frac{1}{4\pi^2}\int\int f(\theta,\tau) g(q,p)\, e^{i\theta(Q-q)+i\tau(P-p)} d\theta\, d\tau\, dq\, dp \ . \tag{14}$$

EIGENVALUE PROBLEM. [1,2] We now illustrate the above by showing how one can derive the energy eigenvalue problem for phase space distributions in a simple way. One of the simplest distributions is obtained by taking $f = e^{i\theta\tau/2}$, which results in

$$F_M(q) = \psi(q)\phi^*(p)e^{-iqp} \ . \tag{15}$$

Now, starting with the time-independent Schrödinger equation

$$\left[-\frac{1}{2m}\frac{\partial^2}{\partial q^2} + V(q) \right] \psi(q) = E\psi(q) \tag{16}$$

and multiplying both sides by $\phi^*(p)e^{-iqp}$, we immediately get that

$$\left\{ \frac{1}{2m}\left(p - i\frac{\partial}{\partial q} \right)^2 + V(q) \right\} F_M(q,p) = E\, F_M(q,p) \ . \tag{17}$$

Now suppose we want the eigenvalue equation for the Wigner distribution, F_W. Using (9) we have that

$$F_W(q,p) = e^{\frac{i}{2}\frac{\partial^2}{\partial q \partial p}} F_M(q,p) \ . \tag{18}$$

Multiply Eq. (17) by $e^{\frac{i}{2}\frac{\partial^2}{\partial q \partial p}}$. The right-hand side then immediately becomes F_W, and for the left-hand side we consider the two terms separately. Consider first the potential term

$$e^{\frac{i}{2}\frac{\partial^2}{\partial q \partial p}}\; V(q)\;\; F_M(q,p)\;, \tag{19}$$

which by using Eq. (3) becomes

$$e^{\frac{i}{2}\frac{\partial}{\partial q_V}\frac{\partial}{\partial p_F}}\; V(q)\;\; e^{\frac{i}{2}\frac{\partial}{\partial q_F}\frac{\partial}{\partial p_F}} F_M(q,p) \;=\; e^{\frac{i}{2}\frac{\partial}{\partial q_V}\frac{\partial}{\partial p_F}}\; V(q)\;\; F_W(q,p) \;=\; V(q + \frac{i}{2}\frac{\partial}{\partial p})\, F_W(q,p)\;. \tag{20}$$

Now consider the first term in Eq. (17),

$$e^{\frac{i}{2}\frac{\partial}{\partial q}\frac{\partial}{\partial p}}\,\frac{1}{2m}\left(p - i\frac{\partial}{\partial q}\right)^2 F_M \;=\; e^{\frac{i}{2}\frac{\partial}{\partial q_F}\frac{\partial}{\partial p_A}}\,\frac{1}{2m}\left(p - i\frac{\partial}{\partial q}\right)^2 e^{\frac{i}{2}\frac{\partial}{\partial q_F}\frac{\partial}{\partial p_F}}\, F_M \tag{21}$$

$$=\; \frac{1}{2m}\left(p + \frac{i}{2}\frac{\partial}{\partial q} - i\frac{\partial}{\partial q}\right)^2 F_W \;=\; \frac{1}{2m}\left(p - \frac{i}{2}\frac{\partial}{\partial q}\right)^2 F_W\;. \tag{22}$$

In the above, the subscript A indicates that the operation is only on the first term. Combining, we have the eigenvalue equation for the Wigner distribution derived in a simple way,

$$\left\{\frac{1}{2m}\left(p - \frac{i}{2}\frac{\partial}{\partial q}\right)^2 + V(q + \frac{i}{2}\frac{\partial}{\partial p})\right\} F_W \;=\; E\,F_W\;. \tag{23}$$

Using Eq. (2), the potential term can be written as

$$V(q + \frac{i}{2}\frac{\partial}{\partial p})\, F(q,p) \;=\; \frac{1}{\pi}\int V(q')\, e^{2i(p-p')(q-q')}\, F(q,p')dp'\,dq'\;. \tag{24}$$

The general eigenvalue problem can be formulated in terms of phase space distributions in a similar fashion. Again it is easy to derive it first for the distribution given by Eq. (15) and then derive it for the Wigner case or for an arbitrary representation,

$$g_M(q, p - i\frac{\partial}{\partial q_F})\, F_M(q,p) \;=\; \alpha\, F_M(q,p) \tag{25}$$

$$g_W(q + \tfrac{1}{2}i\frac{\partial}{\partial p_F}, p - \tfrac{1}{2}i\frac{\partial}{\partial q_F})\, F_W(q,p) \;=\; \alpha\, F_W(q,p) \tag{26}$$

$$\frac{f(i\frac{\partial}{\partial q}, i\frac{\partial}{\partial p}) f(-i\frac{\partial}{\partial q_A}, -i\frac{\partial}{\partial p_F})}{f(i\frac{\partial}{\partial q_F}, i\frac{\partial}{\partial p_F})}\; g(q + \tfrac{1}{2}i\frac{\partial}{\partial p_F}, p - \tfrac{1}{2}i\frac{\partial}{\partial q_F})\, F(q,p) \;=\; \alpha\, F(q,p)\;. \tag{27}$$

ψ REPRESENTABILITY. Not every function of position and momentum is a proper joint distribution because, for an arbitrary $F(q,p)$, a wave function may not exist which generates it. We must therefore constrain solutions to those distributions for which wave functions exist. Such solutions are called ψ-representable. A necessary and sufficient condition is that the right-hand side of Eq. (28) be factorable as indicated

$$\psi(x)\,\psi^*(x') = \left(\frac{1}{2\pi}\right)^3 \int \frac{F(q,p)}{f(\theta, x - x')}\, e^{+i\,\theta(q-(x+x')/2)\,+\,(x-x')p\}}\, dq\, dp\, d\theta\;. \tag{28}$$

For the Wigner case this becomes,

$$\psi(x)\,\psi^*(x') \;=\; \int F\left(\frac{x+x'}{2}, p\right) e^{i\,(x-x')p}\, dp\;. \tag{29}$$

We point out that it is only for ψ-representable distributions that the variational principle and other standard quantum mechanical results will hold when phase space integration is used to calculate expectation values.

REFERENCES.

1. See the following and references therein: a) L. Cohen, J. Math. Phys. 7, 781 (1966); b) J. Math. Phys. 17, 1863 (1976); c) in: *Frontiers of Nonequilibrium Statistical Physics* , edited by G. T. Moore and M. O. Scully (Plenum Press, New York, 1986); d) in : *Density Matrices and Density Functional Theory* , edited by R. Erdahl and V. H. Smith, in print.

2. J.P. Dahl,in *Energy Storage and Redistribution in Molecules*, Edited by J. Hinze, Plenum Publishing Co. (1983); in *Semiclassical Description of Atomic and Nuclear Collision*, Eds. J. de Boer and J. Bang (Elsevier, Amsterdam 1985).

SUMS OVER PATHS ADAPTED TO QUANTUM THEORY IN PHASE SPACE

José M. Gracia-Bondía and J. C. Varilly
Escuela de Matemática, Universidad de Costa Rica
San José, Costa Rica

I. In the WWM formalism[1], it is well-known that information about a quantum system is stored in the "evolution function" or "twisted exponential", this is to say, the solution of the (twisted product) Schrödinger equation:

$$2i\, \partial\chi_H/\partial t = H \times \chi_H \; ; \; \chi_H(0) = 1 \, . \qquad (1)$$

Here H denotes the classical hamiltonian of the system under consideration, χ_H is the corresponding evolution function (then, χ_H is a function of time and phase-space coordinates) and $\times$ denotes the twisted product. We take units with $\hbar = 2$.

A Fourier transformation with respect to t gives us the spectral projectors ("Wigner functions") for each value of the energy E:

$$\Pi_H(E) = (1/4\pi)\int \chi_H(t)\, \exp(itE/2)\,dt \, . \qquad (2)$$

The spectrum of H is simply the support of $\Pi_H(E)$ on the E-axis. We prove the following: the evolution function may be expressed as a Feynman-type integral:

$$\chi_H(u;\, t) = \int \mathcal{D}[x(\tau)]\,\mathcal{D}[y(\tau)]\, \exp[-\tfrac{i}{2}\int (H(x) - 2yJ\dot{y} + 2xJ\dot{y})\,d\tau] \qquad (3)$$

where the phase-space trajectories have to fulfil $x(0) = y(0)$, $y(t) = u$; aJb denotes the symplectic product of the vectors a,b.

Proof: $\chi_H(u;\, t) = \lim_{N\to\infty} \exp(-\frac{it}{2N}H) \times \dots \times \exp(-\frac{it}{2N}H) := \lim_{N\to\infty} \chi_H^{(N)}(u;t)$

$$\chi_H^{(N)}(u;\, t) = \exp(-\,\frac{it}{2N}H) \times \chi_H^{(N-1)}(u;\, t) =$$

$$= \iint \underset{\sim}{dx}_N \underset{\sim}{dy}_N \, \exp[-\tfrac{i}{2}(\tfrac{t}{N}H(x_N) - 2uJx_N - 2x_NJy_N - 2y_NJu)]\; \chi_H^{(N-1)}(y_N) = \dots$$

$$\dots = \int\dots\int \prod_{j=2}^{N} \underset{\sim}{dx}_j\, \underset{\sim}{dy}_j\, \exp\{-\tfrac{i}{2}[\sum_{j=1}^{N} tH(x_j)/N + \sum_{j=2}^{N} 2(x_j - y_j)J(y_{j+1} - y_j)]\} \qquad (4)$$

with $x_1 = y_2$, $u = y_{N+1}$, the limit of which expression we represent by (3). Here $\underset{\sim}{dx} = (2\pi)^{-n}dx$.

II. (A classical interlude). In much the same way as the classical action is selected by the Euler-Lagrange equations for the ordinary lagrangian, we may regard the integrand in (3) as a "Lagrangian" dependent on the variables x,y. The Euler-Lagrange equations give then:

$$\dot{x}_c = J\partial H/\partial x_c \, ; \quad \dot{y}_c = \dot{x}_c/2 \quad . \qquad (5)$$

We remark that the first equations are nothing but Hamilton's equations; then $x_c(\tau)$ is a classical trajectory governed by H and $y_c(\tau) =$

$= 1/2(x(\tau) + x(0))$; the trajectory is chosen in such a way that $y_c(t) = u$. A simple calculation gives for the phase in (3):

$$g_c(u; t) = \int_0^t \left(H(x_c) + 1/2(x_c J \dot{x}_c)\right) d\tau - x_c(0)\, Ju \quad . \tag{6}$$

Contrary to appearances, this function does not depend on $x_c(0)$. In classical mechanics g_c makes a lot of sense.[2,3] It may be directly related to the action associated to the trajectory $x_c(\tau)$. Let us introduce $\{q(\tau), p(\tau)\} = x_c(\tau)$; $(q_i, p_i) = x_c(0)$; $(q_f, p_f) = x_c(t)$; $\rho = q_f - q_i$, $u = (r,k)$. Then we have:

$$g_c(u; t) = g_c(r, k; t) = k\rho - S(r+\rho/2 \ , \ r-\rho/2 \ ; \ t) \ , \tag{7}$$

where $S(q_f, q_i; t)$ is the action and k turns out to be $\frac{1}{2}\left(\frac{\partial s}{\partial q_f} - \frac{\partial s}{\partial q_i}\right)$ (Legendre transformation). Weinstein[2] calls g_c the "Poincare's generating function" and proves its invariance under linear canonical changes of coordinates. Using (7) one gets the following modified Hamilton-Jacobi equation for g_c :

$$\partial g_c/\partial t = H(u + (1/2)\, J\, \partial g_c/\partial u) \ . \tag{8}$$

III. Let us consider quadratic hamiltonians, of the form $H = 1/2({}^t uBu) + {}^t cu + d$ where B is a symmetric $2n \times 2n$ matrix, c is a vector in $\mathbb{R}^{2n}$ and d is a scalar constant. Now, clearly the usual "trick" for quadratic hamiltonians allowing to calculate propagators by factoring out the classical paths, works also in the present context. This will permit us to calculate with ease the evolution function for any quadratic hamiltonian. We write it in the form:

$$\chi_H(u; t) = F(t) \exp\left[\left(-\frac{i}{2}\, g_c(u; t)\right)\right] \tag{9}$$

and calculate in turn $g_c(t)$, $F(t)$. We make the following Ansatz:

$$g_c(u; t) = {}^t u\, G(t) u + {}^t uk(t) + v(t) \ . \tag{10}$$

Let us define $L := JB$, $R := JG$. Replacing expression (12) in (9) gives:

$$\dot{R} = \frac{1}{2}(L - RL + RL - RLR) \ ; \ \dot{k} = \frac{1}{2}(I - GJ)BJk + (I - GJ)c \ ; \ \dot{v} = d - \frac{1}{8}{}^t kJBJk + \frac{1}{2}{}^t cJk \ . \tag{11}$$

The first of these equations is a variant of the "matrix Riccati equation" interesting in its own right.[4] The solution is:

$$R(t) = -\{\Sigma(t) + I\}^{-1} \{\Sigma(t) - I\} \tag{12}$$

where $\Sigma(t)$ solves $\dot{\Sigma} = -L\Sigma = -\Sigma L$, with $\Sigma(0) = I$. From the group property $\chi(t) \times \chi(t') = \chi(t + t')$ one gets:

$$F(t)F(t') \left[\det G(t + t')\right]^{1/2} = F(t + t') \left[\det(G(t) + G(t'))\right]^{1/2} \tag{13}$$

from which we infer: $F(t) = \left[\det\left(\frac{\Sigma(t) + I}{2}\right)\right]^{-1/2}$.

IV. Let us put in what follows $c = 0$, $d = 0$ for simplicity (as long as $\det B \neq 0$, the general case may be brought to that form except for a trivial summand).

In order to calculate the Poincare's generating functions and spectra what remains to be done is, in essence, the calculation of matrix exponentials. This can be greatly simplified using Williamson's[5] classification theorem for normal canonical forms. The solutions for $n = 1$ are well-known :

TABLE 1

Hamiltonian type	Eigenvalues of L	Generating function	Spectrum
1	real	$2\,H\,\mathrm{th}\,\frac{t}{2}$	T.A.C.
3,4	null	Ht	T.A.C.
5	pure imaginary	$2\,H\,\mathrm{tg}\,\frac{t}{2}$	D.P.P.

The solutions for $\mathbb{R}^4$ may be separated in two classes, in an obvious way, according to whether the hamiltonian decomposes in direct sum of two $\mathbb{R}^2$-hamiltonians or not. The indecomposable cases are:

TABLE 2

Hamiltonian type	Eigenvalues of L	Generating function	Spectrum
1	real	$2H_1\mathrm{th}(t/2)+tH_2\mathrm{sech}^2(t/2)$	T.A.C.
2	complex	$\frac{2(H_1\mathrm{senh}(at\ )+H_2\mathrm{sen}(bt))}{\cosh(at\ )+\cos(bt)}$	T.A.C.
4	null	Ht	T.A.C.
6	pure imaginary	$2H_1\mathrm{tg}\frac{t}{2} + tH_2\sec^2\frac{t}{2}$	D.P.P.

In the tables T.A.C. means "transient absolutely continuous spectrum" and D.P.P. "discrete pure point spectrum".[6] Whenever H_1, H_2 appear it is to be reckoned that $H = H_1 + H_2$ and the Poisson bracket of H_1 and H_2 is zero.

We offer two comments: (i) Calculation of g is relatively simpler than calculation of the action S. This is due to its canonical invariance. The simplicity of the evolution function relative to the usual propagator may be traced back to this fact. Also, the pre-exponential factor is computed much more easily. (ii) We know very little about the singular case ($\det B = 0$, $c \neq 0$) when $n > 1$.

REFERENCES

1. J.M. Gracia-Bondía, Phys. Rev. A 30, 691(1984).
2. A. Weinstein, Inv. mat. 16, 202 (1972).
3. M.S. Marinov, J. Phys. A 12, 31 (1979).
4. P. Winternitz "Lie Groups and solutions of nonlinear differential equations", Université de Montréal preprint.
5. P. Broadbridge, Physica A 99, 494 (1979).
6. J.E. Avron and B. Simon, J. Funct. Anal. 43, 1 (1981).

QUANTUM PHASE SPACE DYNAMICS OF HARD ROD SYSTEMS

P.Kasperkovitz, Ch.Foidl, and R.Dirl
Institut für Theoretische Physik
Technische Universität Wien
Karlsplatz 13, A-1040 Wien, Austria

Introduction. For a long time, classical 1-dimensional systems of hard rods have been considered as a simple model of a gas or a fluid /1,2/. It is known that most quantities of physical interest can be calculated exactly in this model /3,4/. Recently it has been shown that the system is integrable and that all calculations of time-dependent quantities are performed by introducing, implicitly or explicitly, action and angle variables /5/. In this note we present the analogous construction in quantum mechanics using a phase space formalism adapted to the present problem. We discuss only the simplest case: the relative motion of two points of unit mass moving on a ring of diameter $\sqrt{2}$. However, it should be emphasized that all results can be generalized to any finite number of rods of equal mass and finite - even varying - diameter.

Classical mechanics. The relative distance of the two particles is given by $(1/\sqrt{2})(x_1-x_2) = x \in (-\pi,\pi)$ and their relative velocity by $(1/\sqrt{2})(p_1-p_2) = p \in \mathbb{R}$. If

$$(z|2\pi) = z \text{ modulo } 2\pi \in (-\pi,\pi) \tag{1}$$

and sgn denotes the usual sign function, then the state (p_t,x_t) evolving from the state (p,x) in time t is given by

$$p_t\{p,x\} = p \operatorname{sgn}(x+pt|2\pi), \quad x_t\{p,x\} = (x+pt|2\pi) \operatorname{sgn}(x+pt|2\pi). \tag{2}$$

Note that the spatial order of the two particles is conserved: If, for instance, $x_1>x_2$ and hence $x \in (0,\pi)$ at t=0, then this holds for all time ("right-localized states").

The evolution of an initially fully specified state, $(p,x)\to(p_t,x_t)$, determines that of all ensembles of such states. Therefore, if W_0 is the distribution function at time t=0, then

$$W_t(p,x) = W_0(p_{-t}\{p,x\},x_{-t}\{p,x\}). \tag{3}$$

Now let W^R be a distribution of right-localized states, i.e., $W^R(p,x) = 0$ for $x \notin (0,\pi)$. We may then pass from the function W^R to a new function $W^{(R)}$ according to

$$T^{(R)}(p,x) = T^{(R)}(p,x+2\pi) =$$
$$= (1/2) \{ W^R(p,x) + W^R(-p,-x)\} \quad \text{for } x \in (-\pi,\pi). \tag{4}$$

Because of (4), (3), and (2), the function $W^{(R)}$ evolves as

$$W_t^{(R)}(p,x) = W_0^{(R)}(p,x-pt). \tag{5}$$

Equations (5) and (4) show that the original problem has been transformed into a free motion on a torus. Now let A be an observable, A^R be defined by $A^R(p,x)=A(p,x)$ for $x \in (0,\pi)$, and $A^R(p,x)=0$ for $x \notin (0,\pi)$, and $A^{(R)}$ be related to A^R by eq.(4); then the expectation value of A for the state W_t^R is given by

$$\langle A\rangle_{W_t^R} = 2 \int_R dp \int_{2\pi} dx \; W_t^{(R)}(p,x) \; A^{(R)}(p,x). \tag{6}$$

Conventional quantum mechanics. The state space is $\mathcal{H} = L^2(-\pi,\pi)$ and the Hamiltonian $\underline{H}$ is given by the differential operator $-(\hbar^2/2)\, d^2/dx^2$ together with the boundary conditions $\psi(-\pi) = \psi(0) = \psi(\pi) = 0$. The Hilbert space $\mathcal{H}$ decomposes into the two orthogonal subspaces $\mathcal{H}^R = L^2(0,\pi)$ and $\mathcal{H}^L = L^2(-\pi,0)$ and contains a subspace $\mathcal{H}^F$, defined by $\psi \in \mathcal{H}^F$ iff $\psi(x) = -\psi(-x)$, where the evolution given by $\underline{H}$ coincides with the usual free evolution ($\underline{H}^F = -(\hbar^2/2)\, d^2/dx^2$ plus $\psi(-\pi) = \psi(\pi)$). Since $\mathcal{H}^R \cong \mathcal{H}^F$ it is possible to find a unitary operator $\underline{U}^R$: $\mathcal{H}^R \to \mathcal{H}^F$ ("odd continuation" and renormalization of right-localized wave functions). Now let $\underline{P}^R$ be the operator projecting onto $\mathcal{H}^R$, $\underline{W}^R=\underline{P}^R\underline{W}^R\underline{P}^R$ be the density operator of a mixture of right-localized states, $\underline{A}$ be an observable, $\underline{A}^R=\underline{P}^R\underline{A}\underline{P}^R$, and $\underline{T}^{(R)}$ be related to $\underline{T}^R$ by

$$\underline{T}^{(R)} = \underline{U}^R \underline{T}^R \underline{U}^{R\dagger}; \tag{7}$$

then the expectation value of $\underline{A}$ for the state $\underline{W}_t^R$ is

$$\langle \underline{A}\rangle_{\underline{W}_t^R} = \text{trace } \underline{W}_t^R \underline{A} = \text{trace } \underline{W}_t^{(R)} \underline{A}^{(R)}, \tag{8}$$

where

$$\underline{W}_t^{(R)} = \underline{U}_t^F \underline{W}_0^{(R)} \underline{U}_t^{F\dagger}, \qquad \underline{U}_t^F = \exp(-it\underline{H}^F/\hbar). \tag{9}$$

Phase space formulation. Following Weyl, Wigner, and others /6/, we can formulate quantum mechanics in such a way that it looks very similar to classical statistical mechanics. In the present case we start from unitary operators $\underline{G}(n,y)$, $n \in Z$, $y \in R$, defined by

$$\{\underline{G}(n,y)\psi\}(x) = \exp\{in(y+2x)/2\} \; \psi(x+y). \tag{10}$$

Here the wave functions are assumed to be periodic with period 2π; therefore $\underline{G}(n,y)=(-1)^n\underline{G}(n,y+2\pi)=\underline{G}(n,y+4\pi)$. An arbitrary operator $\underline{T}$ is expanded according to

$$\underline{T} = (1/2) \sum_n \int_{4\pi} dy\ T(n,y)\ \underline{G}(n,y) \tag{11}$$

and its Weyl transform is defined as

$$\tilde{T}(\hbar k,x) = (1/2) \sum_n \int_{4\pi} dy\ T(n,y)\ \exp\{i(ky+nx)\}. \tag{12}$$

This peculiar definition, entailing the relations $\tilde{T}(\hbar k,x)=\tilde{T}(\hbar k,x+2\pi)=(-1)^{2k}\tilde{T}(\hbar k,x+\pi)$, is needed to obtain a simple time evolution (eq.(14) below), especially for systems with more than two particles. Note that the "phase space" on which the functions $\tilde{T}$ are defined consists here of all pairs (p,x) with $p=\hbar k$, $2k \in Z$, $x \in (-\pi,\pi)$. Like in the familiar scheme, where p is continuous and x unlimited, it is possible to define a binary relation $\tilde{A}*\tilde{B}$ that mimics the operator product $\underline{AB}$. This can be used to obtain $\tilde{T}^{(R)}$ from $\tilde{T}^{R}$. For $\hbar\downarrow 0$ these functions tend weakly to functions $\bar{T}^{(R)}$ and $\bar{T}^{R}$, respectively, that are related by eq.(4) and agree up to constant factors with the corresponding classical phase space functions.

In terms of Weyl transforms the expectation value (8) can be written as

$$\langle\underline{A}\rangle_{\underline{W}_t^R} = (2\pi)^{-1} \sum_k \int_{2\pi} dx\ \tilde{W}_t^{(R)}(\hbar k,x)\ \tilde{A}^{(R)}(\hbar k,x), \tag{13}$$

where $\tilde{W}_t^{(R)}$ is given by

$$\tilde{W}_t^{(R)}(\hbar k,x) = \tilde{W}_0^{(R)}(\hbar k,x-\hbar kt). \tag{14}$$

Redefinition of phase space. Up to now we considered only right-localized states, but left-localized states may be treated in quite a similar way. Since the new functions $F^{(R,L)}$ or $\tilde{F}^{(R,L)}$ are defined on identical domains this approach would result in a doubling of phase space. This can be avoided by restricting $f^{(R)}$ to $p\geq 0$ and $f^{(L)}$ to $p\leq 0$. In doing so no information is lost since all these functions satisfy $f(p,x)=f(-p,-x)$. However, the line $p=0$ has to be properly taken into account, especially in quantum mechanics.

1) F.Zernike and J.A.Prins, Z.Physik 41(1926)184
2) L.Tonks, Phys.Rev. 50(1936)955
3) D.W.Jepsen, J.Math.Phys. 6(1965)407
4) J.L.Lebowitz and J.K.Percus, Phys.Rev. 155(1967)123
5) P.Kasperkovitz and J.Reisenberger, Phys.Rev. A31(1985)2639; J.Math.Phys. 26(1985)2601
6) M.Hillery, R.F.O'Connell, M.O.Scully, and E.P.Wigner, Phys.Rep. 106(1984)121

AN INTRODUCTION TO TOMITA REPRESENTATIONS IN PHYSICS.

Hans H. Grelland
NORKYB
P.O.B.128, N-1362 Billingstad
Norway

The mathematical language made available to physicists through the work of Tomita and Takesaki can be considered as a generalization of the Wigner function formalism. A natural starting point is the abstract W*-algebra, which is an abstraction of the set of operators on a Hilbert space.

We first consider a variety of sets used in the formulation of the theory. For each set of physical entities we have a mathematical set which is its model, and which is a building block in the theory.

First, we have a set of physical quantities. This set has a subset of bounded physical quantities. As a matter of fact, the set of bounded ones is sufficient, since the value of an unbounded quantity can be obtained from the value of a bounded function of it, e.g., the arctan function of $\mathbb{R}$. One model of the set of bounded quantities is the set $\mathcal{R}$ of self-adjoint elements in a W*-algebra $\mathcal{A}$. A state maps $\mathcal{R}$ into $\mathbb{R}$, and hence $\mathcal{A}$ into $\mathbb{C}$. The model of the state space is the norm-one subset of the dual $\mathcal{A}^*$, or, with a lesser degree of idealization permitted, the predual $\mathcal{A}_*$, which can be considered as a subset of the dual. A particular kind of model is the one in which the algebra is presented by an algebra M of bounded operators, a von Neumann algebra. Such a model is called a representation.

Each W*-algebra has a class of unitarily related representations, called standard, or Tomita, or Tomita-Takesaki representations. They are not irreducible, like the representations described by the Dirac bra/ket formulation of quantum mechanics. On the other hand, Tomita representations exist also for classical theories, which are the ones with commutative W*-algebras. The Hilbert space of the Tomita representations is $L^2(S)$, where S is the phase space of the system, or of its classical counterpart. Hence their relevance to this conference. This fact implies that the Tomita representations of classical and quantum mechanics are different sets of operators on the same Hilbert space. This is useful in the study of quantization/classical limit. In the Tomita representations, the positive part $\mathcal{A}_{*+}$ of the predual is represented by a cone of the Hilbert space. Hence all the normal states, including the mixed ones, have vector representations. This is useful in the study of statistical physics. The two advantages mentioned are also known from the Wigner theory. In fact, the Wigner formulation is essentially one of the several possible Tomita representations.

Consider a Newtonian system with a phase space S. $L^{\infty}(S)$ is the set of bounded functions of S, and is a model for the abstract $\mathcal{A}$. In this case, $L^{\infty}(S)_* = L^{\infty}(S)^* = L^1(S)$, the set of (Lebesgue) integrable functions of S. The subset of positive functions, $L^1(S)_+$, has a subset of norm one, which is a model for the set of states. They are interpreted as statistical distributions over S, and are the Wigner functions of the system. In the Tomita representations of this theory, the Hilbert space is $L^2(S)$. One Tomita representation, called the canonical one, is obtained by interpreting the elements a(x,y) of $L^{\infty}(S)$ as multiplication operators on $L^2(S)$: $Af(x,y) = a(x,y)f(x,y)$. Note that this defines a subset of the total operator algebra on $L^2(S)$. The cone representing the positive part of the predual is the set of positive (hence real) functions in $L^2(S)$. They are the square root of the classical Wigner functions. So, this is the main difference between the Tomita representations and the traditional Wigner representations; one works with "amplitudes" instead of "densities" as state representatives. This leads to a simpler description of the dynamics.

A useful way to describe the various representations, both for classical and for quantum mechanics, is in terms of the generators q,p, which are the representatives of the position and momentum operators, respectively. The representation described above has the generators $q=x\cdot$ and $p=y\cdot$, where x,y are the coordinates of S.

In the quantum case, a Tomita representation on $L^2(S)$ is generated by $q=x\cdot$ and $p=-i\hbar\,\partial/\partial x$. Again, we have a cone of vectors representing the normal states. In this representation the vectors $\psi(x,y)$ are directly related to density matrices: $\psi(x,y) = \rho(x,y)$. The Wigner representation is related to this one by the well-known transformation from the density matrix to the Wigner function. The transformation is unitary on $L^2(S)$. In the representation thus obtained, q,p are represented by the Bopp operators $q = x\cdot + (i\hbar/2)\partial/\partial y$ and $p = y\cdot - (i\hbar/2)\partial/\partial x$, which are now interpreted physically as the position and momentum operators. In this representation, $\Psi(x,y) = W(x,y)$, where $\Psi(x,y)$ is the state vector, and $W(x,y)$ is the corresponding Wigner function. Since q,p are non-local operators, $W(x,y)$ cannot be interpreted as a statistical distribution. After these examples, we review the general theory:

DEFINITIONS. Let $\mathcal{H}$ be a Hilbert space, with a set of bounded operators $\mathcal{L}(\mathcal{H})$. A conjugation on $\mathcal{H}$ is an antilinear isometry J such that $J^2=1$. A standard von Neumann algebra is a pair (M,J), where $M \subset \mathcal{L}(\mathcal{H})$ is a von Neumann algebra, J is a conjugation, and the map $A \mapsto j(A) = JA^*J$ is a *-antiisomorphism. A representation of a W*-algebra is an operator algebra $M \subset \mathcal{L}(\mathcal{H})$ which is *-isomorphic to it. A Tomita (Tomita-Takesaki, standard) representation of $\mathcal{A}$ is a pair (M,J) where M is a representation of $\mathcal{A}$, and (M,J) is a standard von Neumann algebra.

THEOREMS.

(1) Every W*-algebra has a Tomita representation.

(2) Two Tomita representations of a W*-algebra are related by a unitary transformation $A \mapsto UAU^{-1}$.

(3) If (M,J) is a Tomita representation, then there exists a unique cone $P \subset \mathcal{H}$ with the properties: (i) $\xi \in P \Rightarrow J\xi = \xi$ (i.e., the vectors representing states are unaffected by J), (ii) $A \in M \Rightarrow AJAJ\, P \subset P$ (this is a way to construct operators leaving P invariant, mapping states onto states; this is how symmetry operators are constructed), (iii) P is self-dual, i.e. $P=P\hat{}$, where $P\hat{} = \{\eta \in \mathcal{H} \mid \langle\eta|\xi\rangle \geqslant 0 \ \forall\ \xi \in P\}$ (this implies that for two states η,ξ, $\langle\eta|\xi\rangle \geqslant 0$), (iv) $\mathcal{H}$ is linearly spanned by P (we can find a basis for $\mathcal{H}$ consisting of state vectors).

(4) There exists a one-one mapping $P \to M_{*+}$, $\xi \mapsto \omega$, where $\omega(A) = \langle\xi|A\xi\rangle$. In particular, normal states corresponds to vectors of norm 1.

The dynamics of the system is constructed by theorem (3.ii) from the Hamiltonian H(p,q). In the case of quantum mechanics:

$$U(t) = \exp(-itH/\hbar)J\exp(-itH/\hbar)J = \exp(-it(H-JHJ)/\hbar)$$

The generator H - JHJ is also known from the work of Bopp on the dynamics of the Wigner function. It is seen here to be a special case of a general construction. The variety of representations thus obtained shows the great potentiality of Wigner's original idea, basing the description on the phase space of the system, both for classical and quantal systems.

A SEMICLASSICAL SCHEME FOR THE DESCRIPTION OF THE STATIC PROPERTIES OF NUCLEI AT FINITE TEMPERATURES

H. G. Miller
Theoretical Physics Division, NRIMS, CSIR,
Pretoria, South Africa

H. Kohl
Institut für theoretische Physik der Universität
Frankfurt, Robert Mayer Str 8-10, D-6 Frankfurt,
West Germany

Most static calculations at finite temperatures are fully quantum mechanical mean field calculations[1-3]. One of the aims of such calculations is to investigate the occurrence of phase transitions in finite systems at which either shape changes occur or the nuclear system evaporates and undergoes a liquid to gas phase transition. In the latter case, the role of the continuum is usually neglected in the mean field calculations although it must play an important role.

Full quantum mechanical mean field calculations are lengthy and difficult to perform for large systems. However, as we will demonstrate, temperature-dependent density functional theory provides the possibility to calculate the local density of the system in a fully consistent manner from a nonlinear differential equation.

The Wigner transform of the density may be obtained from the inverse Laplace transform of the Bloch propagator[4]

$$f(q,p,\mu) = \mathcal{L}^{-1}_{\beta\to\mu}\left[(e^{-\beta\hat{H}})_W \frac{\pi\beta T}{\sin\pi\beta T}\right], \qquad [1]$$

where μ is the chemical potential, which ultimately is fixed by the number of particles, and the index W designates the Wigner transform of the operator. The factor, $\frac{\pi\beta T}{\sin\beta T}$, contains all of the temperature dependence[5].

For a local potential, V, the Wigner function up to terms of order $\hbar^2$ may be obtained from eq. (1) and is given by

$$f(q,p,\mu) = \phi - \frac{\hbar^2}{8m}\Delta V\cdot\phi'' + \frac{\hbar^2}{24m}\left[(\nabla V)^2 + \frac{(\vec{p}\cdot\vec{\nabla})^2}{m}V\right]\phi''', \qquad [2]$$

where $\phi = (1+\exp(\beta(\frac{p^2}{2mc^2} + V-\mu)))^{-1} = (1+\gamma)^{-1}$; $\phi^{(k)} = \frac{\partial^k\phi}{\partial u^k}$ and $\beta = \frac{1}{kT}$.

Near the classical turning point and for $T \to 0$, the Fermi function, ϕ, goes over into a step function.

Integrating the sides of eq. (2) over $\frac{d^3p}{\hbar^3}$ yields an equation for the single particle density

$$\rho - I_1(\mu,\beta) + \frac{d^2V}{dq^2}(I_2(\mu,\beta) + I_4(\mu,\beta)) + (\frac{dV}{dq})^2 I_3(\mu,\beta))$$

$$+ \frac{2}{q}\frac{dV}{dq} I_2(\mu,\beta), \qquad [3]$$

where

$$I_1(\mu,\beta) = \frac{(2mc^2)^{3/2}}{2\pi^2(\hbar c)^3\beta^{3/2}} \int dx \frac{x^2}{\gamma+1}$$

$$I_2(\mu,\beta) = -\frac{(2mc^2)^{1/2}}{8\pi^2(\hbar c)} \beta^{1/2} \int dx\, x^2 \frac{\gamma(\gamma-1)}{(\gamma+1)^3}$$

$$I_3(\mu,\beta) = \frac{(2mc^2)^{1/2}\beta^{3/2}}{24\pi^2(\hbar c)} \int dx\, x^2 \frac{\gamma(\gamma^2-4\gamma+1)}{(\gamma+1)^4}$$

$$I_4(\mu,\beta) = \frac{(2mc^2)^{1/2}\beta^{1/2}}{36\pi^2(\hbar c)} \int dx\, x^4 \frac{\gamma(\gamma^2-4\gamma+1)}{(\gamma+1)^4}$$

and

$$x = \sqrt{\frac{\beta}{2mc^2}}\, \rho .$$

If one now assumes that $V := V(\rho)$ one obtains a second order highly nonlinear differential equation for determining ρ. The following choice of boundary conditions appears to be physically reasonable,

$$\rho(\infty) = \rho'(\infty) = 0 .$$

The above equation must be solved iteratively with μ determined by the subsidiary condition

$$\int \rho dv = N. \qquad [4]$$

Eq. (3) may be reduced to a first order nonlinear differential equation if one assumes, as in the WKB approximation, that V is a reasonably smooth function of q. In this case $\frac{d^2V}{dq^2} \approx 0$.

At low temperatures, one expects to obtain a localized distribution for the density if Eq. (4) is to be satisfied. At higher temperatures, the density should tend toward a constant distribution (particularly if the equations are solved numerically within a box of finite dimension). This behavior would be indicative of a liquid to gas phase transition which has also been observed in the full quantum mechanical treatment of the problem.

References

1. M. Brack and Ph. Quentin, Phys. Sc. A10 (1974) 163; Phys. Lett. 52B (1974) 159.
2. H. G. Miller, R. M. Quick, G. Bozzolo, and J. P. Vary, Phys. Lett. 168B (1986) 13.
3. G. Bozzolo and J. P. Vary, Phys. Rev. C31 (1985) 1909.
4. M. Durand and P. Schuck, preprint 1985.
5. P. Morel and Ph Nozieres, Phys. Rev. 126 (1962) 1909, M. Brack and Ph. Quentin, Nucl. Phys. A361 (1981) 35.

WIGNER-KIRKWOOD EXPANSION AND MANY BODY QUANTUM CORRECTIONS CALCULATIONS

Fabrizio Barocchi
Dipartimento di Fisica, Universita' di Firenze
Largo Enrico Fermi 2, I-50125 Firenze, Italy

Martin Neumann
Institut Für Experimentalphysik Der Universität Wien
Strudlhofgasse 4, A-1090 Wien, Austria

Marco Zoppi
Consiglio Nazionale delle Ricerche
Istituto di Elettronica Quantistica
Via Panciatichi 56/30, I-50127 Firenze, Italy

In recent years, the development of fast computers has permitted a great expansion of the simulation techniques for calculating properties of dense gases, liquids, and solids, in the framework of classical mechanics. However, calculations of quantum properties is still at a preliminary level and some approximate methods are currently under testing (1-7).

The method we have proposed (8), and extensively used (3-5), in order to evaluate the quantum corrections to the classical properties of an N-body system is based on the Wigner's expansion, in series of h, of any quantum mechanical average (9). This method has advantages and disadvantages. In fact, its use is limited to those systems and thermodynamic conditions for which the h-expansion is convergent. However, when this condition is fulfilled, the reliability of the results is total, as no approximation is involved in the calculations. Moreover, since quantum corrections are evaluated by means of special averages on a classical ensemble, the same sampling of the phase space, generated by computer simulation, can be used to calculate a large number of properties.

The basic concepts behind the method are the following. The quantum mechanical average of any operator A(R,P), which corresponds to the observable A of an N-body system, function of the 3N-dimensional operators R and P, is written as

$$\langle A\rangle = \mathrm{Tr}(\rho A) = h^{-3N}\int dr\, dp\, \rho_W(r,p)\, A_W(r,p) \qquad (1)$$

where r and p are the 3N-dimensional variables of the classical phase-space, corresponding to the operators R and P, while ρ_W and A_W are the Weyl-Wigner equivalent functions of the density matrix ρ and the operator A, respectively. Either ρ_W and A_W can be expanded in series of h, the product can be recast as a series in h which, in turn, can be integrated term by term. As a result we get an expression for <A> which is a power expansion in h and whose coefficients are classical averages of functions of r and p. Of course, in practice, the situation is a little more complex than it can appear at present and we refer to our previous papers for the details (3-5,8).

We have developed the series expansion to the third non-zero correction (i.e., the sixth power in h) and we have evaluated a number of properties of a moderately quantum Lennard-Jones fluid in different points of its phase space. We have chosen to work with the pairwise additive Lennard-Jones interaction potential as it is the most widely used potential in classical simulations of monatomic systems.

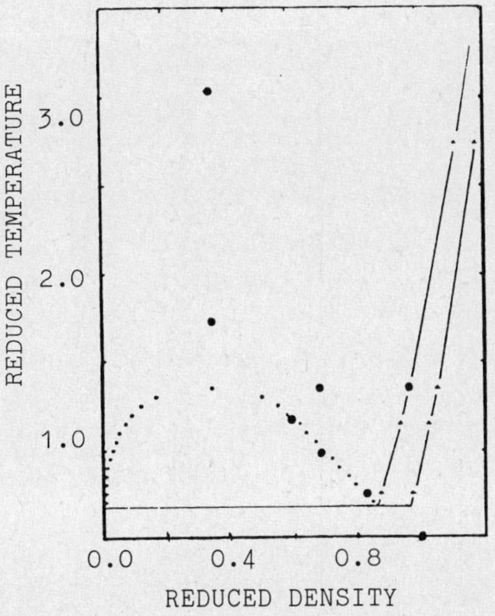

Fig.1 Large dots in the diagram indicate the thermodynamic points of our simulations.

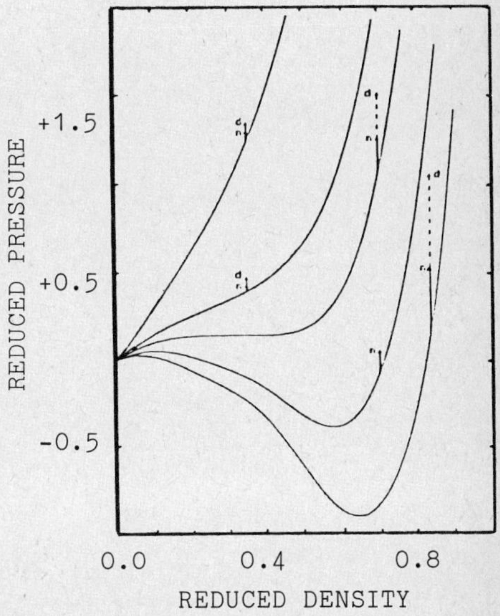

Fig.2 Classical values and quantum corrections to the pressure of neon and deuterium (n,d).

Fig.1 shows the phase diagram for a classical Lennard-Jones system, as it is obtained by the equation of state of Nicholas et al (10), and the dots indicate the thermodinamic points of our simulations. Quantum corrections to the kinetic and potential energy, to the pressure, and to the Helmotz free energy, were calculated to fourth order in h. The pair correlation function was evaluated to the sixth order in h. A detailed report of these results will be given elsewhere (11). As an example, we report here the corrections to the pressure for various systems. It should be mentioned, however, that, since the hamiltonian is always the same (i.e., kinetic energy plus Lennard-Jones potential energy), the same reduced temperature and density corresponds to different temperatures and densities for different systems. One can reverse this concept and obtain, for various systems, different reduced value for h which can be used in the calculations. Fig. 2 shows some results. The full lines are the classical values, while the arrows show the changes due to quantum corrections. Full arrows mean that the second correction term is less than 30% of the first, while broken arrows mean that the second term is within 30% and 50% of the first. This gives an indication of the convergence rate of the series.

To conclude, we would like to show how the calculated quantities compare with the experimental ones when quantum corrections are important. To give an example, the classical calculated value of the pressure of Ne at T=35.1 K and ρ=33.37 nm^{-3} is negative and turns out -2.0 Mpascal. This indicates an unstable thermodinamic state. The experimental value is +2.17 Mpascal. If the quantum corrected value for the pressure is evaluated, the result is +0.98 Mpascal. The inclusion of quantum properties makes the thermodynamic state stable and the remaining discrepancy could be attributed to the not perfect resemblance of the true potential of neon with the Lennard-Jones expression.

REFERENCES

1-N. Corbin and K. Singer, Mol. Phys. 46, 671 (1982)
2-K. Singer and W. Smith, Mol. Phys. 57, 761 (1986)
3-F. Barocchi, M. Zoppi, and M. Neumann, Phys. Rev. A27, 1587 (1983)
4-F. Barocchi, M. Neumann, and M. Zoppi, Phys. Rev. A29, 1331 (1984)
5-F. Barocchi, M. Neumann, and M. Zoppi, Phys. Rev. A31, 4015 (1985)
6-E.L. Pollock and D.M. Ceperley, Phys. Rev. B30, 2555 (1984)
7-D.M. Ceperley and E.L. Pollock, Phys. Rev. Lett. 56, 351 (1986)
8-F. Barocchi, M. Moraldi, and M. Zoppi, Phys. Rev. A26, 2168 (1982)
9-E.P. Wigner, Phys. Rev. 40, 749 (1932)
10-J.J. Nicholas, K.E. Gubbins, W.B. Streett, and D.J. Tildesley, Mol. Phys. 37, 1429 (1979)
11-F. Barocchi, M. Neumann, and M. Zoppi, To be published

A General Approximation Scheme for Quantum Many-Body Dynamics

M. Ploszajczak

Institute of Nuclear Physics, PL-31-342 Krakow, Poland

M.J. Rhoades-Brown and M.E. Carrington

Physics Department SUNY, Stony Brook, N.Y. 11794

1. Introduction

A large variety of phenomena has been observed in heavy ion collisions, calling for a comprehensive theory based on basic principles of quantum many-body dynamics. At low energies the nuclear dynamics is governed by the mean field. At higher energies, as the two body correlations become more effective in distributing particles over the entire phase-space, a proper theory should include collision effects. Finally, at fully relativistic energies the comprehensive theory should also contain the field aspects of nuclear explosion, the production mechanism of a large number of non-conserved quanta and the possibility for the formation of exotic forms of nuclear matter such as the quark-gluon plasma (QGP). In all these energies domains the applications of kinetic theories proved to be very fruitful, leading to an increased interest in the Wigner phase-space representation of quantum mechanics[1–3]. In this contribution, we present a new self-consistent approximation scheme for the quantum mechanics of a many-body system which is valid for the non-relativistic and relativistic domains. This scheme, which was shown to be valid for both pure states and statistical averages does not assume the concept of local equilibrium or the existence of convergent iterative procedure via Green's functions. The Wigner representation[6] is used in our representation only as a tool to approximate the Liouville equation for the evolution of the density matrix and allows to gain an insight into a problem of construction of various useful many-body models valid beyond the classical domain.

2. Non-Equilibrium Approximation Scheme in the Mean Field Limit

The evolution of a quantum system can be described using the Liouville equation:

$$i\hbar\partial_t\hat{\rho}(x,x';t) = [\hat{H},\hat{\rho}(x,x';t)], \tag{1}$$

where $\hat{\rho}$ is the density matrix and $\hat{H}$ is a many-body hamiltonian. In the Hartree approximation, (1) can be rewritten as an equation for a one-body density matrix:

$$i\hbar\partial_t\rho(x,x';t) = [\hat{H}_{MF},\rho(x,x';t)], \tag{2}$$

whose evolution is goverened by the mean-field hamiltonian $\hat{H}_{MF} = T+U$ consisting of the kinetic energy operator T and an effective one body potential U. In general, U depends on the density ρ. (2) can also be expressed in terms of the one-body reduced Wigner distribution:

$$\partial_t f + \frac{p}{m}\cdot\partial_x f + \frac{2}{\hbar}U\, sin[\frac{\hbar}{2}\overleftarrow{\partial_x}\overrightarrow{\partial_p}]f = 0, \tag{3}$$

where, $f(x,p;t) = (2\pi\hbar)^{-3}\int d^3y\; exp(ipy/\hbar)\;\rho(x-\frac{1}{2}y, x+\frac{1}{2}y;t)$ and x, p are the position and momentum coordinates respectively. Let us now introduce the moment function[7]:

$$\rho_o(x;t) < p_{x_i}^{n_i}p_{x_j}^{n_j}p_{x_k}^{n_k} > (x;t) \equiv \int p_{x_i}^{n_i}p_{x_j}^{n_j}p_{x_k}^{n_k} f(x,p;t)d^3p$$

$$= (\frac{\hbar}{2i})^{n_i+n_j+n_k}[(\partial_{x_i}-\partial_{x_i'})^{n_i}(\partial_{x_j}-\partial_{x_j'})^{n_j}(\partial_{x_k}-\partial_{x_k'})^{n_k}\rho(x,x';t]_{x=x'}, \tag{4}$$

where x_i, x_j, x_k, (p_{x_i}, p_{x_j}, p_{x_k}) are x,y,z components of the position (momenta) coordinates respectively, and $\rho_o(x;t)$ is the diagonal density. The moment functions permit us to rewrite off-diagonal elements of the density matrix[4]):

$$\rho(x,x';t) = \rho_o(\frac{1}{2}(x+x');t)e^{(-\frac{i}{\hbar}(x-x')\cdot<p>)} < e^{(-\frac{i}{\hbar}(x-x')\cdot(p-<p>)} >, \tag{5}$$

as an expansion in deviations from the mean momentum value or, equivalently, in the deviations from the current density. The equivalent expansion for the Wigner function reads:

$$f(x,p;t) = \rho_o(x) < exp(-(p- < p >)\cdot\partial_p > \delta(p- < p >). \tag{6}$$

Inserting (5) into (2) one obtains an infinite set of coupled equations. For n=0,1,2, these equations have the form of the usual fluid dynamic equations,

$$n = 0 \quad \partial_t \rho_o + \sum_\alpha \partial_{x_\alpha}(\rho_o < u_\alpha >) = 0, \tag{7a}$$

$$n = 1 \quad \rho_o[\partial_t < u_\beta > + \sum_\alpha < u_\alpha > \partial_{x_\alpha} < u_\beta >]$$

$$= m^{-1}\rho_o \partial_{x_\beta} U - \sum_\alpha \partial_{x_\alpha} P_{\alpha\beta}, \tag{7b}$$

$$n = 2 \quad \partial_t P_{\beta\gamma} + \sum_\alpha (P_{\alpha\beta}\partial_{x_\alpha} < u_\beta > + P_{\alpha\gamma}\partial_{x_\alpha} < u_\gamma >)$$

$$= -\sum_\alpha \partial_{x_\alpha}(W_{\alpha\beta\gamma} + < u_\alpha > P_{\beta\gamma}), \tag{7c}$$

where the indicies α, β, γ, change from 1 to 3, $\rho_o < u_\alpha > \equiv m^{-1} < p_\alpha >$ is the mean velocity in the x_α direction, $P_{\alpha\beta} \equiv \rho_o < (u_\alpha - < u_\alpha >)(u_\beta - < u_\beta >) >$ is the kinetic pressure tensor, $W_{\alpha\beta\gamma} \equiv \rho_o < (u_\alpha - < u_\alpha >)(u_\beta - < u_\beta >)(u_\gamma - < u_\gamma >)$, is a component of a heat current density. However, these coupled equations do not form a closed set, and as such they are useless. To achieve the closure one usually assumes that the quantum system, as a consequence of frequent collisions between its constituents, will always be in local equilibrium. Instead, we prefer to exploit further the possibilities suggested by the form of an expansion (5). Near the classical domain, this expansion should converge rapidly, and in the classical limit $< (p - < p >)^n >= 0$ for all n. The first quantum correction has terms up to $< (p - < p >)^2 > (x' - x)^2/\hbar^2$. Obviously, the density expansion (5) is asymptotically convergent. However, the convergence is not uniform in the whole space. Depending on a separation distance $x' - x$ this expansion may even be strongly convergent at any order in the expansion. This feature makes the density expansion (5) extremely useful for developing practical approximation schemes; it is well known that reactions between composite systems in molecular, nuclear, and particle physics exhibit a large degree of classical behaviour. An introduction of quantum effects into the density matrix should modify $\rho(x,x')$ mainly in the neighbourhood of $\rho_o((x+x')/2;t)$. With this in mind, we suggest that the most physical truncation scheme for moment equations (7) is to put $<(p- <p>)^n>=0$ for all n greater than a given m. It is easy to see that if the truncation scheme is introduced at m=1 then the continuity equation (7a) remains unchanged and the n=1 equation reduces to an analogue of Newton's law. For n=2 one obtains identically zero, thus separating these two equations from the equations for higher moments. For the cutting at m=2 one finds (7a) and (7b) unchanged. (7c)reduces to the quantum analogue of the equation of state,

$$\partial_t P_{\alpha\beta} + \sum_\alpha < u_\alpha > \partial_{x_\alpha} P_{\beta\gamma}$$

$$= -\sum_\alpha [P_{\beta\gamma}\partial_{x_\alpha} < u_\alpha > + P_{\alpha\beta}\partial_{x_\alpha} < u_\gamma > + P_{\alpha\gamma}\partial_{x_\alpha} < u_\beta >], \tag{8}$$

whereas the n=3 fluid dynamical equation is changed into the time-independent consistency condition:

$$P_{\beta\gamma}\partial_{x_\delta} P_{\alpha\delta} + P_{\alpha\beta}\partial_{x_\delta} P_{\gamma\delta} + P_{\alpha\gamma}\partial_{x_\delta} P_{\beta\delta} = -\frac{1}{4m}\rho_o^2\hbar^2\partial^{(3)}_{x_\alpha x_\beta x_\gamma} U, \tag{9}$$

relating the diagonal density and components of the kinetic pressure tensor to the cubic derivatives of the mean-field potential. Again, the coupled equations for n=4 give identically zero. Analogously, for the cutting at m=3, (7a-7c) are unchanged. The continuity equation for components of the heat current density (n=3) now take the form,

$$\partial_t W_{\alpha\beta\gamma} + \sum_\delta < u_\delta > \partial_{x_\delta} W_{\alpha\beta\gamma}$$

$$= \rho_o^{-1} \sum_\delta [P_{\alpha\beta}(\partial x_\delta P_{\gamma\delta}) + P_{\alpha\gamma}(\partial_{x_\delta} P_{\beta\delta}) + P_{\beta\gamma}(\partial_{x_\delta} P_{\alpha\delta})$$

$$-W_{\alpha\gamma\delta}(\partial_{x_\delta} < u_\beta >) - W_{\beta\gamma\delta}(\partial_{x_\delta} < u_\alpha >) - W_{\alpha\beta\delta}(\partial_{x_\delta} < u_\gamma >) - W_{\alpha\beta\gamma}(\partial_{x_\delta} < u_\delta >)]$$

$$+\frac{1}{4}m^{-1}\rho_o^2\hbar^2\partial^{(3)}_{x_\alpha x_\beta x_\gamma}U, \tag{10}$$

for the relations between the density ρ_o and various components of the kinetic pressure tensor and the heat current density. These relations replace constraints imposed by the requirement of local equilibrium and by an equation of state in the ordinary fluid dynamic description. At this level of approximation the equation for n=4 takes the form of the time-independent relation:

$$W_{\alpha\beta\gamma}\partial_{x_\epsilon}P_{\delta\epsilon}+W_{\alpha\beta\delta}\partial_{x_\epsilon}P_{\gamma\delta}+W_{\alpha\gamma\delta}\partial_{x_\epsilon}P_{\beta\delta}+W_{\beta\gamma\delta}\partial_{x_\epsilon}P_{\alpha\epsilon}=0. \tag{11}$$

For reasons of clarity we have introduced our method in the mean field limit. Application of this truncation scheme beyond the Hartree approximation, including collision effects can be found elsewhere[8].

2.1 Conservation Laws

In developing a self-consistent truncation scheme, one has to check whether the simplified theory respects conservation laws for particle number, momentum, and energy which are obeyed by the exact dynamical equations. In the low energy domain the dynamics of heavy-ion collisions is described by the mean-field theory. Conservation laws for this theory were discussed by Bertsch[9]. Here we discuss them in the context of our truncation scheme for the mean field evolution (2). Conservation of particle number,

$$dN/dt=(2\pi\hbar)^{-3}\int d^3p\,d^3x\left(\frac{p}{m}\cdot(\partial_x f)+\frac{2}{\hbar}U\,sin[\frac{\hbar}{2}\overleftarrow{\partial}_x\overrightarrow{\partial}_p]f\right)=0, \tag{12}$$

follows immediately if one changes the volume integrals in (12) into surface integrals. Evaluated at large x, and p values, dN/dt becomes zero. Obviously this property of the Wigner function is not changed by the cutting condition. The conservation of momentum in the mean-field limit follows from the translational invariance of the potential U. Since the self-consistent truncation method leaves the continuity equation, and hence the diagonal density $\rho_o(x)$ unchanged it also does not influence translational properties of the mean-field potential. Consequently, neglecting exchange terms:

$$dp/dt = -Tr(i[\hat{H}_{MF}, \hat{p}]) = -\int d^3x \rho_o(x) \partial_x U = 0. \qquad (13)$$

Analogously, the energy conservation can be traced back to the translational invariance of U(x). Besides the global conservation laws, the mean field theory satisfies also certain local conservation laws. Some of them, like the conservation of particle number and the conservation of momentum, have direct physical meaning and allow for practical evaluation of transport properties in heavy-ion collisions. Particle number conservation is a consequence of the continuity (7a). Local momentum conservation can be derived if the mean-potential U(x) is a functional of the density $\rho_o(x)$ only. This requirement for U is not stringent and, in general, it is satisfied for field-producing short range nucleon-nucleon interactions. Hence, it also does not depend on the relations introduced by the cutting conditions.

3. Covariant Formulation of the Non-equilibrium Truncation Scheme

We begin discussion of the transport properties for the relativistic matter with scalar particles[5)]. The relativistic Liouville equation for scalar field densities is:

$$(\Box_2 - \Box_1) < \phi(x_1)\phi(x_2) > = < \phi(x_1) j(x_2) > - < j(x_1)\phi(x_2) >; \qquad (14)$$

where $\Box$ is taken for the D'Alembertian and j(x) is a source function, and can be expressed in the language of the Wigner phase-space functions. In the mean field limit one obtains the transport equation which resembles closely (3):

$$ip \cdot \partial_x f = -\hbar^{-1} U(x) sin(\frac{\hbar}{2} \overleftarrow{\partial_x} \overrightarrow{\partial_p}) f, \qquad (15)$$

where the "mean-field" potential $U(x) = \lambda(n-1) < \phi^{\dagger n-1}(x)\phi^{n-1}(x) >$ which depends on the centre of mass four vector, is derived from the interaction Lagrangian $L_I = \frac{\lambda}{n}\phi^{\dagger n}\phi^n$ By analogy with (5) we write the relativistic density matrix,

$$\rho(x, x') = \rho_o(\frac{x+x'}{2}) exp(-i(x'-x) \cdot < p >) < exp(-i(x'-x) \cdot (p - < p >) >, \qquad (16)$$

where ρ_o now represents the charge density distribution and $< p >$ is the current density. Inserting (16) into (14) one derives an infinite set of coupled equations which include the continuity equation for the charge and current densities as well as dynamic equations for higher order deviations from the current density. These equations can then be truncated by imposing the condition:

$$< \Pi_{\alpha=1}^{4}(p_\alpha - < p_\alpha >)^{k_\alpha} >= 0, \tag{17}$$

for all $\sum_{\alpha=1}^{4} k_\alpha = n$ greater than a given integer m value[5]. For m=2 one obtains continuity equations describing charge and current density conservations,

$$\sum_\alpha \partial_{x_\alpha}(\rho_o < u_\alpha >) = 0, \tag{18a}$$

$$\sum_\alpha (\partial_{x_\alpha} P_{\alpha\beta}) = -\frac{1}{2}\rho_o \partial_{x_\beta} U - \sum_\alpha \rho < u_\alpha > (\partial_{x_\alpha} < u_\beta >), \tag{18b}$$

as well as the relations between the components of the kinetic pressure tensor and the heat current density which replace the assumption of local equilibrium in ordinary relativistic fluid dynamics.

$$\sum_\alpha < u_\alpha > (\partial_{x_\alpha} P_{\alpha\beta})$$

$$= -\sum_\alpha [P_{\alpha\beta}(\partial_{x_\alpha} < u_\gamma >) + P_{\alpha\gamma}(\partial_{x_\alpha} < u_\beta >) + P_{\beta\gamma}(\partial_{x_\alpha} < u_\alpha >)], \tag{18c}$$

and the consistency relation:

$$\sum_\alpha [P_{\beta\gamma}(\partial_{x_\alpha} P_{\alpha\delta}) + P_{\beta\delta}(\partial_{x_\alpha} P_{\alpha\gamma}) + P_{\gamma\delta}(\partial_{x_\alpha} P_{\alpha\beta})] = -\frac{\rho_o^2 \hbar^2}{8} \partial^{(3)}_{x_\beta x_\gamma x_\delta} U, \tag{18d}$$

between the self-consistent field $U(x)$ and various components of the pressure tensor. It should be emphasised that all global and local conservation laws respected by the non-relativistic mean-field theory are also satisfied in our approximation to the quantum statistical mechanics of the relativistic mean field.

4. Quantum Kinetic Theory of the Quark Gluon Plasma

The collision of heavy ions at ultra-relativistic energies may lead to the formation of QGP. The quantum kinetic theory of such a plasma should respect the spinor character of the quark fields as well as the non-abelian nature of the colour interactions. To date, the attempts to understand the dynamics of QGP have been concentrated on relativistic hydrodynamics[11)] and the classical kinetic theory[12)]. To introduce the quantum effects into these theories, let us write the Wigner distribution function for quark fields,

$$f(x,p) = -(2\pi)^4 \int d^4y \; exp(-ip_\alpha y^\alpha) \; <: \psi(x - y/2) \; \overline{\psi}(x + y/2) :> . \qquad (20)$$

$f(x,p)$ is a solution of the conjugate transport equations resulting from the Dirac equation:

$$i\gamma^\mu \partial_{x_\mu} f = -2(\gamma^\mu p_\mu - m) f + \int d^4y \, exp(-ip_\alpha y^\alpha) < \gamma^\mu Q_a \hat{A}^a_\mu(x'')\psi(x'')\overline{\psi}(x') >,$$

$$if \partial_{x_\mu} \gamma^\mu = 2f(\gamma^\mu p_\mu - m) - \int d^4y \, exp(-ip_\alpha y^\alpha) < \psi(x'')\overline{\psi}(x')\hat{A}^a_\mu(x')\gamma^\mu Q_a >, \quad (21)$$

where the $\hat{A}^a_\mu$ are the non-abelian gluon fields, $Q_a = -\lambda_a/2$ ($a = 1, ...8$) are Gell-Mann matricies, and p is the four momentum. $f(x,p)$ as such is not an observable quantity. Moreover, it does not satisfy the requirement of gauge invariance. Hence, to make (21) useful one should replace them by the coupled infinite hierarchy of equations for the moments of $f(x,p)$. These equations will have a direct value for the descriptions of transport properties of the quark-gluon plasma if the lowest two equations of this hierarchy can be written in the form of continuity equations for locally conserved charge density and current density. This requirement constrains a choice of the momentum operators to be used in this scheme to the kinetic momentum operator:

$$P_\alpha = p_\alpha - Q_b A^b_\alpha . \qquad (22)$$

Utilizing the mean-field limit of (21) one obtains[13],

$$i\partial_{x_\mu} M^{\mu}_{\alpha^n} = i\partial^{(P)}_{x_\mu} M^{\mu}_{\alpha^n} - Tr\Big(\sum_{t=0,2,4...}^{n} \Big(\binom{n}{t}\Big) 2^{-t} (i^t \partial^{(t)}_{x^t_\alpha} A^a_\mu) M^{\mu}_{[\alpha^{n-t},a]-} \Big)$$

$$+Tr\Big(\sum_{t=1,3,5...}^{n} \Big(\binom{n}{t}\Big) 2^{-t} (i^t \partial^{(t)}_{x^t_\alpha} A^a_\mu) M^{\mu}_{[\alpha^{n-t},a]+} \Big), \tag{23}$$

where,

$$M^{\mu}_{\alpha^n}(x) = Tr \int d^4p\, (P_\alpha)^n \gamma^\mu f(X,p), \tag{24}$$

$$M^{\mu}_{[\alpha^n,a]\pm}(x) = Tr \int d^4p (Q^a (P_\alpha)^n \pm (P_\alpha)^n Q^a) \gamma^\mu f(x,p). \tag{25}$$

It is easy to verify that for n=0, (24) defines the current density, and for n=1 it gives the energy-momentum tensor. Hence, the lowest two equations in (23) express the local conservation laws for the baryon density and the energy momentum tensor.

In conclusion, we have introduced a general truncation scheme for kinetic theories that is based on the Wigner function and defined for Schrödinger wave functions or field amplitudes. A general scheme of this kind will be necessary to tackle the wide variety of problems available in heavy-ion physics.

This work is supported by the Department of Energy under contract DE-AC02-76ER13001.

References

(1) J.W. Negele, Rev. Mod. Phys. 54(1982)913

(2) G. Bertsch, H. Kruse and S. Das Gupta, Phys. Rev. C29(1984)673

(3) U. Heinz, Phys. Rev. Lett. 51(1983)351

(4) M. Ploszajczak and M.J. Rhoades-Brown, Phys. Rev. Lett. 55(1985)147

(5) M. Ploszajczak and M.J. Rhoades-Brown, Phys. Rev. D33(1986), in the press

(6) E.P. Wigner, Phys. Rev. 40(1932)749

(7) E. Moyal, Proc. Cambridge Philos. Soc. 45(1945)99

(8) M. Ploszajczak and M.J. Rhoades-Brown, to be published

(9) G. Bertsch, Invited paper presented at School of Heavy-Ion Physics Erice, Sicily, July 17-23, 1984.

(10) P. Carruthers and F. Zachariasen, Rev. Mod. Phys. 55(1983)245

(11) J. Kapusta and A. Mekijan, Phys. Rev. D33(1986)1304

G. Baym et al, Nucl. Phys. A407(1983)541

(12) A. Biatas, W. Czyz, Phys. Rev. D30(1984)2371

T. Matsui, L. McLerran, B. Svetitsky, M.I.T. preprint CTP-1320, CTP-1344

(13) M.E. Carrington and M.J. Rhoades-Brown, to be published

D. OTHER SEMICLASSICAL THEORIES

COHERENT STATES AND THE GLOBAL, UNIFORM APPROXIMATION OF WAVE EQUATION SOLUTIONS

John R. Klauder
AT&T Bell Laboratories
Murray Hill, NJ 07974

INTRODUCTION

Parabolic wave equations arise in numerous physical situations, e.g., in quantum mechanics and unidirectional acoustic wave propagation, to mention just two cases. Each situation is endowed with a dimensional parameter ($\hbar$ in quantum mechanics; $ƛ$, the reduced wavelength, in acoustics), which under suitable circumstances may be regarded as small. Development of the solution with respect to the small parameter yields the semiclassical, or eikonal, approximation, usually composed of a sum of terms each with a phase factor expressed in units of the small parameter and an amplitude factor. Each term in the sum corresponds to a classical ray satisfying the boundary conditions with the phase proportional to the action evaluated for the extremal (classical) path, and the amplitude factor proportional to the square root of the local ray density. Due to singularities in the ray density at caustics, direct estimation of the amplitude factor often breaks down, requiring a more sophisticated treatment.[1] By focusing on a Fourier transformation of the amplitude, Maslov offers a semiclassical approximation that locally captures the effects of caustics.[2] However, this representation generally fails as one moves away from the region of the caustic. As we shall see, an alternative semiclassical approximation based on a coherent-state transformation provides a semiclassical approximation that is both uniformly valid at and near caustics and in addition is globally valid.[3,4]

AMPLITUDES AND WAVE EQUATIONS

As an example with which to illustrate various semiclassical amplitudes we choose a one-dimensional quantum mechanical wave equation given by

$$i\hbar\frac{\partial}{\partial t}\phi(x,t) = \mathscr{H}\left(-i\hbar\frac{\partial}{\partial x}, x,t\right)\phi(x,t)\,,$$

where $\mathscr{H}$ denotes a self-adjoint operator on $L^2(\mathbf{R})$. By changing $\hbar \rightarrow ƛ$ and $t \rightarrow z$, the range, this equation applies to a general unidirectional acoustic wave propagation problem, for example, while analogous substitutions relate to other systems. Interest centers on a solution to the wave equation at time $T > 0$ subject to the initial condition that the solution is a δ function at time 0. We denote the desired solution by

$$J(x'',T;\, x',0)\,,$$

with the property that

$$\lim_{T\rightarrow 0} J(x'',T;\, x',0) = \delta(x''-x')\,.$$

It follows from linearity that the solution for an arbitrary initial condition may be expressed with the aid of the propagator J according to

$$\phi(x'',T) = \int J(x'',T;\, x',0)\; \phi(x',0)\; dx' \,.$$

From the assumption that $\mathcal{H}$ is self adjoint we have, for all $T > 0$,

$$\int |\phi(x,T)|^2 dx = \int |\phi(x,0)|^2 dx,$$

which is physically interpreted as conservation of probability in quantum mechanics or conservation of energy in acoustics, etc.

In most cases a full determination of the propagator J is not possible, and approximate solutions are sought. To that end we shall now introduce two kinds of transformation of the original amplitude and wave equation.

The Fourier Representation

The Fourier transformation

$$\chi(p) = \frac{1}{\sqrt{2\pi\hbar}} \int e^{-ipx/\hbar}\, \phi(x)\, dx$$

and its inverse

$$\phi(x) = \frac{1}{\sqrt{2\pi\hbar}} \int e^{ipx/\hbar}\, \chi(p)\, dp$$

relate the amplitudes $\phi(x)$ and $\chi(p)$. This is a norm-preserving transformation in the sense that

$$\int |\phi(x)|^2 dx = \int |\chi(p)|^2 dp \,.$$

Under such a transformation the basic wave equation is transformed into

$$i\hbar \frac{\partial}{\partial t}\, \chi(p,t) = \mathcal{H}(p,\, i\hbar \frac{\partial}{\partial p})\; \chi(p,t) \,.$$

Our interest centers on a solution of the transformed wave equation expressed in the form

$$\chi(p'',T) = \int M(p'',T;\, x',0)\; \phi(x',0)\, dx',$$

where the propagator M satisfies the initial condition

$$\lim_{T\to 0} M(p'',T;\, x',0) = \frac{1}{\sqrt{2\pi\hbar}}\, e^{-ip''x'/\hbar} \,.$$

Moreover, it follows that

$$J(x'',T;\, x',0) = \frac{1}{\sqrt{2\pi\hbar}} \int e^{ip''x''/\hbar}\, M(p'',T;\, x',0)\, dp'' \,.$$

It is this Fourier-space approach that Maslov has promoted for semiclassical approximations, as we shall discuss shortly. Variations on the Maslov approach have recently been discussed by Littlejohn.[5]

The Coherent-State Representation

We now introduce a second kind of transformation, which we refer to as the coherent-state transformation.[6] Specifically, we introduce

$$\psi(p,x) = \frac{1}{(\pi\hbar)^{1/4}} \int e^{-y^2/2\hbar - ipy/\hbar}\, \phi(y+x)\, dy$$

and the inverse

$$\phi(x) = \frac{(\pi\hbar)^{1/4}}{2\pi\hbar} \int \psi(p,x)\, dp,$$

which relate the amplitudes $\phi(x)$ and $\psi(p,x)$. Like the Fourier transform, the coherent-state transform is a norm preserving transformation in the sense that

$$\int |\phi(x)|^2 dx = \int |\psi(p,x)|^2\, (dp\,dx/2\pi\hbar)\ .$$

However, the coherent-state transform is not a map from $L^2(\mathbf{R})$ to all of $L^2(\mathbf{R}^2)$, but only to a proper subspace composed of very smooth functions, as is evident from the definition of $\psi(p,x)$. In particular, each such ψ satisfies the integral equation

$$\psi(p'',x'') = \int \mathscr{K}(p'',x'';\, p',x')\, \psi(p',x')\, (dp'dx'/2\pi\hbar),$$

where the reproducing kernel $\mathscr{K}$ is given by

$$\mathscr{K}(p'',x'';\, p',x') = \exp\{\frac{i}{2\hbar}(p''+p')(x''-x') - \frac{1}{4\hbar}[(p''-p')^2 + (x''-x')^2]\}\ .$$

Lastly we should note that we have arbitrarily chosen to set a free parameter equal to unity in our description of the coherent-state transformation. This free parameter may be interpreted as the angular frequency Ω of the harmonic oscillator ground state wave function used in defining the coherent-state transformation. This parameter may be restored in the reproducing kernel, for example, simply by scaling all p variables as $p \to p/\Omega$ and all x variables as $x \to \Omega x$. In what follows we shall continue to adopt the choice $\Omega = 1$.

Under the coherent-state transformation it is straightforward to determine that the basic wave equation assumes the form

$$i\hbar\frac{\partial}{\partial t}\psi(p,x,t) = \mathscr{H}(-i\hbar\frac{\partial}{\partial x},\, x+i\hbar\frac{\partial}{\partial p},\, t)\, \psi(p,x,t)\ .$$

Our interest centers on a solution of this wave equation expressed in the form

$$\psi(p'',x'',T) = \int L(p'',x'',T;\, x',0)\, \phi(x',0)\, dx',$$

where the propagator L satisfies the initial condition

$$\lim_{T\to 0} L(p'',x'',T;\, x',0) = \frac{1}{(\pi\hbar)^{1/4}}\, e^{-(x''-x')^2/2\hbar + ip''(x''-x')/\hbar}$$

as follows from the definition of the coherent-state transformation. However, in addition to the propagator L it is also convenient to examine simultaneously the solution expressed as

$$\psi(p'',x'',T) = \int K(p'',x'',T;\, p',x',0)\, \psi(p',x',0)\, (dp'dx'/2\pi\hbar)\ .$$

As initial condition for the propagator K we choose the reproducing kernel, i.e.,

$$\lim_{T\to 0} K(p'',x'',T;\, p',x',0) = \mathscr{K}(p'',x'';\, p',x')$$

the expression for which is given above. It follows from this choice that

$$|K(p'',x'',T;\, p',x',0)| \leqslant 1$$

holds uniformly in all arguments.

From the inverse relation to the coherent-state transformation we observe that

$$J(x'',T;\, x',0) = \frac{(\pi\hbar)^{1/4}}{2\pi\hbar} \int L(p'',x'',T;\, x',0)\, dp'',$$

and

$$J(x'',T;\, x',0) = \frac{(\pi\hbar)^{1/2}}{(2\pi\hbar)^2} \int K(p'',x'',T;\, p',x',0)\, dp''dp' .$$

SEMICLASSICAL APPROXIMATIONS

We now take up the question of semiclassical approximations to the various propagators J,M,L, and K. As we shall see, there are important differences between the first two (J & M) and the last two (L & K) in this respect.

Semiclassical approximations may be readily constructed from a stationary-phase approximation of a path-integral representation of the solution.[1,4] The dominant term is determined by the action evaluated at an extremum, while the sub-dominant term, the amplitude factor, is essentially given as a Gaussian path integral determined by quadratic deviations about the extremal path. If several extrema contribute, the result is a sum over such contributions one for each extremum (assumed isolated). It is particularly noteworthy that the semiclassical expressions for all our propagators (J,M,L, and K) have an essentially similar structure differing principally only with regard to boundary conditions. Thus we initiate the discussion with a sketch of the common structure.

We relate the Hamiltonian operator to an underlying classical Hamiltonian H(p,x,t) by Weyl ordering. Associated with the classical Hamiltonian H are the Hamiltonian equations,

$$\dot{x}(t) = \partial H/\partial p(t) \quad , \qquad \dot{p}(t) = -\, \partial H/\partial x(t) ,$$

which, with suitable boundary conditions, determine an extremal solution (ES). An integral, possibly plus appropriate additional terms, determines the action in each case.

The amplitude factor in all cases is determined by a solution of an *auxiliary*, linear dynamical problem which takes the general form

$$\dot{\tilde{x}}(t) = H_{px}(t)\, \tilde{x}(t) + H_{pp}(t)\, \tilde{p}(t) ,$$

$$\dot{\tilde{p}}(t) = -\, H_{xx}(t)\, \tilde{x}(t) - H_{px}(t)\, \tilde{p}(t) ,$$

where, for ab either pp, px, or xx,

$$H_{ab}(t) \equiv \frac{\partial}{\partial a}\frac{\partial}{\partial b} H(p,x,t)\bigg|_{ES} ,$$

evaluated, as indicated, for the extremal solution. The different amplitude factors are distinguished by different sets of initial conditions to this auxiliary dynamical problem and/or by different functional forms of the solutions.

Listed below in a kind of tabular form are the various semiclassical approximations. Each propagator expression is followed, in order, by the required boundary conditions for the extremal solution, the action, the initial conditions for the auxiliary equations, and finally the expression for the amplitude factor.[4]

The J semiclassical propagator:

$$J_{sc}(x'',T;x',0) = \sum A_J\, e^{iS_J(x'';x')/\hbar};$$

$$x(0) = x' \quad , \qquad x(T) = x'';$$

$$S_J(x'';x') = \int [p\dot{x} - H(p,x,t)]\,dt;$$

$$\tilde{x}(0) = 0 \quad , \qquad \tilde{p}(0) = 1;$$

$$A_J = 1/\sqrt{2\pi i\hbar\,\tilde{x}(T)}\ .$$

The M semiclassical propagator:

$$M_{sc}(p'',T;x',0) = \sum A_M\, e^{iS_M(p'';x')/\hbar};$$

$$x(0) = x' \quad , \qquad p(T) = p'';$$

$$S_M(p'';x') = -\,p''x(T) + \int [p\dot{x} - H(p,x,t)]\,dt;$$

$$\tilde{x}(0) = 1 \quad , \qquad \tilde{p}(0) = 0;$$

$$A_M = 1/\sqrt{2\pi\hbar\,\tilde{x}(T)}\ .$$

The L semiclassical propagator:

$$L_{sc}(p'',x'',T;x',0) = A_L\, e^{iS_L(p'',x'';x')/\hbar};$$

$$x(0) = x' \quad , \qquad x(T) - ip(T) = x'' - ip'';$$

$$S_L(p'',x'';x') = \tfrac{1}{2}[p''+p(T)][x''-x(T)] + \int [p\dot{x} - H(p,x,t)]\,dt;$$

$$\tilde{x}(0) = 0 \quad , \qquad \tilde{p}(0) = 1;$$

$$A_L = 1/\sqrt{\sqrt{\pi\hbar}[\tilde{p}(T) + i\tilde{x}(T)]}\ .$$

The K semiclassical propagator:

$$K_{sc}(p'',x'',T;p',x',0) = A_K\, e^{iS_K(p'',x'';p',x')/\hbar};$$

$$x(0) + ip(0) = x' + ip' \quad , \qquad x(T) - ip(T) = x'' - ip'';$$

$$S_K(p'',x'';p',x') = \tfrac{1}{2}[p''+p(T)][x''-x(T)] - \tfrac{1}{2}[p'+p(0)][x'-x(0)]$$

$$+ \int [p\dot{x} - H(p,x,t)]\,dt;$$

$$\tilde{x}(0) = -i/2 \quad , \qquad \tilde{p}(0) = 1/2;$$

$$A_K = 1/\sqrt{\tilde{p}(T) + i\tilde{x}(T)}\ .$$

Several comments on these expressions are in order. For the J and M solutions the extremal solutions are real, and as a consequence the actions (S_J and S_M) and the auxiliary solutions ($\tilde{x}_J$ and $\tilde{x}_M$) are always real. For the L and K solutions, on the other hand, the extremal solutions are generally complex, and as a consequence the actions (S_L and S_K) and the auxiliary solutions

($\tilde{x}_L, \tilde{p}_L$ and $\tilde{x}_K, \tilde{p}_K$) are also generally complex. It may seem that the boundary conditions are overspecified for the extremal equations in the L and K cases, however this is not the case. In each case one may choose the complex degree of freedom w'' open in the final boundary condition, i.e., $x(T) = x'' + w''$, $\tilde{p}(T) = p'' - iw''$, to ensure that the one complex initial condition is satisfied, i.e. $x(0) = x'$, $p(0)$ arbitrary, for the L case, and $x(0) + ip(0) = x' + ip'$ for the K case. The J and M propagators generally involve a sum over several distinct extrema, while the L and K propagators involve a *single* term; in the latter cases the correct extremum solution to take is the one which, as $T \to 0$, is continuously connected to the unique solution for $T = 0$.

Special and General Points

For certain, "special", points for which $w'' = 0$, the extremal solutions for the L and K cases are *real*. This occurs when the classical evolution extended backwards in time from the point p'', x'' for a time T coincides with the point p', x' (for K) or the point x' (for L), for which $p(0) = p'$ as well. Real extremal solutions lead to real actions (S_L and S_K), and thus for special points

$$|e^{iS_L/\hbar}| = |e^{iS_K/\hbar}| = 1 .$$

Observe for special points that $x(T) = x''$ and $x(0) = x'$, and therefore

$$S_L(p'', x''; x') = S_K(p'', x''; p', x') = S_J(x''; x')$$

for that particular extremal solution labelled by p''.

For "general" points, where $w'' \neq 0$, on the other hand, the extremal solutions are complex, and the actions (S_L and S_K) are likewise complex. However, it follows that $\mathrm{Im}\, S_L > 0$ and $\mathrm{Im}\, S_K > 0$, so that for general points

$$|e^{iS_L/\hbar}| < 1 , \quad |e^{iS_K/\hbar}| < 1 .$$

Thus, in the coherent-state cases (L and K) *the action contains important amplitude information* and for general points it leads to an *exponential damping of the amplitude.* This behavior follows from the uniform bound $|K(p'', x'', T; p', x', 0)| \leqslant 1$. Assuming good behavior of the amplitude factors (A_L and A_K), the qualitative behavior of the coherent-state semiclassical amplitudes (L_{sc} and K_{sc}) for fixed x' and x'' is that of a strongly (exponentially) suppressed amplitude except at and near those special points p'' (for L_{sc}) and p'', p' (for K_{sc}) that characterize real extremal solutions. It is also clear from this structure how an integral over p'' (for L_{sc}) or p'' and p' (for K_{sc}) leads to a sum of contributions, one for each acceptable extremal path, just as in the indicated behavior for J_{sc}.

Amplitude Factors

Now let us discuss the amplitude factors that appear in the various semiclassical approximations. At a caustic, defined as the envelope of ray crossings, it follows that $\tilde{x}_J(T) = 0$ and $A_J = \infty$. This is an error caused by the breakdown of the approximation of including only quadratic deviations about the extremal solution in a path-integral evaluation of the amplitude. Inclusion of cubic deviations, and possibly higher-order terms as well, corrects the error and leads to a finite amplitude distinguished by the fact that it involves the small parameter ($\hbar$) to a smaller power than otherwise, implying thereby an intrinsically increased amplitude.

While perfectly correct, the approach just outlined is unnecessarily complicated. This is just the point where the Maslov approach proves to be important. The Maslov representation of the semiclassical amplitude is

$$J^M_{sc}(x'',T;x',0) = \sum \frac{1}{2\pi\hbar} \int \frac{1}{\sqrt{\tilde{x}_M(T)}} e^{i[p''x''+S_M(p'';x')]/\hbar} dp'' .$$

Normally it suffices to approximate S_M by a quadratic expansion in p″ save for those cases when the coefficient of the quadratic term vanishes. In the latter case it is necessary to include cubic (or higher-order terms) in the expansion of S_M, which inevitably leads to an altered dependence of the amplitude on $\hbar$. Difficulty with this representation arises when $\tilde{x}_M(T) = 0$ for any of the terms in the sum. Since $\tilde{x}_M$ is real it must cross zero if it changes sign, and this will be the generic behavior for any Hamiltonian H with the tendency to confine the rays to a finite region of phase space which includes the origin. Such an error arises, of course, because the quadratic deviation approximation in a path integral again breaks down. Higher-order deviations must be included in the path integral to correct the error; it is not sufficient, for example, simply to make $\hbar$ smaller — the divergence in J^M_{sc} is disconnected from the phase factor.

The amplitude factors for the coherent-state cases (L_{sc} and K_{sc}) have very different properties, and as we shall argue they can actually be chosen as *uniformly bounded.* Since the auxiliary equations are linear, the general solution may be written in the form

$$\begin{pmatrix} \tilde{x}(T) \\ \tilde{p}(T) \end{pmatrix} = \begin{pmatrix} A & B \\ C & D \end{pmatrix} \begin{pmatrix} \tilde{x}(0) \\ \tilde{p}(0) \end{pmatrix} .$$

The coefficients satisfy the constraint

$$AD - BC = 1 ,$$

as follows from the fact that the auxiliary dynamics is represented by a canonical transformation. With $\tilde{x}(0)$ and $\tilde{p}(0)$ as previously given, it follows that

$$\tilde{p}_L(T) + i\tilde{x}_L(T) = D + iB ,$$

$$\tilde{p}_K(T) + i\tilde{x}_K(T) = \tfrac{1}{2}[A+D+i(B-C)] .$$

Now at special points, where the extremal solutions are real, it follows that all four coefficients, A, B, C, and D, are *real.* Thus

$$|\tilde{p}_L(T) + i\tilde{x}_L(T)|^2 = D^2 + B^2 > 0 ,$$

i.e., $|A_L| < \infty$ at all special points thanks to the fact that $D = B = 0$ is incompatible with the constraint on these terms. Near to special points, i.e., where $w'' = O(\sqrt{\hbar})$, it follows by continuity that $|\tilde{p}_L(T) + i\tilde{x}_L(T)|^2 > 0$, i.e., $|A_L| < \infty$, so the amplitude A_L is finite at and nearby all special points. (We shall take up the question of arbitrary general points below.) For the K amplitude we may say even more. At special points it follows that

$$\begin{aligned} |\tilde{p}_K(T) + i\tilde{x}_K(T)|^2 &= \tfrac{1}{4}[(A+D)^2 + (B-C)^2] \\ &= 1 - AD + BC + \tfrac{1}{4}[(A+D)^2 + (B-C)^2] \\ &= 1 + \tfrac{1}{4}[(A-D)^2 + (B+C)^2] \geqslant 1 . \end{aligned}$$

Thus at all special points $|A_K| \leqslant 1$, as in fact must hold in view of the bound $|K| \leqslant 1$. Near to special points $[w'' = O(\sqrt{\hbar})]$ it follows by continuity that $|A_K| \lesssim 1$.

Finally, we take up the question of the amplitudes at arbitrary general points. In such cases, A, B, C, and D are complex and the constraint they satisfy no longer ensures that the amplitudes A_L or A_K will be finite. As usual, if the amplitude diverges it is due to a failure of the quadratic approximation to the deviations in a path-integral representation; the amplitude is rendered finite by including cubic or higher-order terms, leading in addition to an altered dependence on $\hbar$. But this effort is not necessary!

In the case of the K amplitude *the precise value of the finite amplitude at typical general points is irrelevant since* $|e^{iS_K/\hbar}| << 1$ *already leads to a large exponential suppression of the amplitude.* Thus there is no real loss of accuracy in computing

$$J_{sc}^K(x'',T;x',0) \equiv \frac{(\pi\hbar)^{1/2}}{(2\pi\hbar)^2} \int K_{sc}(p'',x'',T;p',x',0)\,dp''\,dp'$$

if we take

$$K_{sc}(p'',x'',T;p',x',0) \equiv A'_K\, e^{iS_K/\hbar}$$

where

$$A'_K \equiv A_K \min(1,\, 100/|A_K|)\ ,$$

thus having a *uniformly bounded amplitude factor*. This prescription therefore yields a *global, uniform semiclassical approximation for* J that becomes ever more accurate as $\hbar$ decreases in magnitude.

For the L_{sc} approximation a divergent amplitude is also made finite by including cubic or higher-order deviations in the path integral, but, just as before, *the precise finite value of the amplitude is irrelevant since any point of trouble always occurs in a region where* $|e^{iS_L/\hbar}| << 1$. Thus in the semiclassical approximation

$$J_{sc}^L(x'',T;x',0) \equiv \frac{(\pi\hbar)^{1/4}}{2\pi\hbar} \int L_{sc}(p'',x'',T;x',0)\,dp''$$

we can adopt a locally bounded expression for L_{sc} defined, for example, by

$$L_{sc}(p'',x'',T;x',0) = A'_L\, e^{iS_L/\hbar},$$

where

$$A'_L = A_L \min(1,\, 100\alpha/|A_L|)\ .$$

Here, as one of several possible choices, we take

$$\alpha = |A_{L,sp}|,$$

the modulus of the amplitude at the nearest special point, where nearest is defined by the least value of $|p''-p''_{sp}|$. When there are only a finite number of special points, then it follows that A'_L is uniformly bounded. This prescription for L_{sc} also yields a *global, uniform semiclassical approximation for* J that becomes ever more accurate as $\hbar$ decreases. It is noteworthy that the amount of computation time involved in computing J_{sc}^L should be comparable to that involved in computing J_{sc}^M, and so the advantages of a global, uniform approximation should be readily accessible.

ACKNOWLEDGEMENTS

This work was largely motivated by discussions with A. H. Carter, R. Holford, and R. S. Patton. It was carried out in part at the Institute for Advanced Studies in Princeton, and the hospitality offered by S. L. Adler and R. F. Dashen was greatly appreciated.

REFERENCES

1. See, e.g., L. S. Schulman, "Techniques and Applications of Path Integration," (Wiley, New York, 1981).
2. V. P. Maslov, "Théorie des Perturbations et Méthodes Asymptotiques," (Dunod, Paris, 1972).
3. J. R. Klauder, Phys. Rev. Lett. *56*, 897 (1986); 1513 (E) (1986). For precursors see J. R. Klauder, in "Path Integrals and Their Application in Quantum, Statistical, and Solid-State Physics," ed. G. J. Papadopoulis and J. T. Devreese (Plenum, New York, 1978), p. 5; Phys. Rev. D *19*, 2349 (1978); in "Quantum Fields-Algebras, Processes," ed. L. Streit (Springer, Vienna, 1980), p. 65. See also S. W. McDonald, Phys. Rev. Lett. *54*, 1211 (1985).
4. J. R. Klauder, "Some Recent Results on Wave Equations, Path Integrals, and Semiclassical Approximations," in the proceedings of the IMA Workshop on Random Media, (Springer, New York, 1986).
5. R. G. Littlejohn, Phys. Rev. Lett. *54*, 1742 (1985).
6. See, e.g., J. R. Klauder and B.-S. Skagerstam, "Coherent States," (World Scientific, Singapore, 1985).
7. In this regard, see J. N. L. Connor and P. R. Curtis, J. Math. Phys. *25*, 2895 (1984), and references therein.

QUANTUM WAVE-FUNCTIONS FROM CLASSICAL PHASE-SPACE MANIFOLDS: AN INTRODUCTION TO MASLOV'S SEMICLASSICAL THEORY

J. B. Delos
College of William and Mary
Williamsburg, VA 23185

The relationship between wave motion and particle motion has been studied for more than 150 years, since the work of Hamilton in the 1830's. Yet new formulations and new discoveries about this relationship continue to be made. Recently a formulation of semiclassical theory has been developed by Maslov and Fedoriuk.[1] We have found this formulation to be a very nice way of thinking about quantum wave-functions and classical trajectories: it is straightforward, logical, mathematically sound, and easy to implement on a computer. In recent papers we have presented the theory in a simplified form,[2] and we have used it to calculate wave-functions for simple bound-state and for collision systems.

The essence of the method can be summarized very simply. Let us consider for a moment a one-dimensional non-linear oscillator with $H = p^2/2+V(q)$ as in Fig. 1a. Suppose the nth quantum state has an energy E_n. Then associated with that state is a classical trajectory represented by a closed curve in phase space $H(p,q) = E_n$ (Fig. 1b). This closed curve is represented in different ways in different regions. Near the highest and lowest parts of the curve, it is represented by smooth functions $p = \varphi_i(q)$, while near the leftmost and rightmost parts of the curve it is represented by functions $q = \mathfrak{f}_i(p)$. (These functions are differentiable only in the regions indicated.)

These functions can be used to construct approximate solutions to the Schroedinger equation,

$$[H(-i\hbar\partial/\partial q, q)-E]\psi = 0.$$

For those values of q such that the function $\varphi_i(q)$ is differentiable, a WKB approximation to $\psi(q)$ can be written as

$$\psi_i(q) = C_i \left| \frac{\partial H}{\partial p} \right|^{-1/2}_{p=\varphi_i(q)} \exp\left[i \int^q \varphi_i(q')\, dq' / \hbar \right].$$

This wave-function is accurate at small q, where $\varphi(q)$ is smooth, but it diverges near the classical turning points, where $\dot{q} = (\partial H/\partial p)$ vanishes, and $\partial p_i(q)/\partial q$ goes to infinity.

On the other hand, there is also a smooth wave-function in momentum space, which satisfies

$$[H(p, i\hbar\,\partial/\partial p) - E]\,\tilde{\Psi}(p) = 0.$$

A WKB approximation to this wave-function is given by

$$\tilde{\Psi}(p) = \tilde{C}_i \left|\frac{\partial H}{\partial q}\right|^{-1/2}_{q = q_i(p)} \exp\left[-i \int^{p} q_i(p')\,dp'/\hbar\right].$$

This function is accurate when p is not too large, but it diverges where $\dot{p} = -\partial H/\partial q$ vanishes. Those points are turning points in momentum space, where the deivative of $q_i(p)$ is infinite.

We now have two approximations, one valid when q is small and the other valid where p is small. Together, these approximations cover the entire classical domain. To combine them, we define on the trajectory a set of "switching functions" e_j (a partition of unity) such that $e_1 + e_2 + \tilde{e}_1 + \tilde{e}_2 = 1$ and

$e_1 \sim 1$	small q, $p<0$,		$e_2 \sim 1$	small q, $p>0$,	
$\tilde{e}_1 \sim 1$	small p, $q<0$,		$\tilde{e}_2 \sim 1$	small p, $q>0$.	

Then a smooth global wave function is given by a combination of the ordinary WKB solution together with the Fourier transform of the momentum-space WKB approximation,

$$\Psi(q) = \sum_i e_i(q)\,\Psi_i(q) + \int \exp(i p q/\hbar)\,\tilde{e}_i(p)\,\tilde{\Psi}_i(p)\,dp.$$

The switching functions select the function which is accurate in the specified region.

This one-dimensional example is quite trivial, but the generalization to n dimensions is not trivial. (1) The trajectory is replaced by a family of trajectories corresponding to a smooth surface in phase-space called a Lagrangian manifold. (2) Each region of the manifold is described by a mixed set of coordinates and momenta p_α, q_β. (3) The WKB approximation in position or momentum space is written as

$$\tilde{\Psi}(p_\alpha q_\beta) = |\tilde{J}(p_\alpha q_\beta)|^{-1/2} \exp[i\,S(p_\alpha q_\beta)/\hbar].$$

(4) The phase function $S(p_\alpha q_\beta)$ satisfies a mixed-space form of

the Hamilton-Jacobi equation. (5) A smooth global wave-function is obtained by tying together such local wave-functions using switching functions.

Results are presented at the conference and in reference 2.

References

1. V. P. Maslov and M. V. Fedoriuk, Semiclassical Approximation in Quantum Mechanics, Reidel, Boston (1981).
2. J. B. Delos, Adv. Chem. Phys. 65, 161 (1986); J. B. Delos, S. K. Knudson and B. Bloom, J. Chem. Phys. 83, 5703 (1985); J. B. Delos, S. K. Knudson and D. W. Noid, J. Chem. Phys. (in press).

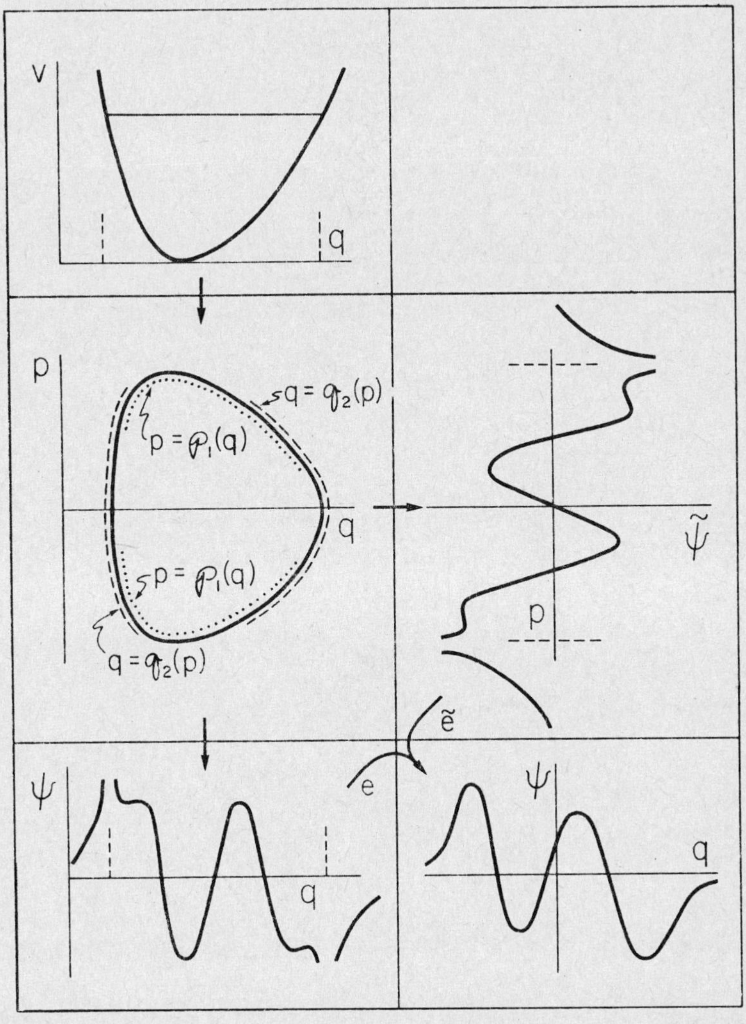

SEMICLASSICAL ANALYSIS OF COUPLED CHANNEL SYSTEM WITH NON-LOCAL INTERACTION

--- Appearance of Berry Phase and Another Additional Phase Accompanying Phase Space Motion ---

K. Yabana and H. Horiuchi

Department of Physics, Kyoto University, Kyoto 606, JAPAN.

The Schrödinger equation for the coupled channel system is, in general, given by

$$\{H_{coll}(\hat{x},\hat{p}) + H_{in}(\hat{\xi}) + V(\hat{x},\hat{p},\hat{\xi})\}\Psi = E\Psi, \tag{1}$$

where $\hat{x}$ and $\hat{p}$ represent the coordinate and momentum operators of the collective motion, respectively, and $\hat{\xi}$ stands for the collection of the operators of internal motion. On condition that the collective motion is sufficiently slower than the internal motion, the adiabatic approximation is often applied to Eq.(1). When the coupling V does not depend on the collective momentum operator $\hat{p}$, the application of the WKB method to Eq.(1) has been studied and, as is well-known, it results essentially in the adiabatic approximation. We here discuss the WKB analysis of Eq.(1) for the case of $\hat{p}$-dependent coupling V, putting emphasis on its relation to the adiabatic approximation[1].

The WKB method for Eq.(1) with p-dependent coupling was recently formulated[2,3]. We first prepare the basis states for internal motion denoted by $|i)$, which are eigenstates of the internal Hamiltonian $H_{in}(\hat{\xi})$. This basis is called the diabatic basis. Utilizing this basis, Equation (1) is rewritten as

$$\sum_j H_{ij}(\hat{x},\hat{p})\psi_j(x) = E\psi_i(x), \tag{2}$$

$$H_{coll}(\hat{x},\hat{p}) + H_{in}(\hat{\xi}) + V(\hat{x},\hat{p},\hat{\xi}) = \sum_{ij} H_{ij}(\hat{x},\hat{p})|i)(j|, \tag{3}$$

$$\Psi = \sum_i \psi_i(x)|i). \tag{4}$$

Then we represent $\psi_i(x)=\exp(iS^i(x)/\hbar)$ and put it into Eq.(2). After expanding $S^i(x)=S_0^i(x)+\hbar S_1^i(x)+\ldots$, the equations for $S_n^i(x)$ can be obtained by regarding the Planck's constant as a small parameter. From the $\hbar^{-1}$ and $\hbar^0$ order, it can be derived that the local momentum $p(x)=dS_0(x)/dx$ does not depend on the channel superfix i and is given by the solution of

$$\det G_{ij}^W\left(x, \frac{dS_0(x)}{dx}\right) = 0, \tag{5}$$

$$G^W_{ij}(x,p) \equiv H^W_{ij}(x,p) - E\delta_{ij}\,, \tag{6}$$

where $H^W_{ij}(x,p)$ is the Wigner transform of the operator $H_{ij}(\hat{x},\hat{p})$. Up to $\hbar^1$ order, the WKB wave function is given by

$$\psi_i(x) = C_0\sqrt{\left(\frac{\Delta_{fi}}{\Delta_{ff}}\right)_{x_0}}\frac{\sqrt{\Delta_{ff}}}{\sqrt{\partial|G^W|/\partial p}}\exp\left[\frac{i}{\hbar}\int_{x_0}^{x}dx\left\{p(x)+\frac{\hbar}{i}\,\frac{P_f(x,p(x))}{2\,\partial|G^W|/\partial p}\right\}\right], \tag{7}$$

where $\Delta_{ij}(x,p)$ stands for the cofactor matrix of $G^W_{ij}(x,p)$, and $|G^W|$ stands for $\det G^W_{ij}(x,p)$. The function $P_f(x,p)$ is defined as

$$P_f(x,p) \equiv \frac{1}{\Delta_{ff}}\sum_{ij}G^W_{ij}\left(\frac{\partial\Delta_{fi}}{\partial p}\frac{\partial\Delta_{jf}}{\partial x} - \frac{\partial\Delta_{fi}}{\partial x}\frac{\partial\Delta_{jf}}{\partial p}\right), \tag{8}$$

which is purely imaginary and is non-zero only when $H^W_{ij}(x,p)$ with $i\neq j$ are complex.

Now let us introduce the adiabatic basis $|\alpha(x,p))$, which is the eigenstates of the operator $\sum_{ij}H^W_{ij}(x,p)|i)(j|$, that is,

$$|\alpha(x,p)) = \sum_i U_{i\alpha}(x,p)|i)\,, \tag{9}$$

$$\sum_j H^W_{ij}(x,p)U_{j\alpha}(x,p) = E^\alpha(x,p)U_{i\alpha}(x,p). \tag{10}$$

Then the local momentum p(x) satisfies

$$E^\alpha(x,p^\alpha(x)) = E\,, \tag{11}$$

for some α because of $\det G^W_{ij}(x,p) = \prod_\alpha(E^\alpha(x,p)-E)$. The WKB wave function of Eq.(7) can be re-expressed by the quantities related to the adiabatic basis as follows

$$\psi_i(x) = C_0'\frac{U_{f\alpha}(x,p^\alpha(x))}{\sqrt{\partial E^\alpha/\partial p}}\exp\left[\frac{i}{\hbar}\int_{x_0}^{x}dx\,p^\alpha(x) + \int_{t_0}^{t}dt\left\{\omega_1^\alpha+\omega_2^\alpha\right\}\right], \tag{12}$$

where the time variable t is related to x by $dx/dt = \partial E^\alpha/\partial p$. The functions ω_1 and ω_2 are defined

$$\omega_1^\alpha(x,p) = -\left\{\frac{\partial E^\alpha}{\partial p}\left(U^\dagger\frac{\partial U}{\partial x}\right)_{\alpha\alpha} - \frac{\partial E^\alpha}{\partial x}\left(U^\dagger\frac{\partial U}{\partial p}\right)_{\alpha\alpha}\right\}, \tag{13}$$

$$\omega_2^\alpha(x,p) = \frac{1}{2}\left\{\left(\frac{\partial U^\dagger}{\partial p}U(E(x,p)-E)U^\dagger\frac{\partial U}{\partial x}\right)_{\alpha\alpha} - (x\leftrightarrow p)\right\}. \tag{14}$$

The full WKB wave function is then represented as

$$\Psi = C_0\frac{1}{\sqrt{\partial E^\alpha/\partial p}}\exp\left[\frac{i}{\hbar}\int_{x_0}^{x}dx\,p^\alpha(x) + \int_{t_0}^{t}dt\left\{\omega_1^\alpha+\omega_2^\alpha\right\}\right]|\alpha(x,p^\alpha(x))) \tag{15}$$

The above expression of the WKB wave function clearly shows the nature of the WKB approximation: The internal motion is described by the adiabatic state $|\alpha(x,p))$ on the classical trajectory in the phase space of collective motion. The collective motion is described by the usual WKB wave function with the adiabatic Hamiltonian $E^\alpha(x,p)$ except for the existence of the additional phases $\int dt(\omega_1^\alpha+\omega_2^\alpha)$ besides

$\int$pdx. The phase of ω_1^α part is known within the general adiabatic theory and is responsible for the change of the internal state $|\alpha(x,p))$ along the classical trajectory. The phase of ω_2^α part has not been known previously and seems to originate from the operator character of the collective variables $\hat{x}$ and $\hat{p}$.

For the bound state problem, the quantization condition is modified from the usual Bohr-Sommerfeld rule by these additional phases and is given by

$$\oint p\,dx + \frac{\hbar}{i}\oint dt\{\omega_1^\alpha + \omega_2^\alpha\} = 2\pi\hbar\left(n+\frac{1}{2}\right), \quad n = \text{integer}. \tag{16}$$

Especially $\oint dt\omega_1^\alpha$ just corresponds to the Berry phase which is recently taken interest in the adiabatic theory. We have applied the quantization condition of Eq.(16) for a simple two channel problem, where it has been shown that the inclusion of both the additional phases is important and indispensable to reproduce the quantum energy spectra.

References

1. K. Yabana and H. Horiuchi, Prog. Theor. Phys. 75 (1986),592.
2. H. L. Berk and D. Pfirsch, J. Math. Phys. 21 (1980),2054.
3. K. Yabana and H. Horiuchi, Prog. Theor. Phys. 71 (1984),1275.
4. M. V. Berry, Proc. Roy. Soc. London A392 (1984),45.

A FUNCTIONAL DENSITY MATRIX FOR QUANTUM ELECTRODYNAMICS AND ITS CLASSICAL LIMIT

J. K. Thurber
Department of Mathematics
Purdue University
West Lafayette, IN 47907

P. Ajanapon
Division of Mathematics and Computer Science
Northeast Missouri State University
Kirksville, Missouri 63501

A functional density matrix for quantum electrodynamics is defined by means of nonstandard analysis [5][8], and a corresponding Wigner functional is also obtained. The density functional satisfies a quantum functional Liouville equation. A classical limit is obtained which satisfies a classical Liouville equation. Special singular solutions of this latter equation are obtained which correspond to the classical deterministic coupled Maxwell-Dirac fields.

A propagation kernel for the density matrix is defined as a Feynman sum over histories [2]. The action is a difference of two corresponding actions which differ only by the labeling of the fields, primed and double primed. The primed fields correspond to rows and the double primed to columns of the density matrix [4]. Nonstandard analysis [5] is used in order to avoid the necessity of taking limits which in fact are too difficult to handle in a sensible way using standard analysis. The photon field variables and the electron-positron field variables are all taken as ordinary C numbers [1]. No Grassmann numbers and Berezin integrals are used [3]. A version of the symmetry requirements for photon fields and anti-symmetry for electron-positron fields is obtained by imposing these as initial conditions. By an infinitesimal perturbation [5] of the Hamiltonian which gives the time evolution of the density matrix, the perturbed Hamiltonian can be made symmetric with respect to an uncountably infinite but star finite [5][8] number of fields such that this star finite number divided by the total number of fields is an infinitesimal [5]. For these fields the density matrix at time t satisfies the same symmetry and anti-symmetry as at time 0, to within an infinitesimal percent error. In an expanded version of this paper we will show how the equal time constraint on the symmetry can be relaxed.

As subsidiary condition we set the time components of the photon fields equal to 0, and corresponding Faddeev-Popov factors [7] are inserted in the integrand in order to guarantee that the propagator for the density matrix is independent of the gauge fixing or subsidiary condition chosen. The density matrix satisfies an equation of the form

$\rho = e^{\left(\frac{H'-H''}{i\hbar}\right)t}\rho_0$ where ρ_0 is density matrix at t=0. H is of the form given by Hoyer [6] with additional terms since our photon fields are quantized. Even though our ψ, Ψ are ordinary C-number fields, we obtain the same form as Hoyer does for Grassmann fields. We also follow the discretization scheme used by Hoyer [6], with the difference that our mesh widths are infinitesimal [5].

Next, a change of field variables is made [4]:

$$A=\frac{1}{2}(A'+A''),\quad B=A'-A'',\quad R=\psi'-\psi'',\quad \bar{R}=\Psi'-\Psi''\quad \left(\Psi=\psi^{+}\gamma^{0}\right).$$
$$\psi=\frac{1}{2}(\psi'+\psi''),\quad \Psi=\frac{1}{2}(\Psi'+\Psi'')$$

A functional Fourier transform in the B_ν fields is taken and the corresponding transform variables are denoted by V_ν. Functional Fourier transforms (analytically continuing where necessary) are taken in the $\bar{R}_\alpha$ fields, the corresponding transform variables are denoted by $\bar{S}_\alpha$. The transform of ρ is denoted by f, and the resulting equation is divided by $i\hbar$. We do not transform in the R_α fields. However, if we did, the resulting representation for ρ would give a Wigner functional. The resulting equation is written in dimensionless form with the aid of a macroscopic length L. An expansion is made in powers of $\epsilon=\frac{\hbar}{mcL}$ and all terms in positive powers of ϵ are ignored. Note however that $\tilde{\gamma}=\hbar\gamma$, when made dimensionless, is not treated as a small quantity. This is due to the fact that $\hbar$ occurs multiplying $\not{\partial}$ in the classical (non-quantized) Maxwell-Dirac equations and ignoring $\hbar\not{\partial}$ would lead to a singular perturbation which we do not consider. There remain terms involving ϵ^{-1} which multiply $R_\beta(y)$. These terms are eliminated by restricting the class of solutions that we consider to be multiples of $\prod_{\beta,y}\delta(R_\beta(y))$.

In the classical limit the functional f satisfies the equation

$$\frac{\partial f}{\partial t}+\sum\left[i\vec{\tilde{\gamma}}\cdot\vec{\Delta}(x,y)c+mc^{2}\delta(x,y)+e\gamma^{r}A_{r}(x)\delta(x,y)\right]^{\alpha\beta}\text{times}$$
$$\left[\psi_\beta(y)\frac{\partial f}{\partial\bar{S}_\alpha(x)}+\left(\tilde{\gamma}_0^{-1}\right)^{\alpha'\alpha}\left(\tilde{\gamma}_0^{-1}\right)^{\beta'\beta}\bar{S}_{\beta'}(y)\frac{\partial f}{\partial\psi_{\alpha'}(x)}\right]$$
$$+\sum\left[e\gamma^{r}\delta(x,y)\right]^{\alpha\beta}\Psi_\alpha(x)\psi_\beta(y)\frac{\partial f}{\partial V_r(x)}$$
$$+\sum c\left(\Delta_r(x,y)A_s(y)-\Delta_s(x,y)A_r(y)\right)\text{times}$$
$$\left(\Delta_r(x,y)\frac{\partial f}{\partial V_s(y)}-\Delta_s(x,y)\frac{\partial f}{\partial V_r(y)}\right)$$
$$+\sum cV_r(x)\frac{\partial f}{\partial A_r(x)}=0\ .$$

The classical deterministic limit is given by the following singular solution

$$C=\prod\left\{\begin{array}{l}\delta(R_\gamma(y))\,\delta(A_r(x)-A_r^c(t,x))\,\delta\left(V_s(x)-\frac{\partial A_s^c(t,x)}{\partial t}\right)\text{ times}\\ \delta\left(\psi_\beta(x)-\psi_\beta^c(t,x)\right)\delta\left((\tilde{\gamma}_0^{-1}S)_\alpha(x)-\bar{\psi}_\alpha^c(t,x)\right)\delta\left(\Psi_\ell(y)-\psi_\lambda(y)\gamma_{0\lambda\ell}\right)\end{array}\right\},$$

where A^c, Ψ^c, and ψ^c are arbitrary solutions of the classical Maxwell-Dirac Equations. A delta functional with complex arguments is defined in terms of nonstandard [5] Kronecker deltas, and the chain rule for functional directional derivatives is obtained.

The solution C is paradoxical in that it is not a limit of a sequence of pure states. The matrix equation $C^2=C$ is not satisfied. The same conclusion for ordinary quantum mechanics was originally noted by the first author, then later derived by the second author [8].

COMMENTS

In the course of this research the first author found errors in the second half of the thesis [9] which forced the abandonment of the procedure used there. This work is entirely different from the thesis [9].

REFERENCES

1 J.R. Klauder,"The Action Option and a Feynman Quantization of Spinor Fields in Terms of Ordinary C-Numbers",Annals of Phys.,11, 1960, 123-168.

2 R.P.Feynman and A.R.Hibbs, Quantum Mechanics and Path Integrals, Mc Graw Hill, 1965.

3 F.A. Berezin, The Method of Second Quantization, Academic Press 1966.

4 H.S.Green, The Molecular Theory of Fluids, Dover, 1969.

5 A.H. Lightstone and A.Robinson, Nonarchimedean Fields and Asymptotic Expansions, North Holland, 1975.

6 P.Hoyer "Time Evolution in Fermion Path Integrals", Acta Physica Polonica, B14, 1983, 203-216.

7 V.N.Popov, Functional Integrals in Quantum Field Theory and Statistical Physics, D.Reidel, 1983.

8 P.Ajanapon, "Classical Limit of the Path-Integral Formulation of Quantum Mechanics in Terms of the Density Matrix", to be pub. in The Amer. Jour. of Phys.

9 P. Ajanapon,"A Derivation Of The Classical Limit Of Quantum Mechanics And Quantum Electrodynamics", thesis, Purdue Univ. Aug. 1985.

STRUCTURAL CONNECTIONS BETWEEN THE WKB AND WIGNER-KIRKWOOD SEMICLASSICAL APPROXIMATIONS

T.A. Osborn and F.H. Molzahn
Department of Physics
University of Manitoba
Winnipeg, Manitoba, Canada R3T 2N2

1. THE PROBLEM

We establish the interdependence and mutual consistency of the WKB and Wigner-Kirkwood semiclassical approximations. In a comparative study of the analytic behavior of the quantum propagator in both mass and Planck's constant, it is shown that the generalized Wigner-Kirkwood approximation is realized as a large mass expansion of the higher-order WKB approximation(1).

Consider the dynamical evolution of an N-particle spinless nonrelativistic system interacting via time-dependent scalar and vector potentials. The propagator K is the fundamental solution of the Schrödinger equation in d = 3N dimensional Euclidean space without boundaries,

$$i\hbar \frac{\partial}{\partial t} K(x,t;y,s) = [\frac{1}{2m}(\frac{\hbar}{i}\nabla_x - A(x,t))^2 + v(x,t)]\, K(x,t;y,s) . \tag{1.1}$$

Here m is particle mass, $h = 2\pi\hbar$ is Planck's constant, s and t are initial and final times, and the vectors y and x describe the initial and final system configurations. The scalar potential $v: R^{d+1} \to R^1$ and the vector potential $A: R^{d+1} \to R^d$ are assumed smooth and bounded. The Cauchy initial data condition is $K \to \delta(x-y)$ as $t \to s$.

The two interesting singular perturbation limits of eq. (1.1) are $\hbar \to 0$ and $m^{-1} \to 0$. For time displacements $\Delta t = t-s$ small enough to avoid caustics, the higher-order multidimensional WKB approximation(2,3) takes the form of an asymptotic expansion for small $\hbar$

$$K \sim (m/ih\Delta t)^{d/2} \exp [\frac{i}{\hbar} S(x,t;y,s) + \sum_{p=1}^{\infty} (i\hbar)^{p-1} \Lambda_p(x,t;y,s)] , \tag{1.2}$$

where S is the classical action

$$S(x,t;y,s) = \int_s^t d\tau\, L(q(\tau),\dot{q}(\tau),\tau) \tag{1.3}$$

for Lagrangian $L(x,\dot{x},\tau) = \frac{m}{2}\dot{x}^2 + A(x,\tau)\cdot\dot{x} - v(x,\tau)$, and q is the unique classical path satisfying the boundary conditions q(s) = y and q(t) = x. The transport method(4) will determine coefficients Λ_p recursively using $q(\tau)$.

By way of contrast, the large mass expansion of K has the analytic form

$$K \sim (m/ih\Delta t)^{d/2} \exp\{\frac{i}{\hbar}\frac{m}{2\Delta t}|x-y|^2 + \sum_{p=0}^{\infty} m^{-p} J_p(x,t;y,s)\} . \tag{1.4}$$

This expansion is the definition of the generalized Wigner-Kirkwood expansion[5]. If potentials $v(x)$ and $A(x)$ are static and β is the inverse temperature, the substitution $\Delta t = \hbar\beta/i$ transforms (1.4) into the traditional Wigner-Kirkwood expansion[6,7] for the heat kernel $\langle x|e^{-\beta H}|y\rangle$. The bridge between (1.2) and (1.4) is implemented by finding a large-mass expansion of the trajectory q and of the transport recurrence identities that define Λ_p.

2. WKB TRANSPORT AND m^{-1} EXPANSION OF q

The small Δt expansion

$$S(x,t;y,s) = \frac{m}{2\Delta t}|x-y|^2 + (x-y)\cdot\int_0^1 d\xi\, A(y + \xi(x-y),s) + O(\Delta t) \qquad (2.1)$$

shows that the initial condition $K \to \delta(x-y)$ is satisfied by (1.2) if one requires $\Lambda_p(x,s;y,s) = 0$, $p \geq 1$. Substituting (1.2) into (1.1) and equating common powers of $\hbar$ provides the differential equations the coefficient functions Λ_p satisfy. The lowest power $\hbar^{-1}$ has zero coefficient because S is a solution of the Hamilton-Jacobi equation. The higher powers of $\hbar$ yield, for $p \geq 1$,

$$[\frac{\partial}{\partial t} + \frac{1}{m}(\nabla_1 S - A)\cdot\nabla_x]\Lambda_p = \frac{1}{2m}[\delta_{p,1}\nabla\cdot A + \Delta_1\Lambda_{p-1} + \sum_{k=1}^{p-1}\nabla_1\Lambda_{p-k}\cdot\nabla_1\Lambda_k] \,. \qquad (2.2)$$

Here the sum is absent if $p = 1$, $\Lambda_o \equiv \frac{m}{2\Delta t}|x-y|^2 - S$, and ∇_1 is a gradient on the first vector argument. Replace x,t with $q(\tau),\tau$ throughout (2.2). By Jacobi's theorem the momentum $p(\tau) = \nabla_1 S$, thus the lhs of (2.2) becomes a total derivative $\frac{d}{d\tau}\Lambda_p(q(\tau),\tau;y,s)$. Integrating over τ subject to $\Lambda_p(x,s;y,s) = 0$ gives the transport representation (which determines Λ_p recursively given q and Λ_o)

$$\Lambda_p(x,t;y,s) = \frac{1}{2m}\int_s^t d\tau\,[\delta_{p,1}\ \nabla\cdot A + \Delta_1\Lambda_{p-1} + \sum_{k=1}^{p-1}\nabla_1\Lambda_{p-k}\cdot\nabla_1\Lambda_k](q(\tau),\tau;y,s). \quad (2.3)$$

The classical path q satisfies the Euler-Lagrange equation associated with L. For each vector component $\alpha = 1,\cdots,d$ of q,

$$\ddot{q}_\alpha = \frac{1}{m}[f_\alpha(q,\tau) + \dot{q}_\beta F_{\alpha\beta}(q,\tau)] \,. \qquad (2.4)$$

Here the repeated index β is summed. The vector and tensor force fields are $f_\alpha = -\nabla^\alpha v - \partial A_\alpha/\partial t$ and $F_{\alpha\beta} = \nabla^\alpha A_\beta - \nabla^\beta A_\alpha$, where ∇^α denotes $\partial/\partial x_\alpha$. The path q is then written in terms of a large mass expansion

$$q(\tau) \equiv \rho_o(\tau) + \eta(\tau) = \rho_o(\tau) + \sum_{j=1}^{\infty} m^{-j}\rho_j(\tau) \qquad (2.5)$$

where η is the displacement between q and the free (linear) classical path ρ_o

$$\rho_o(\tau) = y + \frac{\tau-s}{\Delta t}(x-y) \,. \qquad (2.6)$$

Both paths q and ρ_o obey the two-point boundary conditions specified by (y,s) and (x,t). Thus, η and the coefficients ρ_j ($j \geq 1$) are required to vanish at both s and t. By use of the unit interval $\xi \in [0,1]$ Green function for $d^2/d\xi^2$ which has homogeneous boundary conditions, i.e., $g(\xi,\xi') = \xi_<(\xi_> - 1)$, one may solve for ρ_j recursively. For example, one finds

$$\rho_1(s + \xi\Delta t)_\alpha = \Delta t^2 \int_0^1 d\xi' \; g(\xi,\xi')[f_\alpha(\tilde{\xi}') + \Delta t^{-1}(x-y)_\beta F_{\alpha\beta}(\tilde{\xi}')] \; , \tag{2.7}$$

where $\tilde{\xi}$ denotes the space-time linear path

$$\tilde{\xi} = (y + \xi(x-y), \; s + \xi(t-s)) \; , \quad \xi \in [0,1] \; . \tag{2.8}$$

Similar formulae hold[1] for the higher order coefficients ρ_j.

3. GENERALIZED WIGNER-KIRKWOOD EXPANSION

The analytic character in mass of the Wigner-Kirkwood expansion (1.4) originates via two different mechanisms. First, the transport recurrence relations (2.3) have explicit mass dependence. Second, the classical path $q(\tau)$ which is used to construct the action and coefficients Λ_p has an implicit mass dependence. For example, consider Λ_o. Inserting (2.5) into (1.3) yields

$$\Lambda_o = \int_s^t d\tau \; \{\tfrac{1}{2} \eta_\alpha[f_\alpha(q,\tau) + \dot{q}_\beta F_{\alpha\beta}(q,\tau)] - \dot{q}\cdot A(q,\tau) + v(q,\tau)\} \; . \tag{3.1}$$

This shows that for $p = 0$:

$$\Lambda_p = \sum_{k=p}^{\infty} m^{-k} G_p^k \; , \tag{3.2}$$

where G_p^k are coefficient functions independent of both $\hbar$ and m. That (3.2) holds for $p \geq 1$ is seen from the transport identity (2.3). Using (3.2) and the analytic expansions in (1.2) and (1.4) yields the connecting relationship

$$J_p = \sum_{\ell=0}^{p} (i\hbar)^{\ell-1} \; G_\ell^p \; . \tag{3.3}$$

As an example, we compute $J_o = (i\hbar)^{-1} \; G_o^o$. From (3.1), (3.2) we find

$$i\hbar J_o = \int_s^t d\tau \; [-\dot{\rho}_o\cdot A(\rho_o,\tau) + v(\rho_o,\tau)] = \int_0^1 d\xi \; [\Delta t v(\tilde{\xi}) - (x-y)\cdot A(\tilde{\xi})] \; ,$$

where $\xi = (\tau-s)/\Delta t$. Since the functions G_ℓ^p are available from the large mass expansion (3.2), it is thus possible to calculate all the coefficient functions J_p of the generalized Wigner-Kirkwood expansion. For further details and explicit formulae for J_1 and J_2, we refer to Ref. [1].

REFERENCES

[1] T.A. Osborn and F.H. Molzahn: Phys. Rev. A (1986), to be published

[2] V.P. Maslov and M.V. Fedoriuk: Semiclassical Approximation in Quantum Mechanics (Reidel, Dordrecht, 1981)

[3] D. Fujiwara: Duke Math. J. 47, 559 (1980)

[4] G.D. Birkhoff: Bull. Am. Math. Soc. 39, 681 (1933)

[5] T.A. Osborn: J. Phys. A17, 3477 (1984)

[6] D. Bollé and D. Roekaerts: Phys. Rev. A31, 1667 (1985)

[7] Y. Fujiwara, T.A. Osborn and S.F.J. Wilk: Phys. Rev. A25, 14 (1982)

SIMPLE CONNECTED GRAPH EXPANSIONS OF PROPAGATORS

F. H. Molzahn and T. A. Osborn
Department of Physics
University of Manitoba
Winnipeg, Manitoba, Canada R3T 2N2

1. THE PROBLEM

We describe three formal connected graph representations of the quantum time-evolution operator. These results have great utility because they permit one to expand the propagator in terms of relevant physical parameters such as particle mass m, or Planck's constant $h = 2\pi\hbar$.

Consider an N-body nonrelativistic spinless quantum system, having configurations described by vectors $x,y \in R^d$ $(d = 3N)$. Let the time-dependent Hamiltonian $H_\lambda(t)$, acting in $L^2(R^d_x)$, be determined by the differential operator

$$H_\lambda(x,\frac{\hbar}{i}\nabla_x,t) = -\frac{\hbar^2}{2m}\nabla_x^2 + \lambda\nu(x,t) . \qquad (1.1)$$

Here $\lambda \in R$ is a coupling constant and the scalar potential ν, describing all interparticle and external interactions, is assumed to be a real-analytic function of x. The system's dynamics is described by Schrödinger's equation

$$i\hbar \frac{\partial}{\partial t} U(t,s) = H_\lambda(t)\, U(t,s) \qquad (1.2)$$

for the time-evolution operator $U(t,s)$ satisfying initial condition $U(s,s)$ = identity.

The propagators we consider here are the following Dirac matrix elements of $U \equiv U(t,s)$: $K_\lambda = \langle x|U|y\rangle$ (the coordinate representation) and $K^*_\lambda = \langle x|U|k\rangle$, $K^\#_\lambda = \langle p|U|y\rangle$ where $p,k \in R^d$ are momentum variables (the mixed representations). These propagators will be generically denoted by $K^\bullet_\lambda$ where $\bullet = \ ,*,\#$, with a similar convention for related quantities. This notation is employed both for efficiency and to emphasize the underlying structural similarities which persist in the three propagator representations studied here.

It is easy to see that each free $(\lambda=0)$ propagator $K^\bullet_0$ is everywhere nonzero. Hence, it is possible to define functions F_λ, F^*_λ, $F^\#_\lambda$ by the relation

$$K^\bullet_\lambda = K^\bullet_0\, F^\bullet_\lambda(\ ,t;\ ,s) . \qquad (1.3)$$

The blank vector arguments of $F^\bullet_\lambda$ are those associated with the $\bullet$ case, e.g., they are (x,k) if $\bullet = *$. The algebraic factorization (1.3) is of interest because $K^\bullet_0$ contains all of the essential singularity of the function $K^\bullet_\lambda$ as $m \to \infty$.

It may be shown that $F^\bullet_\lambda$ has an exponentiated coupling constant expansion

$$F^\bullet_\lambda(\ ,t;\ ,s) = \exp\Big[\sum_{j=1}^{\infty} \lambda^j L^\bullet_j(\ ,t;\ ,s)\Big] \qquad (1.4)$$

with coefficient functions $L^\bullet_j$ depending on $\hbar$ and m but not on λ. The $L^\bullet_j$ have a formal representation completely determined by connected simple graphs (clusters). In Ref. [1] these results are derived using the Dyson series expansion of U(t,s), and a combinatoric method based on the cluster expansion of the classical grand partition function. The final form of these results is presented in Sec. 2 below, and some comments about their applications are made.

2. DESCRIPTION OF $L_j^{\bullet}$

We begin by describing the objects which will appear in the formula for $L_j^{\bullet}$. Some of these have a geometrical character, while others are combinatoric.

The first geometrical objects are a trio of linear paths $\xi^{\bullet}$ in R^d, one for each of the three matrix element cases. Put $\Delta t = t-s$ and let $\xi \in [0,1] = I$ be a parameter; then we define

$$[\bullet = \ ] \qquad \tilde{\xi} = y + \xi(x-y) \tag{2.1}$$

$$[\bullet = *] \qquad \xi^{*} = x + (\xi-1)\Delta t u \qquad (u = k/m) \tag{2.2}$$

$$[\bullet = \#] \qquad \xi^{\#} = y + \xi \Delta t v \qquad (v = p/m)\ . \tag{2.3}$$

Path $\tilde{\xi}$ runs from y to x, whereas ξ^{*} runs in the k direction ending at x, and $\xi^{\#}$ starts at y and runs in the p direction. It is useful to think of u as an initial velocity for ξ^{*}, and v as a final velocity for $\xi^{\#}$. The paths (2.1)-(2.3) may then be regarded as the $m \to \infty$ limits of classical paths satisfying the same two-point boundary conditions (cf. Ref. [2] and note $\tilde{\xi} = \rho_o(s+\xi\Delta t)$).

Next, recall that the one-dimensional Green functions $g^{\bullet}(\xi,\xi')$ for the differential operator $d^2/d\xi^2$ satisfy

$$\partial_1^2\, g^{\bullet}(\xi,\xi') = \delta(\xi-\xi') \qquad \xi,\xi' \in I\ .$$

There is a $g^{\bullet}$ for each of the above cases. Specifically, these functions and their associated homogeneous boundary conditions are

$$g(\xi,\xi') = \xi_<(\xi_>-1)\ , \qquad g(0,\xi') = g(1,\xi') = 0\ ,$$

$$g^{*}(\xi,\xi') = \xi_>-1\ , \qquad \partial_1 g^{*}(0,\xi') = g^{*}(1,\xi') = 0\ ,$$

$$g^{\#}(\xi,\xi') = -\xi_<\ , \qquad g^{\#}(0,\xi') = \partial_1 g^{\#}(1,\xi') = 0\ ,$$

where $\xi_< = \min\{\xi,\xi'\}$, $\xi_> = \max\{\xi,\xi'\}$.

Turning to the combinatoric objects, let $\mathcal{C}_j$ be the set of all simple connected graphs[(3)] $G = (V_j,E)$ on $j \geq 1$ vertices. $V_j = \{1,\cdots,j\}$ is the vertex set, while the edge set E consists of unordered pairs $\alpha = \{\ell,\ell'\}$ of distinct vertices ($\ell,\ell' \in V_j$, $\ell \neq \ell'$). That G is simple means there are neither loops (edges of the form $\{\ell,\ell\}$) nor multiple edges (repetitions of a given α) in E. Let $\xi = (\xi_1,\cdots,\xi_j) \in I^j$ be a vector in the unit j-cube. Then we associate with each edge $\alpha \in E$ the differential operators

$$b_\alpha^{\bullet} = g^{\bullet}(\xi_\ell,\xi_{\ell'})\, D_\ell \cdot D_{\ell'}\ .$$

Here D_ℓ provides a spatial gradient on the potential whose argument contains index ℓ, viz.,

$$D_\ell \nu(\xi^\bullet_{\ell'},\tau) = (\nabla\nu)(\xi^\bullet_\ell,\tau)\,\delta_{\ell,\ell'} \quad .$$

Also set

$$c^\bullet_j = \sum_{\ell=1}^{j} b^\bullet_{\{\ell,\ell\}} \quad .$$

Further associate to each edge $\alpha \in E$ a natural number i_α and put $r = \Sigma_{\alpha\in E}\, i_\alpha$ (unless $j = 1$ where $r \equiv 0$). Finally, introduce a summation denoted by $\mathcal{B}_j$ which consists of a sum over all graphs $G \in \mathcal{C}_j$, followed by the sum over $i_\alpha = 1 \sim \infty$ for each $\alpha \in E$:

$$\sum_{\mathcal{B}_j} \equiv \sum_{G\in\mathcal{C}_j} \Big[\prod_{\alpha\in E} \sum_{i_\alpha=1}^{\infty} \Big] \quad .$$

With these ingredients, the formula for $L^\bullet_j$ [which determines propagator $K^\bullet_\lambda$ via (1.3) and (1.4)] may be stated as follows.

$$L^\bullet_j = \sum_{\mathcal{B}_j} \sum_{n=o}^{\infty} \frac{1}{j!n!2^n} (-i\Delta t)^{j+n+r}\, \hbar^{n+r-j}\, m^{-n-r} \int_{I^j} d^j\xi$$
$$\times \Big[\prod_{\alpha\in E} (i_\alpha!)^{-1} (b^\bullet_\alpha)^{i_\alpha} \Big] (c^\bullet_j)^n \prod_{\ell=1}^{j} \nu(\xi^\bullet_\ell, s+\xi_\ell\Delta t). \qquad (2.4)$$

The differential operators $b^\bullet_\alpha$, $c^\bullet_j$ act on the product of j potentials. Two vector arguments from the set $\{x,y,k,p\}$ appear in each $\xi^\bullet_\ell$.

Formula (2.4) may be used to obtain (heuristically) the asymptotic expansions of $K^\bullet_\lambda$ in the physical parameters m^{-1} and $\hbar$. The large mass expansion defines the generalized Wigner-Kirkwood expansion and one recovers formulae like (1.4) and (3.3) of Ref. [2]. The small $\hbar$ expansion gives a graphical representation of the higher-order multidimensional WKB approximation for short time displacements Δt. Consequently, one can thereby construct three explicit complete integrals of the Hamilton-Jacobi equation. A connected graph expression for the momentum representation propagator $\langle p|U(t,s)|k\rangle$ may be derived. However, it turns out to be equivalent to the Fourier transform of a mixed representation, and so will not be presented here. For details on these applications, and for the derivation of (2.4), see Refs. [1,4,5].

REFERENCES

[1] F.H. Molzahn and T.A. Osborn: "Connected Graph Representations of the Quantum Propagator and Semiclassical Expansions", University of Manitoba Preprint (1986).

[2] T.A. Osborn and F.H. Molzahn: previous paper in these proceedings

[3] R.J. Wilson: Introduction to Graph Theory (Academic, New York, 1975)

[4] F.H. Molzahn: Ph.D. Thesis, University of Manitoba, Winnipeg, Canada (1986, unpublished)

[5] F.H. Molzahn and T.A. Osborn: J. Math. Phys. 27, 88 (1986)

THEOREM ON THE SCHWINGER REPRESENTATIONS OF LIE GROUPS AND ITS APPLICATION TO THE COHERENT STATES AND THE VIBRON MODEL

S. K. KIM

Department of Chemistry, Temple University

Philadelphia, Pennsylvania 19122.

1. Introduction

We shall first discuss a general transformation theory of the boson creation and annihilation operators under a Schwinger representation[1] of a Lie group. Its application has two parts. The first part is to describe the generalized coherent states of SU(2) and SU(1,1) groups[2,3]. The present treatment can be easily extended to higher dimensional transformation groups since it does not require Baker-Campbell-Hausdorff (BCH) formula[4] in contrast with the ordinary treatments. This part contains hardly any new result and can claim only a methodological interest. The second part of the application is to construct the most general algebraic Hamiltonian for the U(4) vibron model.[5,6] This will be achieved by classifying the boson creation and annihilation operators into scalars and vectors under orthogonal transformations based on the general transformation theory of the boson operators developed in the beginning.

2. The basic theorems

Let $\{ a_i, a_j^+; i, j = 1, 2, \ldots, n\}$ be a set of Boson annihilation and creation operators satisfying the commutation relations,

$$[a_i, a_j^+] = \delta_{ij}, \quad [a_i, a_j] = [a_i^+, a_j^+] = 0 \quad . \tag{2.1}$$

Let $T = ||t_{ij}||$ be a n x n matrix generator of a given Lie group $L^{(n)}$, then the Schwinger representation of T is defined by

$$\hat{T} = A^+ \cdot T \cdot A = \sum_{i,j} t_{ij}\, a_i^+ a_j \tag{2.2}$$

where A^+ and A are vectors defined by $(a_1^+, a_2^+, \ldots, a_n^+)$ and $(a_1, a_2, \ldots, a_n)$, respectively. The Schwinger representation $\{\hat{T}\}$ of the generator set $\{T\}$ of the Lie group $L^{(n)}$ has following basic properties:

(1) It provides a faithful representation of the Lie algebra of $L^{(n)}$.

(2) Let $U(\theta) = \exp[\theta T]$ be a group element of $L^{(n)}$, where θ is an arbitrary

parameter. Then the transformations of the boson operators under $\check{U}(\theta) = \exp[\theta\hat{T}]$ are described by the original matrix $U(\theta)$ as follows,

$$A^{+'} \equiv \check{U}(\theta)\, A^{+}\, \check{U}(\theta)^{-1} = A^{+}\cdot U(\theta)$$

$$A' \equiv \check{U}(\theta)\, A\, \check{U}(\theta)^{-1} = U(\theta)^{-1}\cdot A\ , \tag{2.3}$$

the proof is elementary; it follows from the commutation relations,

$$[\hat{T}, A^{+}] = A^{+}\cdot T,\quad [\hat{T}, A] = -T\cdot A\ . \tag{2.4}$$

The theorem states that two vectors A^{+} and A transform contragrediently with respect to the tranformation $U(\theta)$. It introduces the Wigner convention for the representations of the Lie group. It also leads to the basic properties of the operator basis which enables us to reconstruct[7] Bargmann's treatment[8] on the representation of the rotation group. A direct consequence of the theorem is that the total boson number $\hat{N} = (A^{+}\cdot A)$ is invariant with respect to $\check{U}(\theta) \in L^{(n)}$. In a special case of the orthogonal group $O(n)$, the generator T becomes antisymmetric and the matrix $U(\theta)$ becomes orthogonal, and thus we have two additional invariants $(A^{+}\cdot A^{+})$ and $(A\cdot A)$. This result is of vital importance in constructing the algebraic Hamiltonian for the U(4) vibron model (Sec. 5).

(3) The transform of a Schwinger generator $\hat{S}$ under a transformation $\check{U}(\theta) \in L^{(n)}$ is given by the matrix transform $S' = U(\theta) S\, U(\theta)^{-1}$ as follows,

$$\hat{S}' = \check{U}(\theta)\hat{S}\,\check{U}(\theta)^{-1} = A^{+}\cdot S'\cdot A\ . \tag{2.5}$$

This theorem is a special case of theorem (1). The examples of this elementary theorem will be given in the following sections on the coherent states.

(4) Let $|0\rangle$ be the boson vacuum state such that $a_i|0\rangle = 0$ for all i. Then a Schwinger operator $\check{U}(\theta) = \exp[\theta\hat{T}]$ leaves the vacuum state invariant, i.e., $\check{U}(\theta)|0\rangle = |0\rangle$. Accordingly, the transformation of a state vector $\psi = Q(A^{+}, A)|0\rangle$ defined by any boson operator $Q(A^{+}, A)$ is completely determined by the similarity tranformation of Q as given by (2.3).

3. The coherent state of SU(2) group.

The Schwinger representation of the generator set (J_1, J_2, J_3) of SU(2) is defined by

$$\hat{J}_i = A^+ \cdot (\sigma_i/2) \cdot A \tag{3.1}$$

with the Pauli spin σ_i and $A^+ = (a_1^+, a_2^+)$. The eigenvector belonging to a spin j and its projection m of $\hat{J}_3$ is given by

$$|j, m> = [(j + m)!(j - m)!]^{-1/2} (a_1^+)^{j+m} (a_2^+)^{j-m}|0> . \tag{3.2}$$

A general rotation which brings the generator J_3 into the direction $n = (\theta,\phi)$ is described in terms of $J_\pm = J_1 \pm iJ_2$ as follows.

$$U(n) = \exp[\alpha J_+ - \bar{\alpha}J_-] = (1 + |\zeta|^2)^{-1/2} \begin{pmatrix} 1 & \zeta \\ -\bar{\zeta} & 1 \end{pmatrix} \tag{3.3}$$

where $\alpha = -(\theta/2)\exp(-i\phi)$, $\zeta=-\tan(\theta/2)\exp(-i\phi)$, and $\bar{\zeta}$ is the complex conjugate of ζ. Then this matrix describes the tranformation of A^+ under $\check{U}(n)$ according to the general theorem (2.3). Thus, the operation of $\check{U}(n)$ on the ground state $|j, -j>$ immediately yields the coherent state of SU(2) in the direction of n,

$$|n> = |\zeta> = (1 + |\zeta|^2)^{-j} \sum_{m=-j}^{j} \binom{2j}{j+m}^{\frac{1}{2}} \zeta^{j+m} |j,m> \tag{3.4}$$

Ordinarily[2], this basic expansion is obtained by writing $\check{U}(n)$ in the normal order using the cumbersome BCH formula[4].

Directly from (3.4) one can derive almost all important properties of the coherent state of SU(2), e.g., the overcompleteness and non-orthogonality and etc. Moreover the characteristic property,

$$(n \cdot \hat{J})|n> = -j|n>, \quad <n|\hat{J}|n> = -jn \tag{3.5}$$

follows directly from $(n \cdot J) = U(n)J_3\,U(n)^{-1}$, $U(n)^{-1}J\,U(n) = J'$ on account of theorem (3).

4. The coherent state of SU(1,1)

The generators of the non-compact Lie group[2,3] SU(1,1) are defined in terms of the generators $\{J_i\}$ of SU(2) as follows,

$$K_1 = iJ_1, K_2 = iJ_2, K_3 = J_3 \tag{4.1}$$

and the Schwinger representation is defined by $A^+ \cdot K_i \cdot A = \hat{K}_i$. Eventhough the noncompact generators $\hat{K}_1$ and $\hat{K}_2$ are antihermitian, the coherent state of SU(1,1) can be defined quite analogously to the case of SU(2). Let the eigenvalues of the Casimir operator $\hat{K}^2$ be denoted as K(K+1). Then the eigenstate |K,m> of the positive discrete representation [2] D^+ may be characterized by K<0 and the eigenvalue m of K_3,

$$m = -K + \mu; \; \mu = 0,\, 1,\, 2,\, \ldots. \qquad (4.2)$$

Let $n = (\mathrm{sh}\tau \cos\phi,\ \mathrm{sh}\tau \sin\phi,\ \mathrm{ch}\tau)$ be a pseudo-Euclidean unit vector. Then a general pseudo-rotation which brings K_3 into $(n \cdot K)$ is described by

$$U(n) = \exp[\alpha K_+ - \bar{\alpha} K_-] = (1 - |\zeta|^2)^{-1/2} \begin{pmatrix} 1 & i\zeta \\ -i\bar{\zeta} & 1 \end{pmatrix}, \qquad (4.3)$$

where $K_\pm = K_1 \pm i K_2$, $\alpha = (\tau/2)\exp(-i\phi)$, $\zeta = \mathrm{th}(\tau/2)\exp(-i\phi)$. Thus, one obtains immediately, for the coherent state of the positive representation D^+,

$$|n\rangle = U(n)|K, -K\rangle = (1 - |\zeta|^2)^{-K} \sum_{m=-K} [(-1)^\mu \binom{2K}{\mu}]^{+1/2} |K, m\rangle, \qquad (4.4)$$

quite analogous to (3.4). Here it is noted that

$$(-1)^\mu \binom{2K}{\mu} = \Gamma(-2K + \mu)/[\mu!\,\Gamma(-2K)] \geq 0 \,. \qquad (4.5)$$

With the basic expansion (4.4) one can obtain almost all the properties of the coherent state of SU(1,1) as in the case of SU(2).

5. The algebraic Hamiltonian for diatomic molecules.

Algebraic approach has been proven to be very effective in describing molecular spectra of vibration and rotation. For a diatomic molecule the spectrum generating algebra has been assumed to be U(4) algebra through the use of the Schwinger representation with 16 generators ($a_i^+ a_j$; i, j = 1,. . ,4). The Hamiltonian of the vibron model must be hermitian, transform as a scalar under rotation, and conserve the total vibron number $\hat{N} = \Sigma\, a_i^+ a_i$. Recently[6], it has been shown by using theorem (2) that $\sigma = a_4$, $\sigma^+ = a_4^+$ behave as scalars and $\pi^+ = (a_1^+, a_2^+, a_3^+)$ and $\pi = (a_1, a_2, a_3)$ as vectors under rotations of $O(3) \subset U(4)$. Thus, one can define additional scalars by $n_\pi = (\pi^+ \cdot \pi)$ $\pi^{+2} = (\pi^+ \cdot \pi^+)$ and $\pi^2 = (\pi \cdot \pi)$. There exist, however, only three elementary hermitian scalars which conserve the vibron number $\hat{N}$;

$$A = n_\pi,\ B = \pi^{+2}\pi^2,\ C = \pi^{+2}\sigma^2 + \sigma^{+2}\pi^2 \,. \qquad (5.1)$$

Thus, the most general Hamiltonian is given by a linear combination of the symmetrized scalars contained in the infinite number of the sets,

$$S_n = \{ [A^{n-2k} B^{k-p} C^p]_+ \} \qquad (5.2)$$

where n = 0,1,. . ., k=0,1,. . . [n/2], p=0,1,. . .k. Hence, the order $|Sn|$ of the set S_n is given by

$$|S_n| = ([n/2]+1)([n/2]+2)/2 \,. \qquad (5.3)$$

Here [n/2] is the integral part of n/2. It can be shown[6] that all terms in the sets can be expressed in terms of the Casimir operators belonging to two different group chains. This helps the actual calculation of the spectra. Recently[6], numerical calculations of the H_2 spectra have been carried out with the Hamiltonian which contains terms up to the fourth order in the generators of U(4). In the ordinary treatment of the vibron model, the boson operators are classified by spherical tensors[5]. This formalism, however, introduces redundant terms in the Hamiltonian such that it seems hardly possible to construct the Hamiltonian which contains more than two-body terms.

Acknowledgement

Part of this work was carried out at the Fritz Haber Research Center, Hebrew University, Jerusalem, Israel. The author is grateful to professor R.D. Levine for the hospitality extended to him. He is also indebted to Professor R. Gilmore, Drexel University, Philadelphia, Pennsylvania for a fruitful suggestion.

References

1. J. Schwinger, "On Angular Momentum," U.S. Atomic Energy Commission, NYO-3071, 1952 (unpublished). It is contained in Quantum Theory of angular Momentum, Ed, L.C. Biedenharen and H. Van Dam, Academic press 1965.
2. A.M. Perelomov, Sov. Phys, Usp. 20, 703(1977).
3. C.C. Gerry, Phys. Rev. A 31, 2721(1985).
4. R. Gilmore, Lie Groups, Lie Algebras, and some of their applications, John Wiley & Sons, New York, 1974.
5. F. Iachello and R.D. Levine, J. Chem. Phys. 77, 3046(1982).
6. S.K. Kim, I.L. Cooper and R.D. Levine, Chem. Phys. To be published.
7. S.K. Kim, To be Published.
8. V. Bargmann, Rev. Mod. Phys. 34, 300(1962).

MULTIPLE-PATH EXPANSION IN QUANTUM MECHANICS AND QUANTUM FIELD THEORY

John P. Ralston
Department of Physics and Astronomy
University of Kansas
Lawrence, Kansas 66045

In this paper we discuss a novel method for constructing approximate Schroedinger wavefunctions. The method has much in common with the finite-element programs of Bender, et al. [1] which approximate the time evolution of operators in quantum mechanics or quantum field theory. Since an operator has so much information, however, we consider the more modest goal of approximating the time evolution of an arbitrary state $|s_o\rangle$. If we have $|s_T\rangle = \exp(-iHT)|s_o\rangle$, a state parametrically depending on time-parameter T, we can project out energy eigenstates $|E\rangle$:

$$|E\rangle = \int dT\, e^{iET}|s_T\rangle \;. \tag{1}$$

The limits on the T integral will be discussed below. Our method will use (1) to construct approximate energy eigenstates from a systematic approximation to the time evolution. The first approximation appears to be as good as WKB, and has much in common with the phase space wave packet superpositions used with striking success for a long time by Heller [2]. We discuss the case of one degree of freedom and a Hamiltonian H(p,q) first.

For the time evolution we borrow path-integral ideas [3], but consider optimizing the "paths". We optimize by replacing the redundant complete sets inserted after a time interval $t - T/M$, $M \to \infty$, by projectors constructed to span the time-evolved state -- at least up to the usual path integral errors of order t^2. The first question is whether one can find a basis in which to implement this idea, besides the unobtainable one made of the time evolved state itself. Surprisingly, the continuous coherent states [4] can be used, as follows.

We expand $|s_o\rangle$ in coherent states $|\alpha_o\rangle$ defined by $a|\alpha_o\rangle = \alpha|\alpha_o\rangle$, where $\alpha = q + ip$ is a complex number. We seek $|\beta_t\rangle$ such that the overlap with each time-evolved $|\alpha_o\rangle$ is unity:

$$\langle\beta_t|1 - iHt|\alpha_o\rangle = 1 + O(t^2) \;\cdot \tag{2}$$

As a first approximation $|\beta_t\rangle$ is still coherent and we solve (2) with the ansatz $\beta_t = \alpha + \dot{\alpha}t$, so the state shifts positions in phase space adiabatically. Along with the constraint that the expected energy

$E_\alpha = <\alpha_o|H|\alpha_o>$ is time independent, we then obtain [5] a contour (Cauchy) prescription for the superpositions given by (1):

$$|\tilde{E}> = \int_C dA\, e^{i2A} |\alpha_o(A)> , \qquad (3)$$

where $dA = (p\dot{q}-q\dot{p})dt/2$ is the area swept out by the state moving in phase space along the contour C s.t. E_α is fixed.

The notation $|\alpha_o(A)>$ indicates $|\alpha_o>$ on C at area parameter A, where $o < A < A_T$ for a closed contour. In obtaining (3) we let $t \to dt$, maintaining a unitary transformation, but made no new approximations. It is interesting that (2) does not necessarily imply classical motion for the β_t parameters, but we are optimizing phases as well as probabilities. For (3) to be consistent with the Hermitian character of H, a surface term (associated with the T limits in (1)) must vanish. That gives the quantization of contour C, if it closes:

$$(e^{i2A_T}-1)|\alpha_o(A_T)> = 0,$$

$$2A_T = 2\oint_C pdq = n\pi \quad n = \text{integer} ; \qquad (4)$$

otherwise, a continuous spectrum is implied.

Numerically, (4) differs from Bohr-Sommerfeld inasmuch as E_α knows all the quantum Hamiltonian's ordering rules, zero-point shifts, etc., which cannot be obtained classically. The wave functions (3) are very similar to Heller's, but more numerical work is required to see how similar. There are, of course, none of the WKB caustics and the procedure is much simpler and more explicit than Maslov's [6].

More significantly, the ansatz (2) is the one-path truncation of something more complicated. It is easy to enlarge the subspace in which $|\beta_t>$ is found by introducing shifted excited states $|\alpha_J> = (a^\dagger - \bar{\alpha})^J|\alpha_o> / \sqrt{J!}$, complete and orthogonal, so $<\alpha_J|\alpha_K> = \delta_{JK}$. These have increasing overlap with higher order, longer-ranged fluctuations beyond the displaced Gaussian ones of the coherent states. In the N path approximation we can project onto N of the $|\alpha_J>$ states, recovering the path integral as $N \to \infty$ and the WKB-like results (3,4) at $N = 1$. As a result we have a systematic series, although it is not at all easy to be precise about what the expansion parameters are: one path is exact for a quadratic Hamiltonian.

The method is also exact in a linearized collective motion problem in field theory [7], although other generalizations to many degrees of freedom are non-trivial. Space does not permit further discussion which will be given elsewhere [5].

ACKNOWLEDGMENT

I thank P. Carruthers, E. Heller, R. Littlejohn, and J. Klauder for useful comments. This work was supported in part under Department of Energy Grant No. DE-FG02-85ER40214.A002 and University of Kansas General Research Allocation No. 3570-0038.

REFERENCES

1. C. M. Bender and D. H. Sharp, Phys. Rev. Lett. 50, (1983)1535; V. Moncrief, Phys. Rev. D28 (1983)2485.
2. E. Heller, J. Chem. Phys. 66 (1977)5777; E. Heller and M. Davis, J. Chem. Phys. 71 (1979)3383; R. G. Littlejohn, Phys. Rep. (to be published).
3. A good review is given by L. Schulman, Techniques and Applications of Path Integrals (Wiley, New York 1981).
4. See, e.g. Coherent States-Applications in Physics and Mathematical Physics, edited by J. R. Klauder and B. Skagerstam (World Scientific, Singapore, 1985).
5. J. P. Ralston, in preparation.
6. V. P. Maslov and M. V. Fedoriuk, Semi-Classical Approximation in Quantum Mechanics (Reidel, Boston 1981).
7. J. P. Ralston, Phys. Rev. D33 (1986)496; D33 (1986)2003.

Strange Semiclassical Phenomena for the Equation $\nabla^2 \partial_t^2 \phi + A(\partial_x^2 + \partial_y^2)\phi + B\, \partial_z^2 \phi = 0$, Describing Waves in Stratified Fluids.

Frank S. Henyey
Center for Studies of Nonlinear Dynamics
La Jolla Institute
3252 Holiday Court, Suite 208
La Jolla, CA 92037

The equation

$$i\,\partial_t \psi = H\,\psi, \tag{1}$$

where H is the operator

$$H = \left[\nabla^{-2}(A\,\partial x^2 + A\,\partial y^2 + B\,\partial z^2)\right]^{1/2}, \tag{2}$$

describes internal gravity waves in the ocean (and atmosphere). (This equation is derived in many textbooks, such as Phillips, 1977.) $A^{1/2}$ is the frequency at which a long thin vertical object would oscillate around its depth of neutral buoyancy. (Deeper fluid is more dense (colder) and shallower fluid is less dense (warmer)). $B^{1/2}$ is the frequency at which a flat object would travel in a horizontal circle due to the Coriolis force on it. Normally $A >> B$.

If the fluid is moving with velocity $\vec{u}$, ∂_t is replaced by the convective derivative

$$\partial_t \rightarrow d_t = \partial_t + \vec{u} \cdot \nabla \tag{3}$$

or equivalently,

$$H = [\nabla^{-2}(A\,\partial x^2 + A\,\partial y^2 + B\,\partial z^2)]^{1/2} - i\,\vec{u} \cdot \nabla. \tag{4}$$

Assuming the flow $\vec{u}(x,y,z,t)$ to be incompressible, H is Hermitian.

An important problem in oceanography is to understand the transport properties of this system, with a stochastic flow field $\vec{u}$. Our group has been attacking this problem, using a (time-dependent) WKB approximation, and doing Monte Carlo calculation (Henyey and Pomphrey, 1983). We find many differences from ''induced diffusion,'' the analog of the Boltzmann equation. These differences are due to overly strong correlations, as decorrelation rates are slower than interaction rates. I think that part of the cause of the disparity of these rates is due to the kinematic structure of the system given by equation 4. In this talk, I will describe some of the unusual features of the classical and semiclassical approximations to this system. This system is a good example on which to test one's intuition. How many of the ''facts'' of mechanics you know are true in general, and how many only apply in familiar cases?

Planck's constant $\bar{h}$, and the number of quanta n, are quantum concepts, but for a bosonic system, their product $n\bar{h}$ is a classical wave quantity, known in oceanography as the wave action. We choose units (or we choose the amount of wave action to follow) so that $n\bar{h}$ is unity. With this choice, deBroglie's expressions are $E = i\,\partial_t$, $p = -i\,\partial_x$. The classical Hamiltonian is, therefore,

$$H = [p^{-2}(Ap_x^2 + Ap_y^2 + Bp_z^2)]^{1/2} + \vec{u}\cdot\vec{p} \ . \qquad [5]$$

First we consider the simple case $\vec{u} = 0$. This Hamiltonian is homogeneous of degree zero in p, very unlike all familiar Hamiltonians. Euler's theorem on homogeneous functions is

$$\vec{p}\cdot\partial H/\partial\vec{p} = 0\,. \qquad [6]$$

The (group) velocity is, by Hamiltonian's equation

$$\vec{v} = \partial H/\partial\vec{p}. \qquad [7]$$

Therefore,

$$\vec{p}\cdot\vec{v} = 0. \qquad [8]$$

The velocity is perpendicular to the momentum! The wave group travels along the wave crests, rather than perpendicular to them. Often to a reasonable approximation,

$$p_x^2 + p_y^2 << p_z^2 << A\,(p_x^2 + p_y^2)/B. \qquad [9]$$

In this case, the Hamiltonian simplifies to

$$H \approx A^{1/2}(p_x^2 + p_y^2)^{1/2}/|p_z| \ . \qquad [10]$$

The behavior in p_x, p_y is not peculiar; $(p_x^2 + p_y^2)^{1/2}$ is exactly what occurs for light and sound in two dimensions. The $1/|p_z|$ is the strange part. We get

$$v_z \approx -sgn(p_z) A^{1/2} (p_x^2 + p_y^2)^{1/2} / p_z^2 . \qquad [11]$$

The vertical velocity has the opposite sign to the vertical momentum, and goes as $1/p_z^2$. Like the traditional mule, the harder you push the slower it goes.

Now, we put the $\vec{u} \cdot \vec{p}$ term back in. For simplicity assume $\vec{u}$ is in the x and y directions, and that it depends only on z. The velocity gets a new term, $\vec{u}$, from $\partial H / \partial \vec{p}$. There is also a force in the vertical

$$dp_z / dt = -\partial H / \partial z = -\partial_z \vec{u} \cdot \vec{p} . \qquad [12]$$

Since we are assuming $\vec{u}$ to be time-independent, the value E of H is conserved. Moreover, p_x and p_y are conserved by the assumption that $\vec{u}$ is x-and y-independent. We can solve equation [5] for p_z,

$$p_z^2 = [A - (E - \vec{u} \cdot \vec{p})^2] (p_x^2 + p_y^2) / [(E - \vec{u} \cdot \vec{p})^2 - B] . \qquad [13]$$

Or, with the approximation of equations [10] and [11],

$$p_z = \pm A^{1/2} (p_x^2 + p_y^2)^{1/2} / (E - \vec{u} \cdot \vec{p}) \qquad [14]$$

which is a first integral of equations [11] and [12]. Assume there is some depth z at which $\vec{u}(z) \cdot \vec{p} = E$. As this depth is approached, p_z becomes infinite. The right side of equation [12] is nearly constant, so $p \propto t$. From equation [11], $z = z_c \propto t^{-1}$ where z_c is the depth at which $E = \vec{u} \cdot p$.

This result can be described as follows: The wave finds itself in a region where the force is strong. Therefore, its momentum increases. Since its velocity is proportional to $1/p_z^2$, it slows down and remains near the same point. The force continues to act, and the wave approaches a ''critical layer.''

This result also holds without the approximation of equation [10], as long as $\vec{u}$ is a function of z alone. More realistically, the time dependence, and possibly the horizontal dependence, eventually cause $\vec{u} \cdot \vec{p}$ to become different than E, so the critical layer does not go to completion. However, our calculations show that the dynamics of waves in random $\vec{u}$ fields are dominated by nearly critical layer events.

We now turn to a peculiar semiclassical phenomenon. The phase of the wave function is, in the semiclassical approximation, the action along the classical trajectories plus the contribution from the Maslov index. (A good review is Ziolkowski and Deschamps, 1984.)

The semiclassical wave function has a factor of the reciprocal of the square root of the Van Vleck determinant. The Morse index (Milnor, 1969) counts how many negative eigenvalues are in the determinant, and the Maslov index tells the difference between the number whose square root should be taken to be i and the number whose square root should be taken to be $-i$. Thus, as each caustic is passed, the Morse index increases by one (this is the Morse index theorem), and the Maslov index either increases or decreases by one. For familiar cases, the Maslov index is equal to the Morse index, which is the number of caustics passed. For our system, this turns out not to be the case.

To find the contribution to the Maslov index from a caustic, a coordinate rotation is made so that the caustic is given by $Q_1 = 0$. The contribution to the Maslov index is

$$\delta I = sgn\left[\delta \frac{dQ_1}{dP_1}\right] = sgn\left[\frac{d}{dt}\,\frac{dQ_1}{dP_1}\right] . \qquad [15]$$

At the caustic, $dQ_1/dt = 0$. Working out $\frac{d}{dt} dQ_1/dP_1$ (for the case $\partial H/\partial t = 0$), we find

$$\delta I = sgn\left[\frac{d}{dt}\,\dot{Q}_1/\dot{P}_1\right] = sgn\left[\ddot{Q}_1/\dot{P}_1\right] = sgn\left[\dot{H}_{P_1}/\dot{P}_1\right] \qquad [16]$$

at the caustic.

To evaluate this expression in general is messy. We will consider a class of examples which are sufficient to show the change in δI. We will assume that Q_1 is a rotated rectangular coordinate, that $\vec{u}$ is perpendicular to Q_1 and depends only on Q_1, and that the family of trajectories we consider depend only on Q_1. In particular, $\vec{u}$ and the family of trajectories are time independent. Since $\dot{Q}_1 = 0$, [Eq. 16] evaluates to, for this case,

$$\delta I = sgn\,(H_{P_1 P_1}) . \qquad [17]$$

We will choose

$$P_1 = \cos\theta\, p_z + \sin\theta\, p_x$$

$$P_2 = \cos\theta\, p_x - \sin\theta\, p_z$$

$$P_3 = p_y \ . \tag{18}$$

We find

$$0 = \dot{Q}_1 = H_{P_1} = H_{p_x} \sin\theta + H_{p_z} \cos\theta$$

$$= (A - B)(p_x p_z^2 \sin\theta - p_z (p_x^2 + p_y^2) \cos\theta) R \ , \tag{19}$$

where R is a positive definite expression, whose form is irrelevant. Using this, we find

$$H_{P_1 P_1} = (A - B)\left[P_2^2(\cos^2\theta - \sin^2\theta) + P_3^2 \cos^2\theta\right] R \ . \tag{20}$$

The usual case is $A > B$. Then if $\theta < 45°$ or if P_3 is too large, $H_{P_1 P_1} > 0$, and the contribution to the Maslov index is the same as the contribution to the Morse index. If $\theta > 45°$ and P_3 is small enough, $H_{P_1 P_1} < 0$, and the Maslov and Morse indices get opposite contributions. This result can be traced to the minus sign in Eq. 19, coming from the strange kinematics. For familiar systems, $H_{P_1 P_1}$ is always positive. As an example, if $H = P^2/2m + V(\vec{x})$, then $H_{P_1 P_1} = \frac{1}{m} > 0$.

In summary, three strange classical and semiclassical phenomena for internal gravity waves have been presented: 1. The velocity is perpendicular to the momentum. 2. The vertical velocity approaches zero as the vertical momentum approaches infinity, leading to critical layers. 3. The Maslov index may differ from the Morse index. All three phenomena are related to the strange dispersion formula, in which the Hamiltonian depends on the direction of the momentum but not on its magnitude.

This work was supported by the Office of Naval Research.

References

Henyey, F. S. and N. Pomphrey, ''Eikonal description of internal wave interactions: a non-diffusive picture of induced diffusion'', Dyn. Atmos. Oceans **7**, 189-219 (1983).

Milnor, J., ''Morse Theory'', Princeton University Press (1969).

Phillips, O. M., ''The dynamics of the upper ocean'' Cambridge Univ. press (1977).

Ziolkowski, R. W. and G. A. Deschamps, ''Asymptotic evaluation of high-frequency fields near a caustic: an introduction to Maslov's method'' Radio Sci. **19**,1001-1025 (1984).

E. SYMPLECTIC GEOMETRY AND QUANTIZATION

QUANTIZATION, TOPOLOGY, AND ORDERING

S. Twareque Ali

Department of Mathematics, Concordia University
Montréal, Canada H4B 1R6

H. D. Doebner

Arnold Sommerfeld Institut für Mathematische Physik
Technische Universität Clausthal, D-3392 Clausthal, FRG.

1. INTRODUCTION

In this report we describe two recently developed approaches to the problem of quantization: Borel quantization (BQ) and prime quantization (PQ). The point of departure in both cases is a notion of localization of the physical system S on a smooth manifold. However, in the one case (BQ) the manifold M in question is the configuration space M_c of S, while in the other case (PQ) it is a space which is eventually identified, by using for example a kinematical group, with the phase space of the system (i.e., the cotangent bundle T^*M_c of the manifold M_c). In any quantization scheme, the first step is the identification of those regions on the manifold M, in which the physical system may be assumed to be localized. These, in the case at hand, are taken to be the Borel sets, i.e., the elements of the Borel field $\mathcal{B}(M)$ of M. Quantization then involves finding a Hilbert space $\mathcal{H}$ and a mapping a of the Borel sets into the set $\mathcal{SA}(\mathcal{H})$ of self adjoint operators on $\mathcal{H}$. In BQ this mapping is effected by means of a projection valued (PV) - measure defined on M_c, while in PQ it is determined by a positive operator-valued measure, having a bounded density, defined on M. In the second step we consider the infinitesimal motions, in other words the momenta, of a localizable physical system. These motions are associated with the infinitesimal motions of the underlying manifold and lead eventually to a dichotomy of the quantized kinematical observables into position and momentum. In BQ these motions are modelled using the flows ϕ_t^X of complete vector fields X on M_c acting suitably on the localization regions. The quantization now involves a mapping p of this set $\mathcal{X}_c(M_c)$ of vector fields to $\mathcal{SA}(\mathcal{H})$ which preserves, on one hand the mapping a and on the other hand some of the algebraic and geometric properties of $\mathcal{X}_c(M_c)$. With the additional assumption (which has no direct 'classical' interpretation) that p(X) be a differential operator on $\mathcal{H}$, we get up to unitary equivalence, a complete classification of all quantizations of the kinematics of the system. This classification is given in an essential way through the specific topology of M_c. It relates the topology directly to physical measurements and opens a gate to the field of topological physics.

In the PQ-approach the dichotomy between (generalized) position and momentum observables is derived through the action on M of a given kinematical group $\mathcal{K}$ such as, for example, the Galilei or the Poincaré group. This ultimately leads to the identification of M with the phase space Γ of the system. Furthermore, the assumption that the POV-measure a has a bounded density, together with a certain covariance condition on it arising as a result of the action of the group $\mathcal{K}$ specifies a class of its unitary representations, whose Lie algebra then yields a quantization of the kinematics. Interestingly, different quantizations of the same classical observable are possible through different choices of a. Thus, a given polynomial in the classical algebra may well be mapped into different operators on $\mathcal{H}$ via different quantizations. This means that the procedure by which an ordering of the operators is carried out in going from the classical to the quantized theory is encoded in the quantization method, and the solution of this problem is linked directly to the kinematical symmetry that prevails and hence indeed to the method of making physical measurements.

2. DIFFERENTIABLE QUANTUM BOREL KINEMATICS AND TOPOLOGICAL PHYSICS

2.1 The mathematical model

To begin with the Borel quantization, or in other words the quantum Borel kinematics (QBK) [1], we consider a class of physical systems S, having a manifold as the configuration space M_c. The momentum of such a system, as well as its localization properties on momentum space, will be given through geometrical objects 'living' on M_c, to be viewed as the 'geometrical arena' for S. A convenient set of localization regions for position is the set $\mathcal{B}(M_c)$ of all Borel sets - i.e., the Borel field - of M_c. We shall denote individual Borel sets by Δ. A motion of the system should then correspond to a motion of the localization regions Δ.

Canonically, such a motion is modelled, at least infinitesimally, with the help of the flow ϕ_t^X of a complete vector field $X \in \mathcal{B}(M_c)$ (the set of all complete vector fields on M_c) with t as a flow parameter. The action of ϕ_t^X on $\mathcal{B}(M_c)$,

$$\Delta \mapsto \phi_t^X(\Delta) = \{ m' \mid m' = \phi_t^X(m),\ m \in \Delta \} \in \mathcal{B}(M_c), \tag{2.1}$$

then gives the flow model. The differential of ϕ_t^X, i.e., X itself is the momentum generated by this flow. This construction, used for all $X \in \mathcal{X}_c(M_c)$, should yield all the physical momenta of S independently of any specific dynamics. Hence we use $(\mathcal{B}(M_c), \mathcal{X}_c(M_c))$ as a classical mathematical model, to be called the Borel kinematics (BK) for a system located and moving (non-relativistically) on M_c, with Δ and X as its (generalized) position and momentum observables, respectively. This set of observables should be complete in the sense that it ought to contain all the information on the physics of the model. In the prime quantization scheme, the flow ϕ_t^X will correspond to one parameter subgroups of the kinematical group $\mathcal{K}$, acting on T*M.

As mentioned in the Introduction, to quantize the Borel kinematics, we have to construct mappings,

$$a: \quad \mathcal{B}(M_c) \longrightarrow \mathcal{SA}(\mathcal{H}) \tag{2.2}$$

$$p: \quad \mathcal{X}_c(M_c) \longrightarrow \mathcal{SA}(\mathcal{H}) \tag{2.3}$$

in a way such that certain characteristic properties of $\mathcal{B}(M_c)$ and $\mathcal{X}_c(M_c)$ survive, namely the Borel structure of $\mathcal{B}(M_c)$ and the Lie structure of $\mathcal{X}_c$ and such that the flow model (2.1), relating momentum to position, acquires a quantum analogue. It will turn out that a quantization calls for additional assumptions, and the structure of M_c as the geometrical arena will be used again to define the so called q-related geometrical objects [2], as possible motivation for defining the map p.

2.2 The quantization procedure

Consider the mapping $\Delta \mapsto a(\Delta)$ in (2.2). If $\psi \in \mathcal{H}$ is a normalized vector, then

$$p_\psi(\Delta) = (a(\Delta)\psi, \psi) \tag{2.4}$$

ought to give the probability for finding the quantum system, when it is in the pure state $T_\psi = |\psi\rangle\langle\psi|$, to be localized in the region Δ of M, and hence $\Delta \mapsto p_\psi(\Delta)$ should be a probability measure on $\mathcal{B}(M)$. Extending this argument to all normalized vectors $\psi \in \mathcal{H}$, we arrive at the result [3] that the mapping (2.2) defines a positive operator-valued (POV)-measure on $\mathcal{B}(M)$. Thus, $\Delta \mapsto a(\Delta)$ must satisfy,

$$a(\emptyset) = 0, \quad a(M) = I, \tag{2.5}$$

$$a\left(\bigcup_{j\in J} \Delta_j\right) = \sum_{j\in J} a(\Delta_j), \tag{2.6}$$

where $\emptyset$ denotes the null set in M, I is the identity operator on $\mathcal{H}$, J is a countable index set, and the Borel sets Δ_j, $j \in J$, are pairwise disjoint. Furthermore, the sum in (2.6) is assumed to converge weakly, and thus we could just as well have arrived at (2.5) - (2.6), had we used density matries ρ instead of pure states in (2.4), writing in that case $p(\Delta) = \mathrm{Tr}[a(\Delta)\rho]$.

The operators $a(\Delta)$, as defined through (2.4) - (2.6), are positive, i.e.,

$$(a(\Delta)\psi, \psi) \geq 0, \quad \forall\ \psi \in \mathcal{H},$$

but otherwise their nature is fairly arbitrary. To select from this wide choice a reasonable class of POV-measures, we assume for the Borel quantization that a is a projection-valued (PV) or spectral measure, and we denote it by a_c. Thus, $\forall\ \Delta \in \mathcal{B}(M)$,

$$a_c(\Delta) = a_c(\Delta)^* = [a_c(\Delta)]^2.$$

There are various physical justifications [10] for this particular choice of a POV-measure. Here we only mention that this choice makes the quantized Borel kinematics analogous to standard Schrödinger quantum mechanics on $M_c = \mathbb{R}^n$, which of course is a special case of our general formalism, and in which case it is well known that $\mathcal{B}(\mathbb{R}^n)$ is represented by projection operators in the quantized theory. To complete the translation of the Borel field into the quantum model, we assume that the von Neumann algebra generated by a_c has an elementary spectral measure, i.e., we neglect spin and internal degrees of freedom. (For a more general formulation, see [5].)

Consider next the map $X \longmapsto p(X) \in \mathcal{SA}(\mathcal{H})$ in (2.3). To construct this map we need a quantum analogue of the flow and of the flow model (2.1). Since for complete vector fields there holds the relation,

$$\phi^X_{t_1} \circ \phi^X_{t_2} = \phi^X_{t_1+t_2}, \qquad t_i = 1,2, \tag{2.7}$$

it is reasonable to model this in $\mathcal{H}$ via a set of continuous one-parameter unitary groups (shift groups). Thus we require that there exist such a shift group V^X_t for any $X \in \mathcal{X}_c(M)$, which acts on $\mathcal{B}(M)$ as (t = group parameter),

$$V^X_t\ a_c(\Delta) V^X_t = a_c(\phi^X_t(\Delta)). \tag{2.8}$$

By analogy with the flow and its momentum, we take for p(X) the generator of V_t^X, which is an essentially self-adjoint operator on a dense domain $\mathcal{D}_{p(X)} \subset \mathcal{H}$, i.e.,

$$ip(X) = \text{s-lim} \frac{V_t^X - I}{t}, \quad \text{on} \quad \mathcal{D}_{p(X)} . \tag{2.9}$$

We note that (2.8) is an imprimitivity relation [6] or a covariance condition, for the set of shift groups on $\mathcal{B}(M)$ (compare (3.12) in Section below). A desirable property of p, connected with the flow model, is the following: Take any pure state T_ϕ, i.e., $T_\phi \psi = \|\phi\|^{-2}(\psi,\phi)\psi$, ϕ, $\psi \in \mathcal{H}$, which is located in Δ, i.e., $\mathrm{Tr}[a_c(\Delta)T_\phi] = \|\phi\|^{-2}(a_c(\Delta)\phi,\phi) = 1$. Furthermore, take any $X \in \mathcal{X}_c(M)$ which vanishes in Δ, i.e., $X\restriction_\Delta = 0$. Then ϕ_t^X acts trivially on Δ and obviously, $(p(0)\phi,\phi) = (p(X)\phi,\phi)$ should hold ($\phi \in \mathcal{D}_{p(0)} \cap \mathcal{D}_{p(X)}$). If this is the case (for any choice of Δ) then p is called local [7]. In addition to the more analytical properties, the map a_c should preserve the partial Lie algebra structure which $\mathcal{X}_c(M)$ inherits from the (infinite dimensional) Lie algebra of smooth vector fields on M, $\mathcal{X}_c(M) \subset \mathcal{X}(M)$. We assume that a_c is a partial Lie homomorphism+) on some domain $\mathcal{D} \subset \mathcal{H}$, i.e.,

$$\left.\begin{aligned} &a_c(X+\alpha Y) = a_c(X) + \alpha a_c(Y) \\ \text{and} \quad &a_c([X,Y]) = [a_c(X), a_c(Y)] \end{aligned}\right\} \tag{2.10}$$

for all $X, Y \in \mathcal{X}_c(M)$, $\alpha \in \mathbb{R}$, where also $X + \alpha Y \in \mathcal{X}_c(M)$ and $[X,Y] \in \mathcal{X}_c(M)$. Because the generators p(X) are in general unbounded, the domain $\mathcal{D}$ has to be properly specified. Moreover, we would also want to apply to $\mathcal{D}$ polynomials of p(X) and $a_c(\Delta)$. To achieve this, it is useful to first smoothen the 'sharp' Borel sets by defining a_c on the functions $f \in C^\infty(M,\mathbb{R})$. This is possible, since by virtue of the spectral theorem a_c yields a map

$$q : C^\infty(M,\mathbb{R}) \longrightarrow \mathcal{SA}(\mathcal{H}),$$

which gives q(f). Thus, we take for $\mathcal{D}$ a common dense domain for all polynomials in p(X) and q(f), $X \in \mathcal{X}_c(M)$, $f \in C^\infty(M,\mathbb{R})$.

+) Note that there exist examples, e.g., connected with magnetic monopoles, in which (2.10) fails.

We summarize with the following definition.

A triple $(\mathcal{H}, a_c, p)$ is called a quantum Borel kinematics (QBK) iff,

$\mathcal{H}$ is a separable Hilbert space.

$a_c : \mathcal{B}(M) \longrightarrow \mathcal{SA}(\mathcal{H})$ gives an elementary spectral measure on $\mathcal{B}(M)$ in $\mathcal{H}$.

$p : \mathcal{X}_c(M) \longrightarrow \mathcal{SA}(\mathcal{H})$ gives $p(X)$ as a generator of a continuous, unitary shift group V^X along X; p is local and a partial Lie homomorphism.

$\mathcal{D} \subset \mathcal{H}$ is a common dense domain for polynomials in $p(X)$, $q(f)$, $f \in C^\infty(M, \mathbb{R})$.

We are thus left with the task of constructing possibly all QBK, up to unitary equivalence, in $\mathcal{H}$. This means having to develop a representation theory for QBK, along with a classification of the different QBK's, depending for example on the topology of M. The corresponding object in the prime quantization scheme is the representation theory of the kinematical group $\mathcal{K}$ on reproducing kernel Hilbert spaces of functions on T*M.

2.3 Results and q-related geometrical objects

The representation of the map a is again unique up to unitary equivalence. If one realizes (standard realization) the abstract Hilbert space $\mathcal{H}$ as the space $L^2(M, \nu)$ of complex functions on M which are square integrable with respect to a smooth Borel measure ν, then $a_c(\Delta)$ acts uniquely as,

$$a_c(\Delta) = \chi_\Delta \psi , \quad \psi \in L^2(M, \nu), \quad \Delta \in \mathcal{B}(M), \tag{2.11}$$

with χ_Δ being the characteristic function of Δ. From a physical point of view, this relation is plausible. It generalizes to arbitrary M the standard results concerning $\mathcal{H}$ and a_c for the case where $M = \mathbb{R}^n$. Since a_c is elementary, ψ cannot be vector valued, thus precluding the appearance of spin or internal degrees of freedom. A generalization of the QBK to the case where a_c is not elementary is possible [5].

For the standard realization of a_c, consider now the inequivalent realizations of p(X) on $L^2(M, \nu)$. No workable classification of these realizations seems to be known, and further physical and geometrical information on p(X) seems necessary in order to proceed further. We consider again the example of $M = \mathbb{R}^n$ and the (unique) irreducible representation of the Heisenberg group in $L^2(\mathbb{R}^n, dx^n)$, together with the representation of the standard vector fields in $\mathbb{R}^n$ as generators of translation. As another example, we consider the Mackey quantization [6] on a homogeneous G-space along with a representation of the associated vector fields on the G-space which span the Lie algebra of G. In both cases, the vector fields are represented through differential operators (of finite order) in a Hilbert space of functions. Hence we assume,

p(X) is a <u>differential operator</u>. (2.12)

Although this assumption appears to be reasonable, its application requires further (geometrical) information, the reason being that $L^2(M,\nu)$ is a complex function space, with elements restricted only by square integrability. One has to define in addition what differentiability means on the set theoretic product $M \times \mathbb{C}$, i.e., one has to equip $M \times \mathbb{C}$ with a <u>differentiable structure D</u> such that it becomes a smooth manifold. There is a trivial method for doing this, viz., the given differentiable structure of M as a smooth manifold and the natural differentiable structure of $\mathbb{C}$ can be used to construct the so called product structure D_0 on $M \times \mathbb{C}$, which is commonly used in mathematical physics. However, there also exist other differentiable structures D on $M \times \mathbb{C}$ which are <u>not</u> isomorphic to D_0. The question then is how to arrive at these latter. A complete answer is not known, but a somewhat sharper formulation of the partial Lie homomorphism (as <u>Lie stability</u> [7]) shows that any $D = D_\eta$ is admissible which turns $M \times \mathbb{C}$ into a hermitian complex line bundle $\eta = (E, \pi, M, \mathbb{C})$, i.e., into a vector bundle with basis M, fibre $\mathbb{C}$ and a hermitian metric. Set theoretically $E = M \times \mathbb{C}$ holds. A complete classification of sets of isomorphic η, i.e. D_η, is known [8], so that a classification of p seems to be possible. Because the set $\mathrm{Sec}_0^\infty(\eta)$ of compactly supported and differentiable (with respect to D_η) sections of η is dense in $L^2(M,\nu$ we can define a <u>differentiable QBK</u> as a QBK with the additional assumption

$$p(X) \text{ is a differential operator on } \mathcal{D} = \mathrm{Sec}_0^\infty(\eta). \quad (2.13)$$

With this assumption, one can prove that, up to unitary equivalence, the set of differentiable QBK is in one-to-one correspondence with $\pi_1{}^*(M) \times \mathbb{R}$ where $\pi_1{}^*(M)$ is the dual (character group) of $\pi_1(M)$ and $\mathbb{R}$ is a parameter space. With this general classification theorem, inequivalent quantizations on M_c are given by (τ, c), $\tau \in \pi_1{}^*(M_c)$, $c \in \mathbb{R}$. Using the Hurewicz theorem, $\pi_1{}^*(M_c) = H_1{}^*(M_c, Z)$, with $H_1(M_c, Z)$ being the 1st homology group which decomposes into a free Abelian (Betti-) group and a discrete torsion part. Different elements of the Betti group yield inequivalent quantizations, and the so called topological potentials $\lambda \cdot \beta$ (X) (i.e., linearly independent logarithmically exact 1-forms) which appear in the expressions for p(X) as additive terms. The constant λ ($0 \leq \lambda \leq 2\pi$) is like a potential strength, similar to what one has in the case of minimal coupling, i.e., (e/c) $\underline{A}$ ($\underline{x}$, t).

Obviously, $\pi_1(M_c)$ is directly related to the topology of M_c. The parameter space $\mathbb{R}$ is independent of the topology and reflects the fact that the dimension of $\mathfrak{X}_c(M_c)$ is not finite. This space reappears in a class of unitary representations of the diffeomorphism group Diff (M_c) (cf. [9] for the case where $M_c = \mathbb{R}^n$) which has (in a certain sense) the set p($\mathfrak{X}_c(M_c)$) as its infinitesimal version. If one introduces a time evolution, i.e., dynamics, via a unitary group on L^2 (M_c, $d\nu$) and a Riemannian structure on M_c (if this is at all possible), one can show that a Hamiltonian having reasonable commutation properties with q(f) and p(X) exists only for c = o. However, c itself is a kind of a quantum number which is of special importance in the description of the kinematical observables of systems of N indistinguishable particles [11].

2.4 Examples

To show how the above developed formalism works, we present examples of non-relativistic systems on topologically non-trivial configuration spaces M_c, i.e., spaces with non-trivial $\pi_1(M_c)$ along with their quantizations. It is possible to imagine various different systems, such as for example, N indistinguishable particles in $\mathbb{R}^3$, a single particle in $\mathbb{R}^3$ with 'holes', systems constrained to submanifolds of $\mathbb{R}^n$ or systems having constrained collective motions. Here the Betti and the torsion parts of $\pi_1(M_c)$ play different roles. We discuss below the first two examples.

A. Consider two particles moving in $\mathbb{R}^3$, which cannot both be at the same point together. The configuration space is $M_c^2 = \mathbb{R}_1^3 \times \mathbb{R}_2^3 - D$, $D = \{(x_1, x_2) \mid x_1 = x_2,\ x_i \in \mathbb{R}_i^3,\ i = 1, 2\}$. Suppose now that the particles are indistinguishable, so that the configurations (x_1, x_2) and (x_2, x_1) become identical. Mathematically this means having to divide the above configuration space by the symmetric group S_2. One gets in this way a factor space (which once again is a smooth manifold),

$$M_c^2 \simeq (\mathbb{R}_1^3 \times \mathbb{R}_2^3 - D) / S_2 .$$

To see its topology, introduce on M_c^2 the coordinates $y = x_1 + x_2$, $z = x_1 - x_2$, to obtain,

$$M_c^2 \simeq \mathbb{R}_y^3 \times \dot{\mathbb{R}}_z^3 \simeq \mathbb{R}_y^3 \times \mathbb{R}^+_{|z|} \times S^2,$$

where $\dot{\mathbb{R}}_z^3 = \mathbb{R}_z^3 - \{0\}$, $\mathbb{R}^+_{|z|} = \{|z| \mid |z| = |x_1 - x_2| > 0\}$ and S^2 is the 2-sphere. Obviously, $\mathbb{R}_y^3$ and $\mathbb{R}^+_{|z|}$ are pointwise S_2 invariant, but the sphere S^2 is not. Here the non-trivial permutations connect antipodal points. Thus, S^2/S_2 is twisted and is the projective space $\mathbb{R}P^2$ (or a half sphere where the two equatorial half circles have been glued together with a twist):

$$M_c^2 \simeq \mathbb{R}_y^3 \times \mathbb{R}^1_{|z|} \times \mathbb{R}P^2.$$

(Configuration spaces of indistinguishable particles are in general not trivial [12]. Thus, for two particles on a circle, one gets a Möbius band.) For the fundamental group we find,

$$\pi_1(M_c^2) = \pi_1(\mathbb{R}P^2) = S_2, \quad \pi_1^*(M_c^2) = S_2,$$

so that π_1 contains only a torsion part.

The classification theorem (c=o) now leads to <u>two</u> inequivalent quantizations, for this system with two indistinguishable particles in $\mathbb{R}^3$. Transforming the kinematics back to $\mathbb{R}_1^3 \times \mathbb{R}_2^3$ one finds (not surprisingly) that they correspond to symmetric or antisymmetric wave functions. This shows that the Pauli principle can be traced back to a topological property of the configuration space, if indistinguishability is formulated appropriately. From systems with more than two particles, one obtains the same result, except that now <u>parastatistics</u> could play an important role. It is tempting to look for a possible connection between <u>spin and statistics</u> along these lines. To do this however, a generalization of QBK to non-elementary measures and a formulation using spinor representations of Diff (M_c^2) [13] would be necessary, in addition to having to modify some conditions, e.g., the spectral conditions used in the proof of the (relativistic) spin-statistics theorem.

B. Consider next the <u>Aharonov-Bohm (AB)</u> configuration space, i.e., a manifold with 'hole': $M_c^{AB} = \mathbb{R}^3 - \{(0,0,x_3) \mid x_3 \in \mathbb{R}\}$ as the geometrical arena for a one particle system. Again, the topology is not trivial and π_1 is a Betti group,

$$\pi_1(M_c^{AB}) \simeq \mathbb{Z}, \quad \pi_1^*(M \quad) \simeq \mathbb{R} \bmod 2.$$

The classification theorem gives (c = o) a one-parameter, $0 \le \lambda \le 1$, family of inequivalent quantizations and correspondingly, the following topological potential

$$\beta\Big(\sum_{j=1}^{3} a_j \frac{\partial}{\partial x_j}\Big) = \lambda \frac{1}{x_1^2 + x_2^2}(-x_2 a_1 + x_1 a_2).$$

The quantization for standard vector fields in $\mathbb{R}^3$ is

$$p\left(\frac{\partial}{\partial x_j}\right) = -i\frac{\partial}{\partial x_j} - A_j(\underline{x}),$$

$$\underline{A}(\underline{x}) = \lambda \frac{1}{x_1^2 + x_2^2}(-x_2, x_1, 0).$$

Hence, the topology of M_c^{AB} produces, through the QBK, just the potential of a $\mathbb{R}^1$- solenoid, with $\Phi \sim \lambda$ where $\hat{\Phi} = \Phi \bmod 2\pi$ and Φ give equivalent theories.

We remark, that this quantization method gives no information on how to realize this topological potential physically - it does not explain the AB-effect. It only shows that the $\mathbb{R}^1$-solenoid potential on M_c^{AB} is topologically the most natural one. This information could be useful for the formulation of quantum mechanics on M_c^{AB} and for the computation of AB-effects for more complicated topologies, e.g., for $\mathbb{R}^3 - T^{2,2}$, where $T^{2,2}$ is an 8-shaped double torus.

2.5 Topological physics

The general classification theorem and the given selection of examples clearly show that the topology of the configuration space M of the system - its geometrical arena - is of direct physical import. Thus, experimental results obtained locally, i.e., in the laboratory, depend on the topology, i.e., globally on M_c. All of this brings us into the domain of (non-relativistic) topological physics. Results obtained in this field can sometimes be rather unexpected. Indeed, quantum systems on manifolds M_c do 'feel' the entire manifold, and this is related to the Hilbert space that is used and the fact that observables have to correspond to self-adjoint operators, in order to ensure a proper probability interpretation of the theory. Quantum Borel kinematics is one way to arrive at topological physics. Here, the topology of M_c becomes enmeshed in the quantization. On the other hand, there are also other routes through which the topology can enter into the physics, such as for example in twisted field theories, in the physics on topologically non-trivial spacetimes, solution varieties of non-linear problems, etc. Each method has its own technical difficulties as well as its unexpected features, at least when compared to our normal experience.

We close this section with a feuilletonistic extension of a remark by Kac. Take a region $B \subset \mathbb{R}^2$ (smooth, no holes) with boundary ∂B. Using B as a drumhead, build a drum in the shape of ∂B. Now beat the drum and listen, and one discovers that one can actually 'hear the shape of the drum'. In other words, one can mathematically analyse the accoustics of the beats and calculate from its spectrum (possibly with some additional assumptions) the boundary ∂B. A similar situation is now seen to prevail in topological physics. Take M to be the universe (possibly of some limited physical experience) and populate it with some physical system. Now conduct (quantum mechanical) experiments and make the relevant local measurements. Analyse the results. It is tempting to imagine that in this way one can (this time quantum mechanically) 'hear the shape of the universe'.

3. PRIME QUANTIZATION AND THE ORDERING PROBLEM

3.1 Phase space as the underlying classical manifold

In this section we turn our attention to the prime quantization procedure [4, 10] . As noted in the Introduction, in this method of quantization the manifold M on which the system is localized is eventually identified with its phase space, denoted by Γ . As a manifold, this phase space could arise in a variety of ways. For example, it could be the cotangent bundle T^*M_c of the manifold M_c considered in the last section; or it could arise as a homogeneous space with a symplectic structure, of a kinematical symmetry group $\mathcal{K}$ of the system; or it could simply be the spectrum of a commutative C*-algebra. The reason for this last possibility is that one can assume that the observables of the classical system to be quantized generate a commutative C*-algebra. This latter, as is well known [14] is isomorphic to an algebra of functions on a certain locally compact space, namely its spectrum, which can then serve as the classical phase space. In every case we shall assume that the phase space Γ comes equipped with a natural measure $d\Gamma$ which has support on the whole of Γ . Indeed, in all the three situations mentioned above, this is in fact the case. In what follows we first lay down the mathematical steps comprising the PQ-procedure and later discuss its physical underpinning as well as its bearing on the ordering problem in quantum mechanics. As a last remark of a general nature, in the transition from the classical to the quantized theory, the concept of a phase space itself has to be modified. To wit, the position and momentum of a particle are not simultaneously measurable with absolute accuracy in quantum mechanics, and hence the local coordinates in Γ lose their classical significance in a quantized theory. However, this altered significance is naturally brought out [10] - without any further assumptions - by the PQ technique.

3.2 The prime quantization procedure

Consider the algebra $\mathfrak{A}_{cl}$ of classical observables. It is a commutative C*-algebra; we denote its spectrum by Γ and identify $\mathfrak{A}_{cl}$ with $C_\infty(\Gamma)$, the C*-algebra under the uniform norm [14] of complex continuous functions on Γ which vanish at infinity. In the case where we begin with Γ itself (arising as some T*M, or as a homogeneous space, etc.) we simply take $C_\infty(\Gamma)$ as the classical algebra. We form next the Hilbert space $\mathcal{H} = L^2(\Gamma, d\Gamma)$, and consider subspaces $\mathcal{H}_K$ of it which are defined by reproducing kernels K. To make this notion more precise, consider the projection operator $\mathbb{P}_K$ on $\mathcal{H}$ for which

$$\mathbb{P}_K \mathcal{H} = \mathcal{H}_K , \tag{3.1}$$

and suppose that the action of $\mathbb{P}_K$ on a vector $\psi \in \mathcal{H}$ is given by means of a kernel, $K : \Gamma \times \Gamma \longrightarrow \mathbb{C}$, such that,

$$\psi_K(\zeta) = (\mathbb{P}_K \psi)(\zeta) = \int_\Gamma K(\zeta, \zeta')\psi(\zeta')\, d\Gamma' \tag{3.2}$$

$\forall \zeta \in \Gamma$. It is important to note that (3.2) defines the projected function ψ_K for all points $\zeta \in \Gamma$, and not just up to a set of measure zero. The property $\mathbb{P}_K = \mathbb{P}_K{}^* = [\mathbb{P}_K]^2$ of the projection operator implies then that the kernel K enjoys the corresponding porperties:

$$K(\zeta, \zeta') = \overline{K(\zeta', \zeta)}, \tag{3.3}$$

$$\int_\Gamma K(\zeta, \zeta'')K(\zeta'', \zeta')\, d\Gamma'' = K(\zeta, \zeta'), \tag{3.4}$$

$\forall \zeta, \zeta' \in \Gamma$, the bar in (3.3) denoting complex conjugation. Eq. (3.4) is the reproducing property of K, a terminology which is self-evident in view of its implication that $\forall \psi_K \in \mathcal{H}_K$,

$$\psi_K(\zeta) = \int_\Gamma K(\zeta, \zeta')\, \psi_K(\zeta')\, d\Gamma'. \tag{3.5}$$

The reproducing kernel Hilbert space $\mathcal{H}_K$ has a canonically associated POV-measure defined on $\mathcal{B}(\Gamma)$. This is obtained by first noting [15] that for each fixed $\zeta \in \Gamma$, the linear evaluation map $E^K_\zeta : \mathcal{H}_K \longrightarrow \mathbb{C}$, defined by

$$E^K_\zeta(\psi_K) = \psi_K(\zeta), \tag{3.6}$$

is bounded. The general theory of reproducing kernel Hilbert spaces can then be used to establish first, that

$$K(\zeta, \zeta') = E^K_\zeta \, E^{K\,*}_{\zeta'} , \tag{3.7}$$

where $E_\zeta^{K*}: \mathbb{C} \longrightarrow \mathcal{H}_K$ is the adjoint of the linear map E_ζ^K , and secondly, that

$$F_K(\zeta) = E_\zeta^{K*} E_\zeta^K \tag{3.8}$$

is a bounded positive operator on $\mathcal{H}_K$. Furthermore, writing

$$a_K(\Delta) = \int_\Delta F_K(\zeta) d\Gamma , \tag{3.9}$$

$\Delta \longmapsto a_K(\Delta)$ is a POV-measure (cf. Eqs. (2.5) - (2.6)) on $\mathcal{H}_K$, with the <u>bounded density</u> F_K .

The prime quantization procedure is now a prescription for mapping $C_\infty(\Gamma)$, linearly as a vector space, into the set $\mathcal{L}(\mathcal{H}_K)$ of bounded operators on $\mathcal{H}_K$, using the POV-measure a_K . Of course, real valued functions in $C_\infty(\Gamma)$ should as a consequence be mapped onto self-adjoint operators in $\mathcal{L}(\mathcal{H}_K)$. Thus, we define the <u>prime quantization map</u>, which is a positive linear map,

$$\pi_K^*: C_\infty(\Gamma) \longrightarrow \mathcal{L}(\mathcal{H}_K), \tag{3.10}$$

such that

$$\pi_K^*(f) = \int_\Gamma f(\zeta) F_K(\zeta) d\Gamma . \tag{3.11}$$

Suppose next that the physical system has an underlying symmetry group G which acts as a transformation group on Γ . Thus, there ought to exist in the Hilbert space $\mathcal{H}_K$ of the quantized system a continuous unitary irreducible representation $g \longmapsto U_K(g)$ of G , and hence, as a consequence of the general theory of group representations on reproducing kernel Hilbert spaces (cf. [15] and references eited therein), U_K should be a subrepresentation of an induced representation [6] . Moreover, under reasonable physical assumptions it is possible to consider G to be a semidirect product, $G = G_0 \circledS T^n$ of a group G_0 and an n-dimensional Abelian group T^n. The phase space Γ , which is now isomorphic to a homogeneous space of $G_0 \circledS T^n$, has a natural local coordinatization into a 'configuration' and a 'momentum' part. Additionally, there exists a non-trivial subgroup G' of G for which

$$U_K(g')a_K(\Delta)U_K(g')^* = a(g'[\Delta]), \tag{3.12}$$

$\forall \quad g' \in G', \Delta \in \mathcal{B}(\Gamma)$, where $g'[\Delta]$ is the translate of the set Δ (considered as a subset of a homogeneous space of G) under g' . Eq. (3.12) is again a <u>generalized imprimitivity</u> or covariance relation, similar to (2.8).

To sum up therefore, the problem of prime quantization reduces to that of finding reproducing kernel subspaces of the Hilbert space $L^2(\Gamma, d\Gamma)$ which carry unitary irreducible representations of the group G .

3.3 The physical interpretation

We start with two mathematical comments. First, given an abstract Hilbert space $\mathcal{H}$ and a POV-measure a which is defined on $\mathcal{B}(\Gamma)$, and which admits a bounded density $\zeta \longrightarrow F(\zeta)$ in the sense of (3.9), one can prove [15] that there exists a reproducing kernel Hilbert space $\mathcal{H}_K$, with canonically associated POV-measure a_K, to which the pair $\{\mathcal{H}, a\}$ is unitarily isomorphic. The second remark is that it is possible, in a fairly straightforward manner, to extend both the domain and the range of the map π_K^* in (3.10) so as to accomodate classical observables which are not necessarily bounded, or even continuous functions on Γ.

In view of the second remark, we see that π_K^* can be obtained as the adjoint of a map $\pi_K: \mathcal{J}(\mathcal{H}_K) \longrightarrow L^1(\Gamma, d\Gamma)$, where $\mathcal{J}(\mathcal{H}_K)$ is the Banach space of all trace class operators on $\mathcal{H}_K$. Thus, every physical state, i.e., density matrix ρ, is mapped by the prime quantization procedure to a probability measure

$$d\mu_\rho = f_\rho(\zeta)d\Gamma \tag{3.13}$$

with $f_\rho \in L^1(\Gamma, d\Gamma)$. Hence, $f_\rho(\zeta)$ denotes the probability density of finding the system localized at the phase space point ζ. Actually, the localization point ζ is specified only to within a volume $\hbar$ in Γ, in complete consonance with the uncertainly principle. (For a complete discussion of this point cf. [10].) Furthermore, the probability density f_ρ satisfies an equation of continuity when the appropriate time evolution is applied to the system. To show that the interpretation is consistent, it is possible to prove, using the first remark, that any quantum system which is localized in phase space, in the sense that it possesses a POV-measure a on $\mathcal{B}(\Gamma)$, having a bounded density, can be realized on a reproducing kernel Hilbert space and hence is achievable by means of a prime quantization of a classical system.

Finally, every reproducing kernel Hilbert space $\mathcal{H}_K$ admits an <u>overcomplete family</u> of vectors Φ_ζ, $\zeta \in \Gamma$, obtained as

$$\Phi_\zeta(\zeta') = K(\zeta', \zeta) \tag{3.14}$$

and hence there is a close connection [10, 16] between the PQ-technique and the use of <u>generalized coherent states</u> to describe quantum mechanical systems.

3.4 The ordering problem and examples

For free non-relativistic and relativistic problems, a classification of all covariant (with respect to the Galilei and the Poincaré groups) quantizations has been given in [10] and [16] . Denoting either one of these two groups by $\mathcal{K}$ and by H the subgroup $SO(3) \otimes T$, of spatial rotations and time translations of both these groups, we write as the homogeneous space

$$\Gamma = \mathcal{K}/H .$$

Thus, in both cases, points $\zeta \in \Gamma$ can be (globally) parametrized by $(\underline{q}, \underline{p}) \in \mathbb{R}^6$, and in fact the invariant measure on Γ is then $d\Gamma = d\underline{q}\, d\underline{p}$. Hence, the Hilbert space $L^2(\Gamma, d\Gamma)$ consists of all square integrable functions $\psi(\underline{q}, \underline{p})$ of the (3-) position and (3-) momentum variables, with respect to the usual Lebesgue measure. Every reproducing kernel Hilbert space $\mathcal{H}_K \subset L^2(\Gamma, d\Gamma)$ which carries a unitary irreducible representation U_K of $\mathcal{K}$ (corresponding to mass-m and spin-j) is characterized by a single vector η . Denoting by β the canonical surjection

$$\beta : \Gamma \longrightarrow K ,$$

the overcomplete family (generalized coherent states) of vectors in $\mathcal{H}_K$ is defined as

$$= U_K \ (\beta(\underline{q},\underline{p}))\eta \quad , \ (\underline{q},\underline{p}) \in \Gamma . \tag{3.15}$$

Furthermore, the kernel K itself and the density F_K are given by

$$K(\underline{q}, \underline{p}; \underline{q}', \underline{p}') = (\eta_{\underline{q},\underline{p}} , \eta_{\underline{q}',\underline{p}'}) , \tag{3.16}$$

$$F_K(\underline{q}, \underline{p}) = |\eta_{\underline{q},\underline{p}}\rangle\langle\eta_{\underline{q},\underline{p}}| . \tag{3.17}$$

The different quantizations correspond to different ordering possibilities in a sense we now make precise. Defining the classical position and momentum operators by means of the phase space functions,

$$\left. \begin{aligned} f_{pos}(\underline{q}, \underline{p}) &= \underline{q} \\ f_{mom}(\underline{q}, \underline{p}) &= \underline{p} \end{aligned} \right\} \tag{3.18}$$

we find for their quantized versions two operators

$$\begin{aligned} \underline{Q}_K &= \pi_K^*(f_{pos}) \\ \underline{P}_K &= \pi_K^*(f_{mom}) \end{aligned} \tag{3.19}$$

whose components satisfy the canonical commutation relations,

$$[Q_K^j , P_K^k] = i \hbar \delta_{jk} . \tag{3.20}$$

Consider now a classical observable f which is a finite polynomial in the components of $\underline{q}$ and $\underline{p}$. Its quantized version $\pi_K^*(f)$ will in general also be a polynomial in the components of Q_K^j and P_K^j. However, the order in which these products of Q_K^j and P_K^j appear in any term of $\pi_K{*}(f)$ is completely determined by the particular K, and hence by the specific $\eta \in \mathcal{H}_K$, that is chosen. Different kernels K give rise to different orderings. For example, in the non-relativistic case, if one takes for η the function $A\,exp[-B(q^2 + p^2)]$ where A and B are fixed constants (determined by the theory), one gets the well-known antinormal ordering which is so often used in quantum mechanics.

ACKNOWLEDGEMENTS

One of us (STA) is grateful to the Alexander von Humboldt-Stiftung for financial support during the period of time when the find draft of this manuscript was completed. He would also like to thank the Arnold Sommerfeld Institut, Clausthal, for hospitality.

REFERENCES

1. B. Angermann, H.-D. Doebner and J. Tolar, Lecture Notes in Math. 1037, 171-208 (1984).

2. H.-D. Doebner, Czech. J. Phys. (1987), in press.

3. J.M Jauch, Foundations of Quantum Mechanics, Addison-Wesley, Reading, Mass. (1968).

4. S.T. Ali and H.-D. Doebner 'The ordering problem in quantum mechanics: Prime quantization and a physical interpretation' to appear.

5. H.-D. Doebner and U.-A. Gehringer, in preparation.

6. G.W. Mackey, Induced Representations of Groups and Quantum Mechanics, Benjamin, New York (1968).

7. B. Angermann, Ph.D. dissertation, Clausthal (1983).

8. See, e.g., B. Kostant, Lecture Notes in Math. 170, 87-208 (1970).

9. G.A. Goldin, R. Menikoff and D.H. Sharp, Phys. Rev. Lett. 51, 2246-49 (1983)

10. S.T. Ali, Rivista del Nuovo Cim. 8, 1-128 (1985).

11. H.-D. Doebner and G.A. Goldin, preprint.

12. E. Fadell, L. Neuwirth, Math. Scand. 10, 111-118 (1962).

13. G.A. Goldin and D.H. Sharp, Commun. Math. Phys. 92, 217-228 (1983).

14. S. Sakai, C*-Algebras and W*-Algebras, Springer, Berlin (1971).

15. S.T. Ali, J. Math. Phys. to appear.

16. S.T. Ali and E. Prugovečki, Acta Appl. Math. 6, 1-18 (1986); 6, 19-45 (1986) and 6, 47-62 (1986).

THE DIFFERENTIAL GEOMETRY OF PHASE SPACE AND QUANTISATION

D.J. Simms
School of Mathematics
Trinity College, Dublin

In the geometric approach, phase space is taken to be a smooth manifold M carrying a non-degenerate closed differential two-form ω. Such a space is called a symplectic manifold. With respect to a class of local coordinates $p_1, \ldots, p_n, q_1, \ldots, q_n$, called canonical, ω is $\sum_{i=1}^{n} dp_i \wedge dq_i$. Here we consider the case where M has finite dimension 2n. Thus, ω_m is a non-degenerate skew-symmetric bilinear form on the tangent space M_m, and M_m is called a symplectic vector space. The word symplectic was coined by Hermann Weyl to describe a geometry based on a skew-symmetric scalar product rather than the usual Euclidean symmetric scalar product.

Each smooth scalar field H on the phase space defines, by contraction of the covector field dH with the two-form ω, a vector field X_H called the Hamiltonian vector field generated by H. The set of smooth scalar fields on M forms a Lie algebra under the Poisson bracket: $[H,F] = X_H(F)$.

Now the tangent space M_m is itself a linear symplectic manifold, with the constant symplectic form ω_m, and may be considered as the linearisation of the phase space at m. Of special interest is the (2n+1)-dimensional Lie algebra N_m of polynomial functions on M_m of degree at most one. This is a Lie algebra under the Poisson bracket and is called the Heisenberg algebra at m since it has Heisenberg commutation relations relative to a suitable basis.

The simply connected Lie group N_m generated by the Lie algebra N_m is called the Heisenberg group at m. For each m we can select an irreducible unitary representation W_m of N_m on a Hilbert space such that $W_m(1) = \exp(i/\hbar)1$, this being the Weyl integrated form of the canonical commutation relations. However, such a choice is only unique up to unitary equivalence. The geometric approach to quantisation constructs as an initial step, a fully covariant theory which incorporates at each point of M a representation of the CCR.

This is done as follows. A basis for M_m with respect to which ω_m has component matrix $\begin{pmatrix} O & I \\ -I & O \end{pmatrix}$ is called a symplectic frame. Consider $\mathbb{R}^{2n}$ as a symplectic vector space with the usual basis as a symplectic frame. Denote by $N(\mathbb{R}^{2n})$ the Poisson bracket Lie algebra of polynomial functions of degree at most one on $\mathbb{R}^{2n}$. Then the usual coordinates $x^1, \ldots, x^n, y^1, \ldots, y^n$ satisfy the Heisenberg realations $[x^i, y^j] = \delta^{ij}$. Denote by $N(\mathbb{R}^{2n})$ the simply connected Lie group having $\mathcal{N}(\mathbb{R}^{2n})$ as Lie algebra, and fix an irreducible unitary representation W of $N(\mathbb{R}^{2n})$ on the Hilbert space $L_2(\mathbb{R}^n)$ with $W(1) = \exp(i/\hbar)1$.

A choice of symplectic frame for M_m gives a linear isomorphism of M_m with $\mathbb{R}^{2n}$, a Lie algebra isomorphism of $\mathcal{N}_m$ with $\mathcal{N}(\mathbb{R}^{2n})$ and a Lie group isomorphism of N_m with $N(\mathbb{R}^{2n})$.

The group of transformations between symplectic frames is called the symplectic group $Sp(n,\mathbb{R})$. It is the group of 2n x 2n real matrices g such that $g^t\begin{pmatrix} 0 & I \\ -I & 0\end{pmatrix}g = \begin{pmatrix} 0 & I \\ -I & 0\end{pmatrix}$. It is a non-compact Lie group, and it plays the same role in phase space as the Lorentz group does in space-time. The symplectic group acts on $\mathbb{R}^{2n}$ in the natural way and hence on the Lie algebra $\mathcal{N}(\mathbb{R}^{2n})$ and on the Lie group $N(\mathbb{R}^{2n})$. We denote by g.a the effect of $g \in Sp(n,\mathbb{R})$ acting on $a \in N(\mathbb{R}^{2n})$. If an element of N_m is represented by $\underline{a}$ with respect to one symplectic frame, then under a change of frame by g it is represented by g.a.

We know that, for some unitary operator U on $L_2(\mathbb{R}^n)$ that

$$W(g.a) = U\, W(a)\, U^{-1}$$

for all $a \in N(\mathbb{R}^{2n})$. However U is only unique up to a U(1) phase factor. Thus, the natural group to implement a change of frame is the Weil-metaplectic group $Mp^c(n,\mathbb{R})$ which is defined as the set of all pairs (U,g) such that U is a unitary operator on $L_2(\mathbb{R}^n)$, and $g \in Sp(n,\mathbb{R})$, and $W(g.a) = U\, W(a)\, W^{-1}$ for all $a \in N(\mathbb{R}^{2n})$.

$Mp^c(n,\mathbb{R})$ is a group containing U(1) as a subgroup and whose quotient by this subgroup is the symplectic group $Sp(n,\mathbb{R})$. Thus, $Mp^c(n,\mathbb{R})$ may be thought of as the symplectic group together with an extra U(1) gauge freedom.

To get a fully covariant theory which incorporates the CCR at each point of phase space we abandon the use of symplectic frames and instead use Weil-metaplectic frames which incorporate the additional U(1) gauge freedom. This is analogous to the procedure used in introducting spinors in space-time, where the usual Lorentz frames are replaced by spin frames which incorporate an additional Z_2 freedom.

More formally, the situation may be described by saying that we have a central extension

$$1 \to U(1) \to Mp^c(n,\mathbb{R}) \to Sp(n,\mathbb{R}) \to 1$$

and we replace the symplectic frame bundle of M by a principal $Mp^c(n,\mathbb{R})$

bundle P which we call a Weil-metaplectic frame bundle for M. The point to note is that this can always be done, and the number of ways it can be done is naturally parametrised by the second integral cohomology group of M.

The group $Mp^c(n,\mathbb{R})$ acts naturally on $N(\mathbb{R}^{2n})$, on $L_2(\mathbb{R}^n)$, and on the space $S'(\mathbb{R}^n)$ of tempered distributions. The corresponding bundles associated to P have fibres N_m, H_m, S'_m (say) respectively. The representation W of $N(\mathbb{R}^{2n})$ on $L_2(\mathbb{R}^n)$ gives a well-defined representation of N_m on H_m for each m. Sections of S' are the symplectic spinors of Kostant.

The Hamiltonian vector fields on M preserve ω and hence define vector fields on the symplectic frame bundle which are invariant under the symplectic group. Quantisation, in this approach, requires lifting these to vector fields on the Weil-metaplectic frame bundle P which are invariant under the Weil-metaplectic group. Such a lifting is fixed by choosing a suitably normalised differential one-form γ on P which is invariant under the Weil-metaplectic group and such that $d\gamma$ corresponds to $\omega/i\hbar$. The value of γ on the lift of the Hamiltonian vector field X_H is then fixed to correspond to $H/i\hbar$. This ensures that γ itself is invariant under the lifted vector fields.

Such a one-form γ on P exists if and only if the cohomology class of ω/h plus half the first Chern class of the symplectic frame bundle is an integral de Rham class. This is a quantisation condition on phase space, due to Hess, which refines the original work of Kostant and Souriau.

The action of the lifted Hamiltonian vector field on the symplectic spinors gives a Lie algebra representation of the complete Poisson bracket Lie algebra $C^\infty(M)$. This is called prequantisation, and it is the basis of the geometric approach to constructing the quantum operators. Further geometric properties of phase space are required to complete the construction.

Literature: P. Robinson, Mp^c structures and applications. Warwick University thesis, 1984.

N. Woodhouse, Geometric Quantization. Oxford Univ. Press 1980.

THE STRUCTURE OF PHASE SPACE AND QUANTUM MECHANICS

M. Moshinsky
Instituto de Física, UNAM
Apdo. Postal 20-364
México, D.F. 01000 México

The author and his collaborators[1-6] have been interested for a number of years in the structure of phase space as a carrier of canonical transformations, and in the representations of the latter in quantum mechanics. This has led to their conviction that some quantum features, such as the nature of the spectra of Hamiltonians i.e., continuous, discrete, mixed or of bands, are already implicit in the classical picture. The main new development in this field was the recent discussion of the canonical transformation to action and angle variables for periodic potentials and their representation in quantum mechanics.[4] In this note though, we shall try to give a general overview of our program and indicate the references where the reader can find a more systematic analysis.

We shall only be discussing problems with one degree of freedom and thus our phase space is actually a (q,p) plane with q being the coordinate and p the momentum. To be a carrier of canonical transformations this plane should have a structure similar to the Riemann surface of the complex plane,[5] which allows the latter to be a carrier of conformal transformations in a bijective (i.e., one-to-one onto) fashion. This Riemann surface structure for the phase plane turns out not to be as convenient as it is for the complex plane, because we also want to discuss the representation in quantum mechanics of the canonical transformations. Thus an alternative structure was developed,[1-6] using the concepts of ambiguity group and ambiguity spin, where the latter can also be introduced in quantum mechanics.

We shall begin by showing through the example of a simple conformal transformation that the concepts of ambiguity group and spin present an alternative to the Riemann surface also in the case of the complex plane. Let us consider two complex variables z and $\bar{z}$ related by the conformal transformation

$$z=\bar{z}^{k}, \text{ k integer}. \qquad (1)$$

Clearly when $\bar{z}$ is in a sector of angle $(2\pi/k)$, the variable z will cover the whole plane and we need to introduce k sheets in z plane joined, for example, along the real axis from 0 to ∞ to have a one—to—one mapping of the $\bar{z}$ plane to the k sheeted Riemann surface associated with z. An alternative way of introducing the mapping in a bijective way is to note that in the $\bar{z}$ plane we have an ambiguity group, as when we carry out the operation

$$\bar{z} \to \bar{z} \exp (i2\pi r/k), \; r=0,1,2,... k-1 \qquad (2)$$

we get from (1) the same value of z. The group of these operations is clearly C_k, the cyclic one of order k.

We wish now to find a way of mapping entire functions of $\bar{z}$, which we denote by

$$\phi(\bar{z}) = \sum_{\nu=o}^{\infty} \alpha_\nu \bar{z}^\nu, \qquad (3)$$

on the corresponding ones of z, without the need of using the Riemann surface structure of the z plane. For this purpose we note that $\phi(\bar{z})$ can be decomposed in its irreducible parts $\phi^\lambda(\bar{z})$, $\lambda=0,1,2,...$ k-1, with respect to C_k as

$$\phi(\bar{z}) = \sum_{\lambda=o}^{k-1} \phi^\lambda(\bar{z}) \;, \qquad (4a)$$

$$\phi^\lambda(\bar{z}) = k^{-1} \sum_{r=o}^{k-1} \exp (i2\pi\lambda r/k) \; \phi[\bar{z} \exp (-i2\pi r/k)] \;, \qquad (4b)$$

where

$$\exp (i2\pi\lambda r/k) \;, \; r=0,1,2,... k-1 \;, \qquad (5)$$

are the irreducible representations (irreps) characterized by $\lambda=0,1,..., k-1$ of the abelian cyclic group C_k.

Clearly then, if we write $\nu \equiv \lambda (\bmod k)$, i.e., $\nu=nk+\lambda$; $\nu=0,1,2,...,$ $\lambda=0,1,...$ k-1; n=0,1,2,..., the $\phi^\lambda(\bar{z})$ become

$$\phi^\lambda(\bar{z}) = \bar{z}^\lambda \Phi^\lambda(\bar{z}^k) \qquad (6a)$$

$$\Phi^\lambda(\bar{z}^k) = \sum_{n=o}^{\infty} \alpha_{nk+\lambda} (\bar{z}^k)^n. \qquad (6b)$$

From (1) and (6b) we see that $\Phi^\lambda(z)$ are entire functions of z and we can then associate with a scalar function $\phi(\bar{z})$ a vector Φ in the z plane of the form

$$\phi(\bar{z}) \leftrightarrow \underline{\Phi} = \begin{bmatrix} \Phi^0(z) \\ \Phi^1(z) \\ \cdot \\ \cdot \\ \cdot \\ \Phi^{k-1}(z) \end{bmatrix}. \tag{7}$$

If we now operate on the function $\phi(\bar{z})$ by multiplication with $\bar{z}$, differentiation with respect to $\bar{z}$ or in any other way, the corresponding operation on the vector Φ will be a matrix one. For example

$$\psi(\bar{z}) \equiv \bar{z}\phi(\bar{z}) = z^{1/k} \sum_{\lambda=o}^{k-1} z^{\lambda/k} \Phi^{\lambda}(z)$$

$$= z\Phi^{k-1}(z) + \sum_{\lambda=1}^{k-1} z^{\lambda/k} \Phi^{\lambda-1}(z)$$

$$= \sum_{\lambda=o}^{k-1} z^{\lambda/k} \Psi^{\lambda}(z), \tag{8}$$

which implies the correspondance

$$\bar{z} \leftrightarrow \begin{bmatrix} 0 & 0 & \dots & 0 & z \\ 1 & 0 & \dots & 0 & 0 \\ \dots & \dots & \dots & \dots & \dots \\ 0 & 0 & \dots & 1 & 0 \end{bmatrix} \tag{9}$$

between the two complex variables.

Turning now our attention to canonical transformations, we shall first discuss the simple one relating the Hamiltonians of two oscillators of frequency 1 and k^{-1}, where k is integer, i.e.,

$$\frac{1}{2}(p^2+q^2) \ , \ \frac{1}{2}(\bar{p}^2+k^{-2}\bar{q}^2). \tag{10a,b}$$

In (10b) we can carry the point transformation $\bar{q} \to k\bar{q}$, $\bar{p} \to k^{-1}\bar{p}$ and then our canonical transformation is defined by the implicit equations[5)]

$$\frac{1}{2}(p^2+q^2) = (1/2k)(\bar{p}^2+\bar{q}^2) \tag{11a}$$

$$\arctan(p/q) = k \arctan(\bar{p}/\bar{q}), \tag{11b}$$

where (11b) involves the canonically conjugate variables to the Hamiltonians in (11a).[5)]

Introducing the observables η,ξ by the definition

$$\eta=(1/\sqrt{2})(q-ip) \quad , \quad \xi=(1/\sqrt{2})(q+ip) \quad , \tag{12a,b}$$

and similar expressions $\bar{\eta},\bar{\xi}$, we see that (11) implies the relations

$$\eta=k^{-1/2}(\bar{\eta}\bar{\xi})^{(1-k)/2}\bar{\eta}^{k} \tag{13a}$$

$$\xi=\bar{\xi}^{k}k^{-1/2}(\bar{\eta}\bar{\xi})^{(1-k)/2} \quad , \tag{13b}$$

which look similar to the conformal transformation (1) and in fact admits the same ambiguity group C_k since

$$\bar{\eta}\to\bar{\eta}\ \exp\ (i2\pi r/k) \quad , \quad \bar{\xi}\to\bar{\xi}\ \exp\ (-i2\pi r/k)\ \ r=0,1,\ldots\ k-1 \tag{14}$$

leave η,ξ invariant.

A parallel analysis[5] to the one that led to (9) indicates that to the scalar observables $\bar{\eta},\bar{\xi}$ correspond the matrices

$$\bar{\eta}\leftrightarrow\begin{bmatrix} 0 & 0 \;\ldots\; & 0 & k^{1/2}\eta \\ (k\eta\xi)^{1/2} & 0 \;\ldots\; & 0 & 0 \\ \cdots & \cdots & \cdots & \cdots \\ 0 & 0 \;\ldots\; & (k\eta\xi)^{1/2} & 0 \end{bmatrix} \tag{15a}$$

$$\bar{\xi}\leftrightarrow\begin{bmatrix} 0 & (k\eta\xi)^{1/2} & \ldots & 0 \\ \cdots & \cdots & \cdots & \cdots \\ 0 & 0 & \ldots & (k\eta\xi)^{1/2} \\ k^{1/2}\xi & 0 & \ldots & 0 \end{bmatrix} . \tag{15b}$$

The rows and columns in the matrices on the right-hand side of (15) are labeled by the ambiguity spin indices, i.e., the irreps $\lambda=0,1,\ldots k-1$ of the C_k group. We can rewrite these matrices in component form if we introduce the indices λ',λ'' associated respectively with the row and column.

So far our discussion has been entirely classical despite the appearance of the matrix observables in (15). We can pass to the quantum picture by asking about the representation of the <u>operators</u> $\bar{\eta},\bar{\xi}$ in, for example, a basis in which the oscillator Hamiltonian of unit frequency is diagonal. Our bras and kets are then characterized by the eigenvalues n' and n" of the number operator $\eta\xi$ and also by the values λ' and λ'' of the ambiguity spin. Thus, from the operator character of our observables in quantum mechanics[6] we have now for $\bar{\eta},\bar{\xi}$ a matrix representation of the form

$$< n'\lambda'|\bar{\eta}|n''\lambda'' >= k^{1/2}(n''+1)^{1/2}\delta_{n',n''+1}\delta_{\lambda',0}\delta_{\lambda'',k-1} + (kn''+\lambda''+1)^{1/2}\delta_{n',n''}\sum_{s=0}^{k-2}\delta_{\lambda',s+1}\delta_{\lambda'',s} \tag{16a}$$

$$< n'\lambda'|\bar{\xi}|n''\lambda'' >= k^{1/2}(n'')^{1/2}\delta_{n',n''-1}\delta_{\lambda',k-1}\delta_{\lambda'',0} + (kn''+\lambda'')^{1/2}\delta_{n',n''}\sum_{s=0}^{k-2}\delta_{\lambda',s}\delta_{\lambda'',s+1}. \tag{16b}$$

From (16) we see that the matrix elements for the barred number operator $\bar{\eta}\bar{\xi}$ associated with the observables $\bar{q},\bar{p}$ take the form

$$<n'\lambda'|\bar{\eta}\bar{\xi}|n''\lambda''>=(kn''+\lambda'')\delta_{n'n''}\delta_{\lambda'\lambda''}, \tag{17}$$

which precisely reflect the fact that a jump of n quanta in the oscillator of unit frequency at the left-hand side of Eq. (11a) requires a jump of kn quanta in the oscillator of frequency k^{-1} on the right-hand side of the same equation. We note the appearance of the ambiguity spin λ in (17) which indicates the correspondance of k levels in the interval nk to (n+1)k for the oscillator of frequency k^{-1} with single level n for the oscillator of unit frequency.

We see that to represent the operators $\bar{\eta},\bar{\xi}$, or functions of them such as the $\bar{q},\bar{p}$, in the Hilbert space in which $\eta\xi$ is diagonal, we need to supplement in the bra and ket the eigenvalue n of $\eta\xi$ with the ambiguity spin λ. This is a very general result that applies to all types of observables obtained by non-bijective canonical transformations,[1-6] and for whatever operator we choose to diagonalize in the Hilbert space we associate with our original q,p observables. What does change in each case is the type of ambiguity group, and the indices required to characterize the correponding ambiguity spin.

Among the more interesting non-bijective canonical transformations are those that take us from our original (q,p) to the $(\bar{q},\bar{p})$ which are respectively the action and angle variables (or simple functions of them) associated with definite Hamiltonians. For reasons of space, we just state the ambiguity groups associated with the different Hamiltonians, the Riemann surface character of the mappings, and the type spectra that we have in the corresponding quantum mechanical problem.

If the potential in the Hamiltonian H is monotonically decreasing from ∞ to $-\infty$ when q goes from $-\infty$ to ∞, there is of course no way to define action and angle, but we can instead consider $\bar{q}=H$, $\bar{p}=T$ where T is the canonically conjugate variable to H. In this case the canonical transformation is bijective[3] and thus there is no ambiguity group. The quantum mechanical spectrum is continuous in the range $-\infty$ to ∞.

If the potential is $+\infty$ at $q=-\infty$, decreases monotonically to 0 at, for example, q=0 and then increases monotonically to $+\infty$ at $q=+\infty$, then the action $\bar{q}$ and angle $\bar{p}$ variables can be defined for all energies. The ambiguity group is then the semidirect product $T \wedge I$ where T is a finite translation group $\bar{q}\to q$, $\bar{p}\to\bar{p}+2\pi m$, $m=0,\pm1,\pm2\ldots$, and I is the inversion operation $\bar{q}\to-\bar{q}$, $\bar{p}\to-\bar{p}$. From a Riemann surface standpoint, the single sheeted plane $(\bar{q},\bar{p})$ is associated with an ∞ number of sheets in (q,p). The quantum mechanical spectra is discrete, and this feature is closely related with the ambiguity spin associated with the above group.[1,3]

If the potential is periodic, the ambiguity group is the $T \wedge I$ of the previous paragraph supplemented by a translational one T associated with the periodicity of the potential, i.e., $q\to q+ma$, $p\to p$. From the Riemann surface standpoint both planes (q,p) and $(\bar{q},\bar{p})$ have now an infinite number of sheets. The quantum mechanical spectra is in the form of bands and this fact is closely related with the ambiguity spin associated with the above group.[4]

Similar results hold for other types of potentials, e.g., those that give rise to mixed spectra, and thus we can conclude that a connection exists between the type of spectra in the quantum mechanical Hamiltonian and the ambiguity spin for the canonical transformation to action and angle variables. As this spin is a classical concept this corroborates the initial observation of this note, that some quantum features, such as the type of spectra of Hamiltonians, are already implicit in the classical picture.

REFERENCES

1. M. Moshinsky and T.H. Seligman, Ann. Phys. (N.Y.) 114, 243 (1978)
2. M. Moshinsky and T.H. Seligman, Am. Phys. (N.Y.) 120,402 (1979)
3. J. Deenen, M. Moshinsky and T.H. Seligman, Ann. Phys. (N.Y.) 127, 458 (1980)
4. J. Flores, G. López, G. Monsivais and M. Moshinsky, Ann. Phys. (N.Y.)(submitted for publication)
5. M. Moshinsky and T.H. Seligman, J. Math. Phys. 22, 1338 (1981)
6. P. Kramer, M. Moshinsky and T.H. Seligman, J. Math. Phys. 19, 683 (1978)

PHASE SPACE FORMULATION OF GENERAL RELATIVITY WITHOUT A 3+1 SPLITTING

Abhay Ashtekar, Luca Bombelli and Rabinder Koul
Physics Department
Syracuse University, Syracuse, NY 13244-1130

1. INTRODUCTION

There exist in the literature two Hamiltonian formulations of general relativity, one based on the space of initial data on spacelike hypersurfaces [1,2], the second one on the space of radiative modes at null infinity [3]. These have brought out the role of conserved quantities at spacelike infinity and fluxes of certain quantities at null infinity, respectively, as generators of the asymptotic symmetry groups.

Our purpose here is to discuss the present status of a manifestly 4-dimensional Hamiltonian formulation of general relativity, based on the space of asymptotically flat solutions of Einstein's equation. This formulation has been used in establishing a relationship between the two previous ones [4], and gives a framework in which the role of the full diffeomorphism group of spacetime can be studied.

2. THE SYMPLECTIC STRUCTURE

Fix a 4-dimensional manifold M, topologically $\Sigma \times R$, where Σ is topologically flat, at least outside some compact region C, and fix a flat reference metric η_{ab} outside $C \times R$. Consider now the space Γ of all metrics g_{ab} on M which are globally hyperbolic solutions of the vacuum Einstein equations and approach η_{ab} sufficiently fast to satisfy the asymptotic conditions of [5,6], including the requirement that the magnetic part of the Weyl tensor vanish at spatial infinity.

The tangent space $T_g\Gamma$ at some point g_{ab} of Γ is the space of solutions of the linearized Einstein equation around g_{ab}. We give on Γ the symplectic structure $\Omega : T\Gamma \times T\Gamma \to R$, defined at any $g_{ab} \in \Gamma$, with corresponding ∇_a and ϵ_{abcd}, by

$$\Omega(h,h') := \frac{1}{16\pi} \int_\Sigma (h_{ab}\nabla_n h'_{cd} - h'_{ab}\nabla_n h_{cd})\, \epsilon^{anc}{}_e \, dS^{ebd}, \qquad (1)$$

where Σ is any Cauchy surface in M. Our choice of boundary conditions ensures that the integral is finite, and independent of Σ.

This symplectic form is, however, degenerate. To characterize the degenerate directions, we discuss first the action of spacetime diffeomorphisms on Γ. Let ξ^a be a smooth vector field on M which preserves the boundary conditions; then the spacetime tensor field

$$X_{ab} := \mathcal{L}_\xi g_{ab} \equiv 2\nabla_{(a}\xi_{b)}$$

automatically satisfies the linearized Einstein equation off g_{ab}, whence X_{ab} defines a vector field on Γ. Furthermore, the motions X_{ab} generates on Γ are canonical transformations, i.e., they preserve the symplectic structure. Consider now the action of Ω on this X_{ab} and any other $h_{ab} \epsilon T\Gamma$. We have

$$\Omega(h,X) = \frac{1}{16\pi}\int_{\partial\Sigma}(v^a\epsilon_{amn} + \epsilon^{abc}{}_m\xi_c\nabla_b h_{an})ds^{mn} \tag{2}$$

where

$$v^a := \epsilon^{bac}{}_p\epsilon^{pqrs}t_s h_{br}\nabla_{[q}\xi_{c]},$$

$\partial\Sigma$ is the 2-sphere at infinity representing the boundary of Σ, t^a is the unit normal to Σ, and ϵ_{abc} the volume element on it.

Consider first the case when ξ^a goes to zero at spatial infinity. Then expression (2) vanishes for all h_{ab}, and the corresponding vector X_{ab} on Γ is a degenerate direction for Ω. We will call gauge the diffeomorphisms generated by such ξ^a.

3. CONSERVED QUANTITIES AND HAMILTONIANS

Consider now vector fields ξ^a on spacetime which represent (nonzero) asymptotic symmetries at spatial infinity. We wish to discuss those functions on Γ whose hamiltonian vector fields are given by $X_{ab} = \mathcal{L}_\xi g_{ab}$. We notice that, because of the degeneracy of Ω, the generating functions will not distinguish between vector fields ξ^a which differ by terms vanishing at infinity, and there will be a corresponding ambiguity in all Hamiltonian vector fields.

We say that X is an infinitesimal canonical transformation with Hamiltonian or generating function H_ξ if, for any $h \epsilon T\Gamma$ [7],

$$\mathcal{L}_h H_\xi \equiv \lim_{\epsilon\to 0}\frac{1}{\epsilon}[H_\xi(g+\epsilon h) - H_\xi(g)] = \Omega(h,X).$$

For our choice of boundary conditions, the symmetry group at spatial infinity admits a preferred Poincaré subgroup, and one can talk about asymptotic translation vector fields and asymptotic rotation or boost vector fields, without supertranslation ambiguities [5]. When ξ^a is an asymptotic translation, the first term in (2) vanishes, and using the remaining term one can show the following. If ξ^a is an asymptotic time translation, the Hamiltonian H_ξ is precisely the ADM [1] energy:

$$H_\xi = \frac{1}{16\pi} \int_{\partial\Sigma} (\partial_a q^{bc} - \partial_b q_{ac}) e^{ac} ds^b$$

where q_{ab} is the metric induced by g_{ab} on Σ, e_{ab} is the flat metric on Σ to which q_{ab} is asymptotic, and ∂_a is the derivative operator of e_{ab}. Next, if ξ^a approaches a space translation, the generating function is the ADM 3-momentum in the direction corresponding to ξ^a:

$$H_\xi = \frac{1}{8\pi} \int_{\partial\Sigma} P_{ab}\xi^a ds^b,$$

where P_{ab} is the canonical momentum conjugate to the 3-metric on Σ. An analogous result is expected to hold when ξ^a is an asymptotic rotation vector field, in which case both terms in (2) will contribute. Work is in progress to check this, as well as for the case of boosts.

Consider now asymptotic BMS [8,9] vector fields ξ^a, i.e., fields which preserve the structure at null infinity. If one conformally completes spacetime along null directions by adding the surface $\mathcal{I}$ (either future or past null infinity) [10] and uses $\mathcal{I}$ instead of Σ in equations (1)and (2), one can show that the generating function is

$$H_\xi = \int_{\mathcal{I}} N^{ab}[({}_\xi D_a - D_a {}_\xi) l_b + 2 l_{(a} D_{b)} k] d^3\mathcal{I},$$

where N_{ab} is the news tensor of g_{ab}, D_a is the derivative operator induced on $\mathcal{I}$ by ∇_a, ℓ_a is any vector such that $n^a \ell_a = 1$, with n^a the null normal to $\mathcal{I}$, and k is the scalar field on $\mathcal{I}$ such that $\mathcal{L}_\xi q_{ab} = 2kq_{ab}$ (ξ is a conformal Killing vector on $\mathcal{I}$).

This last expression is just the flux through $\mathcal{I}$ associated with the BMS generator corresponding to ξ^a [3]. So, if ξ^a is an asymptotic translation or rotation, we get, respectively, the flux of Bondi [8,9] energy-momentum or angular momentum.

4. CONCLUDING REMARKS

A question to study in our framework is that of recovering the

Bondi "charge integrals", defined on cross sections of $\mathcal{I}$, that are expected to play the role of generating functions of canonical transformations when one chooses, as a surface to evaluate Ω on, a Cauchy surface including a null or asymptotically null surface in spacetime that meets $\mathcal{I}$, and part of $\mathcal{I}$ itself. Related to this, one can expect to get new insights on the characteristic initial value formulation of general relativity.

One can also extend this formalism in a fairly straightforward way to include spacetimes with matter fields: Γ would then be the space of solutions of the coupled Einstein-matter equations.

This work was supported in part by the National Science Foundation, under Grants PHY-8310041 and PHY-8318350.

REFERENCES

[1] R. Arnowitt, S. Deser and C.W. Misner, in: Gravitation, an introduction to current research, L. Witten ed., Wiley 1962.

[2] P.A.M. Dirac, "Lectures on quantum mechanics", Yeshiva University Press 1964.

[3] A. Ashtekar and M. Streubel, Proc. R. Soc. Lond. A376 (1981) 585-607.

[4] A. Ashtekar and A. Magnon-Ashtekar, Comm. Math. Phys. 86 (1982) 55-68.

[5] A. Ashtekar, in General relativity and gravitation, vol. 2, A. Held ed., Plenum 1980.

[6] R. Beig and B.G. Schmidt, Comm. Math. Phys. 87 (1982) 65-80.

[7] P.R. Chernoff and J.E. Marsden, "Properties of infinite dimensional hamiltonian systems", Springer-Verlag 1974.

[8] H. Bondi, M.G.J. van der Burg and A.W.K. Metzner, Proc. R. Soc. Lond. A269 (1962) 21-52.

[9] R.K. Sachs, Phys. Rev. 128 (1962) 2851-2864.

[10] R. Penrose, Proc. R. Soc. Lond. A284 (1965) 159-203.

DIFFEOMORPHISM GROUPS, COADJOINT ORBITS, AND THE QUANTIZATION OF CLASSICAL FLUIDS*

Gerald A. Goldin
Departments of Mathematics and Physics
Rutgers University
New Brunswick, New Jersey 08903

Ralph Menikoff and David H. Sharp
Theoretical Division
Los Alamos National Laboratory
Los Alamos, New Mexico 87545

The configuration space for a classical, inviscid, incompressible fluid in R^n is $G = \mathrm{sDiff}(R^n)$, the set of volume-preserving diffeomorphisms. Indistinguishability of fluid configurations means that G acts on configuration space as a symmetry group under composition. The fluid is assumed stationary at ∞; i.e. for $\varphi \in G$, $\varphi(\vec{x}) \to \vec{x}$ rapidly as $x \to \infty$. Classical phase space is the cotangent bundle $T^*(G)$, and the action of G "lifts" to $T^*(G)$. Specification of values for all conserved quantities--not just total energy and momentum, but local observables such as circulation of the fluid around closed loops--restricts the motion to an orbit in the coadjoint representation of G.[1]

We quantize the theory by constructing continuous unitary representations of G associated with such coadjoint orbits or "reduced" phase spaces.[2] The main steps are: (1) selection of a satisfactory orbit and description of the stability group K associated with a point in it; (2) construction of a polarization group H, thereby selecting half the phase-space coordinates to describe observable configurations; (3) inducing a representation of G from a character of H; and (4) construction of a measure on configuration space quasi-invariant for G. The aspect of quantization highlighted here is the question of existence of a polarization. We mention some examples and their physical interpretations, including orbits for which representations can be found and others for which they cannot.

For n = 2, the Lie algebra g is the set of divergenceless velocity fields in R^2 under the Lie bracket operation. For $\vec{v} \in g$, we have $\vec{v} = \mathrm{curl}\,\vec{\chi}$, where $\vec{\chi} = \chi\hat{z}$. As $x \to \infty$, $\vec{v} \to 0$ and $\chi \to$ constant. The choice $\chi = 0$ at ∞ determines a unique stream function $\chi_{\vec{v}}$ for any given $\vec{v}$. Then a useful identity is $\vec{\chi}_{[\vec{v}_1,\vec{v}_2]} = \vec{v}_1 \times \vec{v}_2$. In the adjoint representation of G, $\varphi \in G$ acts on g by $\vec{v}' = \varphi\vec{v}$, where $(v')^j = [(\partial_k\varphi^j)v^k]\circ\varphi^{-1}$. The corresponding action on χ is given by $\chi' =$

---

*Work supported by the U.S. Department of Energy.

$\chi \circ \varphi^{-1}$. Let g' be the dual space of g; $\vec{A} \in g'$ is a momentum density field whose components are generalized functions admitting jump discontinuities, Dirac δ-functions and their derivatives, etc. We write $\langle \vec{A}, \vec{v} \rangle = \int \vec{A}(\vec{x}) \cdot \vec{v}(\vec{x}) d^2x$. Supposing $\vec{A}$ to be bounded at ∞, we have $\langle \vec{A}, \vec{v} \rangle = \langle \nabla \times \vec{A}, \chi_{\vec{v}} \rangle$. Thus, the value of $\vec{A}$ as a functional depends only on the vorticity density $\vec{B} = \text{curl } \vec{A} = B\hat{z}$, a generalized function. The coadjoint representation of $G = \text{sDiff}(R^2)$ is given by $\vec{A}' = \varphi^* \vec{A}$, where $\langle \vec{A}', \vec{v} \rangle = \langle \vec{A}, \vec{v}' \rangle$. Then $B' = B \circ \varphi$. Consider now some specific orbits.

<u>The Point Vortex</u>. In polar coordinates (r, θ), let $\vec{A}$ be given by $A_r = 0$, $A_\theta = c/2\pi r$. Then $B = c\delta^{(2)}(\vec{x})$, and we have a pure point vortex at the origin. Let $B_{\vec{a}} = c\delta^{(2)}(\vec{x} - \vec{a})$; then $\varphi: B_{\vec{a}} \to B_{\varphi^{-1}(\vec{a})}$. The coadjoint orbit is 2-dimensional, and parameterized by $\vec{a}$. The stability group $K_{\vec{a}}$ is $\{\varphi \mid \varphi(\vec{a}) = \vec{a}\}$, which is a maximal subgroup of G. Thus, no polarization is possible! There is no consistent way to designate one of the two components of $\vec{a}$ as "position-like" and the other as "momentum-like." We are thus led to conclude that the existence of a quantum point vortex is incompatible with a theory of an ideal incompressible fluid in R^2 incorporating the local momentum densities as phase-space coordinates. This is in sharp contrast with conclusions drawn from the classical equations for a point vortex, where it is consistent to let the phase space coordinates x and y be canonically conjugate.

<u>N Point Vortices</u>. Here we have $B = c_1 B_{\vec{a}_1} + \ldots + c_N B_{\vec{a}_N}$, with B' specified by $\vec{a}_j{}' = \varphi^{-1}(\vec{a}_j)$. The stability group K contains diffeomorphisms φ satisfying $\varphi(\vec{a}_j) = \vec{a}_j$, or permuting those $\vec{a}_j$ for which the values of c_j are equal. The components of K map naturally onto elements of the braid group. The coadjoint orbit is 2N-dimensional, but no polarization group exists.

<u>The Vortex Dipole at a Point</u>.[3] Let $B_{(\vec{\lambda}, \vec{a})} = \vec{\lambda} \cdot \nabla_{\vec{x}} \delta^{(2)}(\vec{x} - \vec{a})$; i.e. $\langle B, \chi \rangle = -(\lambda^j \partial_j \chi)(\vec{a})$. We think of B as describing two point vortices of equal and opposite (infinite) vorticity, at infinitesimal separation, centered at $\vec{a}$ with finite (vorticity) dipole moment $\vec{\lambda}$. In the coadjoint representation $B' = B_{(\vec{\lambda}', \vec{a}')}$, where $(\lambda')^k = \partial_j (\varphi^{-1})^k (\vec{a}) \lambda^j$ and $\vec{a}' = \varphi^{-1}(\vec{a})$. The coadjoint orbit is thus 4-dimensional, with the plane $\vec{\lambda} = 0$ excluded. The stability group $K_{(\vec{\lambda}, \vec{a})}$ contains diffeomorphisms φ such that $\varphi(\vec{a}) = \vec{a}$ and the Jacobian matrix $\partial_j (\varphi^{-1})^k (\vec{a})$ is of the form $\begin{bmatrix} 1 & \kappa \\ 0 & 1 \end{bmatrix}$ (in an appropriate basis). A polarization group is obtained by relaxing the constraints on $\partial_j (\varphi^{-1})^k (\vec{a})$, so that $H = \{\varphi \mid \varphi(\vec{a}) = \vec{a}\}$. For $\vec{v} \in g$, $\langle B_{(\vec{\lambda}, \vec{a})}, \chi_{\vec{v}} \rangle = \hat{z} \cdot \vec{\lambda} \times \vec{v}(\vec{a})$, which is 0 if $\vec{v}$ is

in the Lie algebra h of H. So we have trivially the important condition $\langle \vec{A}, [\vec{v}_1, \vec{v}_2] \rangle = 0$ for $\vec{v}_1, \vec{v}_2 \in h$, and obtain the character which is identically 1. The induced representation V of G is particle-like: $(V(\varphi)\Psi)(\vec{a}) = \Psi(\varphi(\vec{a}))$ for $\Psi \in L^2(R^2)$.

The Rotating Vortex Dipole. Glue a point vortex to a vortex dipole by setting $B^c_{(\vec{\lambda},\vec{a})} = c\,\delta^{(2)}(\vec{x} - \vec{a}) + \vec{\lambda}\cdot\nabla_{\vec{x}}\delta^{(2)}(\vec{x} - \vec{a})$. Then $\vec{a}'$, $\bar{\lambda}'$, $K_{(\vec{\lambda},\vec{a})}$ and H are exactly as in the previous example. For $\vec{v} \in g$, $\langle B^c_{(\vec{\lambda},\vec{a})}, \chi_{\vec{v}} \rangle = c\chi_{\vec{v}}(\vec{a}) + \hat{z}\cdot\vec{\lambda}\times\vec{v}(\vec{a})$, which is $c\chi_{\vec{v}}(\vec{a})$ for $\vec{v} \in h$. And for $\vec{v}_1, \vec{v}_2 \in h$, we have $\langle \vec{A}, [\vec{v}_1, \vec{v}_2] \rangle = c\chi_{[\vec{v}_1,\vec{v}_2]}(\vec{a}) = c\hat{z}\cdot\vec{v}_1\times\vec{v}_2(\vec{a}) = 0$. Now the functional $c\chi_{\vec{v}}(\vec{a})$ may be written as $c\int_\Gamma^{\vec{a}} \vec{v}\cdot\hat{n}\,ds$, the flux through a curve Γ from ∞ to $\vec{a}$ (which is independent of Γ because the fluid is incompressible). Let $m(\varphi)$ be the area between Γ and $\varphi\Gamma$ (also independent of Γ). Then $m(\varphi_2\circ\varphi_1) = m(\varphi_1) + m(\varphi_2)$, and $c\chi_{\vec{v}}(\vec{a})$ exponentiates to give the character $\exp[icm(\varphi)]$ of H. Finally, the induced representation V of G becomes: $(V(\varphi)\Psi)(\vec{a}) = \exp[icm(\varphi,\vec{a})]\Psi(\varphi(\vec{a}))$, where $m(\varphi,\vec{a})$ is the cocycle on G associated with $m(\varphi)$ on H. Explicitly, $m(\varphi,\vec{a}) = \int_{\varphi\Gamma}^{\varphi(\vec{a})} \vec{w}\cdot d\vec{s} - \int_\Gamma^{\vec{a}} \vec{w}\cdot d\vec{s}$, where curl $\vec{w} = z$; the value being independent of Γ. Note that we cannot consistently let $\vec{\lambda} \to 0$ relative to c in this orbit to recover a point vortex, since area-preserving diffeomorphisms enlarge $\vec{\lambda}$ without bound, leaving c fixed.

The Uniform Vortex Patch. As the idealization of a pure point vortex is too severe, one may investigate the coadjoint orbit containing a uniform circular vortex patch: $B(r,\theta) = c$ for $r \leq R$, and $B(r,\theta) = 0$ for $r > R$. The stability group includes all diffeomorphisms leaving the circle $r = R$ fixed (as a set). Again we are unable to find a polarization. But in the next paper we see that an extended object of different internal structure, the vortex filament in R^2, has a natural polarization that permits quantization of the theory.

References

1. J. Marsden and A. Weinstein, "Coadjoint Orbits, Vortices, and Clebsch Variables for Incompressible Fluids," in Procs. of the Los Alamos Conference "Order in Chaos", ed. by D.K. Campbell, H.A. Rose, and A.C. Scott, Physica 7D (1983), 305-323.

2. L. Auslander and B. Kostant, Inventiones Math. 14 (1971), 255. A.A. Kirillov, Ser. Math. Sov. 1 (1981), 351. G.A. Goldin, R. Menikoff, and D.H. Sharp, Phys. Rev. Letts. 51 (1983), 2246-2249. G.A. Goldin, "Diffeomorphism Groups, Semidirect Products, and Quantum Theory," in Fluids and Plasmas: Geometry and Dynamics, ed. by J.E. Marsden, Contemp. Math. 28 (1984), 189-207.

3. See also G.A. Goldin and R. Menikoff, J. Math. Phys. 26 (1985), 1880-1884, for quantum dipoles, etc., in a different context.

QUANTIZED VORTEX FILAMENTS IN INCOMPRESSIBLE FLUIDS*

Gerald A. Goldin
Departments of Mathematics and Physics
Rutgers University
New Brunswick, New Jersey 08903

Ralph Menikoff and David H. Sharp
Theoretical Division
Los Alamos National Laboratory
Los Alamos, New Mexico 87545

Among the solutions $\vec{A}_t(\vec{x})$ to the Euler equations for a classical incompressible fluid are those describing vortex filaments. Here we discuss quantum analogues of such solutions in 2 and 3 dimensions.

Within the framework of our accompanying paper[1], we expect vortex filaments to generate coadjoint orbits in the classical phase space $T^*(\mathrm{sDiff}(R^n))$, $n = 2$ or 3. These orbits can be used as reduced phase spaces describing simple classes of fluid flow.

Consider such an orbit for $n = 2$. Let $\vec{C}(\alpha)$, for $0 \leq \alpha \leq 2\pi$, be a parametrized curve in R^2 (an arc or a loop). For $\vec{C}(\alpha)$ smooth, we can introduce an <u>unparametrized</u> curve $\Gamma = \{\vec{C}(\alpha)\}$ and a function $\gamma = d\alpha/ds$. We define $\vec{A}_{\vec{C}} \in T^*$ by its value on $\vec{v} \in T$; that is

$$\langle \vec{A}_{\vec{C}} \, , \vec{v} \rangle = \int_\Gamma ds\, \gamma(s)\, \chi_{\vec{v}}(\vec{C}(\alpha)) \ , \text{ where } \vec{\nabla} \times (\chi_{\vec{v}}\, \hat{z}) = \vec{v} \ . \quad (1)$$

It follows from Eq. (1) that the vorticity of the generalized momentum field $\vec{A}_{\vec{C}}$ is $\vec{\nabla} \times \vec{A}_{\vec{C}} = \gamma\, \delta_\Gamma \hat{z}$. Thus Γ represents the vortex filament and γ the vortex density on this filament.

The action of $\phi \in \mathrm{sDiff}(R^2)$ on $\vec{A}_{\Gamma,\gamma}$ is given by $\phi^*\vec{A}_{\Gamma,\gamma} = \vec{A}_{\Gamma',\gamma'}$ where $\Gamma' = \phi\Gamma$ and $\gamma' = (||\hat{s} \cdot \nabla\phi||)^{-1}\, \gamma \circ \phi^{-1}$. Here $\hat{s}$ is the unit vector tangent to Γ. Therefore, a coadjoint orbit corresponds to a collection of vortex filaments satisfying a number of additional constraints. For example: (a) the total vorticity $\int_\Gamma ds\, \gamma(s)$ is an invariant of the orbit; (b) the smoothness class of Γ (e.g., the number of kinks or jumps in derivatives of Γ) is preserved;

*Work supported by the U.S. Department of Energy.

(c) all topological properties of Γ are invariants including whether Γ is bounded or unbounded, whether it is an arc or a closed loop, etc.; (d) if Γ is a loop, then the area it encloses is an invariant.

For specificity, consider the coadjoint orbit Δ containing the closed loop filament $\vec{A}_{\vec{C}_o}$, where $\vec{C}_o(\alpha) = (\cos\alpha, \sin\alpha)$, or equivalentl $\vec{A}_{\Gamma_o,\gamma_o}$, where Γ_o is the unit circle and $\gamma_o = 1$. The stability group (or little group) K_{Γ_o,γ_o} of $\vec{A}_{\Gamma_o,\gamma_o}$ is the group of all area-preserving diffeomorphisms ϕ which merely rotate $\vec{C}$; i.e., such that $\phi(\vec{x}) = R\vec{x}$ for all $\vec{x} \in \Gamma_o$, where R is a rigid rotation in R^2. The orbit, regarded as the reduced phase space for the classical system, is thus identified with the quotient space $sDiff(R^2))/K_{\Gamma_o,\gamma_o}$. An element of this space corresponds to a pair (Γ,γ), where Γ is a C^∞ closed loop of area π and γ is a positive C^∞ vorticity density function on Γ with total vorticity 2π.

Next we look more closely at the physical interpretation of this coadjoint orbit as describing a vortex filament. The general solution to the Euler equations for an incompressible fluid (with $\vec{A}$ proportional to the velocity) reduces in the vortex filament case to

$$\vec{A}(\vec{x},t) = \nabla \times \int_0^{2\pi} d\alpha \; \ell n|\vec{x} - \vec{C}(\alpha,t)|\hat{z} \quad , \tag{2}$$

where

$$\frac{d\vec{C}(\alpha,t)}{dt} = \vec{A}_{av}(\vec{C}(\vec{\alpha},t)), \tag{3}$$

and $\vec{A}_{av}$ is defined as follows. The generalized momentum field $\vec{A}$ has a shear, i.e., a discontinuity in its tangential component along Γ, given by $\gamma = \hat{s} \cdot (\vec{A}^+ - \vec{A}^-)$. Then $\vec{A}_{av} = (\vec{A}^+ + \vec{A}^-)/2$. Thus, the coadjoint orbit Δ describes the fluid flow of a vortex ring. It can also be shown that Eqs. (2) and (3) can be written in Hamiltonian form.

Now we turn to the quantum theory. The first step in geometric quantization is to define the polarization group, K_p:

$$K_p = \{\phi \in sDiff(R^2) \;|\phi\{\Gamma\} = \{\Gamma\}\} \quad . \tag{4}$$

The notation $\phi\{\Gamma\} = \{\Gamma\}$ means simply $\phi(\vec{x}) \in \Gamma$ for $\vec{x} \in \Gamma$. Clearly K_p is a subgroup of $sDiff(R^2)$ and $K_p \supset K_{\Gamma_o,\gamma_o}$. The generators of K_p are elements of the algebra of velocity fields $\mathcal{A}_p = \{\vec{v} | \hat{n} \cdot \vec{v} = 0$ on $\Gamma\}$.

The essential property required of K_p is that $(\vec{A}_{\Gamma_o,\gamma_o}, [\xi,\eta]) = 0$ for $\xi,\eta \in \mathcal{A}_p$, which is easily shown.

This result allows us to decompose phase space into coordinates and momenta. In particular, coordinate space is isomorphic to $sDiff(R^2)/K_p \approx \{\Gamma\}$, i.e. to curves $\vec{C}(\alpha)$ modulo parametrization. Thus, Γ is the coordinate and γ is the associated momentum variable.

The next step is to find the character $\chi_p(\phi)$ on K_p determined by $\vec{A}_{\Gamma_o,\gamma_o}$. It can be shown[2] that $\chi_p(\phi) = \exp i\Omega(\omega,\phi)$, where ω is any curve from infinity to a point on the curve Γ_o, and $\Omega(\omega,\phi)$ is the area enclosed by the curves ω, $\phi\omega$, and Γ_o.

Thus the algebraic part of the geometric quantization program has been accomplished. To obtain a unitary representation of $sDiff(R^2)$ describing vortex filaments, i.e., to fully construct a quantum theory of vortex filaments, one still needs a measure on the set of curves Γ, quasi-invariant for $sDiff(R^2)$. Unlike the finite dimensional case, the required measure does not follow simply from the existence of the canonical 2-form on the coadjoint orbit. An approach to obtaining such measures will be outlined in a future paper.

In conclusion, we note that these methods can be applied to vortex filaments in 3 dimensions. However, in this case the analysis shows that the little group is maximal and no polarization exists. Thus, a 3-dimensional vortex filament can not be quantized. If the vortex filament is smeared out so as to form a vortex tube, one again expects a polarization to exist so that it will be possible to carry out the quantization program.

References

1. G. A. Goldin, R. Menikoff and D. H. Sharp, "Diffeomorphism Groups, Coadjoint Orbits and the Quantization of Classical Fluids" (These proceedings.)

2. G. A. Goldin, R. Menikoff and D. H. Sharp, (in preparation).

SYMPLECTIC GEOMETRY OF THE RELATIVISTIC CANONICAL COMMUTATION RELATIONS *

J. A. Brooke
Department of Mathematics
University of Saskatchewan
Saskatoon, Canada, S7N 0W0

1. Relativistic Canonical Commutation Relations (RCCR)

Let $M^{\mu\nu}$, P^{μ} be (Lie algebra) generators of the Poincaré group:

$$[M^{\mu\nu},M^{\rho\sigma}] = i\hbar(g^{\mu\sigma}M^{\nu\rho} + g^{\nu\rho}M^{\mu\sigma} - g^{\mu\rho}M^{\nu\sigma} - g^{\nu\sigma}M^{\mu\rho}) \qquad (1.1a)$$

$$[M^{\mu\nu},P^{\lambda}] = i\hbar(g^{\nu\lambda}P^{\mu} - g^{\mu\lambda}P^{\nu}) \qquad (1.1b)$$

$$[P^{\mu},P^{\nu}] = 0 \qquad (1.1c)$$

with $(g_{\alpha\beta})= \mathrm{diag}(1, -1,-1,-1)$; $\alpha,\beta,\ldots,\lambda,\mu,\nu,\ldots\varepsilon\{0,1,2,3\}$ and $M^{\mu\nu} + M^{\nu\mu} = 0$.

Suppose that $M^{\mu\nu}$, P^{μ} generate a strongly continuous unitary representation U of the Poincaré group in some Hilbert space and form a tensor operator with respect to U:

$$M'^{\mu\nu} \equiv U(g)^{-1}M^{\mu\nu}U(g) = \Lambda^{\mu}_{\alpha}\Lambda^{\nu}_{\beta}M^{\alpha\beta} + a^{\mu}\Lambda^{\nu}_{\rho}P^{\rho} - a^{\nu}\Lambda^{\mu}_{\rho}P^{\rho} \qquad (1.2a)$$

$$P'^{\mu} \equiv U(g)^{-1}P^{\mu}U(g) = \Lambda^{\mu}_{\rho}P^{\rho}, \qquad (1.2b)$$

where $g=(a,\Lambda)$ is an element of the Poincaré group, a being a spacetime translation and Λ a Lorentz transformation.

Assuming that $P^{\mu}P_{\mu}$ is positive definite, define M as the positive square root of $P^{\mu}P_{\mu}$, i.e.,

$$M^2 \equiv P^{\mu}P_{\mu} \qquad (1.3)$$

and define the intrinsic spin operator $S^{\mu\nu}$,

$$S^{\mu\nu} \equiv (\delta^{\mu}_{\alpha} - \frac{P^{\mu}P_{\alpha}}{M^2})(\delta^{\nu}_{\beta} - \frac{P^{\nu}P_{\beta}}{M^2})\, M^{\alpha\beta}. \qquad (1.4)$$

Noting that $S^{\mu\nu}P_{\nu} = 0$, $S^{\mu\nu}$ is the "projection of $M^{\mu\nu}$ into the spacelike hyperplane orthogonal to P^{μ} ".

It is possible to express $M^{\mu\nu}$ as follows:

$$M^{\mu\nu} = Q^{\mu}P^{\nu} - Q^{\nu}P^{\mu} + S^{\mu\nu} \qquad (1.5)$$

where Q^{μ} is an operator transforming as a spacetime coordinate:

$$Q'^{\mu} \equiv U(g)^{-1}Q^{\mu}U(g) = \Lambda^{\mu}_{\rho}Q^{\rho} + a^{\mu} \qquad (1.6)$$

with $g=(a,\Lambda)$.

Then (Brooke and Prugovečki, (1985)), as a consequence of the fore-

* Supported in part by NSERC grant A 8943

going assumptions:

$$Q^{\mu} = \tau\frac{P^{\mu}}{M} + \frac{M^{\mu\lambda}P_{\lambda}}{M^2} \qquad (1.7)$$

where τ transforms as a Lorentz-invariant time coordinate:

$$\tau' \equiv U(g)^{-1}\tau U(g) = \tau + a\cdot\frac{\Lambda P}{M} \qquad (1.8)$$

and moreover the RCCR hold:

$$[Q^{\mu},Q^{\nu}] = \frac{i\hbar}{M^2}S^{\mu\nu} \qquad [P_{\mu},Q^{\nu}] = i\hbar\delta^{\nu}_{\mu} \qquad [P_{\mu},P_{\nu}] = 0 \qquad (1.9a)$$

$$[S^{\mu\nu},Q^{\lambda}] = \frac{i\hbar}{M^2}(S^{\nu\lambda}P^{\mu} - S^{\mu\lambda}P^{\nu}) \qquad [S^{\mu\nu},P_{\lambda}] = 0 \qquad (1.9b)$$

$$[S^{\mu\nu},S^{\rho\sigma}] = i\hbar(\Pi^{\mu\sigma}S^{\nu\rho} + \Pi^{\nu\rho}S^{\mu\sigma} - \Pi^{\mu\rho}S^{\nu\sigma} - \Pi^{\nu\sigma}S^{\mu\rho}) \qquad (1.9c)$$

where $\Pi^{\alpha\beta} \equiv g^{\alpha\beta} - \frac{P^{\alpha}P^{\beta}}{M^2}$.

Equivalently, in terms of

$$S^{\mu} \equiv \tfrac{1}{2}\varepsilon^{\mu\nu\rho\sigma}S_{\nu\rho}\frac{P_{\sigma}}{M} \qquad (1.10a)$$

so that

$$S_{\mu\nu} = -\varepsilon_{\mu\nu\rho\sigma}S^{\rho}\frac{P^{\sigma}}{M}, \qquad (1.10b)$$

we may replace (1.9b,c) by

$$[Q^{\mu},S^{\nu}] = \frac{i\hbar}{M^2}S^{\mu}P^{\nu} \qquad [P_{\mu},S^{\nu}] = 0 \qquad (1.9b')$$

$$[S^{\mu},S^{\nu}] = i\hbar S^{\mu\nu}. \qquad (1.9c')$$

Note also that $[M,\tau] = i\hbar$ holds in addition to the RCCR.

2. RCCR - An Underlying Classical Model

Let (V,g) denote a spacetime with Lorentz metric g, and local coordinates q^{μ}.

If $s>0$, let X_s denote the subbundle of $T^*V \oplus TV$ consisting of triples $\{q^{\mu},p_{\mu},s^{\mu}\}$ for which (q^{μ}) denotes a point of V, (p_{μ}) is a future-pointing timelike covector and (s^{μ}) is a vector orthogonal to (p^{μ}) of fixed length ($s^{\mu}s_{\mu} = -s^2$). Let us define $m = \sqrt{p^{\mu}p_{\mu}}$, a positive function on V.

Now, if $\{e_a\}$, $a=0,1,2,3$, denotes a Lorentz frame on V, and $\{\theta^a\}$ its dual coframe, one may describe X_s as follows:

$$X_s = \bigcup_{q\in V}\{(m\theta^0,\ se_3) \in T^*_qV \oplus T_qV : m>0, \text{ and } \{e_a\},\{\theta^a\} \text{ are Lorentz frame and dual coframe, resp., at } q\}.$$

Clearly, X_s is a 10-dimensional manifold with local coordinates q^{μ}, $p_{\mu} = m\theta^0_{\mu}$, $s^{\mu} = se^{\mu}_3$. If V possesses a global Lorentz frame (i.e., if V supports a spin structure by a result of Geroch (1968)) then X_s is topologically $T^*_+V \times S^2$, the timelike cotangent bundle of V Cartesian product with the 2-sphere.

On the Lorentz frame bundle P over V, let $\omega^a{}_b = \theta^a_\alpha\,(de^\alpha_b + \Gamma^\alpha_{\beta\gamma} e^\gamma_b\, dq^\beta)$ denote the connection form with $\omega_{ab} + \omega_{ba} = 0$. The corresponding curvature form is $R^a{}_b = d\omega^a{}_b + \omega^a{}_c \wedge \omega^c{}_b$ and the torsion form, which we set equal to zero, is $T^a = d\theta^a + \omega^a{}_b \wedge \theta^b$. Let $\tilde{P}$ denote the extended Lorentz bundle, $\bigcup_{q\in V}$ {Lorentz frames at q}$\times\mathbb{R}^+$ with local coordinates $\{q^\alpha, e^\alpha_a, m\}$. Following Künzle (1972), who considers the case of <u>fixed</u> mass, define on $\tilde{P}$ the 1-form $\tilde{\theta} = m\overset{0}{\theta} + s\overset{1}{\omega}{}_2$ and define on $\tilde{P}$ the presymplectic 2-form $d\tilde{\theta}$. Then $d\tilde{\theta}$ has a 1-dimensional kernel, and therefore $\tilde{P}/\ker(d\tilde{\theta})$ is a 10-dimensional symplectic manifold. The pullback to X_s of this symplectic structure under the obvious mapping of X_s to $\tilde{P}$ is denoted by Ω.

Turning to Minkowski spacetime where the connection is flat, we find

$$\Omega = dp_\mu \wedge dq^\mu - \frac{s^{\mu\nu}}{2m^2} dp_\mu \wedge dp_\nu + \frac{s_{\mu\nu}}{2s^2} ds^\mu \wedge ds^\nu \tag{2.1}$$

from which it easily follows that (X_s, Ω) is a symplectic manifold. When restricted to the mass-m subbundle of X_s, the 2-form (2.1) becomes the presymplectic 2-form of Souriau (1974).

If f is a smooth function on X_s, the Hamiltonian vector field ξ_f defined by $\xi_f \lrcorner \Omega = -df$ allows us to define Poisson brackets accordingly:

$$\{f,g\} = \xi_f(g). \tag{2.2}$$

One finds that:

$$\xi_{q^\mu} = \frac{1}{m^2} s^{\mu\alpha} \frac{\partial}{\partial q^\alpha} - \frac{\partial}{\partial p_\mu} + \frac{s^\mu p^\alpha}{m^2} \frac{\partial}{\partial s^\alpha} \tag{2.3a}$$

$$\xi_{p_\mu} = \frac{\partial}{\partial q^\mu} \tag{2.3b}$$

$$\xi_{s^\mu} = -\frac{p^\mu s^\alpha}{m^2} \frac{\partial}{\partial q^\alpha} + s^{\mu\alpha} \frac{\partial}{\partial s^\alpha}. \tag{2.3c}$$

As a result, the Poisson algebra of (X_s, Ω) is mutatis mutandis, (1.9):

$$\{q^\mu, q^\nu\} = \frac{1}{m^2} s^{\mu\nu}, \qquad \{p_\mu, q^\nu\} = \delta^\nu_\mu, \qquad \{p_\mu, p_\nu\} = 0, \tag{2.4a}$$

$$\{s^{\mu\nu}, q^\lambda\} = \frac{1}{m^2}(s^{\nu\lambda} p^\mu - s^{\mu\lambda} p^\nu), \qquad \{s^{\mu\nu}, p_\lambda\} = 0, \tag{2.4b}$$

$$\{s^{\mu\nu}, s^{\rho\sigma}\} = \pi^{\mu\sigma} s^{\nu\rho} + \pi^{\nu\rho} s^{\mu\sigma} - \pi^{\mu\rho} s^{\nu\sigma} - \pi^{\nu\sigma} s^{\mu\rho}. \tag{2.4c}$$

The RCCR qualify then as the quantum mechanical version of the relativistic, variable-mass, fixed-spin classical particle.

3. <u>References</u>

Brooke, J.A. and E. Prugovečki (1985): Nuovo Cimento <u>A89</u>,126-148

Geroch, R. (1968): J. Math. Phys. <u>9</u>,1739-1744

Künzle, H.P. (1972): J. Math. Phys. <u>13</u>,739-744

Souriau, J.-M. (1974): Ann. Inst. Henri Poincaré <u>A30</u>,315-364

THREE PHYSICAL QUANTUM MANIFOLDS FROM THE CONFORMAL GROUP+

V.Aldaya*, J.A.de Azcárraga* and J. Bisquert
Departamento de Física Teórica, Facultad de Ciencias Físicas
Universidad de Valencia, Burjasot (Valencia), Spain.

I.- Introduction

We wish to report in this talk about the construction of three quantum manifolds from the conformal group. Each one gives rise to a different quantum phase space and is characterized by a different principal bundle structure of SO(4,2), whose fiber is one of the three one-parameter subgroups of its SL(2,ℝ) subgroup. This contribution follows a general group manifold approach to Geometric Quantization which has been developed in recent years [1-3], and which only requires a group law $\tilde{G}$ with a principal bundle structure as the starting point.

The principal bundle structure of $\tilde{G}$ allows us to define a connection 1-form Θ in a canonical way: Θ is the vertical component of the (say) left canonical 1-form on $\tilde{G}$. The pair $(\tilde{G},\Theta)$ can be transformed (if desired) into a Quantum Manifold in the usual sense by taking its quotient by the characteristic module $C_\Theta = \{ X \text{ on } \tilde{G} / i_X\Theta = i_X d\Theta = 0 \}$. C_Θ unambiguously defines, up to equivalence, the notion of a Full Polarization $\mathcal{P}$ on the whole group $\tilde{G}$ as a maximal left horizontal subalgebra containing C_Θ. The wavefunctions ψ are U(1)-functions on $\tilde{G}$ for which $X.\psi = 0 \;\; \forall X \in \mathcal{P}$, and the quantum operators are the right vector fields acting on ψ as usual derivation [see 1,3 for details].

A special class of what might be called Dynamical Groups is provided by the central extensions $\tilde{G}$ of semi-invariance groups G of classical lagrangians by U(1); in this case the symplectic cohomology group of G parametrizes the different quantum dynamics whose classical counterparts are obtained by substituting ℝ for U(1). Note that this last operation of "opening U(1)" is not possible in general if the group $\tilde{G}$ is not a central extension, and in this case a classical limit does not exist. Also, the possibility exists of considering trivial central extensions or "pseudo extensions" of Lie groups with trivial cohomology but obtained as dilatations of groups G with non trivial cohomology; this kind of group admits classes of coboundaries (i.e., admit a "pseudocohomology" group) which in the contraction process tranform into true cocycles of the contracted group G [4] . This last possibility was considered in making a first treatment of relativistic quantization [5], the result being the usual Klein-Gordon equation but for the rest mass energy.

II.- Structure of the Relativistic Quantization

In the case of the free relativistic particle, a pseudo-extension of the Poincaré group leads to a quantization 1-form Θ given by

$$\Theta = -\vec{x}\cdot d\vec{p} - (p^0 - mc)\, dx^0 + \frac{d\zeta}{i\zeta},$$

where x^μ are the four-translation parameters , $\vec{p} = 2mc\vec{\alpha}\sqrt{1+\vec{\alpha}^2}$

+Paper partially supported by a CAICYT research grant.
*also IFIC, Centro Mixto Universidad de Valencia-CSIC.

are the boost parameters, and ζ parametrizes the U(1) group.

In order to restore the rest mass energy in $\circledast$, as well as in the left vector field $\tilde{X}^L_{(x^0)} \in \mathcal{C}_{\circledast}$, which provides the wave equation

$$i\frac{\partial}{\partial x^0}\psi = \left\{\sqrt{\vec{p}^2+m^2c^2} - mc\right\}\psi ,$$

a linear combination of the Lie algebra generators is required. However, the new generators $\tilde{X}^{L\prime}_{(g^a)}$ - among which we find, in particular, the evolution generator $\tilde{X}^{L\prime}_{(x^0)} = \tilde{X}^L_{(x^0)} + mc\,(1+2\vec{\alpha}^2)\tilde{X}^L_{(\zeta)}$ - do not correspond to a canonical chart at the unity of the group, thus indicating the presence of some anomaly. They rather look like the generators of an orbit of the Poincaré group on a bigger manifold.

The most appropriate way to recover the missing rest mass energy seems to be to enlarge the group and to take some constraint after the group quantization procedure has been applied. This was in fact done [6] by taking as the bigger group the following one which is provided by a contraction of the conformal group:

$$\begin{aligned}
x''^\mu &= (\mathrm{ch}\,\Omega\tau)x'^\mu + \frac{\mathrm{sh}\,\Omega\tau}{\Omega}u'^\mu + \Lambda'^\mu_{\cdot\nu}x^\nu \\
u''^\mu &= (\mathrm{ch}\,\Omega\tau)u'^\mu + \Omega(\mathrm{sh}\,\Omega\tau)x'^\mu + \Lambda'^\mu_{\cdot\nu}u^\nu \\
\Lambda(\vec{\alpha}'',\vec{\epsilon}'') &= \Lambda(\vec{\alpha}',\vec{\epsilon}')\cdot\Lambda(\vec{\alpha},\vec{\epsilon}) \qquad (\Lambda \in SO(3,1)) \\
\Omega\tau'' &= \Omega\tau' + \Omega\tau \qquad (\Omega = \mathrm{const} \in R,\ [\Omega] = T^{-1}).
\end{aligned}$$

In these expressions the parameters $x^\mu, u^\nu, \alpha^i, \epsilon^j$ and $\Omega\tau$ are identified with the conformal parameters $m^{\mu 6}$, $m^{\nu 5}$, m^{0i}, m^{ki} and m^{65}, respectively (the metric tensor is $g^{AB} = (+,-,-,-;\,-,+)$. The correct Klein-Gordon equation ($i\partial_0\psi = \sqrt{\vec{p}^2+m^2c^2}\,\psi$) is obtained from the off-shell theory associated with this fifteen-parameter group after imposing the mass-shell condition,

$$u^i = 2c\sqrt{1+\vec{\alpha}^2}\,\alpha^i\ ,\quad u^0 = c\,(1+2\vec{\alpha}^2) \quad (\Rightarrow u^\mu u_\mu = c^2).$$

III.- Three Quantum Manifolds from SO(4,2)

In the previous construction we have obtained a dynamical system having (m^{i6} , m^{05}) as coordinates, (m^{j5} , m^{06}) as momenta, and m^{65} as the absolute (proper) time. The corresponding Quantum Manifold, to be called $Q^{(65)}$, is obtained from the contraction of SO(4,2) with respect the subgroup (m^{65}, m^{0i}, m^{jk}), which has non trivial cohomology. This cohomology arises from the coboundary generated by the parameter m^{65}, considered as a real function on SO(4,2) much in the same way as x^0 generates the pseudocohomology of Poincaré. We point out, by the way, that the characteristic module turns out to be precisely the subgroup associated with the contraction procedure.

Generalizing the scheme, three quantum manifolds $Q^{(65)}$, $Q^{(05)}$ and $Q^{(06)}$ arise in a natural way by applying the group quantization formalism to the groups $G^{(65)}$, $G^{(05)}$ and $G^{(06)}$, which are obtained by contracting SO(4,2) with respect to the subgroups which contain the rotation generators and each one of the three SL(2,$\mathbb{R}$) generators, i.e., M^{65}, M^{05} and M^{06}, respectively.

We present in a schematic manner the features of these three Quantum Manifolds (the m's indicate the parameters associated with the generators M).

$\tilde{G}^{(65)}$ | coordinates $m^{\mu 6}$, velocities $m^{\nu 5}$, absolute time m^{65};
characteristic module (M^{0i}, M^{ij}, M^{65}), Polarization($M^{\mu i}$, M^{65}; $M^{\mu 6}$),
System described: Free particle in Minkowski space.

$\tilde{G}^{(06)}$ | coordinates (m^{i6}, m^{05}), velocities (m^{0i}, m^{65}), absolute time m^{06};
characteristic module (M^{i5}, M^{ij}, M^{06}), polarization (M^{i5}, M^{ij}, M^{06}; $M^{i6}+iM^{0i}$, $M^{05}+iM^{65}$),
System described: Free particle in non-relativistic anti-de Sitter space.

$\tilde{G}^{(05)}$ | coordinates (m^{i5}, m^{06}), velocities (m^{0i}, m^{65}), absolute time m^{05};
characteristic module (M^{i6}, M^{ij}, M^{05}), polarization (M^{i6}, M^{ij}, M^{05}; $M^{i5}+M^{0j}$, $M^{06}+M^{65}$),
System described: Free particle in non-relativistic de Sitter space.

Finally, the <u>relativistic</u> dynamics associated with the M^{06} and M^{05} fibrations may be obtained by considering the contraction of SO(4,2) with respect to the subgroups formed by each SL(2,$\mathbb{R}$) generator with the whole set of Lorentz generators.

References

1 V. Aldaya and J.A.de Azcárraga, J.Math.Phys. 23,1297(1982)

2 V. Aldaya, J.A.de Azcárraga and K.B. Wolf, J.Math. Phys.25,506(1984)

3 V. Aldaya and J.A.de Azcárraga, Fortschr.der Phys.(1987 to appear)

4 E.J. Saletan, J.Math.Phys. 2, 1 (1961)

5 V. Aldaya and J.A.de Azcárraga, Int.J.Theor.Phys. 24, 141(1985)

6 V. Aldaya and J.A.de Azcárraga, Ann.Phys. (N.Y.)165,484(1985)

REDUCTION OF DEGENERATE LAGRANGIANS AND THE SYMPLECTIC REDUCTION THEOREM

Luis A. Ibort
University of California
Department of Mathematics
Berkeley CA 94720

The symplectic reduction theorem [1] has been one of the most useful tools in the study of Hamiltonian systems with symmetry. Many of the symmetries appearing in physical systems are related with the degenerate character of the Lagrangian function describing them. Therefore, it is important understanding the relationship between the geometrical structure of degenerate Lagrangians and the symplectic reduction provided by gauge groups of symmetries.

The following discussion shows that for an interesting class of Lagrangians it is possible to trace the origin of the gauge symmetry group of the Lagrangian from its geometrical structure, and we will prove for them the equivalence of two differents reduction procedures: The reduction of degenerate Lagrangians as it is described in [2] and the well-known symplectic reduction theory.

Let L be an almost regular [3] degenerate Lagrangian defined on the tangent bundle TQ of some differentiable manifold Q. We will assume that the characteristic sheaf of the Lagrange 2-form ω_L of L, denoted by $\mathrm{Char}\,\omega_L$, as well as its vertical part $V(TQ)\cap\mathrm{Char}\,\omega_L$, denoted by $\mathrm{VChar}\,\omega_L$, are both subbundles of TTQ. Denoting by FL the Legendre transformation defined by L, notice that $\mathrm{KerT(FL)}=\mathrm{VChar}\,\omega_L$. It turns out that the dimension of the subbundle $\mathrm{VChar}\,\omega_L$ has to be at least half of the dimension of the characteristic bundle $\mathrm{Char}\,\omega_L$ [4]. Degenerate Lagrangians having minimal vertical characteristic bundle, this is $\dim\mathrm{VChar}\,\omega_L=\frac{1}{2}\dim\mathrm{Char}\,\omega_L$, will be called semiregular Lagrangians (or type II, using the terminology in [2]) and among its remarkable properties let us mention that there always exists a SODE solution of the dynamical equation $X\lrcorner\omega_L=-dE_L$, in the points for which this equation has sense [4].

In what follows we will focus our attention in semiregular Lagrangians for which there **exists a** global solution X of the dynamical equation. This solution projects along FL and it defines a solution for the dynamical equation $Y \lrcorner \omega_1 = -dH_1$ on the primary constraint submanifold $M_1 = \mathrm{Im}FL$, where ω_1 denotes the pullback of the canonical symplectic structure Ω_Q of T^*Q, and H_1 is the projection of the energy function E_L under FL. That implies that there are only primary constraints and the final constraint submanifold is M_1. In addition, L being semiregular is equivalent to M_1 being coisotropic [2], or equivalently, in Dirac's terminology, all primary constraints are first class.

Simultaneously, the subbundle $\mathrm{Char}\,\omega_L$ enjoys very special properties. It is invariant under the canonical (1,1) tensor field S of TQ [5], its dimension is twice **the dimension of its vertical** part, and if we assume that S is projectable under $\mathrm{Char}\,\omega_L$, i.e., $\mathrm{Im}(\mathcal{L}_Z S) \subset \mathrm{Char}\,\omega_L$ for every Z in $\mathrm{Char}\,\omega_L$ [2], that suffices to prove that there exists an integrable distribution D in Q, such that $\mathrm{Char}\,\omega_L$ is spanned by the complete and vertical liftings of vector fields in D. This distribution is denoted by TD and called the tangent distribution of D. Assuming that Q is a fibration over Q/D, the presymplectic manifold (TQ, ω_L) can be reduced taking the quotient by $\mathrm{Char}\,\omega_L$, and we get a symplectic manifold diffeomorphic to T(Q/D). The original Lagrangian would pass to the quotient T(Q/D) if and only if it were invariant under the vertical and complete liftings of vector fields in D. Lagrangians such that $\mathrm{Char}\,\omega_L = TD$ for some distribution D in Q and smooth in the zero section of TQ are invariant under $\mathrm{Char}\,\omega_L$ if and only if they are invariant under vertical liftings, i.e., $\mathcal{L}_{X^v} L = 0$ for every X in D. We will call D the infinitesimal gauge distribution of L.

Assuming that L is invariant under the vertical liftings of vector fields in the infinitesimal gauge distribution D, it is easy to check that M_1 is an open submanifold of the annihilator of D defined by $\mathrm{Ann}D = \{\alpha_q \in T_q^*Q \mid \langle X_q, \alpha_q \rangle = 0 \ \forall X \in D\}$. The characteristic distribution of ω_1 is spanned by the complete **cotangent liftings** of vector fields of D because, as an straightforward computation shows, the complete liftings X^c and the cotangent liftings $\tilde{X}^c$ are FL-related. Defining the momentum map of D as $J: T^*Q \to D^*$ by $J(\alpha_q)(X_q) = \langle X_q, \alpha_q \rangle$ for every α_q in T^*Q and every vector field X in D, it is easy to prove that J is infinitesimally equivariant and $J^{-1}(0) = \mathrm{Ann}D$. Then finally the reduced phase space $M_1/\mathrm{Char}\,\omega_1$ is identified with $T^*(Q/D)$ endowed with its canonical symplectic structure $\Omega_{Q/D}$.

We have essentially proved the following theorem:

Theorem. Let L be an almost regular Lagrangian on TQ such that the dynamical equation admits a global solution. If L is semiregular and S is reducible under Charω_L, then there exists an integrable distribution D in Q such that Charω_L=TD. If $\mathcal{L}_{X^v}L=0$ for every vector field X in D, the final constraint manifold for L is a coisotropic submanifold of T*Q and an open submanifold of the zero level set of the momentum map of D. Finally, the Legendre transformation FL induces a symplectomorphism between the reduced regular Lagrangian system $(T(Q/D),\omega_{\tilde{L}})$ and the reduced Hamiltonian system $(T^*(Q/D),\Omega_{Q/D},H_o)$ obtained by the symplectic cotangent reduction theorem.

Next result is a sort of converse of the theorem before. Basically it shows that we can construct a degenerate Lagrangian which reproduces the usual symplectic reduction procedure.

Theorem. Let L be an hyperegular Lagrangian invariant under the lifted action of a Lie group G acting freely and properly in the configuration space Q. Then there exists a degenerate Lagrangian L', such that its characteristic bundle Char$\omega_{L'}$ is generated by the action of TG in TQ. Besides, the image of FL' is $J^{-1}(0)$ where J is the momentum map of the cotangent action of G and the reduced regular Lagrangian induced in T(Q/D) is equivalent to the reduced Hamiltonian system $(T^*(Q/G),\Omega_{Q/G},H_o)$.

Proof. Consider the reduced Hamiltonian H_o. The inverse Legendre transformation FH_o will be a global diffeomorphism because L is hyperegular and H is G-invariant. Then there exists a regular Lagrangian $\tilde{L}$ defined in T(Q/G) such that $E_{\tilde{L}}=H_o\circ F\tilde{L}$. The quotient of TQ under TG is diffeomorphic to T(Q/G) and the pullback of $\tilde{L}$ under this action provides the desired Lagrangian L'.

REFERENCES.

[1] R. Abraham & J. Marsden. Foundations of Mechanics, 2nd ed. Benjamin/Cummings Publ. Comp., Reading MA (1978).

[2] F. Cantrijn, J.F. Cariñena, M. Crampin & L.A. Ibort. Reduction of degenerate Lagrangians, Preprint (1986).

[3] M.J.Gotay & J.M.Nester. Ann. Inst. H. Poincaré, A30, 129-142 (1979)

[4] J.F.Cariñena & L.A.Ibort. J.Phys.A:Math. Gen., 18, 3335-3341 (1985)

[5] M. Crampin. J. Phys. A:Math. Gen., 16, 3755-3772 (1983).

FORMAL QUANTIZATION OF QUADRATIC MOMENTUM OBSERVABLES

Mark J. Gotay
Mathematics Department
United States Naval Academy
Annapolis, Maryland 21402

The problem of quantizing classical observables is a venerable one. It is well known that it is impossible to fully quantize all classical observables [1-3]; in fact, it is not even possible to consistently quantize the algebra of polynomials on a Euclidean phase space [1,3-6]. Consequently, research has centered on both quantizing restricted classes of observables [1,7,8] and weakening the notion of a "full" quantization [2,8-10]. To these ends, numerous quantization schemes have been developed (see, e.g., [7,10-15]).

In this paper, I study the quantization of quadratic momentum observables using the geometric quantization procedure of Kostant and Souriau [12-14] in the Schrödinger representation.

Consider a classical system with an oriented n-dimensional configuration space M. It is convenient (but certainly *not* necessary) in what follows to assume that M comes equipped with a semi-Riemannian metric g. Let $V = ker\ T\pi$ be the vertical polarization on phase space T^*M, where $\pi : T^*M \to M$ is the projection.

I take the prequantization line bundle to be trivial and use the orientation on M to induce a metaplectic structure for V on T^*M. Relative to this data, the quantum Hilbert space is $L^2(M, |det\ g|^{\frac{1}{2}})$.

Turning now to the classical observables, denote by $C_N(V)$ the subspace of $C^\infty(T^*M)$ consisting of functions which are at most N^{th}-degree polynomials along the fibers of π. Then $C_0(V) = \pi^* C^\infty(M)$ is the set of *configuration observables*. The space $C_1(V)$ consists of *linear momentum observables* $P_X + f$, where $f \in C_0(V)$ and P_X is the momentum in the direction of the vector field X on M.

The linear momentum observables play a distinguished role in quantization theory since they form a Schrödinger subalgebra [2,16] of the Poisson algebra $C^\infty(T^*M)$. In particular, this means that elements of $C_1(V)$ can be unambiguously and equivalently quantized in all viable quantization schemes [2,8]. This is so in geometric quantization because the flow of (the Hamiltonian vector field of) each such observable preserves the vertical polarization. Specifically, $P_X + f \in C_1(V)$ is quantized as the operator

$$Q(P_X + f) = -i(X + \tfrac{1}{2} div\ X) \tag{1}$$

on $L^2(M, |det\ g|^{\frac{1}{2}})$, where $\hbar = 1$.

However, trouble arises when quantizing momentum observables of degree two or higher. Even when applicable, different quantization procedures will typically yield ambiguous and/or inequivalent results. Geroch [17] gives a succinct discussion of the problems that can occur. For example, due to factor-ordering ambiguities, the canonical quantization of an element of $C_N(V)$ is only determined up to the addition of a differential operator of order $N-2$; this quantization is therefore well defined only in the semiclassical limit (see [15] for details and [8] for an illustration). In geometric quantization, on the other hand, such observables present difficulties because their flows do not preserve the vertical polarization. By suitably pairing V with some transverse polarization $V^{\perp}$, Kostant [12] is able to avoid this problem and can actually quantize all polynomial momentum observables. But $V^{\perp}$ may not exist and, when it does, the quantizations resulting from different choices of $V^{\perp}$ agree only in the semiclassical limit. For this reason, I work solely with the vertical polarization, i.e., in the Schrödinger representation. Quantization then requires "moving" V and employs the Blattner-Kostant-Sternberg kernels [13,14].

Regarding $C_2(V)$, the main result is

Theorem: *Suppose that $\{X_1,\ldots,X_K\}$ is a collection of linearly independent commuting vector fields on M. Let γ and ζ be any homogeneous quadratic and linear polynomials, respectively. Then*

$$Q\left[\gamma(P_{X_1},\ldots,P_{X_K}) + \zeta(C_1(V))\right]$$
$$= \gamma(QP_{X_1},\ldots,QP_{X_K}) + \zeta(Q[C_1(V)]). \tag{2}$$

Remark: There are several strong analytic and geometric conditions, not made explicit here, which must be satisfied before (2) rigorously follows. If any of these conditions fail, as is commonly the case, then (2) must be interpreted in a formal sense. These technical matters are discussed in [13,14,18].

Sketch of Proof: Write

$$\gamma(P_{X_1},\ldots,P_{X_K}) = \gamma^{ij}P_{X_i}P_{X_j}$$

where the γ^{ij} are constants. Without loss of generality, it may be assumed that γ^{ij} has rank K (for otherwise it would be possible to eliminate some constant linear combinations of the P_{X_i}'s). Then a constant linear transformation $X_i \to Y_i$ brings γ into the form

$$\gamma = \textstyle\sum_{i=1}^{K} \varepsilon_i \left(P_{Y_i}\right)^2 ,$$

where $\varepsilon_i = \pm 1$. Furthermore, by the assumptions on the X_i and the construction of the Y_i, there exists a coordinate system $\{q^a\}$, $a = 1,\ldots,n$, such that $Y_i = \partial_i$, $i = 1,\ldots,K$. Finally, since any element of $C_1(V)$ can be written as $P_Z + f$ for some Z and f, the observable to be quantized becomes

$$\textstyle\sum_{i=1}^{K} \varepsilon_i p_i{}^2 + Z^a(q)p_a + f(q),$$

where $p_a = P_{\partial_a}$.

With this normal form, it is now straightforward to compute the LHS of (2) via the BKS transform (see §5-7 of [13] for the details of such calculations). The desired result then follows from a comparison of this with the RHS of (2), which is easily computed in these coordinates using (1). ■

I now examine some of the implications of this result.

First of all, note that (2) yields

$$QF^2 = (QF)^2 \tag{3}$$

for all $F \in C_1(V)$. This may or may not be surprising, but in any case is at variance with the predictions of most other quantization schemes (e.g., [9-12,15]). On the other hand, (3) *is* consistent with von Neumann's rule (I) [19].

Secondly, simple examples show that (2) does not respect the "Poisson bracket → commutator" rule. This is unfortunate, but not entirely unexpected [1-7].

Thirdly, the conditions on the X_i's indicate that the Theorem is capable of quantizing only a limited subset of $C_2(V)$. But these requirements -- involutivity in particular -- are essential; the Theorem will fail without them.

For instance, it is not difficult to see that quantization cannot be multiplicative over noncommuting observables in general [7,17]. Indeed, suppose that X is nonvanishing and that g is a positive function on M chosen in such a way that $[gX, g^{-1}X] \neq 0$. Then

$$(QP_X)^2 = QP_X{}^2 = Q(P_{gX}P_{g^{-1}X})$$

is not equal to the symmetrized RHS of (2), which in this case is just the anti-commutator of QP_{gX} with $QP_{g^{-1}X}$. Thus the Theorem is also incompatible with the "product → anti-commutator" rule.

More significant, however, is the fact that quantization is not necessarily *additive* over noncommuting observables. In particular, it follows from (1) and (3) that $QP_X{}^2$ depends only upon $X \otimes X$. But $QP_X{}^2 + QP_Y{}^2$ will not usually depend only upon $X \otimes X + Y \otimes Y$ unless $[X,Y] = 0$. A simple example is provided by the free particle Hamiltonian on $M = \mathbf{R}^2$ in polar coordinates.

Thus geometric quantization violates von Neumann's additivity axiom (II) [19] and again this is at odds with most other quantization techniques.

The lack of additivity is really what prevents one from quantizing general elements of $C_2(V)$. On the other hand, the Theorem shows that one need only consider observables which are homogeneous quadratic in the momenta. Such an observable may be written

$$\gamma = \gamma^{ab}(q)p_a p_b.$$

Only in special cases can γ be easily quantized.

For example, if $rk\ \gamma = 1$ everywhere, then up to sign there exists a unique globally defined vector field X on M such that $\gamma^{ab} = X^a X^b$. Then $\gamma = P_X^2$ can be directly quantized.

When $rk\ \gamma > 1$ and the hypotheses of the Theorem are not satisfied, one must quantize γ by brute force. There is no problem with this in principle, but the BKS computations are intractable. To simplify them, it is necessary to find a suitable normal form for γ (note that the conditions on the X_i's in the Theorem also serve to guarantee the existence of such a normal form).

Other than the cases covered by the Theorem, apparently the only instance in which a suitable normal form exists is when $rk\ \gamma = n$. In particular, consider the Hamiltonian

$$H = \tfrac{1}{2}g^{ab}p_a p_b + V(q)$$

which can be formally quantized in Riemann normal coordinates [13,14]. The result is

$$QH = -\frac{1}{2}(\Delta - \frac{R}{6}) + V(q)$$

where Δ is the Laplace-Beltrami operator and R is the scalar curvature of g.

Acknowledgments

I would like to thank I. Vaisman, T. Mahar and J. Sniatycki for useful discussions. This work was supported in part by a grant from the United States Naval Academy Research Council.

References

1. L. Van Hove, Acad. Roy. Belg. Bull. Cl. Sci., (5), 37, 610 (1951); Mem. Acad. Roy. Belg., (6), 26, 1 (1951).
2. M. J. Gotay, Int. J. Theor. Phys., 19, 139 (1980).
3. R. Abraham and J. E. Marsden, *Foundations of Mechanics* (Benjamin-Cummings, Reading, PA, 1978).
4. H.J. Groenewold, Physica, 12, 405 (1946).
5. A. Joseph, Commun. Math. Phys., 17, 210 (1970).
6. P. Chernoff, Had. J., 4, 879 (1981).
7. F. Bayen, M. Flato, C. Fronsdal, A. Lichnerowicz and D. Sternheimer, Ann. Phys., 111, 111 (1978).

8. P. B. Guest, Rep. Math. Phys., 6, 99 (1974).
9. W. Arveson, Commun. Math. Phys., 89, 77 (1983).
10. J. Underhill, J. Math. Phys., 19, 1932 (1978).
11. F. J. Bloore, in *Géométrie Symplectique et Physique Mathématique*, Coll. Int. C.N.R.S. 237, 299 (1978).
12. B. Kostant, in *Géométrie Symplectique et Physique Mathématique*, Coll. Int. C.N.R.S. 237, 187 (1978).
13. J. Śniatycki, *Geometric Quantization and Quantum Mechanics*, Springer Appl. Math. Ser. 30 (Springer, Berlin, 1980).
14. N. M. J. Woodhouse, *Geometric Quantization* (Clarendon, Oxford, 1980).
15. I. Vaisman, J. Austral. Math. Soc., 23, 394 (1982).
16. R. Hermann, *Lie Algebras and Quantum Mechanics*, (Benjamin, New York, 1970).
17. R. Geroch, *Geometrical Quantum Mechanics*, mimeographed lecture notes (Univ. of Chicago, 1975).
18. R. J. Blattner, in LNM 570, 11 (1977).
19. J. von Neumann, *Mathematical Foundations of Quantum Mechanics*, (Princeton, 1955).

THE USE OF GHOST VARIABLES IN THE DESCRIPTION OF CONSTRAINED SYSTEMS

David McMullan
Department of Physics
University of Utah
Salt Lake City, Utah 84112

The principle of local gauge invariance is an important ingredient in the construction of realistic models for the interactions observed in nature. It is well-known that a consequence of this type of symmetry is the occurrence of first class constraints in the canonical analysis of such systems. The classical dynamics, and associated symplectic geometry, of constrained theories is now well understood. One finds that it is the Lie algebra of the symmetry group which supplies the important phase space structures.

Upon quantization the situation appears very different. Indeed, the standard methods employed in high energy physics to deal with such theories seem totally unrelated to the known symplectic descriptions. In particular, Yang-Mills theory requires the introduction of scalar fields with fermionic statistics in order to maintain unitarity. These ghost fields play a very important role in the renormalization of such theories. Also, one finds that now the Lie algebra cohomology supplies the important quantum structures [1].

In order to gain a deeper understanding of these quantum aspects of constrained systems it is important to develop an extension of the standard symplectic techniques which exposes these structures at the classical level. So let us consider the following situation. We have a Lie group G which has a smooth free action on some configuration space M, i.e., there is a Lie algebra homomorphism $\gamma : E \to \mathrm{Vect}M$ where $(E^*)E$ is the (dual) Lie algebra of G and VectM are the vector fields on M. To model the traditional approach to constrained theories one lifts the G action to a symplectic action on the phase space T^*M. The momentum map $\Phi: T^*M \to E^*$ is then constructed, and $\Phi_a := \langle \Phi, \rho_a \rangle$ (where ρ_a is a basis of E and $\langle , \rangle$ is the pairing between E^* and E) is identified with the constraints. The physical observables then correspond to the equivalence classes of weakly invariant functions on T^*M, where the equivalence class structure is that given by weak equivalence.

In terms of the configuration space M, the invariant functions are the physically relevant ones. These can be given a straightforward characterization in terms of the Lie algebra cohomology of E taking values in $C^\infty(M)$. We recall that if $\Gamma^p(E,M)$ is defined as the space of all p-linear, continuous, skew mappings from ExEx...xE (p-times) to $C^\infty(M)$, then there is a unique coboundary operator $\delta': \Gamma^p(E,M) \to \Gamma^{p+1}(E,M)$ such that $(\delta' f)(\lambda) = \gamma(\lambda)f$, where $f \varepsilon \Gamma^0(E,M) \approx C^\infty(M)$ and $\lambda \varepsilon E$, and $\delta'^2 = 0$. Thus the zeroth cohomology group of δ' corresponds to the invariant functions on M.

Using standard arguments found in the construction of graded manifolds [2], one can view $\Gamma^{\cdot}(E,M) := \Sigma_p \Gamma^p(E,M)$ as the module of functions on a graded manifold $\mathcal{M}$, whose body is M. Then δ' is an odd vector field on $\mathcal{M}$. One can now construct $T^*\mathcal{M}$, which comes equipped with a graded symplectic structure. If we identify the module of functions on $T^*\mathcal{M}$ with $\Gamma^{\cdot}(ExE^*, T^*M)$, then we can describe the induced graded Poisson algebra as that derived from the Poisson bracket on T*M and the duality between E and E*.

One finds [3] that δ' can be lifted to an odd Hamiltonian vector field δ on $T^*\mathcal{M}$, which is nilpotent. If we denote the Maurer-Cartan form on G by θ, then

$$\delta\omega = \{Q, \omega\} = \{\Phi + \tfrac{1}{2}[\theta, \theta], \omega\} \quad ,$$

where $\omega \varepsilon \Gamma^{\cdot}(ExE^*, T^*M)$, [,] are the Lie brackets on E and {,} are the graded Poisson brackets. Q is the odd momentum map for this action. In terms of the basis ρ_a of E and dual basis η^a of E*, we find that

$$Q = \Phi_a \eta^a - \frac{1}{2} C^a_{bc} \eta^b \eta^c \rho_a \quad ,$$

where C^a_{bc} are the structure constants of E, and multiplication of basis vectors is the exterior product. This expression for Q allows us to identify it with the phase space version of the Becchi-Rouet-Stora-Tyutin charge introduced by Fradkin and Vilkovisky [4] in their phase space path integral approach to constrained systems. Thus we can identify the ghost fields with η^a.

Given a weakly invariant function on T*M, one can construct a δ-closed even function on $T^*\mathcal{M}$. Then the equivalence classes of physical observables can be identified with the zeroth cohomology group of δ.

The above construction shows us how to understand the introduction of ghosts and cohomology into field theory using methods which are very similar in spirit to the traditional approach to such constrained systems. However, whereas we might expect, in retrospect, to be able to develop a cohomological description for constraints related to a Lie group action, it is not at all obvious that a similiar analysis can be carried out for more general first class systems. This is not a totally academic problem since many important constrained theories are of this more general type, the paradigm example being gravity. Surprisingly though, a cohomological description can be given for these systems [5].

In order to explain how this is achieved, and also to expose an important property of the graded phase space we are using, consider the following invariance of a constrained theory. If we have canonical coordinates (q,p) on T*M (suppressing any indices) and $\Lambda_a^b(q,p)$ is a non-singular matrix (at least in a neighbourhood of the constrained surface) of smooth functions on T*M, then $\tilde{\Phi}_a := \Lambda_a^b(q,p)\Phi_b$ can also serve as the constraints of the theory and should not change any physical result. This transformation of the constraints though, does not correspond to any canonical transformation on T*M. However, on $T^*\mathcal{M}$ we can construct an even canonical transformation implementing this change of constraints. Indeed, if we have canonical coordinates (q,p,η,ρ) on $T^*\mathcal{M}$ which transform to canonical coordinates $(Q,P,\tilde{\eta},\tilde{\rho})$ under the even transformation $K^{-1}: T^*\mathcal{M} \to T^*\mathcal{M}$ which has generating function

$$F(q,\ P,\tilde{\eta},\rho) = qP - \Lambda_a^b(q,\ P)\ \tilde{\eta}^a\ \rho_b\ ,$$

then $\tilde{Q} := K^*Q$ is the odd charge generating $\tilde{\delta}$, the coboundary operator constructed from the constraints $\tilde{\Phi}_a$. Now since K^{-1} is a canonical transformation, it is easy to show that $\tilde{\delta}K^* = K^*\delta$. So K^* is also a chain mapping between the complexes defined by $\tilde{\delta}$ and δ. Thus they have the same cohomology and hence describe the same observables.

Here we see an unexpected intertwining between the symplectic and cohomology structures on $T^*\mathcal{M}$ which can prove useful in developing new insights into the quantization of such theories [6].

This work was partially supported by the S.E.R.C. (UK) and the NSF under grant number PHY8503653.

[1] L.D. Faddeev, Phys. Lett. 145B (1984) 81.

[2] B. Kostant in Differential Geometric Methods in Mathematical Physics, Lecture Notes in Mathematics, 570, Springer, (1977), 177-306.

[3] D. McMullan, University of Utah preprint (June 1985).

[4] E.S. Fradkin and G.A. Vilkovisky, CERN Report TH-2332 (1977).

[5] A.D. Browning and D. McMullan, University of Utah preprint (October 1985).

[6] D. McMullan, Phys. Rev. D33 (1986) 2501.

Geometric Quantization of Particles in Quark Model

En-Bing Lin
Department of Mathematics
University of California
Riverside, CA 92521

A glance at a table of particle masses shows that the proton and neutron masses are amazingly close in value. They are in fact two manifestations of one and the same particle called the "nucleon," in very much the same way as the two states of an electron with spin up and down are thought of as one, not two, particles. The mathematical structure used to discuss the similarity of neutron and the proton is almost a carbon copy of spin, and is called "isospin". Isospin arises because the nucleon may be viewed as having an internal degree of freedom with two allowed states, the proton and the neutron, which the nuclear interaction does not distinguish. We therefore have an SU(2) symmetry in which they form the fundamental representation. Classical dynamics of particles with internal degrees of freedom is described in terms of a principal SU(2) bundle over the cotangent bundle of the space-time manifold. Geometric Quantization of particles with isotopic spin was exhibited. [2]

One can carry through the same calculation for the quark model.[1] Namely, we consider hadrons are made up of a small variety of more basic entities, called quarks, bound together in different ways. The fundamental representation of SU(3), the multiplet from which all other multiplets can be built, is a triplet. In other words, it is described in terms of a principal SU(3) bundle over the cotangent bundle of the space-time manifold.

Let Z be the principal SU(3) bundle over the space-time manifold Y. We introduce a SU(3)-invariant metric K on Z which gives rise to a metric N on Y. The connection form on Z is defined by

$$\alpha(u) = K(\eta(z), u),$$

where η is the fundamental vector field on Z and $u \varepsilon T_z Z$, $z \varepsilon Z$.

We denote the vertical distribution on Z tangent to the fibers of $Z \xrightarrow{P} Y$ by ver TZ . The horizontal distribution on Z is the K-orthogonal complement of the vertical distribution:

$$\text{hor } TZ = \left\{ u \varepsilon \ TZ \mid \alpha(u) = 0 \right\} .$$

Let g be the Lie algebra of SU(3), J be the moment map on T^*Z

$$J : T^*Z \longrightarrow g^* ,$$

$$dJ_\xi = \xi \lrcorner \omega, \qquad \xi \varepsilon g.$$

For each $x \varepsilon T_z{}^*Z$,ver $x = Q(x)\alpha_z$.
In fact, Q(x) is the moment map $J_\xi(x)$. [2]
We therefore have

$$K(x,x) = \text{hor } K(x,x) + \text{ ver } K(x,x)$$
$$= N(p(x),p(x)) + J_\xi{}^2(x) .$$

The dynamics of particles in T^*Z is described by

$$\xi_K = \xi_{N\ p} + \xi_{J^2}$$

The metric induces an isomorphism between T^*Z and TZ. we use the same notation for metric on each bundle.

Let Z be a principal G-bundle over the space time manifold Y, where G =SU(3) . We imbed T^*G in T^*R^9 via the identification of G with S^8.
Thus ,

$$T^*G = \left\{ (X,Y) \mid (X,Y) \varepsilon R^9 x R^9 , \|X\| = 1, X \cdot Y = 0 \right\}$$

The canonical one-form on T^*G is given by $\vartheta_G = \Sigma\ y_i\ dx_i$ and the canonical one-form on T^*Y is ϑ_Y.
By pulling back position-type functions and momenta from T^*Y to T^*Z denoted by

$$(q_1, q_2, q_3, q_4, p_1, p_2, p_3, p_4)$$

Let $\pi_1 : T^*Z \longrightarrow T^*Y$

$\pi_2 : T^*Z \longrightarrow T^*G$

be the canonical projections.

We denote $u_+, u_-, v_+, v_-, w_+, w_-$ the induced coordinates on T^*Z which are defined by using quaternion structure as follows.

$$u_+ = (bx_1 + iy_1) + (bx_2 + iy_2)j + (bx_3 + iy_3)k ,$$
$$u_- = (bx_1 + iy_1) - (bx_2 + iy_2)j - (bx_3 + iy_3)k ,$$
$$v_+ = (bx_4 + iy_4) + (bx_5 + iy_5)j + (bx_6 + iy_6)k ,$$
$$v_- = (bx_4 + iy_4) - (bx_5 + iy_5)j - (bx_6 + iy_6)k ,$$
$$w_+ = (bx_7 + iy_7) + (bx_8 + iy_8)j + (bx_9 + iy_9)k ,$$
$$w_- = (bx_7 + iy_7) - (bx_8 + iy_8)j - (bx_9 + iy_9)k ,$$

where $b^2 = \sum_{i=1}^{9} y_i^2$.

The condition for T^*G can be replaced by $u_+ u_- + v_+ v_- + w_+ w_- = 0$.

We then pull back $\vartheta_G = \Sigma\, y_i dx_i$ to T^*Z and follow the technique of geometric quantization. [2],[3],[4]

Reference

[1] Halzen, F. , Martin, A. D. : Quarks & Leptons, John Wiley & Sons,1984.

[2] Lin, E. B. , Geometric Quantization of Particles with Isotopic Spin, Hardronic Journal, vol. 5, no. 6, pp. 2041-2068, 1982.

[3] Lin, E. B. , On Prequantization Operators and Schrodinger Operators, to appear.

[4] Sniatycki, J. , Geometric Quantization and Quantum Mechanics, Springer-Verlag, 1980.

COHOMOLOGY AND LOCALLY-HAMILTONIAN DYNAMICAL SYSTEMS

J.F. Cariñena*, M.A. del Olmo** and M.A. Rodríguez***
*Dept. Física Teórica, Facultad de Ciencias, Universidad de Zaragoza, 50009-Zaragoza, Spain.
**Centre de recherches mathématiques, Université de Montréal, C.P. 6128, Succ. A, Montréal, Canada.
***Dept. Métodos Matemáticos de la Física, Facultad de Físicas, Universidad Complutense, 28040-Madrid, Spain.

The Symplectic Geometry has been shown to be the appropriate framework for the description of autonomous regular mechanical systems, providing a way for dealing on the same footing with Lagrangian and Hamiltonian formalisms. The importance of continuous symmetries of classical dynamical systems leads to the study of the realizations of Lie groups as groups of diffeomorphisms of symplectic manifolds. It is the aim of this short note to point out some interesting remarks concerning this problem which, as far as we know, have been forgotten till now.

We will use the notation of Ref. 1 : (M, ω) is a connected symplectic manifold, G is a connected Lie group acting on M and $\tilde{G}$ denotes its Lie algebra. The $C^\infty(M)$-map, $\hat{\omega} : \chi(M) \to \Lambda^1(M)$, defined by contraction $\hat{\omega}(X)Y = \omega(X, Y)$, is an isomorphism and the subalgebras of hamiltonian and locally hamiltonian vector fields will be denoted by $\chi_H(M, \omega)$ and $\chi_{LH}(M, \omega)$ respectively. We recall that the Poisson bracket of two real functions in M is given by $\{f, g\} = \omega(\hat{\omega}^{-1}(df), \hat{\omega}^{-1}(dg))$, and then the map $\sigma = \hat{\omega}^{-1} \circ d : C^\infty(M) \to \chi_H(M, \omega)$ is a homomorphism of Lie algebras, the kernel of σ being the constant functions, i.e., the sequence $0 \to \mathbf{R} \to C^\infty(M) \to \chi_H(M, \omega) \to 0$ is exact. Let G be a Lie group of transformations of M and suppose that the action of G is symplectic. The fundamental vector fields X_a defined by $(X_a f)(m) = (d/dt)\, f(\exp(-ta)\, m)\ (t = 0)$, will be locally hamiltonian for every $a \in G$ and we restrict us to study the case in which X_a is hamiltonian for any $a \in G$ (the action of G on M is said to be a strongly symplectic or hamiltonian action). It is well known that the map $X : G \to \chi_H(M, \omega)$ is a homomorphism of Lie algebras and that, if G is finite-dimensional, there is a (non-uniquely defined) map $f : G \to C^\infty(M)$ such that $\sigma \circ f = X$; it will be called a comomentum map, the corresponding momentum map $P : M \to G^*$ being $P(m)\, a = f_a(m)$.

The important point is now whether or not the comomentum map f can be chosen to be a homomorphism, this fact corresponding to that of equivariance of P when G acts on $\tilde{G}^*$ by the coadjoint action : in the affirmative case, the action of G on M is called a strongly hamiltonian action and the study of such actions can be carried out by using the well known method of the orbits of the coadjoint action as developed by Kirillov, Kostant and Souriau (2). The answer to the question can be given by using homological tools. In fact, the map $X : \tilde{G} \to \chi_H(M)$, supplies a new exact sequence of Lie algebras, the pull-back $0 \to \mathbf{R} \to G_X \to G \to 0$, as well as a map $\bar{X} : G_X \to C^\infty(M)$ such that the following diagram is commutative :

$$\begin{array}{ccccccccc} 0 & \longrightarrow & \mathbf{R} & \longrightarrow & G_X & \xrightarrow{\pi} & G & \longrightarrow & 0 \\ & & \| & & \downarrow & & \downarrow & & \\ 0 & \longrightarrow & \mathbf{R} & \longrightarrow & C^\infty(M) & \longrightarrow & \chi_H(M) & \longrightarrow & 0 \end{array}$$

We recall that G_X is the subset of pairs $(f, a) \in C^\infty(M) \times G$ such that $X_a = \sigma(f)$, the composition law being that of direct sum. Every section $s : G \to G_X$ for the projection π enables us with a comomentum map $f = \overline{X} \circ s$. For instance, if $s' : \chi_H(M) \to C^\infty(M)$ is a section for σ, we can take $s(a) = (s'(X_a), a)$. If the section s' defines the cocycle $\zeta \in H^2(\chi_H(M), \mathbf{R})$, i.e., $\zeta(X, Y) = \{s'(X), s'(Y)\} - s'([X, Y])$, the cocycle corresponding to the section s will be denoted $X^*\zeta \in H^2(G, \mathbf{R})$ and it is given by $X^*\zeta(a, b) = \{s' \circ X_a, s' \circ X_b\} - s'([X_a, X_b])$ and if we take into account that X is a homomorphism and $\overline{X} \circ s = s' \circ X = f$, we will obtain that $X^*\zeta(a, b) = \{f_a, f_b\} - f_{[a, b]}$, then $X^*\zeta$ is measuring the obstruction for a choice of f such that f be a homomorphism : had we chosen a different section s', we would obtain a cocycle $\bar{\zeta} \in H^2(\chi_H(M), \mathbf{R})$ equivalent to ζ and an image $X^*\bar{\zeta}$ equivalent to $X^*\zeta$: more accurately, if s' differs from s' in a function $\tau : \chi_H(M) \to \mathbf{R}$, i.e., $\bar{s}' = s' + \tau$ then $X^* \bar{\zeta}(a, b) = X^* \zeta(a, b) - \tau([X_a, X_b])$. The new comomentum function $\bar{f} : G \to C^\infty(M)$ is given by $\bar{f}(a) = f(a) + \tau(X_a)$. The cocycle $X^*\bar{\zeta}$ being cohomologous to $X^*\zeta$ we see that f can be chosen to be a homomorphism if and only if $X^* \zeta$ is cohomologous to zero.

The action of G of M induces an action of G_X on M, by $\widetilde{X} = X \circ \pi$ which is a strongly hamiltonian action, the subgroup **R** acting in an ineffective way. Therefore, in order to study in this way the hamiltonian actions of a given Lie group G (or its corresponding Lie algebra) it is necessary to find the set of all the Lie algebra extensions of G by **R** and then, the strongly hamiltonian actions of the middle group of every such extension in which **R** acts trivially. Another alternative procedure was proposed by Martínez Alonso (3) following ideas similar to those of Ref. 4 for the problem of reducing the study of the projective representations of a connected Lie group G, to the linear ones of a related group $\hat{G}$, the Lie algebra of such group being a particular central extension of G by the abelian algebra : $\mathbf{R}^{\dim H^2(G, \mathbf{R})}$. The point we want to stress here is that not every factor system of G can arise as a lifting of a symplectic action of G on a given manifold M and therefore, in order to solve the problem for a Lie group of transformations of M it is possible to use a lower dimensional Lie algebra, obtained in a similar way but with the substitution of $H^2(G, \mathbf{R})$ by the subgroup $H_M{}^2(G, \mathbf{R})$ of factor systems arising in that way. We will study the case of an exact symplectic manifold, as for instance the case of a regular Lagrangian system, following the results of Ibort (5). In this case the action is hamiltonian if and only if the 1-form $\alpha_a = L_{X_a} \theta$ is exact, where θ is any 1-form such that $\omega = -d\theta$.

The study of $H_M{}^2(G, \mathbf{R})$ is based on the Chevalley cohomology of G associated to the action of G on $B^1(M)$. We recall that k-cochains are the k-linear skewsymmetric maps $c : G \times G \times G \ldots G \times G \to B^1(M)$, and the ∂ operator is given by

$$\partial c(a_1,\ldots, a_{k+1}) = \sum_{i=1,k} (-1)^{i+1} L_{X_{a_i}} c(a_1,\ldots, \hat{a}_i ,\ldots, a_{k+1})$$

$$+ \sum_{i<j} (-1)^{i+j} c([a_i, a_j] ,\ldots, \hat{a}_i ,\ldots, \hat{a}_j ,\ldots, a_{k+1}) \quad ,$$

the supercaret denoting omission of the corresponding symbol. Then the 1-cochain $\alpha : G \to B^1(M)$

given by $\alpha(a) = \alpha_a = L_{X_a}\theta$ is a 1-cocycle, i.e., it satisfies : $\alpha([a, b]) = d(\alpha_b(X_a) - \alpha_a(X_b))$ and the substitution of θ by $\theta' = \theta + \xi$ with $\xi \in B^1(M)$ would lead to an equivalent cocycle; in other words, the action of G on M determines an element of $H^1(G, B^1(M))$. On the other hand, the exact sequence $0 \to \mathbf{R} \to C^\infty(M) \to B^1(M) \to 0$ induces another exact sequence :
$\ldots \to H^1(G, C^\infty(M)) \to H^1(G, B^1(M)) \xrightarrow{\delta} H^2(G, \mathbf{R})$ and it was shown by Ibort (5) that the cocycle $X^*\zeta$ defining the lifting $0 \to \mathbf{R} \to G_X \to G \to 0$ lies in the image of $[\alpha]$ by the connecting homomorphism δ, and therefore, only the 2-cocycles in the classes of $\delta(H^1(G, B^1(M))$ can arise as liftings of symplectic actions. Moreover, not every such cocycle can arise in this way, as one of the following examples.

Example 1. The 2-dimensional euclidean group acting in the usual way of the plane. The fundamental vector fields are $P_1 = -\partial/\partial x$, $P_2 = -\partial/\partial y$, $J = y\partial/\partial x - x\partial/\partial y$, with nonvanishing defining relations $[J, P_1] = P_2$, $[J, P_2] = -P_1$. Then $H^2(G, \mathbf{R})$ is one-dimensional, the only new defining relation being $[P_1, P_2] = -\lambda$. On the other hand, $H^1(G, B^1(M))$ is also one-dimensional, the simplest representative of the class [c] being $\alpha_{P_1} = 0$, $\alpha_{P_2} = \lambda dx$, $\alpha_J = \lambda(xdx-ydy)$. This cocycle arises from the symplectic form : $\omega = \lambda dy \wedge dx$.

Example 2. The 1+1 Galilei group with the usual action on $\mathbf{R}^2$. The generators of G become : $P = -\partial/\partial t$, $H = -\partial/\partial x$, $K = -t\partial/\partial x$ with defining relations $[K, H] = P$, $[K, P] = 0 = [H, P]$. Then $H^2(G, \mathbf{R})$ is two-dimensional, and a generic extension (λ, μ) being given by the nonvanishing relations $[K, H] = P$, $[K, P] = \mu I$, $[P, H] = \lambda J$. On the other hand $H^1(G, B^1(M))$ is also two-dimensional, the simplest representative in the class $[c_1, c_2]$ being $\alpha_P = 0$, $\alpha_H = -c_1 dx$, $\alpha_K = c_2 dx - c_1\, tdt$, the image of $[c_1, c_2]$ being $\delta([c_1, c_2]) = (\lambda = c_1, \mu = c_2)$. The Lie algebra $\tilde{G}$ is five-dimensional. It is noteworthy that only the 2-forms $\omega = k dx \wedge dy$ with a constant k can be invariant under G and in this case the associate cocycle would be in [k, 0]. It is also possible to check that only elements in [0, 0] will be obtained as associated to degenerated closed 2-forms and therefore the true group $\tilde{G}_M$ is four-dimensional, because we can forget the c_2 ambiguity.

Example 3. The rotor : a free nonrelativistic particle constrained to move in a ring. The symmetries are time translations, angular translations and change to a new frame which is in a constant angular speed motion, with a Lie algebra isomorphic to that of 1+1 Galilei group. However, $H^1(G, B^1(M))$ is now different because the periodicity conditions eliminate the 1-forms with $c_2 \neq 0$ and this fact gives a difference with the previous case, the Lie algebra $\tilde{G}_M$ will be four-dimensional.

References.

1. Cariñena, J.F., and Ibort, L.A., Nuovo Cimento 87B, 41 (1985).
2. Abraham, R, and Marsden, J., Foundations of Mechanics, Benjamin, New York, 1978.
3. Martínez Alonso, L.J., Math. Phys., 17, 1177, (1976).
4. Cariñena, J.F. and Santander, M., J. Math. Phys. 16, 1416, (1975).
5. Ibort, L.A., "Estructura Geométrica de los sistemas con simetría en Mecánica Clásica y Teoría Clásica de Campos" Ph.D. Thesis, University of Zaragoza, 1984.

The Third Quantization of Phase Space and Bilocal Lattice Fields

A. Das

Department of Mathematics, Simon Fraser University

Burnaby, British Columbia V5A 1S6, Canada

Ever since the advent of quantum field theory, divergence difficulties have plagued its mathematical foundations. In spite of the formal successes of cancellations of "infinities", many mathematicians cannot accept this method as the ultimate solution of the divergence problems. In the last decade little effort has been given to this question, but in the author's opinion the solution of the divergence problem is of vital importance to a deeper understanding of elementary particles.

The most reasonable approach to this problem is to introduce a fundamental length. Attempts were made to achieve this end either by introducing lattice structures in space-time[1)] or by considering non-local interactions[2)] of local fields or non-local fields.[3),4)] But the appproach of formally quantizing space-time (or extended phase-space[5)]) with linear operators seems to be the most logical answer. This procedure would be called *the third quantization*. Thirty years ago Snyder[6)] introduced a third quantization of space-time. But in his formulation simultaneous measurements of all four coordinates were not possible. In a different approach, following the idea of reciprocity[7)] (that position and momentum variables should have equal footings), Yukawa[3)] put forward a third quantization rule in the extended phase-space. This quantization was formally identical with the basic postulate of quantum mechanics viz., $[P_a, Q^b] = -i\hbar\delta_a{}^b I$. However, Yukawa considered only the *continuous* spectrum of these operators, which did not resolve the divergence difficulties. In a still different approach, the present author introduced[8)] complex space-time partly to incorporate iso-groups and partly to remove divergences. He also formulated a third quantization in complex space-time and considered only the *discrete* spectrum of the operators. Furthermore, in that quantization four lattice space-time coordinates could be measured simultaneously. In this paper the last two approaches are unified by identifying complex space-time with extended phase-space[9] through the relation $\sqrt{2}Z_a \equiv (Q_a - iP_a)$. Third quantization is then carried out with the *discrete* spectrum. Furthermore, the third quantized field equations for spin-0 and spin-$\frac{1}{2}$ particles are written in both operator and partial *difference* languages. The resulting partial difference equations coincide exactly with the new bilocal lattice field equations published recently.[10]

We now describe the notation and give the definitions used in this paper. The extended phase-space V_8 of space-time-momentum-energy is defined to be the collection of points $(q,p) \equiv (q^k,p_k)$ where k and other Roman indices take the values 1, 2, 3, 4. For the third quantization q^k and p_k are replaced by the corresponding eight linear operators Q^k and P_k, which act on a *separable* Hilbert space. Indices can be lowered or raised by metric tensor $[\eta_{ab}] = [\eta^{ab}] \equiv \mathrm{diag}[-1^3,1]$. The summation convention is followed wherever possible. Born[7] units are chosen so that $a = b = c = 1$, where a, b, c are the fundamental length, momentum, and velocity, respectively. At the present moment the exact values of a and b are not known, but $ab = h$ and $a < 10^{-13}$ cm.

The <u>third quantization</u> is the basic quantum postulate, viz.,

$$[P_k,Q^l] = -i\hbar\delta_k{}^l I. \tag{1}$$

Here only the *discrete* spectrum of the operators is allowed. The commutator relation (1) defines the mathematical structure of the third-quantized V_8. This structure is independent of the possible occupation of V_8 by any field quanta. In this respect Eq. (1) is interpreted *differently* than in the usual quantum particle mechanics.

To obtain a convenient matrix representation of the operators P_k and Q_k, let $|n)$ denote the orthonormal basis of eigenvectors for the Hamiltonian of the usual oscillator problem; i.e.,

$$|n) \equiv |n_1,n_2,n_3,n_4);\ n_a = 0, 1, 2, \ldots;$$

$$(m|n) = \delta_{m_1n_1}\delta_{m_2n_2}\delta_{m_3n_3}\delta_{m_4n_4},$$

$$(1/2)(R_k)^2|n) \equiv (1/2)(P_k{}^2 + Q_k{}^2)|n) = (1/2)r_k{}^2|n),\ r_k = \sqrt{2n_k+1},$$

$$Z^+_\alpha|n) = (\sqrt{2}/2)(Q_\alpha - iP_\alpha)|n_\alpha) = \sqrt{n_\alpha+1}|n_\alpha+1),$$

$$Z^-_\alpha|n) = (\sqrt{2}/2)(Q_\alpha + iP_\alpha)|n_\alpha) = \sqrt{n_\alpha}|n_\alpha-1),\ \alpha = 1,\ 2,\ 3;$$

$$Z^+_4|n_4) = (\sqrt{2}/2)(Q_4 - iP_4)|n_4) = \sqrt{n_4}|n_4-1),$$

$$Z^-_4|n_4) = (\sqrt{2}/2)(Q_4 + iP_4)|n_4) = \sqrt{n_4+1}|n_4+1). \tag{2}$$

Here the vector $|n)$ stands for the tensor product (represented by the Kronecker product) of the vectors $|n_1)$, $|n_2)$, $|n_3)$, $|n_4)$.

For the physical interpretation of Eqs. (2) one can write for the unquantized variables $q_k + ip_k \equiv r_k e^{i\theta}$, where k is not summed. After the third quantization, r_k takes quantized values $\sqrt{2n_k+1}$, whereas the measurement of θ_k becomes *completely uncertain* in the representation diagonalizing r_k. Therefore it can be concluded that each of the four p_k-q_k phase-planes of the quantized V_8 exhibits annular phase cells.

Analyzing the annular cells, one finds that the minimum permissible radius is $r_1 = 1$. Therefore, there exits a forbidden hole around the origin. Every annular cell has the same area, $\pi[2(n_1 + 1) + 1] - \pi[2n_1 + 1] = 2\pi$. Moreover, the thickness of the cells $\sqrt{2(n_k+1)+1} - \sqrt{2n_k+1}$ tends to zero as $n_1 \to \infty$. As the annular cells cannot be further refined or subdivided, it follows that field quanta, which possibly occupy a phase cell, cannot be localized any further. Thus, intracellular causality becomes a meaningless concept whereas intercellular causality is still retained in a probabilistic sense. Thus, interpretation reaffirms the uncertainty principle in a concrete model of quantized extended phase-space.

Regarding the questions of covariance, one notices that the annular cells are isotropic in each phase-plane. This feature is consistent with *generalized reciprocity*.[7),9)] However, the homogeneity of extended phase-space is lost on two accounts. First, there appears a finite hole around the origin and second, the thickness of the cells tends to zero indicating some evolution. Furthermore, measurable space-time coordinates $\hat{q}_a$ must take values $\pm r_a = \pm\sqrt{2n_a+1}$, which produces a strange set of lattice events. None of the translational, rotational, and reflection symmetries in space-time is present. Naturally, one wonders if Lorentz covariance is still relevant in the third quantization! It turns out that the apparent difficulties regarding relativity can be completely resolved by the following arguments. The set of operators P_a and Q^b (and their functions $\Phi(P,Q)$) and the set of basis vectors $|n)$ in Eq. (2) are associated with one observer in a Lorentz frame of reference. For a second observer in a different Lorentz frame of reference

$$
\begin{aligned}
|n') &= |n), \\
P'_a &= U^\dagger P_a U, \\
Q'^b &= U^\dagger Q^b U
\end{aligned}
\tag{3}
$$

in the Heisenberg picture. The unitary operator U in Eqs. (3) and (4) is given by

$$
U = \exp[i l^a P_a + (1/4)\theta^{ab}(Q_b P_a - Q_a P_b + P_a Q_b - P_b Q_a)], \tag{4}
$$

where the parameters l^a, $\theta^{ab} = -\theta^{ba}$ are associated with a finite Poincaré transformation.

Suppose one considers the Heisenberg picture as implied by (3) and (4). In that case $R'_k = U^\dagger R_k U$, where the operator R_k has been defined in (2). It is a mathematical fact that the eigenvalues r'_k of the operator R'_k are exactly the same as the eigenvalues of the operator R_k. Furthermore, an eigenvector $|n)$ of the operator R_k is not necessarily an eigenvector of R'_k. Thus, for the first observer in the first Lorentz frame of reference, the set of lattice events is given by the eigenvalues $\hat{q}_k = r_k = \sqrt{2n_k+1}$ of the operator R_k. For the second observer in the Lorentz-transformed frame, the lattice events of the first observer will be completely smeared or blurred, however, an exactly similiar, though different, set of lattice events corresponding to eigenvalues $r'_k = \sqrt{2n'_k+1}$, $n'_k = 0, 1, 2, \ldots$, will result. Hence the lattice space-time structure is compatible with the principle of special relativity.

Although the transformation of lattice events under a Lorentz mapping is explained here, it must be emphasized that cellular extended phase-space or lattice space-time is not directly observable through experiments. These structures can only be verified indirectly by studying the field operators $\Phi(P,Q)$ or the expectation values $(m|\Phi(P,Q)|n)$ defined over quantized spaces.

The usual Klein-Gordon and Dirac equations have to be replaced by operator equations in the third quantized phase-space. The equations for spin-0 and spin-$\frac{1}{2}$ fields are taken to be

$$-\eta^{ab}\left[\frac{\partial^2\Phi}{\partial Q^a\partial Q^b} + \frac{\partial^2\Phi}{\partial P^a\partial P^b} + \mu^2\Phi\right] \equiv \eta^{ab}\{[P_a,[P_b,\Phi]] + [Q_a,[Q_b,\phi]]\} - \mu^2\Phi = 0; \quad (5a)$$

$$-i\left\{v^k\frac{\partial\Psi}{\partial Q^k} + a^k\frac{\partial\Psi}{\partial P^k}\right\} \equiv v^k[P_k,\Psi] - a^k[Q_k,\Psi] + M\Psi = 0, \quad (5b)$$

$$v^kv^l + v^lv^k = a^ka^l + a^la^k = 2\eta^{kl}\mathrm{I}, \quad v^ka^l + a^lv^k = 0,$$

where μ and Φ are the mass parameter and field operator for the spin-0 particle and M, Ψ, v^k, a^k are the mass parameter, the field operator, velocity, and acceleration matrices, (16×16 size) for the spin-$\frac{1}{2}$ particle. These equations are combinations of Yukawa's operator equations[3] and allow a group of invariance larger than the Lorentz group. Equation (5a) will be called the Klein-Gordon-Yukawa (in short, K.G.Y.) operator equation. Other versions of this equation have already been written.[9] Equation (5b) is the Boltzmann-Dirac-Yukawa (in short, B.D.Y.) operator equation.[9] Multiplying (5a) and (5b) by the operator $U^\dagger$ from the left and U from the right, and noting (3), one can conclude that the operator equations (5a) and (5b) are Lorentz covariant.

The expectation values of the field operators are defined as follows:

$$\phi(m,n) \equiv \phi(m_1,m_2,m_3,m_4;n_1,n_2,n_3,n_4) \equiv (m|\Phi(P,Q)|n),$$

$$\psi(m,n) \equiv \psi(m_1,m_2,m_3,m_4;n_1,n_2,n_3,n_4) \equiv (m|\psi(P,Q)|n).$$

Taking the expectation values of operator equations (5a) and (5b) between eigenvectors $(m|$ and $|n)$, and using (2) and (6) one obtains the following partial difference equations:

$$\sum_{k=1}^{4} \epsilon(k)\{(m_k+n_k+1)\phi(m,n) - \sqrt{(m_k+1)(n_k+1)}\phi(\ldots,m_k+1,\ldots;\ldots,n_k+1,\ldots)$$
$$- \sqrt{m_kn_k}\phi(\ldots,m_k-1,\ldots;\ldots,n_k-1,\ldots)\} + (1/2)\mu^2\phi = 0, \quad (7a)$$

$$\sum_{k=1}^{4} \{(i\epsilon(k)v^k-a^k)\sqrt{m_k+1}\psi(\ldots,m_k+1,\ldots;n) - (i\epsilon(k)v^k+a^k)\sqrt{m_k}\psi(\ldots,m_k-1,\ldots;n)$$
$$+ (i\epsilon(k)v^k+a^k)\sqrt{n_k+1}\psi(m;n_k+1,\ldots) - (i\epsilon(k)v^k-a^k)\sqrt{n_k}\psi(m;n_k-1,\ldots)\}$$
$$+ M\psi(m,n) = 0, \quad (7b)$$

where $\epsilon(1) = \epsilon(2) = \epsilon(3) = -\epsilon(4) = -1$.

The above equations have already been proposed[10] and Lorentz covariance under infinitesimal transformations has been shown directly without recourse to operators. The Green's functions of the difference equations (7a) and (7b) should be nonsingular[8] and therefore these equations are good candidates for a convergent bilocal lattice field theory. A detailed investigation is in progress.

References

1) A.Das Nuovo Cim. **18** (1960), 482.

2) E.Arnous and W.Heitler, Nuovo Cim. **11** (1959), 443.
E.Arnous, W.Heitler, and Y.Takahashi, Nuovo Cim. **16** (1960), 671.

3) H.Yukawa, Phys. Rev. **76** (1949),300; **77** (1950),219; **80** (1950), 1047.

4) A.Pais and G.E.Uhlenbeck, Phys. Rev. **79** (1950), 145.

5) C.Lanczos, *The Variational Principles of Mechanics* (University of Toronto Press, Toronto, 1966), p.186.

6) H.Snyder, Phys. Rev. **71** (1947),38; **72** (1948), 68.

7) M.Born, Proc. Roy. Soc. **A165** (1938), 291; **A116** (1938), 552; Nature **163** (1949), 207; Rev. Mod. Phys. **21** (1949), 463.

8) A.Das, J. Math. Phys. **7** (1966), 45, 52.

9) A.Das, J. Math. Phys. **21** (1980), 1506, 1513, 1521.

10) A.Das, Prog. Theor. Phys. **68** (1982), 336, 341.

PAULI-FORBIDDEN REGION IN THE PHASE SPACE OF THE INTER-NUCLEUS RELATIVE MOTION

H.Horiuchi and K.Yabana
Department of Physics, Kyoto University, Kyoto 606, JAPAN.

In the RGM (resonating group method)[1], the many-body wave function for the system of two interacting nuclei is given by

$$\mathcal{A}[\chi(\vec{r})\phi_1\phi_2] \tag{1}$$

where ϕ_i (i=1,2) is the given internal wave function of the nucleus C_i, $\chi(\vec{r})$ the wave function of the inter-nucleus relative motion to be determined, and $\mathcal{A}$ the anti-symmetrizer. The equation of motion which determines $\chi(\vec{r})$ is

$$\int [H(\vec{\rho},\vec{\rho}\,') - EN(\vec{\rho},\vec{\rho}\,')]\,\chi(\vec{\rho}\,')d\vec{\rho}\,' = 0\,, \tag{2}$$

$$\left\{ \begin{matrix} H(\vec{\rho},\vec{\rho}\,') \\ N(\vec{\rho},\vec{\rho}\,') \end{matrix} \right\} \equiv \langle \delta(\vec{r}-\vec{\rho})\phi_1\phi_2 | \left\{ \begin{matrix} \hat{H}-E_1-E_2 \\ 1 \end{matrix} \right\} | \mathcal{A}[\delta(\vec{r}-\vec{\rho}\,')\phi_1\phi_2] \rangle, \tag{3}$$

where $\hat{H}$ stands for the many-body Hamiltonian, E the energy of the relative motion, and E_i the binding energy of C_i calculated by ϕ_i. In a concise operator notation, Eq.(2) is expressed as $(H - EN)\chi = 0$.

When we apply the WKB approximation[2] to Eq.(2) by expressing $\chi(\vec{r})$ as $\chi(\vec{r}) = \exp\left[(i/\hbar)S_0(\vec{r}) + iS_1(\vec{r}) + \cdots\right]$, we get in the zeroth order of the Planck constant $\hbar$ the following Hamilton-Jacobi equation

$$H^W(\vec{\rho},\vec{p}(\vec{\rho})) - E\,N^W(\vec{\rho},\vec{p}(\vec{\rho})) = 0\,, \quad \vec{p}(\vec{\rho}) \equiv \vec{\nabla}S_0(\vec{\rho}). \tag{4}$$

Here we use the notation $O^W(\vec{\rho},\vec{p})$ which expresses the Wigner transform of the non-local operator $O(\vec{\rho},\vec{\rho}\,')$;

$$O^W(\vec{\rho},\vec{p}) \equiv \int d\vec{s}\, e^{(i/\hbar)\vec{s}\cdot\vec{p}}\, O(\vec{\rho}-\tfrac{\vec{s}}{2}, \vec{\rho}+\tfrac{\vec{s}}{2})\,. \tag{5}$$

Within the WKB approximation, the phase of $\chi(\vec{\rho})$ is given by $\int d\vec{\rho}\,\vec{p}(\vec{\rho})/\hbar$. Once the local momentum $\vec{p}(\vec{\rho})$ is obtained from Eq.(5), we get the equivalent local potential (ELP) $V^{eq}(\rho)$ by

$$(\vec{p}(\vec{\rho}))^2/2\mu + V^{eq}(\rho) = E\,. \tag{6}$$

We can regard that Eq.(6) is the Hamilton-Jacobi equation derived by applying the WKB approximation to the Schrödinger equation $[-(\hbar^2/2\mu)\vec{\nabla}^2 + V^{eq}(\rho)]\,\chi^L(\vec{\rho}) = E\chi^L(\vec{\rho})$. Therefore, within the WKB approximation, the phase of $\chi^L(\vec{\rho})$ is given by $\int d\vec{\rho}\,\vec{p}(\vec{\rho})/\hbar$ which is the same as that of $\chi(\vec{\rho})$. In this sense, $V^{eq}(\rho)$ is the phase-equivalent local potential; namely, the phase of $\chi^L(\vec{\rho})$ is (approximately) equal to that of $\chi(\vec{\rho})$. The Perey factor $F(\rho)$, which relates the amplitudes between $\chi^L(\vec{\rho})$ and $\chi(\vec{\rho})$ as $\chi(\vec{\rho}) = F(\rho)\cdot\chi^L(\vec{\rho})$, is obtained from the first order of $\hbar$ in the WKB expansion and for the details of this point the reader is referred to Ref.3).

The ELP, $V^{eq}(\rho)$, were calculated for many kinds of two-nucleus systems and we confirmed that the phase-shifts evaluated quantum-mechanically by using $V^{eq}(\rho)$ reproduce quite accurately those obtained by solving Eq.(2) quantum-mechanically[3]. An important general feature of the calculated ELP is that they are quite deep like in the double-folding model[3]. This fact is independent of the details of the adopted effective two-nucleon force. It is a result of the existence of the Pauli-forbidden region in the phase space of the relative coordinate $\vec{\rho}$ and its conjugate momentum $\vec{p}$.[3][4]

Below we explain this Pauli-forbidden region which is basically important for the understanding of the ELP.

It is well-known[1] in the RGM theory that there exist the so-called Pauli-forbidden states which the inter-nucleus relative motion can not occupy. The Pauli-forbidden states denoted as $\chi_F(\vec{r})$ are defined as the wave functions which satisfy $\mathcal{A}[\chi_F(\vec{r})\phi_1\phi_2] \equiv 0$ and therefore they have the property $H\chi_F = N\chi_F = 0$. The Pauli-forbidden states are the redundant solutions of Eq.(2) since there holds at any energy $(H - EN)\chi_F = 0$. The physical relative wave function $\chi(\vec{r})$ should be orthogonal to $\chi_F(\vec{r})$; $\langle \chi(\vec{r}) | \chi_F(\vec{r}) \rangle = 0$. The Pauli-forbidden states are expressed by the harmonic oscillator wave functions $R_{n\ell}(r,\gamma)Y_{\ell m}(\hat{r})$ with $N=2n+\ell \leqq N_F$ where the oscillator parameter γ is given by $\gamma = \mu\omega/2\hbar$ with ω standing for the harmonic oscillator angular frequency. The integer N_F is the largest number of the harmonic oscillator quanta of the Pauli-forbidden states. For example, in the case of the $\alpha + {}^{40}$Ca system, N_F is 10 for even ℓ and 11 for odd ℓ. When the RGM equation of motion (Eq.(2)) is treated by the WKB method, the harmonic oscillator wave function $R_{n\ell}(r,\gamma)Y_{\ell m}(\hat{r})$ is transformed into the harmonic oscillator trajectory in the phase space $p^2/2\mu + \rho^2 \cdot \mu\omega^2/2 = \hbar\omega(2n+\ell+3/2)$. Therefore, the functional space spanned by the Pauli-forbidden states $R_{n\ell}(r,\gamma)Y_{\ell m}(\hat{r})$ with $2n+\ell \leqq N_F$ is transformed into such region in the phase space that is defined by $p^2/2\mu + \rho^2 \cdot \mu\omega^2/2 \leqq \hbar\omega(N_F+3/2)$. We call this region in the phase space the Pauli-forbidden region. The quantum-mechanical requirement that the physical relative wave function can not have any component of the Pauli-forbidden wave functions is transformed into the semi-classical requirement that the physical trajectory can not enter into the Pauli-forbidden region in the phase space. Any point $(\vec{\rho},\vec{p}_F)$ in the Pauli-forbidden region satisfies the relation $H^W(\vec{\rho},\vec{p}_F) \approx N^W(\vec{\rho},\vec{p}_F) \approx 0$, which means that there holds $H^W(\vec{\rho},\vec{p}_F) - EN^W(\vec{\rho},\vec{p}_F) \approx 0$ at any energy E. This fact that any trajectory inside the Pauli-forbidden region is the redundant solution of the Hamilton-Jacobi equation of Eq.(4) at any energy E corresponds to the fact that the Pauli-forbidden states are the redundant solutions of the RGM equation of motion Eq.(2) at any energy E.

As explained above, the physical local momentum $\vec{p}(\vec{\rho})$ should satisfy the inequality

$$(\vec{p}(\vec{\rho}))^2/2\mu + \rho^2\mu\omega^2/2 > \hbar\omega(N_F + 3/2) . \qquad (7)$$

When we use Eq.(6), this inequality can be equivalently rewritten in terms of $V^{eq}(\rho)$ as follows:

$$V^{eq}(\rho) < E - \hbar\omega(N_F + 3/2) + \rho^2\mu\omega^2/2 . \qquad (8)$$

Equation (8) means that $V^{eq}(\rho)$ is necessarily deep for small ρ except in such high energy regions where E is comparable with $\hbar\omega(N_F+3/2)$. This is the reason why the numerical calculations always gave us the deep ELP.

The ELP, $V^{eq}(\rho)$, we have obtained provide us with the basis for the microscopic investigation of the real part of the optical potential between nuclei. This point is important because, as is well-known, in many cases the inter-nucleus optical potentials have not been determined unambiguously by experiments. In the cases of light-ion-nucleus projectiles, ^{3}He and α,[6)7)] for which the absorption is not strong, the so-called discrete ambiguity of the optical potential has been successfully removed mainly by utilising the nuclear rainbow scattering in the high scattering energy region. For these cases we found that the real part of the optical potentials are quite similar to our ELP.[3)7)]

The existence of the Pauli-forbidden region in the phase space can be also derived by the path-integral approach. In order to apply the path-integral method we need to use the many-body projection operator onto the functional space spanned by the RGM wave functions of Eq.(1), which is given as[1)]

$$P = \int d\vec{a}\,|\Phi(\vec{a})\rangle\langle\Phi(\vec{a})|\ ,$$
$$\Phi(\vec{a}) \equiv \left(1/\sqrt{\binom{N_1+N_2}{N_1}}\right)\mathcal{A}[N^{-1/2}(\vec{r},\vec{a})\phi_1\phi_2]\ , \qquad (9)$$

where $N(\vec{r},\vec{a})$ is the norm kernel defined in Eq.(2). The overlap of Φ's is just the projection operator Λ onto the space spanned by the Pauli-allowed inter-nucleus relative wave functions. (The Pauli-allowed states are defined to be the states orthogonal to all the Pauli-forbidden states.)

$$\langle\Phi(\vec{a})|\Phi(\vec{b})\rangle = \Lambda(\vec{a},\vec{b}) = 1-\textstyle\sum_F |\chi_F(\vec{a})\rangle\langle\chi_F(\vec{b})|. \qquad (10)$$

The propagator in the path-integral form is calculated to be

$$\langle\Phi(\vec{q}'')|\,e^{(\hat{H}/i\hbar)(t''-t')}\,|\Phi(\vec{q}')\rangle$$
$$= \int\mathcal{D}[\omega(\vec{p}(t),\vec{q}(t))]\exp\Big[\int_{t'}^{t''}dt\Big\{\vec{p}(t)\dot{\vec{q}}(t)-\frac{\tilde{H}^W(\vec{p}(t),\vec{q}(t))}{\Lambda^W(\vec{p}(t),\vec{q}(t))}\Big\}\Big]\ ,$$
$$d\omega(\vec{p},\vec{q}) \equiv \Lambda^W(\vec{p},\vec{q})\,d\vec{p}\,d\vec{q}\ ,\quad \tilde{H} \equiv \frac{1}{\sqrt{N}}H\frac{1}{\sqrt{N}}\ . \qquad (11)$$

Since Λ is the projection operator onto the Pauli-allowed states, $\Lambda^W(\vec{p},\vec{q})$ vanishes for the point $(\vec{p},\vec{q})$ inside the Pauli-forbidden region in the phase space. Therefore, Eq.(11) shows clearly that the propagator is constructed by summing only over the physical paths which are outside of the Pauli-forbidden region. Accordingly, the semi-classical trajectory obtained by the stationary phase method from Eq.(11) lies necessarily outside of the Pauli-forbidden region of the phase space.

Now, Eq.(7) indicates that for non-large ρ the local momentum $\vec{p}(\vec{\rho})$ can not be small even for zero incident energy. This fact teaches us that the often-used adiabatic treatment which neglects the higher power terms of the inter-nucleus relative velocity than quadratic has no a priori justification even at zero incident energy. Namely, the existence of the Pauli-forbidden region in the phase space in the case of the inter-nucleus relative motion invalidates the adiabatic assumption which is plausible in many other kinds of nuclear collective motion. In this respect, the derivation by Saraceno et al[8)] of the existence of the Pauli-forbidden region in the phase space of the inter-nucleus relative motion in the framework of the time-dependent variational theory is quite instructive, because we can explicitly recognize in a common theoretical framework of the time-dependent variational approach how the inter nucleus relative motion differs from the other kinds of nuclear collective motion. For the details of the derivation by Saraceno et al, the reader is referred to Ref.8).

References

1) K.Ikeda et al, Prog.Theor.Phys. Supplement No.62 (1977).
2) H.Horiuchi, Prog.Theor.Phys. 64 (1980), 184.
3) H.Horiuchi, Proc.IV Int.Conf. on Clustering Aspects of Nuclear Structure and Nuclear Reactions (Chester,1984),p.35.
4) H.Horiuchi, Prog.Theor.Phys. 69 (1983), 886.
5) M.Hyakutake et al, Nucl.Phys. A311 (1978), 161.
6) D.A.Goldberg and S.M.Smith, Phys.Rev.Letters 29 (1972), 500.
7) Th.Delbar et al, Phys.Rev. C18 (1976), 1237; F.Michel et al, Phys.Rev. C28 (1983), 1904.
8) M.Saraceno,P.Kramer and F.Fernandez, Nucl.Phys. A405 (1983), 88.

QUANTUM CORRECTIONS TO TIME-DEPENDENT MEAN-FIELD METHOD
--- Case of a Spin System ---

Shinji IIDA*
Department of Physics, Kyoto University, Kyoto 606, JAPAN
*) Present Address: Research Center for Nuclear Physics, Osaka University, Osaka 567, JAPAN

A phase-space method is extended to the case where a system is described by a generalized coherent state (specifically, a spin coherent state), which gives a procedure to calculate quantum corrections to the motion of a mean field for a spin system.

§1. INTRODUCTION: The time-dependent Hartree-Fock (TDHF) method has been known as one of the powerful methods to treat a strongly interacting Fermion system, such as the nucleus, in a non-perturbative way. In spite of its many successes, TDHF method itself does not tell us to what extent it approximates the exact quantum mechanical treatment and it is difficult to evaluate its validity. Therefore, it is highly desired to clarify the position of the TDHF method as an approximation scheme to the exact quantum theory of many body problem, i.e., to formulate quantum mechanics such that the 0-th order term of a certain approximate series coincides with the TDHF method and we can successively evaluate higher-order terms.

In this report, I investigate the above problem by using a phase-space method[1]. In the TDHF method, in place of an operator $\hat{O}$, we treat a c-number function $O(z^*,z)=(z|\hat{O}|z)/(z|z)$, where $|z)$ denotes a Hartree-Fock (HF) state parameterized by z. This results in the classical nature of the TDHF method. If we replace a HF state $|z)$ by a Boson coherent state, $O(z^*,z)$ becomes a classical representation of the operator on the phase space ((z,z^*)-space) with a normal ordering classical-quantal corresponding rule. The HF state is known as a generalized coherent state (GCS) associated with a certain Lie group (a group of unitary transformations of single particle and hole states)[2]. Therefore, it is convenient to use the GCS representation to treat the above subject, because the GCS has properties similar to the usual Boson CS and thus, discussions can be made in parallel way to the case of the phase-space method for a Boson system with a normal ordering rule. Since the simplest example of the GCS is a spin CS associated with the SU(2) group, I here discuss an extension of the phase-space method to the case of a spin system and the quantum corrections to the time-dependent mean-field method.

§2. PHASE-SPACE METHOD FOR A SPIN SYSTEM: We first recapitulate the result of the phase-space method for a Boson system[1].

2.1 The case of a Boson system: A classical-quantal mapping Θ is defined as

$$O(z^*,z) = \Theta(\hat{O})(z^*,z) = (z|\hat{O}|z)/(z|z), \tag{1}$$

where $|z)$ is a Boson CS whose overlap function is $(z|z)=\exp(z^*z)$. By using Θ and Θ^{-1}, we obtain a one to one correspondence between operators and c-number function in the (z,z^*)-space. In order to represent non-commutativity of operators, we introduce a $*$-product $A*B$ which corresponds to the product of operators $\hat{A}\cdot\hat{B}$ as

$$(A*B)(z^*,z) = \Theta(\hat{A}\hat{B})(z^*,z)=(z|\hat{A}\hat{B}|z)/(z|z) = \textstyle\sum_{N=0}^{\infty} g^{(N)}(A,B),$$
$$g^{(N)} = \partial^N A/\partial z^N \cdot \partial^N B/\partial z^{*N}/N!. \tag{2}$$

The expression $z=(q+ip)/\sqrt{2\hbar}$ shows that $g^{(N)}$ is of order $\hbar^N$ and we find that $N\geq 1$ terms express the quantum corrections to the classical limit (the $N=0$ term), i.e., a simple product $A(z^*,z)B(z^*,z)$. By using this $*$-product, the quantity $[A*B]$ corresponding to the commutator $[\hat{A},\hat{B}]$ is defined as

$$[A*B](z^*,z)=\Theta([\hat{A},\hat{B}]/i\hbar)(z^*,z) = \textstyle\sum_{N=0}^{\infty} F^{(N)}(A(z^*,z),B(z^*,z)),$$
$$F^{(N)}(A,B) = (g^{(N+1)}(A,B) - g^{(N+1)}(B,A))/i\hbar. \tag{3}$$

We can see that the $\hbar^0$-term, $F^{(0)}$, is nothing but the usual Poisson Bracket. Various relations in quantum mechanics can be equivalently expressed in the (z,z^*)-space by using $*$ or $[*]$. In particular, we can get the $\hbar$-expansion of the solution of the Heisenberg's equations of motion, $O(z^*,z;t) = \Theta(\hat{O}(t))(z^*,z)$, as[3]

$$O(z^*,z;t)= \textstyle\sum_{N=0}^{\infty} O^{(N)}(z^*,z;t),$$
$$O^{(0)}(z^*,z;t) = (W|\hat{O}(0)|W)/(W|W)\big|_{W=W(t;z,z^*)},$$
$$O^{(N\geq 1)}(z^*,z;t) = \textstyle\sum_{j+k=N,\ j\geq 1} \int_0^t ds,$$
$$\left[F^{(j)}(O^{(k)}(Z^*,Z;s),H(Z^*,Z))\right]_{Z=W(t-s;z,z^*)}, \tag{4}$$

where W and W^* express a classical trajectory with the initial condition $W(0;z,z^*) = z$, $W^*(0;z,z^*) = z^*$. The above equations provide us with a procedure to successively evaluate the higher order terms (quantum corrections) of $O(z^*,z;t)$ starting from the classical approximation $O^{(0)}(z^*,z;t)$.

2.2 The case of a spin system: The above result shows that if $A*B$ (i.e., $g^{(N)}$) can be calculated, we can evaluate quantum corrections by using Eq.(4). The spin CS is written as $|z) = \exp(z\hat{S}_+)|S,-S\rangle$ where $|S,S_0\rangle$ is an eigenstate of $\hat{\vec{S}}^2$ and $\hat{S}_0$ with eigenvalues $S(S+1)$ and S_0, respectively. Noting that all the operators of a spin system are mapped into the "physical space" spanned by $\{ z^m (z^*)^n/(1+x)^{2S}$; $m,n=0,\ldots,2S \}$ where $x=z^*z$, we obtain

$$(A{*}B)(z^*,z) = \textstyle\sum_{N=0}^{2S} g^{(N)}(A(z^*,z),B(z^*,z)),$$
$$g^{(N\geq 1)} = \textstyle\sum_{i+j=N} \sum_{k=1}^{j} f(j,k;i)(1+x)^{j+k} x^i ,$$
$$\{ z^{j-k} \partial^j A/\partial z^j \cdot \partial^k B/\partial z^{*k} + (z^*)^{j-k} \partial^k A/\partial z^k \cdot \partial^j B/\partial z^{*j} \}, \quad (5)$$

where $f(j,k;i)$ is a constant independent of z and z^*. In contrast with the case of a Boson system, the order of A*B for a spin system has an upper limit, 2S. Since explicit forms of $g^{(N)}$ are lengthy, I, here, write down only the lowest order quantum correction term

$$g^{(2)} = 1/S(2S-1) \cdot [(1+x)^2 x \partial A/\partial z \cdot \partial B/\partial z^* + (1+x)^3/2 \cdot \{ z \partial^2 A/\partial z^2 \cdot \partial B/\partial z^*$$
$$+ z^* \partial A/\partial z \cdot \partial^2 B/\partial z^{*2} \} + (1+x)^4/4 \cdot \partial^2 A/\partial z^2 \cdot \partial^2 B/\partial z^{*2}]. \quad (6)$$

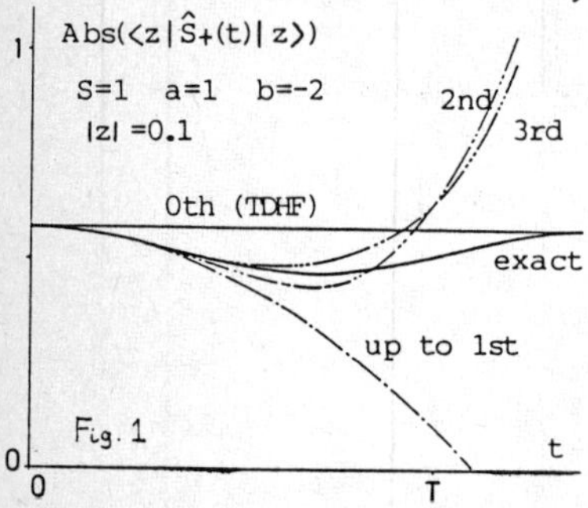

Using a canonical coordinate representation, $z = w/(2S-w^*w)^{1/2}$, we can see that $g^{(2)}$ becomes that of Boson system $g^{(2)} = \partial^2 A/\partial w^2 \cdot \partial^2 B/\partial w^{*2}/2$ when w^*w/S is negligible. Fig.1 shows lower order quantum corrections to the time development of $|(z|\hat{S}_+(t)|z)/(z|z)|$ by using Eq.(4), where Hamiltonian is given by $\hat{H} = a\hat{S}_0{}^2 + b\hat{S}_0 + c$.

§3. REMARKS: A few remarks are in order: (i) We here discuss a one-dimensional case for simplicity. The generalization to multi-dimensional cases is straightforward. (ii) For a Boson system, we can naturally order all terms in A*B according to the degree of the power of $\hbar$ and can fix the expression of $g^{(N)}$. On the other hand, for a spin system, we do not know such an *a priori* "small parameter" $\hbar$ and there is ambiguity in fixing the expression of $g^{(N)}$ except for $g^{(0)}$ and $g^{(1)}$. In order to assign the order to each term in A*B, I use the following : If $\hat{A}$ (or $\hat{B}$) is a generator, $\hat{g}$, of the relevant Lie group ($\hat{S}_0$ and $\hat{S}_\pm$ in the present case), all the higher-order ($N\geq 2$) terms cancel out for an arbitrary physical operator $\hat{B}$. In Eq.(5), the expression of $g^{(N)}$ is determined such that the above cancellations occur between the terms having the same order, i.e., $g^{(N)}(g,B)=0$ should hold for $N\geq 2$.

The author is indebted to Japan Society for the Promotion of Science for financial support.

References

1) G.S. Agarwall and E. Wolf, Phy. Rev. D2 (1970) 2161.
2) J.R. Klauder, J. Math. Phys. 4 (1963) 1055,1058.
R. Gilmore, Ann. of Phys. 74 (1972) 391.
3) S. Iida, Prog. Theor. Phys. 76 (1986) No.1, to be published.

THE REMARKABLE PHASE SPACE OF THE RADIATING ELECTRON

A.O. Barut
Department of Physics
University of Colorado, Box 390
Boulder, CO 80309

I begin with the elementary case of a three-dimensional classical spin vector $\vec{S}$ precessing in a magnetic field $\vec{B}$ according to the equation $d\vec{S}/dt = a\,\vec{S}\times\vec{B}$. A priori, $\vec{S}$ has three dynamical coordinates. But it follows from the equations of motion, because $\vec{S}.d\vec{S}/dt = 0$, that the magnitude of S is a constant of motion. This leaves us with two degrees of freedom. But since the equations of motion are of first order, this two-dimensional space is already the phase space, i.e., spin is its own conjugate variable. More precisely, selecting out the third component S_3, for example, we have the following symplectic system and Poisson brackets:

$$H = a\sqrt{S^2-S_3^2}\,(B_1\cos\varphi + B_2\sin\varphi) + aS_3B_3, \text{ with } \varphi = \arctan g(S_2/S_1) \text{ and}$$

$$\dot{S}_3 = \partial H/\partial\varphi = a\sqrt{S^2-S_3^2}\,(-B_1\sin\varphi + B_2\cos\phi) = \{H,S_3\};\ \dot{\varphi} = -\partial H/\partial S_3$$

$$\dot{\varphi} = -\frac{aS_3}{\sqrt{S^2-S_3^2}}\,(B_1\cos\varphi + B_2\sin\varphi) + 2B_3 = \{H,\varphi\}.$$

These equations coincide with the original equations $d\vec{S}/dt = a\vec{S}\times\vec{B}$. Here $\{f,g\}$ is the usual Poisson bracket. There are of course two other similar equations, one for S_1 and one for S_2 and their conjugate angles. It is interesting that, even classically in a canonical formalism one spin component at a time is meaningful as in quantum mechanics.

The above Poisson bracket $\{S,\varphi\} = 1$ cannot be realized by quantum-mechanical operators in finite-dimensional representations of spin. Therefore, I give a second form of the symplectic relations for the classical spin using a curved space, namely the homogeneous space $S^2 \sim SU(2)/U(1)$, the 2-sphere. Parametrizing the spin with the coordinates S_1 and S_2, for example, we have the system of equations,

$$\dot{S}_1 = \{H,S_1\} = -\sqrt{S^2-S_1^2-S_2^2}\,\frac{\partial H}{\partial S_2} = a(S_2B_3 - S_3B_2)$$

$$\dot{S}_2 = \{H,S_2\} = \sqrt{S^2-S_1^2-S_2^2}\,\frac{\partial H}{\partial S_1} = a(S_3B_1 - S_3B_1)$$

with the Poisson bracket definition $\{f,g\} \equiv \sqrt{S^2-S_1^2-S_2^2}\left(\frac{\partial f}{\partial S_1}\frac{\partial g}{\partial S_2} - \frac{\partial f}{\partial S_2}\frac{\partial g}{\partial S_1}\right)$ and $H = a(S_1B_1 + S_2B_2 + \sqrt{S^2-S_1^2-S_2^2}\,B_3)$. Thus, $\{S_1,S_2\} = (S^2-S_1^2 - S_2^2)^{1/2}$. This definition can now be taken over to quantum theory and shows again the curved phase space of finite quantum systems. A third Poisson bracket can be given in terms of the coordinates θ,φ on the sphere S_2, namely

$$\{\theta,\varphi\} = \frac{1}{S\sin\theta},\ \{f,g\} \equiv \frac{1}{S\sin\theta}\left\{\frac{\partial f}{\partial\theta}\frac{\partial g}{\partial\varphi} - \frac{\partial f}{\partial\varphi}\frac{\partial g}{\partial\theta}\right\},\ \dot{\varphi} = -\frac{1}{S\sin\theta}\frac{\partial H}{\partial\theta} = \{H,\varphi\},$$

$$\dot{\theta} = \frac{1}{S\sin\theta}\frac{\partial H}{\partial\varphi} = \{H,\theta\}.$$

If the magnetic field is inhomogeneous, the spin motion is coupled to the space-time motion even for the nonrelativistic Pauli electron. The phase space is now eight-dimensional, $(\vec{x},\vec{p},S_1,S_2)$, and we have the dynamical system

$$d\vec{x}/dt = \{H,\vec{x}\} = \frac{1}{m}(\vec{p}-e\vec{A}(\vec{x}))$$

$$d\vec{p}/dt = \{H,\vec{p}\} = \frac{e}{m}\vec{A},\ \nu\vec{x}^{\nu} + a\vec{S}.\nabla\vec{B}$$

Thus, in a problem like in the Stern-Gerlach experiment the motion of the spin and the deflection of the beam are intricately and nonlinearly coupled, which to my knowledge have never been fully analyzed. There is another example of a nonlinear spin motion in which the spin direction asymptotically approaches two attractors for large times, the north pole or the south pole relative to the magnetic field B, just to indicate the nontriviality of the spin precession [3]. Classical spin models are also interesting because they give the same Einstein-Podolsky-Rosen spin correlation functions as the quantum mechanics [8].

Now I come to the relativistic electron [1,2]. For the Dirac electron the phase space of the spin is much more intrinsically coupled with the space-time phase space, even for a free particle. In order to see this we summarize the equations of the "dynamical system electron" for the classical Lorentz electron, the classical Lorentz-Dirac electron which contains the radiation of self-energy effects, and on the quantum side, the Dirac electron in Heisenberg representation (Table 1). We also give a new classical model of the Dirac electron whose equations are in one-to-one correspondance with the Dirac electron [4]. There is a remarkable similarity between the radiative effects in Lorentz-Dirac equation and the quantum effects in Dirac theory [2]. In both cases the velocity and momentum are independent dynamical variables. The same is true for the classical model. The equations in Table 1 have been written with respect to an invariant time parameter τ. This is the most convenient way of exhibiting the equations. The classical phase space with respect to invariant time τ is $R_4 \otimes R_4 \otimes C_4 \otimes C_4$. Instead of the internal dynamical conjugate pair z and $\bar{z}$, we can also introduce the spin variables, as shown in the Table. Coherent states and Wigner functions can now be defined on this generalized phase space of the relativistic radiating electron so that a solution can be given to two of the main problems in Wigner function theory, namely, accounting for relativity and spin.

[1] A.O. Barut, Lecture Notes in Math., Vol. 905, 90-98 (1982).
[2] A.O. Barut and A.J. Bracken, Phys. Rev. D23, 2454-63 (1981) and Phys. Rev. D24, 3333-34 (1981).
[3] A.O. Barut, in Symposium on the Foundations of Quantum Mechanics, Japan Physical Society, Tokyo 1984, p. 321-26.
[4] A.O. Barut and N. Zanghi, Phys. Rev. Lett. 52, 2009 (1984).
[5] A.O. Barut and I.H. Duru, Phys. Rev. Lett. 53, 2355 (1984).
[6] A.O. Barut, A.J. Bracken and W.D. Thacker, Lett. Math. Phys. 8, 477 (1984).
[7] A.O. Barut and W.D. Thacker, Phys. Rev. D31, 1386 (1985).
[8] A.O. Barut and P. Meystre, Phys. Lett 105A, 458 (1984).

Table I - Phase space and dynamical equations for the electron

Lorentz electron	$\dot{x}_\mu = \frac{1}{m}(P_\mu - eA_\mu)$ $\dot{p}_\mu = eA_{\nu,\mu}\dot{x}^\nu$
Lorentz-Dirac electron	$\dot{x}_\mu = u_\mu$ $\dot{u}_\mu = \frac{3}{2\alpha}(mu_\mu + eA_\mu - p_\mu)$ $\dot{p}_\mu = eA_{\nu,\mu}u^\nu + \frac{3}{2\alpha}(mu_\nu + eA_\nu - p_\nu)^2 u_\mu$ $\mathcal{H} = u^\mu(p_\mu - eA_\mu), \ \dot{\mathcal{H}} = 0$
Dirac electron	$\dot{x}_\mu = \gamma_\mu$ $\dot{\gamma}_\mu = \frac{2i}{h}(\gamma_\mu \mathcal{H} + eA_\mu - p_\mu)$ $\dot{p}_\mu = eA_{\nu,\mu}\gamma^\nu$ $\mathcal{H} = \gamma^\mu(p_\mu - eA_\mu), \ \dot{\mathcal{H}} = 0$
Classical Dirac electron	$\dot{x}_\mu = \bar{z}\gamma_\mu z = u_\mu$ $\dot{p}_\mu = eA_{\nu,\mu}u^\nu$ $\dot{z} = -i\pi_\mu\gamma^\mu z$ $\dot{\bar{z}} = i\bar{z}\pi_\mu\gamma^\mu, \quad \pi_\mu = p_\mu - eA_\mu$ $\mathcal{H} = \bar{z}\gamma^\mu z(p_\mu - eA_\mu), \ \dot{\mathcal{H}} = 0$
For both Classical and Quantum Dirac electron ($u_\mu \leftrightarrow \gamma_\mu$)	$\dot{x}_\mu = u_\mu$ $\dot{u}_\mu = 4S_{\mu\rho}\pi^\rho$ $\dot{\pi}_\mu = eF_{\mu\rho}u^\rho$ $S_{\mu\nu} = \pi_\mu u_\nu - \pi_\nu u_\mu$

THE QUANTIZATION OF SYMMETRIC SPACES AND ITS APPLICATIONS

A. Unterberger
Mathématiques, Université de Reims
BP 347, F 51062 Reims Cedex (France)

Our point of view in quantization theory is that of a pseudo-differential operator user : it is our feeling that it gives new insights on the subject ; also, it will turn out that the most difficult problems which have to be solved to carry out the quantization program below would not even have been raised had it not been for the demands in this field of applications. The following, not an axiomatization, is meant only as a list of possibly desirable properties of the quantization scheme.

A) Find a phase space Π acted upon by a Lie group G, a unitary representation V of G in a Hilbert space H, and a correspondence Op from functions on Π (" symbols") to (possibly unbounded) operators on H ; the covariance requirement is that the formula

(1) $$V(g)\,Op(a)\,V(g)^{-1} = Op(a \circ g^{-1})$$

should hold for every symbol a and $g \in G$.

B) One should be able to derive estimates for the operator Op(a) in terms of natural estimates for its symbol a : a typical result could be that C^{∞} symbols a such that Da is bounded whenever D is a G-invariant differential operator on Π give rise to bounded operators.

C) Estimates in the reverse direction should hold too. A fixture of the theory is the existence of a natural map Symb from operators to symbols, in some sense the transpose of Op. Then the problem is essentially the spectral analysis of the (G-invariant, but not differential) operator Symb o Op on Π .

D) Is it true that natural classes of symbols (e.g. those described in B) give rise to operator algebras ? Also, does one have Op(a) Op(b) = Op (a # b), where the symbol a # b admits an expansion comparable to the one familiar in the Weyl calculus of operators ?

E) A measure of the usefulness of the whole calculus (= quantization scheme) could be its power to provide parametrices (approximate inverses) for interesting classes of partial differential operators, in the way standard pseudo-differential operators work for elliptic operators.

It turns out that there exists a plethora of calculi satisfying (A) : most have to be thrown away as they will fail in most other respects. If Π is a symmetric space whose group of automorphisms is G , if V is a unitary representation of G, one may, as suggested in [3] , define

(2) $$Op(a) = \int_{\Pi} a(X)\, \sigma_X \, dm(X)$$

where σ_X is the unitary operator on H corresponding to the symmetry of Π around X, and dm is the G-invariant measure on Π . A related suggestion, actually a generalization of the Wick calculus of operators (ours generalizes Weyl's) had been made earlier by Berezin [1] : however, it fails as far as (C) and (D) are concerned [4]. Let us now specialize to the case when $G = SO_o(2, n+1)$ and $\Pi = G/K$ is the complex tube $C + i\mathbb{R}^{n+1}$ above the solid forward light-cone C in $\mathbb{R}^{n+1}$ (when n = 3 , G is the invariance group of Maxwell equations ; when n = 0 , $G \simeq SL(2,\mathbb{R})/\{\pm 1\}$ and C is a half-line). Let $r(t) = t_o^2 - t_1^2 - \ldots - t_n^2$ if $t \in \mathbb{R}^{n+1}$. Given $\lambda > \max(0, n-1)$, let H_λ be the space of complex-valued functions u on C that are square-summable with respect to the measure $(r(t))^{-\lambda/2} dt$. The Laplace transformation makes it possible to realize H_λ as a space of holomorphic functions on Π , on which we can let a representation V_λ taken from the (projective) holomorphic discrete series of G act : when n = 0 , the kernel of σ_X can be made explicit in terms of Bessel functions [4] . It requires rather extensive work, but it is true at least for large λ , that (B) and (C) are valid for general n : (B) was proved in [6] , a joint work with J. Unterberger. One word about the crux of the proof is in order. For every $X = x + i\xi \in \Pi$, let

$$(3) \qquad \varphi_X(t) = c_\lambda \quad r(x)^{1/4(\lambda + n + 1)} \; r(t)^{\lambda/2} \; e^{-2\pi \langle Jt, X\rangle}$$

where J is the matrix of the quadratic form r and c_λ is a constant. Given X and $X' \in \Pi$, let us call Wigner - function of φ_X and $\varphi_{X'}$ the function on Π which is the image under Symb of the operator $u \mapsto (u, \varphi_X)\, \varphi_{X'}$: in this case, it is just a power of some rational function in the coordinates of X and X'. Then everything hinges on proving relevant estimates and (invariant) differential equations for the Wigner functions.

Now one does not get operator algebras in this way : the reason for this is that estimates in the C^k- topology of the symbols a have clear implications as estimates for the operator Op(a) only if $k < \lambda$: thus, though we never get algebras, things improve as λ increases. Also, though, at least for n = 0, one has a nice integral formula for $a \# b$, no meaningful asymptotics can be given for it. To complete the scheme, we now "renormalize" Op so that it will have a non-trivial limit Op^F as $\lambda \to \infty$ (F stands for "Fuchs" : more in a moment). This is tantamount to replacing G by a "contraction" Γ : observing that the restriction of V_λ to the subgroup G_o of G that consists of all affine transformations of Π is , up to equivalence, independent of λ , we substitute for the infinitesimal generators of the more exotic one-parameter subgroups of $V_\lambda(G)$ (actually those in $V_\lambda(K)$) their (renormalized) limits as $\lambda \to \infty$. The net result is the quantization formula

$$(4) \qquad (Op^F(f)u)(t) = 2^{n+1} \int_{C\times\mathbb{R}^{n+1}} f(y,\eta)\, e^{2i\pi \langle \eta, t - S_y t\rangle}\, u(S_y t)\, dy\, d\eta ,$$

where S_y is the symmetry around $y \in C$ (a symmetric space in its own right, isometric to $\mathbb{R}$ times the mass hyperboloid). The Hilbert space H is just H_{n+1} . The covariance group Γ is generated by G_o and a certain involutive transformation of $\mathbb{C} \times \mathbb{R}^{n+1}$. The fundamental properties (B) and (C) can be proved to be true, at the price of considerable work. The full C^∞ topology on symbols is now involved: as a consequence, we do this time get operator algebras, as well as asymptotics for a $\#$ b. The easier case n = 0 has been described in full in [5]. We called it the "Fuchs calculus" for, as an answer to (E), it provides, when n = 0, parametrices for ordinary differential operators of Fuchs type; similar features, leading to a generalization of the notion of Fuchs type, appear for general n.

We conjecture that the whole procedure works for general hermitian symmetric spaces of tube type, though the complexity of some of the proofs is an increasing function of the rank; abandonment of the complex structure, on the other hand (we made extensive computations with $G = SL(2,\mathbb{C})$, in which case Π is the 3-dimensional mass hyperboloid) creates severe troubles as far as (C) and (D) are concerned.

Besides giving rise to a symmetric space of rank 2, the group SO(2,4) of which the present paper describes a quantization has possible relevance in theoretical physics in view of its role in I.Segal's cosmology: one may also note, in this context, the likely significance of its contractions as was shown in a geometric study by S.Sternberg [2]. Now do you find it somewhat humiliating, for a tentative model of the universe, to be at the same time a good tool for a special class of partial differential equations?

REFERENCES

[1] F.A.Berezin, Quantization in Complex Symmetric Spaces, Math.U.S.S.R.Izvestija 9 (1975), 341-379.

[2] S.Sternberg, Chronogeometry and Symplectic Geometry, Coll.Intern.CNRS 237 (1975), Paris, 45-57.

[3] A.Unterberger, Quantification de certains espaces hermitiens symétriques, Séminaire Goulaouic-Schwartz 1979-80, Ecole Polytechnique, Paris.

[4] A. Unterberger, Symbolic Calculi and the Duality of Homogeneous Spaces, Contemporary Mathematics 27 (1984), 237-252 .

[5] A. Unterberger, The calculus of pseudo-differential operators of Fuchs type, Comm. in Part. Diff. Equ. 9(12), (1984), 1179-1236.

[6] A. and J. Unterberger, A quantization of the Cartan Domain BD I (q = 2) and Operators on the Light-Cone, to appear in J. Funct. Anal.

A CLOSED FORM FOR THE INTRINSIC SYMBOL OF THE RESOLVENT PARAMETRIX OF AN ELLIPTIC OPERATOR

S. A. Fulling* and G. Kennedy
Mathematics Department
Texas A & M University
College Station, Texas, 77843

At a previous College Park conference [1], one of us pointed out that the intrinsic symbolic calculus of pseudodifferential operators [2–3] offers a way to formulate Wigner distribution functions (and the related Weyl calculus of functions of the noncommuting operators $\mathbf{q}$ and $\mathbf{p}$) in a manifestly covariant way in the presence of external gauge fields (possibly non-Abelian) and gravitational fields. (Recent independent work toward these goals appears in [4–6].)

Practical calculations with pseudodifferential operators are usually based on the asymptotic expansion of the symbol of the resolvent of a differential operator A. From this one can obtain expansions of other symbols and kernels associated with A, such as the celebrated heat kernel, which is exploited in index theory, in renormalization theory, and (in the guise of the partition function) in statistical mechanics. (The symbol of an operator is essentially the classical function $A(\mathbf{q},\mathbf{p})$ which is associated with $A = A(\mathbf{q}, -i\nabla)$ under Weyl's correspondence.)

We have determined this expansion to arbitrary order, for a very general class of elliptic operators [7]. This work deals not with the (midpoint-based) Weyl calculus, but with the intrinsic version of the (endpoint-based) Kohn-Nirenberg calculus more common in the mathematical literature. (This fact creates the necessity for the word "essentially" in the paragraph above.) An intrinsic Weyl calculus has not yet been worked out in detail, but when it becomes available our methods should be easily adaptable to it.

Before stating the theorem, we briefly explain the notation. The operator A acts on sections of a vector bundle E over a manifold M. As in the mathematical literature, the basic variables are called (x,ξ) rather than $(\mathbf{q},\mathbf{p})$. (Thus $\xi \in T^*_x(M)$.) $\hat{\nabla}$ denotes a symmetrized covariant derivative with respect to x. A multi-index notation is used, where, for example,

$$q_j = (q_{j1},\ldots,q_{jT_j}) \in \mathbf{Z}_+^{T_j} \quad \text{for all} \quad 1 \le j \le J,$$

$$q = (q_{jt}) \quad \text{where} \quad 1 \le j \le J \quad \text{and} \quad 1 \le t \le T_j,$$

$$|q| = \sum_{j=1}^{J} |q_j| = \sum_{j=1}^{J} \sum_{t=1}^{T_j} q_{jt},$$

$$q! = \prod_{j=1}^{J} (q_j!) = \prod_{j=1}^{J} \prod_{t=1}^{T_j} (q_{jt}!),$$

and

$$n_j = (n_{j0},\ldots,n_{jk_j}) \in \mathbf{Z}_+^{k_j+1} \quad \text{for all} \quad 1 \le j \le J,$$

$$n = (n_{j\kappa}) \quad \text{where} \quad 1 \le j \le J \quad \text{and} \quad 0 \le \kappa \le k_j,$$

*speaker

$$|n| = \sum_{j=1}^{J} |n_j| = \sum_{j=1}^{J} \sum_{\kappa=0}^{k_j} n_{j\kappa},$$

$$n_+! = \prod_{j=1}^{J} \Big(n_{j0}! \prod_{\kappa=1}^{k_j} (1 + n_{j\kappa})!\Big).$$

Moreover,

$$N_j = \overline{n}_j + k_j + |n_j| \quad \text{and} \quad M_j = k_j + |m_j| \quad \text{for all } 1 \le j \le J.$$

$\otimes_R$ denotes a tensor product in reversed order. Finally, τ^E is the parallel transport in the bundle (defined by the gauge field), and Φ is the inverse of the exponential mapping (defined by the gravitational field) — i.e., $\Phi(x, y)$ is the tangent vector to the geodesic joining x to y. In the formula, the derivatives of Φ and τ^E are evaluated at coinciding arguments; these are then the quantities known to some physicists as DeWitt's $[-\sigma_{;\alpha'\beta\gamma\ldots}]$ and $[I_{;\alpha\beta\ldots}]$. The recursion relations determining these objects have *not* been solved in closed form, but several methods for calculating them recursively to high order are under investigation.

Theorem: Let $A = \sum_{r=0}^{\ell} A_r \hat{\nabla}^{\ell-r}$, where $A_r \in \Gamma^\infty(\mathrm{End}(E) \otimes S(\otimes^{\ell-r}(T(M))))$, be a differential operator of order $\ell > 0$ which is elliptic with respect to a ray Γ in $\mathbf{C}$, let $\lambda \in \Gamma$, and let $B_\lambda \in L^{-\ell}(M, E, E)$ be a resolvent parametrix of A (i.e., an inverse of $A - \lambda$ modulo operators with C^∞ integral kernels). The intrinsic symbol of B_λ has asymptotic expansion $b \sim \sum_{s \ge 0} b_s$, where $b_0 = (A_0(\otimes^\ell(i\xi)) - \lambda)^{-1}$ and, for all $s \ge 1$,

$$b_s = \sum_{J=1}^{s} \sum_{\substack{\sum_{i=1}^{j} T_i \le s-J+j \\ T_j \ge 0 \\ |T| \ge \frac{s}{\ell} - J}} \sum_{\substack{|r+\overline{n}+k|+|n+m|=s \\ r_j+M_j \le \ell \\ r_j\ge0,\overline{n}_j\ge0,k_j\ge0,n_{j\kappa}\ge0,m_{j\kappa}\ge0 \\ r_j+\overline{n}_j+k_j+|n_j|+|m_j|\ge1 \\ \sum_{i=1}^{j} N_i \ge \sum_{i=1}^{j} T_i}} \sum_{\substack{|\overline{p}+p|+|q|=|N| \\ \sum_{i=1}^{j}(\overline{p}_{i+1}+p_{i+1}+|q_i|) \le \sum_{i=1}^{j} N_i \\ \overline{p}_j \le \ell - r_j - M_j, p_j \le k_j, q_{jt} \le \ell \\ \overline{p}_j\ge0,p_j\ge0,q_{jt}\ge1 \\ \overline{p}_1=p_1=0}}$$

$$i^{(\ell+2)(J+|T|)-s} (\ell!)^{|T|} \overline{N}!(\ell - r)! \left(\overline{n}!n_+!m_+!\overline{p}!p!q!(\overline{N} - N)!(\ell - q)!(\ell - r - M - \overline{p})!(k - p)!\right)^{-1}$$

$$\Big[\bigotimes_{R\, j=1}^{J} \Big[\Big(\bigotimes_{t=1}^{T_j} b_0 A_0\Big) b_0 (\hat{\nabla}^{\overline{n}_j} A_{r_j}) (\hat{\nabla}^{n_{j0}} \hat{\nabla}^{m_{j0}} \tau^E)$$

$$\Big(S_{k_j}\Big(\bigotimes_{\kappa=1}^{k_j} \hat{\nabla}^{1+n_{j\kappa}} \hat{\nabla}^{1+m_{j\kappa}} \Phi\Big)\Big)\Big]\Big] b_0(\otimes^{\ell(J+|T|)-s}\xi).$$

Here $\ell - r = (\ell - r_1, \ldots, \ell - r_J) \in \mathbf{Z}_+^J$, $\ell - q = (\ell - q_{jt})$, where $1 \le j \le J$ and $1 \le t \le T_j$, and S_{k_j} indicates symmetrization in the k_j contravariant indices of $\bigotimes_{\kappa=1}^{k_j} \hat{\nabla}^{1+n_{j\kappa}} \hat{\nabla}^{1+m_{j\kappa}} \Phi$. The (inner) covariant derivatives of the form $\hat{\nabla}^{m_{j0}}$ and $\hat{\nabla}^{1+m_{j\kappa}}$ are contracted with M_j of the $\ell - r_j$ contravariant indices of $\hat{\nabla}^{\overline{n}_j} A_{r_j}$ for all $1 \le j \le J$. The (outer) covariant derivatives of the form $\hat{\nabla}^{\overline{n}_j}$, $\hat{\nabla}^{n_{j0}}$, and $\hat{\nabla}^{1+n_{j\kappa}}$ have much more complicated contractions, which are described inductively as follows: For all $1 \leq j \leq J-1$ the N_j covariant derivatives of the form $\hat{\nabla}^{\overline{n}_j}$, $\hat{\nabla}^{n_{j0}}$, and $\hat{\nabla}^{1+n_{j\kappa}}$ are combined with the $\sum_{i=1}^{j-1}(N_i - \overline{p}_{i+1} - p_{i+1} - |q_i|)$ as yet uncontracted covariant derivatives of the form $\hat{\nabla}^{\overline{n}_i}$, $\hat{\nabla}^{n_{i0}}$, and $\hat{\nabla}^{1+n_{i\kappa}}$ for all $1 \leq i \leq j-1$, and the total of $\overline{N}_j = N_j + \sum_{i=1}^{j-1}(N_i - \overline{p}_{i+1} - p_{i+1} - |q_i|)$ covariant derivatives is symmetrized. Of these, q_{jt} are contracted with q_{jt} of the ℓ contravariant indices of the tth factor of A_0 in $\bigotimes_{t=1}^{T_j} b_0 A_0$ for all $1 \leq t \leq T_j$, p_{j+1} are contracted with p_{j+1} of the k_{j+1} contravariant indices of $S_{k_{j+1}}(\bigotimes_{\kappa=1}^{k_{j+1}} \hat{\nabla}^{1+n_{j+1\,\kappa}} \hat{\nabla}^{1+m_{j+1\,\kappa}} \Phi)$, and $\overline{p}_{j+1}$ are contracted with $\overline{p}_{j+1}$ of the $\ell - r_{j+1} - M_{j+1}$ as yet uncontracted contravariant indices of $\hat{\nabla}^{\overline{n}_{j+1}} A_{r_{j+1}}$. The remaining $\overline{N}_j - \overline{p}_{j+1} - p_{j+1} - |q_j| = \sum_{i=1}^{j}(N_i - \overline{p}_{i+1} - p_{i+1} - |q_i|)$ are combined with the N_{j+1} covariant derivatives of the form $\hat{\nabla}^{\overline{n}_{j+1}}$, $\hat{\nabla}^{n_{j+1\,0}}$, and $\hat{\nabla}^{1+n_{j+1\,\kappa}}$, and the process is repeated. After the final such operation (when $j = J-1$), the remaining $\overline{N}_{J-1} - \overline{p}_J - p_J - |q_{J-1}| = \sum_{i=1}^{J-1}(N_i - \overline{p}_{i+1} - p_{i+1} - |q_i|)$ covariant derivatives are combined with the N_J covariant derivatives of the form $\hat{\nabla}^{\overline{n}_J}$, $\hat{\nabla}^{n_{J0}}$, and $\hat{\nabla}^{1+n_{J\kappa}}$, and the total of $\overline{N}_J = N_J + \sum_{i=1}^{J-1}(N_i - \overline{p}_{i+1} - p_{i+1} - |q_i|) = |q_J|$ covariant derivatives is symmetrized. Of these, q_{Jt} are contracted with q_{Jt} of the ℓ contravariant indices of the tth factor of A_0 in $\bigotimes_{t=1}^{T_J} b_0 A_0$ for all $1 \leq t \leq T_J$. Finally, $\otimes^{\ell(J+|T|)-s}\xi$ is contracted with the remaining $\ell(J+|T|) - s$ uncontracted contravariant indices.

References

1. S. A. Fulling, *How can the Wigner–Weyl formulation of quantum mechanics be extended to manifolds and external gauge fields?*, in *XIIIth International Colloquium on Group Theoretical Methods in Physics*, ed. by W. W. Zachary, World Scientific, Singapore, 1984, pp. 258–260.
2. H. Widom, *A complete symbolic calculus for pseudodifferential operators*, Bull. Sci. Math. **104**, 19–63 (1980).
3. L. Drager, *On the Intrinsic Symbol Calculus for Pseudo-Differential Operators on Manifolds*, Ph.D. Dissertation, Brandeis University, 1978.
4. U. Heinz, *Kinetic theory for plasmas with non-Abelian interactions*, Phys. Rev. Lett. **51**, 351–354 (1983).
5. J. Winter, *Wigner transformation in curved space-time and the curvature correction of the Vlasov equation for semiclassical gravitating systems*, Phys. Rev. D **32**, 1871–1888 (1985).
6. O. T. Serimaa, J. Javanainen, and S. Varró, *Gauge-independent Wigner functions: General formulation*, Phys. Rev. A **33**, 2913–2927 (1986).
7. S. A. Fulling and G. Kennedy, *The resolvent parametrix of the general elliptic linear differential operator: A closed form for the intrinsic symbol*, in preparation.

QUANTUM MECHANICS IN COHERENT ALGEBRAS ON PHASE SPACE

B. Lesche and T.H.Seligman
Instituto de Física, Laboratorio de Cuernavaca
National University (UNAM), México D.F., México

Using the so-called * - product defined between functions on phase space (1,2,3) one can obtain the algebra of quantum mechanics. Our aim is to show that we can introduce physically relevant concepts such as states, coherence, etc., without making explicit reference by means of a Wigner-Weyl map (4,5) to Hilbert-space quantum mechanics. This also allows us to construct quantum mechanics in spaces and coordinate systems other than cartesian coordinates in $\mathbb{R}^{2n}$. The essentially new concept we introduce (6) is that of a coherent algebra, closely related to the superposition principle.

We shall start with a brief introduction of the basic concepts; for simplicity we consider a two-dimensional phase space. The generalization to n dimensions is straight forward. Let

$$\mathcal{P} = \frac{\overleftarrow{\partial}}{\partial q}\frac{\overrightarrow{\partial}}{\partial p} - \frac{\overleftarrow{\partial}}{\partial p}\frac{\overrightarrow{\partial}}{\partial q}$$

be the position operator. Here the arrows $\leftarrow \rightarrow$ indicate action to the right or left of the partial derivatives such that

$$f\mathcal{P}g = \{f,g\} .$$

The * - product is defined as

$$* = \exp\left\{1/2\ i\hbar\mathcal{P}\right\} = \sum_{k=0}^{\infty} \frac{1}{k!}\left(\frac{i\hbar}{2}\mathcal{P}\right)^k .$$

While $\mathcal{P}$ only obeys the Jacobi identity, the *- product is associative (7). Clearly we have

$$q * p - p * q \equiv [q\ ,\ p] = i\hbar$$

and further more in the limit $\hbar \rightarrow 0$,

$$A * B = A\,B\ ,\ 1/i\hbar\,[A,B] = \{A,B\}$$

for functions A and B that are defined on phase space and have all required derivatives. We can write the von Neumann equation

$$\frac{\partial \varrho}{\partial t} = \frac{1}{i\hbar} (H * \varrho - \varrho * H) \equiv \frac{1}{i\hbar} \left[H, \varrho \right]$$

which goes into the classical equation of motion. Finally, complex conjugation is an adjoining operation as $\hbar \longrightarrow 0$

$$\overline{A * B} = \overline{B} * \overline{A} \ .$$

In order to have a complete image of quantum mechanics we need a trace operation and a concept of " state " compatible with each other. Usually these are obtained through the Wigner-Weyl map (4,5) or through embedding (7) in a larger space $\mathbb{R}^{2n}$ where this map is well established. We suggest an algebraic approach within the phase space formulation instead. But we shall not go the abstract route of identifying C* and W* algebras, although this is no doubt possible. Rather we shall try to introduce on physical grounds the relevant conditions, basing ourselves on classical analogy and the superposition principle.

We shall make use of two mathematical results proven in (6):
Theorem 1: Let A and B be two functions on phase space such that the product of these functions and of all their k^{th} partial derivatives vanish at infinity (in all directions where phase space extends to infinity) then $\int A * B \, d\Omega = \int A B \, d\Omega$. Note that this implies

$$\int A * B * C \, d\Omega = \int C * A * B \, d\Omega ,$$

but $\int A * B * C \, d\Omega \neq \int A B C \, d\Omega$.

Let an algebra $\mathcal{A}$ be defined as a linear space closed with respect to the * - product. A real functions $P \in \mathcal{A}$, is called pure if for any real function $F \in \mathcal{A}$ the equation $P * F = F$ is equivalent to $F = cP$. $c \in \mathbb{R}$. This equivalence tells us that P is a minimal idempotent.

$\mathcal{A}$ shall be called coherent if it contains pure functions and if for any two pure functions P_1, P_2 in $\mathcal{A}$, there exists a third one P_3 such that $P_1 * P_3 \neq 0$, $P_2 * P_3 \neq 0$.
We can now formulate
Theorem 2: If $\mathcal{A}$ is a coherent algebra whose pure functions fulfil theorem 1, then all pure functions have the same integral: $\int P_i \, d\Omega = \text{const.}$

Based on these results we can fill the frame work of quantum mechanics on phase space:
The phase space integral of classical mathematics can be extended to be the trace

operation of quantum mechanics; theorem 1 ensures the cyclic invariance. The pure functions can readily be associated to density operators of pure states, and the superposition principle requires us to demand a coherent algebra. A general state is then described by a positive linear combination of pure states rather than by a positive function; the former property is conserved under unitary time evolution whereas the latter would not fulfil this criterion.

Thus quantum mechanics on phase space is given by a coherent algebra of functions on this space with the ⋆ - product as the product of the algebra. Real functions in this algebra are observables and the phase space integral is the trace operation.

On $\mathbb{R}^{2n}$ this leads to the standard Wigner picture, but it can be implemented on other manifolds. If we choose a cylinder as phase space (9), the coherence condition can only be fulfilled by admitting step functions (for which the ⋆ - product is defined through a Fourier transform). But once we allow them, the (angular) momentum quantization follows from the coherence condition. Similarly, we can directly quantize in polar coordinates (6). Note that classical canonical coordinates will in general not appear as observables in the coherent algebra. Summarizing, we may say that the coherence gives us a tool to quantize in other than cartesian frames though implementation may be non-trivial.

References

(1) Groenewold H J 1946 Physica 12 405
(2) Moyal J E 1949 Proc. Camb. Phil. Soc. 45 99
(3) Amiet JP and Huguenin P Mécaniques classique et quantique dans l'espace de phase (unpublished manuscript)
(4) Wigner E 1932 Phys. Rev. 40 749
(5) Weyl H 1928 Z. Phys. 46 1
(6) Lesche B and Seligman T H J. Phys. A 1986 (January)
(7) Plebanski J F 1968 Preprint, Poisson Brackets and Commutators, Institute of Physics of Nicolaus Copernicus University, Torun
(8) Bayen F, Flato M, Fronsdal C, Lichnerowicz A and Sternheimer D 1978 Ann. Phys. 111 61
(9) Lesche B 1982 KINAM 4 81

F. RELATIVITY, QUANTUM MECHANICS AND THERMODYNAMICS

MAXIMAL-ACCELERATION INVARIANT PHASE SPACE

Howard E. Brandt
Harry Diamond Laboratories, Adelphi, MD 20783, USA

Maximal-acceleration invariant phase space is the eight-dimensional product manifold of ordinary spacetime with four-velocity space, and its line element is invariant under the maximal acceleration group.[1] The form of the line element follows directly from the requirement that the identical maximal proper acceleration apply invariantly in all accelerated frames. The existence of a maximal proper acceleration of the order of one Planck-length per Planck-time squared was shown to follow[2,3] from arguments that at such an extreme acceleration, the temperature of the vacuum radiation reaches Sakharov's maximal temperature of thermal radiation, spacetime foam is copiously produced, and the classical spacetime structure breaks down.

The physical basis for maximal proper acceleration is elucidated by elementary heuristic reasoning. By the time-energy uncertainty principle, virtual particles of mass m occur in the vacuum during a time $\hbar/2mc^2$ and over a distance $\hbar/2mc$, the Compton wavelength of a particle. In an accelerated frame the inertial force acts on such a virtual particle, and if energy equaling its rest energy is imparted to it, it becomes real. The inertial force, ma, on a virtual particle in a frame having proper acceleration a, acts over a distance within which creation of the particle can occur, namely a Compton wavelength $\hbar/2mc$, thereby doing work $(ma)(\hbar/2mc)$. Equating this work to the rest energy mc^2 of the particle, it follows that for $a = 2mc^3/\hbar$, particles of mass m will be copiously produced out of the vacuum. If the acceleration is sufficiently large, the created particles will be black holes. A particle of mass m is a black hole if its size, namely its Compton wavelength $\hbar/2mc$, is less than its Schwarzschild radius $2Gm/c^2$, or equivalently if the particle mass exceeds $(\hbar c/G)^{\frac{1}{2}}/2$. The minimum possible mass of a black hole is therefore of the order of the Planck mass.[2] It follows that for proper acceleration $a_0 = 2\pi\alpha(c^7/\hbar G)^{\frac{1}{2}}$, where α is a number of order unity, there will be copious production of Planck-mass black holes out of the vacuum, resulting in the formation of a spacetime foam and breakdown of the classical spacetime structure, as well as breakdown of the very concept of acceleration. This is the maximal proper acceleration relative to the vacuum. The temperature of the vacuum radiation at maximal proper acceleration is given by $T_{max} = \hbar a_0/2\pi kc = \alpha(\hbar c^5/G)^{\frac{1}{2}}/k$, which is Sakharov's maximum possible temperature of thermal radiation. In the language of metaphor, at maximal proper acceleration, spacetime undergoes a topological phase transition and begins to "boil" while the temperature maintains its constant maximal value, analogously to a liquid maintaining a constant temperature while boiling away.

In a frame with proper acceleration a, less than maximal, a host of effects can arise, consistent with invariant maximal proper acceleration, such as time dilatation due to proper acceleration.[1,4] By the equivalence principle, local time dilatation due to a gravitational field also follows.[1,5] For a static observer in a stationary spacetime, the proper time dilatation $\delta\tau/\tau$ may be expressed in terms of the local gravitational field with metric tensor $g_{\mu\nu}$ as follows:

$$\delta\tau/\tau = -(c^4/8a_0^2)g^{\mu\nu}(\ln g_{00})_{|\mu}(\ln g_{00})_{|\nu} \tag{1}$$

to lowest order in a/a_0, where a is the local acceleration due to gravity. For a Schwarzschild field this reduces to $(G^2M^2/2a_0^2r^4)(1-(2GM/c^2r))^{-1}$ outside the horizon.

On the basis of the maximal acceleration group, a differential geometry of the eight-dimensional phase space can be formulated.[1,6] The metric tensor for the phase space is given by

$$G_{AB} = \left[\begin{array}{c|c} g_{\mu\nu} + \rho_0^2\Gamma^{\alpha}{}_{\lambda\mu}g_{\alpha\beta}\Gamma^{\beta}{}_{\delta\nu}v^{\lambda}v^{\delta} & \rho_0 g_{\alpha\nu}\Gamma^{\alpha}{}_{\lambda\mu}v^{\lambda} \\ \hline \rho_0 g_{\alpha\mu}\Gamma^{\alpha}{}_{\lambda\nu}v^{\lambda} & g_{\mu\nu} \end{array}\right], \tag{2}$$

where A,B = 0,1,2,..7, all Greek indices range from 0 to 3, $\rho_0 = c^2/a_0$ is the minimum radius of curvature of world lines, $\Gamma^{\mu}{}_{\alpha\beta}$ is the ordinary spacetime affine connection, and v^{μ} is the four-velocity.

A key step in the development of a complete metric field theory on the phase space is to define the appropriate connection. The eight-dimensional Christoffel connection consists of eight parts, corresponding to whether its indices lie in the spacetime or velocity-space submanifolds. For example, the part of the Christoffel connection having spacetime indices only reduces to[6]

$${}^{(8)}\Gamma^{\mu}{}_{\alpha\beta} = \Gamma^{\mu}{}_{\alpha\beta} - \tfrac{1}{2}\rho_0^2\Gamma^{\gamma}{}_{(\alpha\delta}R^{\mu}{}_{\beta)\gamma\lambda}v^{\delta}v^{\lambda}, \tag{3}$$

for that subclass of manifolds for which $\partial g_{\mu\nu}/\partial v^{\beta}$ is vanishing, where $R^{\mu}{}_{\beta\gamma\lambda}$ is the ordinary spacetime Riemann curvature tensor, and $(\alpha\ldots\beta)$ is the symmetrization notation of reference 1 (Eq.(3) corrects Eq.(13) of Ref. 1). As another example, the spacetime-velocity-spacetime part is given by

$${}^{(8)}\Gamma^{\mu}{}_{a\beta} = -\tfrac{1}{2}\rho_0 v^{\lambda}R^{\mu}{}_{\beta a\lambda}, \tag{4}$$

where a = 4,5,6,7, and $R^{\mu}{}_{\beta a\lambda} \equiv R^{\mu}{}_{\beta a-4\lambda}$.

Under an infinitesimal parallel displacement of an eight-vector $\{A^M;\ M=0,\ldots 7\} = \{A^\mu,\ A^m;\ \mu=0,1,2,3;\ m=4,5,6,7\}$, in which the displacement δx^β is restricted to the spacetime submanifold, it follows that the change in the spacetime part of the vector is given by

$$\delta A^\mu = (\Gamma^\mu{}_{\alpha\beta} - \tfrac{1}{2}\rho_0^2 v^\delta v^\lambda \Gamma^\gamma{}_{(\alpha\delta} R^\mu{}_{\beta)\gamma\lambda}) A^\alpha \delta x^\beta - \tfrac{1}{2}\rho_0 v^\lambda R^\mu{}_{\beta a\lambda} A^a \delta x^\beta . \tag{5}$$

The first term contains the familiar form with an added part depending on the four-velocity, as well as the ordinary affine connection and curvature tensor. The second term depends on the four-velocity and curvature tensor, and mixes the four-velocity part of the eight-vector with its spacetime part. These terms involving ρ_0 are only important for metric variations occurring on the scale of the Planck length. Also, for a flat spacetime with vanishing Riemann curvature tensor, they are vanishing. The eight-dimensional Riemann tensor constructed from the eight-dimensional Christoffel connection leads naturally to a field Lagrangian depending on powers of the ordinary Riemann tensor, and geodesic equations of motion with terms depending explicitly on the curvature tensor and higher powers of the four-velocity.

The metric tensor of maximal-acceleration invariant phase space, equation (2), is of the same form as that for an eight-dimensional Kaluza-Klein-type gauge theory,[7] where the coordinates of the inner-space group-manifold are four-velocity (with a factor of ρ_0), the length scale of the inner space is given by ρ_0, the metric tensor of the group manifold is the ordinary spacetime metric, and the gauge potentials are given by the inner product of the ordinary affine connection and the four-velocity. Here, however, spacetime is restricted to four dimensions, and the higher dimensions are in phase space.

1. H.E. Brandt, XIIIth International Colloquium on Group Theoretical Methods in Physics, W.W. Zachary, editor, World Scientific, Singapore, 519 (1984).
2. H.E. Brandt, Lett. Nuovo Cimento <u>38</u>, 522 (1983); <u>39</u>, 192 (1984).
3. H.E. Brandt, Bull. Am. Phys. Soc. <u>30</u>, 341 (1985).
4. H.E. Brandt, Bull. Am. Phys. Soc. <u>29</u>, 651 (1984).
5. H.E. Brandt, Bull. Am. Phys. Soc. <u>30</u>, 717 (1985).
6. H.E. Brandt, Bull. Am. Phys. Soc. <u>31</u>, 804 (1986).
7. L.N. Chang, K.I. Macrae, F. Mansouri, Phys. Rev. D <u>13</u>, 235 (1976).

OCTONIONIC HADRONIC SUPERSYMMETRY AND LINEARLY RISING REGGE TRAJECTORIES

Sultan Catto
Baruch College of the City University of New York,
New York, N.Y. 10010

The parallelism of baryonic and mesonic Regge trajectories provides a realization of supersymmetry in nature. Recently we have shown that this hadronic supersymmetry is derived from QCD, the gauge field theory of strong interactions in the elongated bag or lattice gauge theory approximations. The supersymmetry group that leaves a semi-relativistic Hamiltonian invariant was shown to be U(6/21). In this note we present an extension of this symmetry that uses the algebra of octonions and operates at the quark-diquark level, and show incorporation of color through an octonionic structure that follows from the absence of diquarks in six-dimensional representation of the color group.

The octonion algebra has eight dimensions and its base vectors, e_0, e_i, $i=1,\ldots,7$, satisfy the product law

$$e_0 e_i = e_i e_0 = e_i , \qquad e_i e_j = -\delta_{ij} e_0 + \epsilon_{ijk} e_k ,$$

where ϵ_{ijk} is completely antisymmetric with $\epsilon_{ijk}=1$ for ijk=123, 246, 435, 651, 572, 714, 367. This algebra is neither commutative nor associative, but belongs to the class of alternative algebras, with the property that for any three octonions the associator of a,b,c, is given by

$$\left[a, b, c\right] = (a\, b)\, c - a\, (b\, c) ,$$

where a,b,c, are called Cayley numbers. The Cayley algebra with the case given above belongs to the class of division algebras, but it can also be represented as a split algebra if we use a new basis defined on the complex field. From one set of octonion units we therefore construct three new Grassmann numbers u_m (m=1,2,3) and their complex conjugates

$$u_m = \tfrac{1}{2}(e_m + i e_{m+3}) , \qquad u_m^* = \tfrac{1}{2}(e_m - i e_{m+3}) .$$

We have

$$u_m u_n + u_n u_m = 0, \quad u_m^* u_n^* + u_n^* u_m^* = 0, \quad u_m u_n^* + u_n^* u_m = \delta_{mn},$$

with

$$u_0 = \tfrac{1}{2}(1+ie_7), \quad u_0^2 = u_0, \quad u_0 u_j = u_j u_0^* = u_j, \quad u_0^* u_j = u_j u_0 = 0 .$$

Unlike ordinary Grassmann numbers, these are non-associative. We can call them exceptional Grassmann numbers, whose automorphism group is SU(3). The octonionic quark field Q_A is now written as

$$Q_A = u_i Q_A^i = \vec{u} \cdot \vec{Q}_A ,$$

where $\vec{Q}$ transforms like a triplet under SU(3). The diquark $D_{AB} = \epsilon_{ijk} u_k^* Q_A{}^i Q_B{}^j = \vec{u}^* \cdot \vec{D}_{AB}$ comes from $D_{AB} = Q_A Q_B = Q_B Q_A$. The SU(3) assignments of u_i is 3, u_j^* is $\overline{3}$, and of u_0 and u_0^* is 1, so that $3 \times 3 = \overline{3}$ (no 6), and $3 \times \overline{3} = 1$ (no 8). The Q_A is in 6, and D_{AB} is in 21 representation. The baryons which are in the (56) symmetrical representation are written as

$$B_{\underline{ABC}} = \tfrac{1}{2} Q_A (Q_B Q_C) + \tfrac{1}{2} (Q_B Q_C) Q_A .$$

We now consider 28x28 octonionic U(6/22) supermultiplet X given by

$$X = \begin{pmatrix} u_0 M & u_0 B & \vec{u} \cdot \vec{Q} \\ u_0 B^\dagger & u_0 N & \vec{u} \cdot \vec{D}^* \\ \epsilon \vec{u}^* \cdot \vec{Q}^\dagger & \epsilon \vec{u}^* \cdot \vec{D}^T & \vec{u}_0^* L \end{pmatrix}$$

where mesons $M(6 \times \overline{6})$, and exotics $N(21 \times \overline{21})$, are Hermitian; $B(6 \times 21)$, $\overline{B}(\overline{21} \times \overline{6})$, $Q(6 \times 1)$, $\overline{Q}(1 \times \overline{6})$, $D(1 \times 21)$, $\overline{D}(\overline{21} \times 1)$, $L(1 \times 1)$, and where ϵ can be taken to be 1, -1, or 0. Closure properties of X matrices are such that

$$[X, X'] = iX'' , \quad \{X, X'\} = X''' ,$$

and in general they are nonassociative, except in the case when $\epsilon = 0$ we have

$$[[X, X'], X''] + [[X', X''], X] + [[X'', X], X'] = 0 .$$

Then we have a true superalgebra (non-semisimple) which is a contraction of a simple algebra that closes but does not satisfy the Jacobi identity. If we now consider the element of the algebra given by

$$Z=\begin{pmatrix} u_0 m & u_0 b & \vec{u}\cdot\vec{q} \\ u_0 b^{\dagger} & u_0 n & \vec{u}\cdot\vec{d} \\ \epsilon\,\vec{u}^{*}\cdot\vec{q}^{\dagger} & \epsilon\,\vec{u}^{*}\cdot\vec{d}^{T} & u_0^{*} l \end{pmatrix}$$

where $m \in SU(6)$, $\begin{pmatrix} u_0 m & u_0 b \\ u_0 b^{+} & u_0 n \end{pmatrix}$ are the color singlet parameters $\left[U(6/21)\right]$, $\begin{pmatrix} \vec{u}\cdot\vec{q} \\ \vec{u}\cdot\vec{d} \end{pmatrix}$ are the colored parameters, and $(u_0 b\,,\ \vec{u}\cdot\vec{q})$ are the fermionic parameters, then the change in X given by $\delta X= i[Z,X]$ leads to (in the case $\vec{q}=0$, $\vec{d}=0$, and $l=0$)

$$i\,\delta\vec{Q}= m\vec{Q}+b\vec{D}^{*}, \qquad \text{and } i\,\delta D^{*}= b^{\dagger}\vec{Q}+n\vec{D}^{*}\,.$$

This subgroup is valid for a semirelativistic Hamiltonian describing $q(x_1)$ and $\bar{q}(x_2)$, $q(x_1)$ and $D(x_2)$, $\bar{D}(x_1)$ and $q(x_2)$, and $\bar{D}(x_1)$ and $D(x_2)$ interacting through a scalar potential $V(\vec{x}_1,\vec{x}_2)=b|\vec{x}_1-\vec{x}_2|$, as was shown by us in an earlier publication. The fundamental representation of $U(6/21)\times[SU(3)^c$ triplet$]$ and the adjoint representation of $U(6/21)\times[SU(3)^c$ singlet$]$ fits in the adjoint representation of $U(6/22)$ denoted by X. The automorphism group of this algebra includes $SU(6)\times SU(21)\times SU(3)^c$. If $\vec{Q}$ is Majorana and $\vec{D}$ real, then the group becomes $OSp(6/21)\times SU(3)^c$, with subgroup $Sp(12,R)\times O(21)\times SU(3)^c$.

Thus, we have shown that this algebra generalizes $U(6/21)$ algebra and puts quarks and diquarks in the same multiplet with the hadrons. It also naturally suppresses quark configuration that are symmetrical in color space and anti-symmetrical in the remaining flavor, spin, and position variables.

ROTATIONS AND GAUGE TRANSFORMATIONS

Y. S. Kim, University of Maryland, College Park, Maryland 20242
D. Han, SASC Technologies, Inc., Hyattsville, Maryland 20784
D. Son, Kyungpook National University, Taegu 635, Korea

The energy-momentum relation for a nonrelativistic particle is $E = P^2/2m$, while that for a massless particle is $E = P$. It was Einstein who unified these two relations through his formula $E = [P^2 + m^2]^{1/2}$. In addition to the four-momentum, a relativistic particle has its internal space-time degrees of freedom. When the particle is slow or at rest, the space-time symmetry manifests itself through by the spin. On the other hand, if the particle is massless, it has the helicity and gauge degrees of freedom. The question is whether the internal space-time symmetries for massive and massless particles can be unified. In order to study this problem, we use Wigner's little groups.[1]

The little group is the maximal subgroup of the Lorentz group which leaves the four-momentum of the particle invariant. In order to construct a transformation which leaves the four-momentum invariant, let us consider a particle moving along the z direction with velcity parameter α. We can rotate this particle around the y axis by θ. We can then make a boost to come back to the original momentum. The four-momentum is unchanged under this boost preceded by the above-mentioned rotation.[2] The matrix which performs this transformation is

$$D(\alpha,\theta) = \begin{bmatrix} 1 - (1-\alpha^2)u^2/2T & 0 & -u/T & \alpha u/T \\ 0 & 1 & 0 & 0 \\ u/T & 0 & 1 - u^2/2T & \alpha u^2/2T \\ \alpha u/T & 0 & -\alpha u^2/2T & 1 + \alpha u^2/2T \end{bmatrix} , \qquad (1)$$

where $\quad u = -2(\tan\frac{\theta}{2}) \quad , \quad$ and $\quad T = 1 + (1 - \alpha^2)(\tan\frac{\theta}{2})^2 \quad .$

This matrix is clearly an element of Wigner's O(3)-like little group, and should take the form

$$D(\alpha,\theta) = A(\alpha)\ W(\theta^*)\ [A(\alpha)]^{-1} \quad , \qquad (2)$$

where $A(\alpha)$ is the matrix which boosts along the z direction the particle at rest to that of the velocity parameter α. W represents the rotation around the y axis by θ^*, where

$$\theta^* = \cos^{-1}\left[\frac{1 - (1-\alpha^2)(\tan\frac{\theta}{2})^2}{1 + (1-\alpha^2)(\tan\frac{\theta}{2})^2}\right] \quad . \qquad (3)$$

Indeed, the $D(\alpha,\theta)$ a Lorentz-boosted rotation matrix.

For massless particles, we can take the limit $\alpha \to 1$, so that the expression becomes

$$D(u) = \begin{bmatrix} 1 & 0 & -u & u \\ 0 & 1 & 0 & 0 \\ u & 0 & 1 - u^2/2 & u^2/2 \\ u & 0 & -u^2/2 & 1 + u^2/2 \end{bmatrix} . \qquad (4)$$

This matrix performs a gauge transformation when applied to the four-potential for a free photon.[3]

Let us next study the above kinematics for spin-1/2 particles, using the algebra of SL(2,c). The generators of SL(2,c) satisfy the commutation relations:

$$[S_i, S_j] = i\varepsilon_{ijk} S_k , \quad [S_i, K_j] = i\varepsilon_{ijk} K_k , \quad [K_i, K_j] = -i\varepsilon_{ijk} S_k . \qquad (5)$$

where S_i and K_i generate rotations and boosts respectively. Since these relations remain invariant under the sign change in K_i, the boost generators can take two different signs: $K_i = (\pm)\frac{i}{2}\sigma_i$, the D matrix can have two different forms:

$$D^{(\pm)}(\alpha,\theta) = \begin{bmatrix} 1/\sqrt{T} & (1 \pm \alpha)u/2\sqrt{T} \\ -(1 \mp \alpha)u/2\sqrt{T} & 1/\sqrt{T} \end{bmatrix} , \qquad (6)$$

where $D^{(+)}$ and $D^{(-)}$ are constructed from $K_i = \frac{i}{2}\sigma_i$ and $K_i = -(\frac{i}{2}\sigma_i)$ respectively. It can be shown that the above form is also a Lorentz-boosted rotation.

If we use $\chi_\pm$ and $\dot{\chi}_\pm$ as the spinors to which $D^{(+)}$ and $D^{(-)}$ are applicable respectively, the angle between the momentum and the directions of the spins represented by χ_+ and $\dot{\chi}_-$ is $\tan^{-1}((1 - \alpha)\tan(\theta/2))$, which becomes zero as $\alpha \rightarrow 1$. However, in the case of χ_- and $\dot{\chi}_+$, the angle becomes $\tan^{-1}((1 + \alpha)\tan(\theta/2))$. In the limit of $\alpha \rightarrow 1$, this angle becomes $\tan^{-1}(2 \tan(\theta/2))$. Indeed, the spins represensted by χ_- and $\dot{\chi}_+$ refuse to align themselves with the momentum.

In the limit $\alpha \rightarrow 1$, $D^{(+)}$ and $D^{(-)}$ become

$$D^{(+)} = \begin{bmatrix} 1 & u \\ 0 & 1 \end{bmatrix} , \qquad D^{(-)} = \begin{bmatrix} 1 & 0 \\ -u & 1 \end{bmatrix} . \qquad (7)$$

These matrices should perform gauge transformations when applied to the spinors. The SL(2,c) spinors are gauge-invariant in the sense that

$$D^{(+)}(u)\ \chi_+ = \chi_+ , \qquad D^{(-)}(u)\ \dot{\chi}_- = \dot{\chi}_- . \qquad (8)$$

On the other hand, the SL(2,c) spinors are gauge-dependent in the sense that

$$D^{(+)}(u)\ \chi_- = \chi_- + u\chi_+ , \qquad D^{(-)}(u)\ \dot{\chi}_+ = \dot{\chi}_+ - u\dot{\chi}_- . \qquad (9)$$

The gauge-invariant spinors of Eq.(8) appear as polarized neutrinos in the real world.[4] However, where do the above gauge-dependent spinors stand in the physics of spin-1/2 particles? In order to see that these spinors are responsible for the gauge dependence of the four-potential, let us take binlienar combinations:

$$-\chi_+\dot{\chi}_+ = (1,\ i,\ 0,\ 0)\ , \qquad \chi_-\dot{\chi}_- = (1,\ -i,\ 0,\ 0)\ ,$$
$$\chi_+\dot{\chi}_- = (0,\ 0,\ 1,\ 1)\ , \qquad \chi_-\dot{\chi}_+ = (0,\ 0,\ 1,\ -1)\ . \tag{10}$$

If we boost a massive particle initially at rest along the z direction, $\chi_+\dot{\chi}_+$ and $\chi_-\dot{\chi}_-$ remain invariant. However, $\chi_+\dot{\chi}_-$ and $\chi_-\dot{\chi}_+$ acquire the constant factors $\sqrt{(1 + \alpha)/(1 - \alpha)}$ and $\sqrt{(1 - \alpha)/(1 + \alpha)}$ respectively. As $\alpha \to 1$, we can drop $|\chi_-\dot{\chi}_+>$ while replacing the coefficient $\sqrt{(1 + \alpha)/(1 - \alpha)}$ by 1. The D(u) matrix for the above spinor combinations should take the form:

$$D(u) = D^{(+)}(u)\ D^{(-)}(u)\ , \tag{11}$$

where $D^{(+)}$ and $D^{(-)}$ are applicable to the first and second spinors of Eq.(10) respectively. Then

$$D(u)\ (-|\chi_+\dot{\chi}_+>) = -|\chi_+\dot{\chi}_+> + u|\chi_+\dot{\chi}_->\ ,$$
$$D(u)\ |\chi_-\dot{\chi}_-> = |\chi_-\dot{\chi}_-> + u|\chi_+\dot{\chi}_->\ , \tag{12}$$
$$D(u)\ |\chi_+\dot{\chi}_-> = |\chi_+\dot{\chi}_->\ .$$

The first two equations of the above expression correspond to the gauge transformations on the photon four-vector.

We can now summarize the above result into the following table.

	Massive Slow	between	Massless Fast
Energy Momentum	$E = P^2/2m$	Einstein's $E = (m^2 + P^2)^{1/2}$	$E = P$
Spin, Gauge, Helicity	S_3 S_3, S_3	Wigner's Little Group	S_3 Gauge Trans.

While Einstein's $E = [P^2 + m^2]^{1/2}$ unifies the energy-momentum relations for massive and massless particles, Wigner's little group unifies the internal space-time symmetries. We are very grateful to Professor Eugene P. Wigner for illuminating discussions.

1. E. P. Wigner, Ann. Math. 40, 149 (1939); Rev. Mod. Phys. 29, 255 (1957).
2. J. Kuperzstych, Nuovo Cimento 31B, 1 (1976); Phys. Rev. D 17, 629 (1978); D. Han, Y. S. Kim, and D. Son, J. Math. Phys. 22, 2228 (1986).
3. S. Weinberg, Phys. Rev. 134, B882 (1964); 135, B1049 (1964); D. Han and Y. S. Kim, Am. J. Phys. 49, 348 (1981).
4. D. Han, Y. S. Kim, and D. Son, Phys. Rev. D 26, 3717 (1982).

HEISENBERG ALGEBRAS IN THE THEORY OF SPECIAL FUNCTIONS

Philip Feinsilver
Department of Mathematics
Southern Illinois University
Carbondale, Illinois 62901

Introduction. In this presentation we will show how Lauricella polynomials of types F_A, F_B (see [1]) arise naturally in terms of representations of the Heisenberg-Weyl (HW) algebra/group. We indicate two approaches. The first follows [2], extending to the multivariate case. The second explicates the results of [3]. This latter method depends on solving a Riccati system of projective type, independently studied in the more general researches of [4].

Remarks. 1) The one-variable hypergeometric polynomials were studied from a Lie-theoretic viewpoint by Weisner [5]. Also see Meixner [6]. 2) Miller studied the F_D functions via Lie-theoretic methods in [7]. 3) As late as 1984, Biedenharn [8] remarked on the absence of a Lie-theoretic approach to Appell F_2 and F_3 functions.

Notational Remarks. The summation convention is: repeated Greek indices are summed from 1 to N. In general, m, n, etc., will denote multi-indices, $|m| = m_1+\ldots+m_N$, e.g., $x^n = x_1^{n_1}\ldots x_N^{n_N}$, $n! = n_1!\ldots n_N!$, etc.

I. Matrix Elements of the HW Group. Consider the Lie algebra of dimension 2N+1 having basis x_j, δ_j, h satisfying the commutation relations $[\delta_j, x_k] = h\delta_{jk}$, $[h,\delta_j] = [h,x_j] = 0$. A typical group element $g(\alpha,\beta,\gamma)$ is of the form $\exp(\alpha\cdot x)\exp(\beta\cdot\delta)\exp(\gamma h)$, the dot denoting the usual scalar product, $a\cdot b = \sum_1^N a_j b_j$.

Compute the action of the HW group on the monomials $X(\ell mn) = x^\ell\delta^m h^n$ forming a basis for the universal enveloping algebra $\mathcal{U}$. We want $g(\alpha\beta\gamma)\,X(\ell mn) = \sum_{L,M,N} M(\alpha\beta\gamma)_{\ell mn}^{LMN}\,X(\ell+L,\ m+M,\ n+N)$, where the sum is such that all components of the indices $\ell+L$, $m+M$, $n+N$ are ≥ 0.

Theorem 1. The matrix elements for the representation of the HW group on $\mathcal{U}$ are given by F_A functions.

Proof (sketch): Expand the group law

$$\exp(\alpha\cdot x)\exp(\beta\cdot\delta)\exp(\gamma h)\cdot\exp(A\cdot x)\exp(B\cdot\delta)\exp(Ch) =$$
$$\exp((\alpha+A)\cdot x)\exp((\beta+B)\cdot\delta)\exp((\gamma+C+\beta\cdot A)h)$$

in power series. I.e., expand the element g(A,B,C) in terms of the basis $X(\ell,m,n)$, and then expand the right-hand side to find the coefficient of $X(\ell+L,\ m+M,\ n+N)$. After collecting terms one finds, for $L \geq 0$,

$$\frac{\alpha^L}{L!}\frac{\beta^M}{M!}\frac{\gamma^N}{N!}\sum_k \frac{(-\ell)_k(-N)_{|k|}}{(L+1)_k\, k!}\left(\frac{\alpha\beta}{\gamma}\right)^k.$$

The latter sum can be expressed as $F_A(-N, -\ell; L+1; \alpha\beta\gamma^{-1})$.

II. Fock-type Representations (see [3]). We are looking for operators ξ_j, V_j, $[V_j, \xi_k] = \delta_{jk}$, where $V = V(D)$, $D = (\partial/\partial x_1, \ldots, \partial/\partial x_N)$, on an underlying space with variables $x_1, \ldots, x_N$. One wants to find a "Hamiltonian" function $L(D)$ such that under the Hamiltonian flow generated by L, the induced representation of the HW algebra generated by ξ, V yields orthogonal polynomials. Specifically, with $\xi(t) = e^{-tL}\xi e^{tL}$, we want $J_n(x,t) = \xi(t)^n 1$ to satisfy $V_j J_n = n_j J_{n-e_j}$, e_j having just a 1 in the $j^{\underline{th}}$ spot, $((\partial/\partial t)+L)J_n = 0$, with J_n orthogonal with respect to the measure $p_t(dx)$ on $\mathbb{R}^N$ such that $tL(z) = \log \int e^{z\cdot x} p_t(dx)$ is analytic for z in some neighborhood of 0 in $\mathbb{C}^N$. It turns out that there are constants $\gamma_j \neq 0$ such that $V_j = \gamma_j L_j$, $L_j = \partial L/\partial z_j$. We state the following theorems from [3]:

Theorem 0. The functions V_j (suitably normalized) satisfy the system

$$V_{jk} = \delta_{jk} + a^{\lambda}_{jk}\ V_\lambda + b_k V_j V_k\,,$$

where $V_{jk} = \partial V_j/\partial z_k$, a^{ℓ}_{jk} and b_k are constants.

Theorem 00. The generating function for the orthogonal system $J_n(x,t)$ is (canonically) of the form

$$\begin{aligned} G(x,t;v) &= \sum v^n J_n/n! \\ &= \prod_j (1+\gamma_\lambda v_\lambda B_{\lambda j})^{X_j+c_j t}\cdot(1-\gamma_\lambda v_\lambda B_{\lambda\mu} c_\mu/\bar{c})^{\bar{c}t-u\cdot X}, \end{aligned}$$

where $B_{\lambda i}\gamma_\lambda B_{\lambda j} = c_j^{-1}\delta_{ij} - 1$ for given constants $c_1, \ldots, c_N$ and matrix B. ($\bar{c} = 1-u\cdot c$, $u = (1, 1, \ldots, 1)$, N ones).

Now we have the following theorems.

Theorem 1. A generating function for the F_B polynomials

$$F_B(-r,\ b;\ t;\ s) = \sum(-r)_n(b)_n z^n/(t)_{|n|}n!$$

is $(1-u\cdot v)^{u\cdot b-t}\prod_j(1-u\cdot v+s_j v_j)^{-b_j} = \sum v^r(t)_{|r|}F_B(-r,b;t;s)/r!$.

Theorem 2. The polynomials $J_n(x,t)$ satisfy

$$\sum(Bv)^n J_n(x,t)/n! = \sum(v^r/r!)(-t)_{|r|}F_B(-r,ct+X;-t;c^{-1}).$$

Remarks. 1) $x_j = c_\lambda{}^{-1}X_\lambda C_{\lambda j}$ where $C = B^{-1}$, inverse matrix to B. 2) Thus, the J_n are linear combinations of the F_B functions. 3) We omit the proofs. They follow by series expansion (1) and by comparison (2) after substituting $v \to Bv$ in Theorem 00.

III. Group-theoretic and Differential-Geometric Properties[3]. The Riccati system in Theorem 0, is of projective type, i.e., it can be linearized in projective space with coordinates (V,y), where we identify V with V/y. Thus, a subgroup of GL(N+1) acts on the solutions as fractional linear transformations: $V \to EV/(1+g\cdot V)$ for matrix E and vector g.

The differential-geometric properties apply to general Riccati systems $V_{jk} = \varepsilon_{jk} + a^{\lambda}_{jk}V_{\lambda} + B^{\lambda\mu}_{jk} V_{\lambda}V_{\mu}$ with, in general, variable coefficients. The differential forms $\varepsilon_j = \sum \varepsilon_{jk}dz_k$, $a^{\ell}_j = \sum a^{\ell}_{jk}dz_k$, $B^{\ell m}_j = \sum B^{\ell m}_{jk}\, dz_k$ satisfy

Theorem 0. (1) $d\varepsilon = A \wedge \varepsilon$ (2) $B\times\varepsilon = dA - A \wedge A$ (3) $dB = B \wedge A$.

Remarks. 1) ε, A, B have components ε_j, a^{ℓ}_j, $B^{\ell m}_j$ respectively. 2) The product × denotes anti-symmetric contraction of indices. 3) Thus one has canonically an associated geometric structure to every Riccati system. Essentially B is a gauge field (curvature) with potential (connection form) A.

References

1. P. Appell, "Sur les fonctions hypergéométriques de plusieurs variables," Mem. de Sci. Math., 3, 1925.
2. B. Gruber, H.D. Doebner, P. Feinsilver, Kinam 4(1982) 241-278.
3. P. Feinsilver, Springer Lect. Notes in Math. 1064(1984) 86-98.
4. R.L. Anderson, J. Harnad, P. Winternitz, J. Math. Phys. 24, 5(1983) 1062-1072.
5. L. Weisner, Pac. J. Math. 5(1955) 1033-1039.
6. J. Meixner, J. London Math. Soc. 9(1934) 6-13.
7. W. Miller, J. Math. Phys. 13(1972) 1393-1399.
8. L.C. Biedenharn, in "Special Functions, Group Theoretical Aspects and Applications," Reidel, 1984, 130.

EXPERIMENTAL AND PHILOSOPHICAL FOUNDATIONS OF THE FORMALISM OF STOCHASTIC QUANTUM MECHANICS

F.E. Schroeck, Jr.
Florida Atlantic University
Boca Raton, Florida 33431

This is a report on joint work with P. Busch.

We have analyzed the neutron interferometer, Stern-Gerlach, single slit, double slit, Michelson interferometer, and amplitude modulation of Mossbauer quanta experiments and find a direct link to the formalism of stochastic quantum mechanics.[1] We review the results for phase space and spin space here.

In neutron interferometry[2] a neutron beam is split, one leg is directed through a retarding material, and interference fringes are observed as a function of the delay (thickness). For d = the relative shift in the center of the neutron wave function ψ, for U a representation of the translation group, for T_d the projection on $U(d)\psi$, then the "contrast" or "visibility", $V(d)$, of the interference fringes equals the autocorrelation[3] which we reexpress:

$$V(d) = \left|\int\psi^*(q)\psi(q-d)dq\right| = |\langle\psi,U(d)\psi\rangle| = [Tr(T_dT_0)]^{1/2}. \quad (1)$$

More generally, we define the transition probability density

$$\beta(d,b) = Tr(T_dT_b) \quad (2)$$

for observing a neutron centered at d when it is in fact centered at b. Then

$$0 \leq \beta(d,b) = \beta(d-b,0) = \beta(0,b-d) \leq 1; \quad (3)$$

so β is directly observable through the contrast. The fine width[4] of ψ is defined by choosing $0 < \alpha < 1$ and then defining

$$w_\psi(\alpha) = \min\{\|d\| \,\big|\, \beta(d,0) = \alpha^2\}. \quad (4)$$

We may generalize this to translations in phase space (Weyl or Heisenberg group). Then in irreducible square integrable representations and for μ the left invariant group measure

$$\int T_d d\mu(d) = \lambda\mathbb{1},\ \lambda > 0, \quad (5)$$

$$\lambda^{-1}\int\beta(d,b)d\mu(d) = 1; \quad (6)$$

so β describes a confidence function for a point in space marked with a neutron. Thus, points correspond to distributions, yielding a stochastic (or quantum) geometry. For low velocities (4) does not change value significantly in this extension.

The gross width[4] is given by choosing confidence level $0 < N < 1$ and defining

$$2W_{\psi}(N) = \text{Min}\{\text{diameter } (\Delta) \mid \Delta \in \mathcal{B}(\mu),\ \int_{\Delta} \beta(d,0)d\mu(d) = N\}\ . \quad (7)$$

W.O.L.O.G. Δ may be chosen to be a sphere. We have used $\beta(d,0)$ rather than $|\psi(d)|^2$ to reflect the stochastic nature of Δ. Then

$$\int_{\Delta} \beta(d,0)d\mu(d) = \langle\psi, A(\Delta)\psi\rangle; \quad (8)$$

$$A(\Delta) = \int_{\Delta} T_d d\mu(d); \quad (9)$$

$$U_d A(\Delta) U_d^{-1} = A(d[\Delta]), d[\Delta] = \Delta \text{ translated,} \quad (10)$$

shows that we may express W_{ψ} in terms of a positive operator valued measure intrinsic to a system of covariance (9), (10). These fine and gross widths are also the major items in a more careful discussion of uncertainty relations.[4,5]

Next consider a Stern-Gerlach device used to prepare a spin up beam (or a polarizer for helicity). For a realistic device oriented in the z direction, the beam density is

$$E(\lambda z) = \tfrac{1}{2}(\mathbb{1} + \lambda z\cdot\sigma)$$
$$= \tfrac{1}{2}(1+\lambda)T_z + \tfrac{1}{2}(1-\lambda)T_{-z},\ 0 \le \lambda \le 1,\ \|z\| = 1, \quad (11)$$

$$T_z = \tfrac{1}{2}(1 + z\cdot\sigma) = \text{a projection.} \quad (12)$$

$E(\lambda z)$ is a positive operator representing a mixture, unless $\lambda = 1$.

Measuring the up-beam with a second apparatus alligned with z', efficiency λ', the collapsed state for the new "up" beam is taken to be $E^{1/2}(\lambda'z')E(\lambda z)E^{1/2}(\lambda'z')$ which, taking trace, yields the correlation $\frac{1}{2}(1 + \lambda\lambda' \cos\theta_{zz'}) = c(\theta)$. The visibility for this peak is $\lambda\lambda'$ and for $\lambda = \lambda'$ we have

$$c(\theta) = \beta(\theta,0) = \text{Tr}(E(\lambda z)U(\theta)E(\lambda z)U(\theta)^{-1})\ , \quad (13)$$

where $U(\theta)$ is the rotation on the sphere taking z into z'. This agrees with (2). Given a particle/beam known to be in some pure state of unknown orientation, if it passes through a Stern-Gerlach device and ends in the z' up-beam, we may only infer a confidence region of its prior direction. For cones of angular halfwidth θ, center z', we have

$$\theta = \cos^{-1}([N\lambda + N - 1]\lambda^{-1}) \quad (14)$$

which for level of confidence $= 1/2 = N$ gives $\theta > \pi/2$.

Interpreting μ now as the rotation invariant measure on the sphere normalized to 2, then (5), (9), and (10) again hold with T_d replaced by $E(\lambda d)$, d a direction, and λ in (5) = 1; so, we again have motivated a system of covariance. Then

$$K(x,y) = T_y T_x \quad \text{or} \quad E^{1/2}(\lambda y)E^{1/2}(\lambda x) \tag{15}$$

defines a reproducing kernel on

$$\int^{\oplus} d\mu(x) T_x \mathcal{H} \quad \text{or} \quad \int^{\oplus} d\mu(x) E^{1/2}(\lambda x)\mathcal{H}$$

which then yields a Hilbert bundle structure.

Defining

$$\rho(x) = \mathrm{Tr}(T_x \rho), \text{ resp. } \mathrm{Tr}(E(\lambda x)\rho) \ , \tag{16}$$

$$A(f) = \int d\mu(x) f(x) T_x, \text{ resp. } \int d\mu(x) f(x) E(\lambda x) \ , \tag{17}$$

where f is a classical observable, yields a Kolmogorov (classical) probability density $\rho(x)$ and a quantization (17) in which the classical and quantum expectations agree, and commutators and Poisson brackets are correctly related.[8]

References

1. S.T. Ali, "Stochastic localization, quantum mechanics on phase space and quantum space-time", La Rivista del Nuovo Cimento 8, No. 11, 1985

2. H. Kaiser, S.A. Werner, E.A. George, Phys. Rev. Lett. 50 (1983) 560-563

3. A.G. Klein, G.I. Opat, W.A. Hamilton, Phys. Rev. Lett. 50 (1983) 563-

4. These definitions generalize those of J.B.M. Uffink, J. Hilgevoord, Found. Phys. 15 (1985) 925; Phys. Lett. 108A (1985) 59-62

5. J.F. Price, Phys. Lett. 105A (1984) 343-345

6. F.E. Schroeck, Jr., Found. Phys. 12 (1982) 479-497

7. P. Busch, Phys. Rev. D 33 (1986) 2253-2261

8. F. E. Schroeck, Jr., J. Math. Phys. 26 (1985) 306-310

EXPLICIT MULTIDIMENSIONAL SOLITARY WAVE SOLUTIONS TO NONLINEAR EVOLUTION EQUATIONS

Henry A. Warchall

Department of Mathematics
University of Rochester
Rochester, NY 14627

Recently, Deumens and Warchall [5] found explicitly all spherically symmetric standing wave solutions to certain semilinear Klein-Gordon and Schrödinger equations in multiple spatial dimensions, and analyzed properties of the associated localized traveling waves. This report outlines heuristic considerations underlying that analysis, and shows that the structure of the infinite family of standing waves is governed by the phase space portraits of a nonlinear dissipative dynamical system with one degree of freedom.

We study solitary wave solutions $u: \mathbf{R}^{n+1} \to \mathbf{C}$ to the nonlinear evolution equations

$$-i\, u_t - \Delta u = g(u) \qquad \text{(NLS)} \qquad (1)$$

$$u_{tt} - \Delta u = g(u) \qquad \text{(NLKG)} \qquad (2)$$

with $n \geq 2$, where the nonlinear function $g: \mathbf{C} \to \mathbf{C}$ has the form $g(z) = h(|z|)\, z/|z|$ with $h: \mathbf{R} \to \mathbf{R}$ a continuous odd piecewise linear function. If we look for solutions u to either of the above two equations of the form $u(x,t) = e^{i\omega t} v(x)$ with $v: \mathbf{R}^n \to \mathbf{R}$, where ω is real, we find that v must satisfy the elliptic equation

$$-\Delta v = f_\omega(v), \qquad (3)$$

where the nonlinear function f_ω is given by

$$f_\omega(s) = \begin{cases} h(s) - \omega s & \text{for NLS} \\ h(s) + \omega^2 s & \text{for NLKG}. \end{cases}$$

Note that if v is a nontrivial solution to equation (3) with $v(x) \to 0$ as $|x| \to \infty$, then the corresponding solutions $u(x,t)$ to equations (1) and (2) are localized standing solitary waves. Because equation (1) (respectively (2)) is invariant under the Galilean (respectively Poincaré) group, the action of this group of symmetries on u generates traveling solitary wave solutions.

Under certain conditions (see [1], [2], [3] and [10]) on f_ω that are easily satisfied by appropriate choices of h and ω, equation (3) has an infinite sequence of distinct solutions, each of which is spherically symmetric and twice continuously differentiable on $\mathbf{R}^n$, and which, together with its first two derivatives, decays exponentially. Furthermore, under certain other compatible conditions (see [4], [8], [9] and [11]), the lowest-energy of the standing wave solutions corresponding to these is stable under the time development given by the evolution equation (1) or (2).

Now consider the particular case of $n = 3$ spatial dimensions. Looking for the spherically symmetric solutions to equation (3) whose existence is asserted above, we substitute $v(x) = w(r)$ where $w: [0, \infty) \to \mathbf{R}$, and $r = |x|$, and find that w must satisfy the ordinary differential equation

$$w'' = -2\,r^{-1}\,w' - f_\omega(w) \tag{4}$$

on the half line, subject to the boundary conditions

$$w'(0) = 0 \qquad \text{and} \qquad \lim_{r\to\infty} w(r) = 0.$$

We choose a nonlinear term that satisfies the criteria for the existence of an infinite family of spherically symmetric solitary waves. Consider continuous piecewise linear functions h and values of ω such that

$$f_\omega(s) = \begin{cases} s + 1 + \lambda^2 & \text{if } s \le -1 \\ -\lambda^2 s & \text{if } -1 \le s \le 1 \\ s - 1 - \lambda^2 & \text{if } 1 \le s \end{cases}$$

where λ is a positive number. Equation (4) is then linear in each of the three amplitude regions $(-\infty, -1]$, $[-1, 1]$, and $[1, \infty)$, and it is possible to construct solutions explicitly by patching together the known closed-form expressions for the solution at various amplitudes. This is done in [5].

Equation (4) may be reinterpreted ([6], [7]) as the equation of motion for a nonlinear dissipative system with one degree of freedom if we think of r as the independent (time) variable and of w as the dependent (position) variable. The term $-2\,r^{-1}\,w'$ then gives a time-dependent damping force acting on a particle that moves in the potential well

$$F_\omega(s) = \int_0^s f_\omega(\tau)\,d\tau\,.$$

Below is shown the potential F_ω corresponding to the piecewise linear f_ω above.

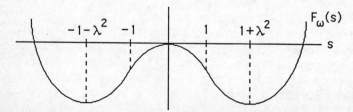

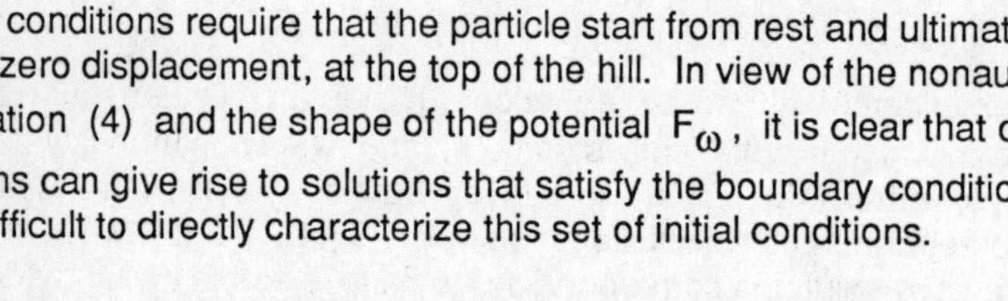

The boundary conditions require that the particle start from rest and ultimately reach equilibrium at zero displacement, at the top of the hill. In view of the nonautonomous nature of equation (4) and the shape of the potential F_ω, it is clear that only nongeneric initial conditions can give rise to solutions that satisfy the boundary condition at $r = \infty$. It is furthermore difficult to directly characterize this set of initial conditions.

To find the solutions of interest, it is simpler to first restrict consideration to that set Ω of nontrivial solutions to equation (4) satisfying the boundary condition at $r = \infty$. If $w(r)$ is an element of Ω, then it is completely characterized by the latest time $r = T_0$ that w had amplitude unity, and the sign of w at that time, that is, by $T_0 = \sup\{ r \in \mathbf{R} : |w(r)| = 1 \}$

and sgn $[w(T_0)]$. Furthermore, for every positive T, there are two solutions to (4) having $T_0 = T$. We henceforth suppose that $w(T_0)$ is positive to obviate this multiplicity. As T_0 is increased from small values, the solution $w \in \Omega$ changes shape, and as T_0 becomes just larger than those values for which the boundary condition $w'(0) = 0$ is satisfied, w acquires additional amplitude-threshold crossings, say at values of r equal to $T_1, T_2, \ldots, T_K$. Below are shown, from left to right, the shapes of w as T_0 increases through the smallest value of T_0 for which $w'(0) = 0$ is satisfied.

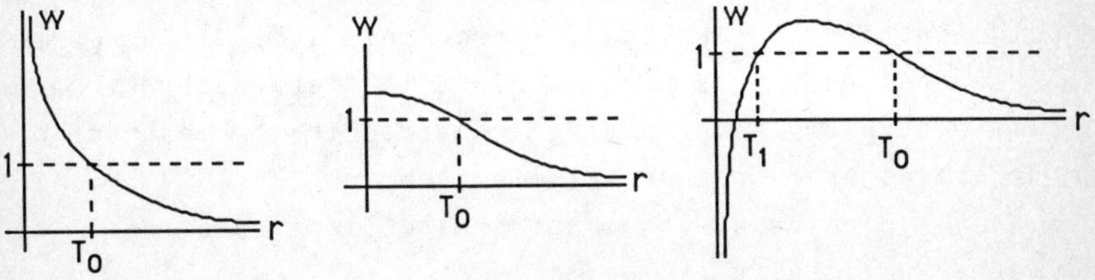

A central result of the analysis in [5] is that the values $T_1, T_2, \ldots, T_K$ of r at amplitude threshold crossings depend continuously and monotonically on T_0. Below are shown the phase space portraits of the three solutions w above.

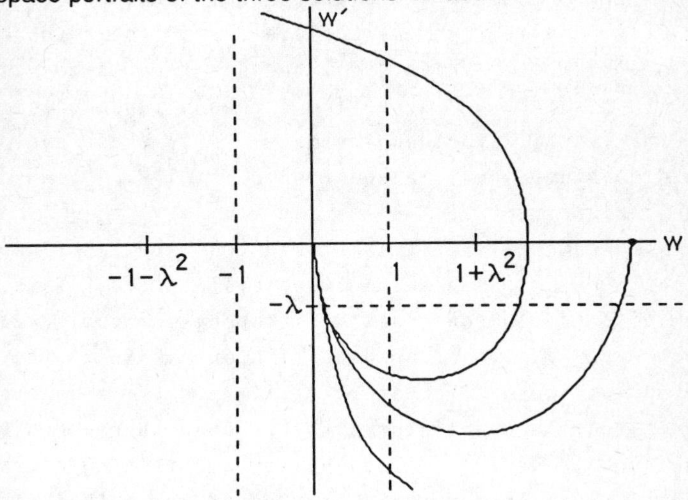

[1] Berestycki, H. and P.-L. Lions, Arch. Rat. Mech. Anal. **82** (1983) 313-375

[2] Berger, M., Nonlinearity and Functional Analysis, Academic Press (1977)

[3] Brezis, H. and E.H. Lieb, Comm. Math. Phys. **96** (1984) 97-113

[4] Cazenave, T. and P.-L. Lions, Comm. Math. Phys. **85** (1982) 549-561

[5] Deumens, E. and H. Warchall, Explicit construction of all spherically symmetric solitary waves for a nonlinear wave equation in multiple dimensions. Preprint.

[6] Jones, C. and T. Küpper, On the infinitely many solutions of a semilinear elliptic equation. To appear in SIAM J. Math. Anal.

[7] Lieb, E.H., Talk at the SIAM-AMS Summer Seminar on Systems of Nonlinear Partial Differential Equations in Applied Mathematics (1984: College of Santa Fe)

[8] Shatah, J., Comm. Math. Phys. **91** (1983) 313-327

[9] Shatah, J. and W. Strauss, Comm. Math. Phys. **100** (1985) 173-190

[10] Strauss, W., Comm. Math. Phys. **55** (1977) 149-162

[11] Weinstein, M., Comm. Pure Appl. Math. **39** (1986) 51-67

ENTROPY, FREQUENCY MIXING, AND PARTICLE CREATION[1]

Henry E. Kandrup
Center for Theoretical Physics
University of Maryland
College Park, MD 20742

Recent work by the author has served to demonstrate the possibility of constructing a sensible and consistent theory of nonequilibrium statistical mechanics in classical general relativity.[2] The object of the research reported here is to reformulate that approach in the context of quantum field theory, emphasizing in particular the connection between entropy generation and particle creation.[3]

As a prototypical example, consider a collection of oscillators described by a Hamiltonian

$$H = \sum_k \frac{1}{2}(p_k{}^2 + \omega_k{}^2 q_k{}^2) + \sum_{k'\neq k} \frac{1}{2} c_{kk'} q_k q_{k'} \equiv \sum_k H_k^o + \sum_{k'\neq k} H_{kk'} \quad ,$$

where $\omega_k > 0$ and $c_{kk'} = c_{k'k}$ are arbitrary real functions of time t. This is, e.g., the generic form appropriate for a linear source-free scalar field in a Mixmaster Universe. The idea then is to view the p_k's and q_k's as operators and to define as one's fundamental statistical object a many-oscillator density matrix $\rho(t)$ whose evolution is governed by the quantum Liouville equation $\dot{\rho} = -i[H,\rho]_-$, where $\cdot$ denotes a time derivative.

Given this fundamental ρ, it is straightforward to define reduced density matrices, such as the one- and two-oscillator $\rho_1(k)$ and $\rho_2(k,k')$, via partial traces, and to construct a hierarchy of coupled equations relating the time derivative of an N-oscillator matrix to the N- and (N+1)-oscillator matrices. Alternatively, if, e.g., one is concerned principally with the one-oscillator $\rho_1(k)$'s, one can use projection operator techniques to derive a "subdynamics," obtaining a closed (albeit nonlocal) equation for the evolution of the $\rho_1(k)$'s which contains no explicit reference to the higher order matrices $\rho_2(k,k')$.

The linearity of the Liouville equation implies that any measure of the "total entropy" constructed from the full ρ (e.g., Tr $\rho \,\ell og\, \rho$) cannot change with time. To the extent, however, that the $\rho_1(k)$'s are considered especially fundamental, one might seek instead to define a time-dependent "reduced entropy" in terms of these $\rho_1(k)$'s. Thus, e.g., in analogy with particle dynamics, one can choose to define

$$S(t) = -\sum_k \mathrm{Tr}_k\, \rho_1(k)\, \ell og\, \rho_1(k) \quad ,$$

where Tr_k denotes a trace over the variables for the kth oscillator. The real question is whether this, or any similar, quantity has a useful physical interpretation. One indication that this S is meaningful is that, at least in one limit, it satisfies an H-theorem inequality: if, at some time t_0, the system is free of "correlations," so that $\rho_2(k,k') = \rho_1(k)\rho_1(k')$, it follows rigorously that $dS(t_0+\Delta t)/dt = |a|^2\,\Delta t > 0$, where the quantity $|a|^2$, which involves the form of the initial $\rho_1(k)$'s, is quadratic in the $c_{kk'}$'s.[4] This demonstrates that an increase in the entropy provides a measure of correlations induced by mode-mode couplings. Indeed, if $c_{kk'} \equiv 0$, it follows that $dS/dt \equiv 0$ for all times. If there is no coupling between modes to generate correlations, the reduced entropy cannot change. It is natural to conjecture that this H-theorem inequality is in fact significantly more general, but that remains to be proved.

There exists also a clear connection between this entropy and particle creation. In the usual way, one can define the operator $N_k + 1/2 = (p_k^2 + \omega_k^2 q_k^2)/(2\omega_k)$ and interpret $\langle N_k\rangle = \mathrm{Tr}\,\rho N_k$ as the number of particles in the kth mode. One then concludes that

$$\partial_t \langle N_k\rangle = -\frac{\dot{\omega}_k}{2\omega_k^2}\langle p_k^2 - \omega_k^2 q_k^2\rangle - \sum_{k'\neq k}\frac{c_{kk'}}{\omega_k}\langle p_k\rangle\langle q_{k'}\rangle - \sum_{k'\neq k}\frac{c_{kk'}}{\omega_k}\,\underset{k}{\mathrm{Tr}}\,\underset{k'}{\mathrm{Tr}}\,\nu(k,k')\,p_k q_{k'} \quad ,$$

where $\nu(k,k') \equiv \rho_2(k,k') - \rho_1(k)\rho_1(k')$ provides a measure of the correlations in the system. The important point is that if ν vanishes at time t_0, the piece of the total $\partial_t\langle N_k\rangle$ generated by that evolving ν must be positive at time $t_0 + \Delta t$.

Quite generally, one calculates that, for small Δt,

$$\begin{aligned}\partial_t \langle N_k(t_0+\Delta t)\rangle = {} & \sum_{k'\neq k}[(c_{kk'})^2/\omega_k](\langle q_{k'}^2\rangle - \langle q_{k'}\rangle^2)\Delta t \\ & + (\dot{\omega}_k^2/\omega_k)\langle q_k^2\rangle\,\Delta t \\ & + \sum_{k'\neq k}(c_{kk'}/\omega_k)\{-\langle p_k\rangle\langle q_{k'}\rangle[1+(\dot{c}_{kk'}/c_{kk'} - \dot{\omega}_k/\omega_k)\Delta t] \\ & \qquad - \Delta t[\langle p_k\rangle\langle p_{k'}\rangle + \omega_k^2\langle q_k\rangle\langle q_{k'}\rangle - \sum_{\ell} c_{k\ell}\,\langle q_{k'}\rangle\langle q_{\ell}\rangle]\} \\ & + (\dot{\omega}_k/2\omega_k^2)\{-\langle p_k^2 - \omega_k^2 q_k^2\rangle[1+(\ddot{\omega}_k/\dot{\omega}_k - 2\dot{\omega}_k/\omega_k)\Delta t] \\ & \qquad + \Delta t[2\omega_k^2\langle p_k q_k + q_k p_k\rangle + 2\sum_{k'} c_{kk'}\langle q_{k'}\rangle\langle p_k\rangle]\} \quad ,\end{aligned}$$

where the first term, arising from the $\nu(k,k')$ contribution, is intrinsically positive.

The total $\partial_t\langle N_k(t_0+\Delta t)\rangle$ admits in fact to a simple interpretation. For many initial conditions of physical interest, including the vacuum or any eigenstate of the N_k's, all but the first two terms in this expression vanish identically, so that $\partial_t\langle N_k\rangle$ is the sum of two positive contributions. One obtains a net particle creation regardless of the sign of $\dot{\omega}_k$ or $c_{kk'}$. These contributions may in turn be interpreted as "spontaneous particle creation" induced, respectively, by "correlations" and "parametric amplification." The former is reflected by an increasing entropy; the latter entails instead a type of "frequency mixing."

The remaining more complicated terms will be nonvanishing only for nontrivial initial conditions (they vanish even if the $\rho(k)$'s are chosen initially to correspond to one-oscillator thermal matrices!). These, therefore, may be interpreted as reflecting "stimulated particle creation" induced by "correlations" or "parametric amplification" in the presence of initial conditions which manifest a "coherence of initial phase."

Note also that, unlike "spontaneous" creation, the "stimulated" contributions to $\partial_t \langle N_k \rangle$ are not necessarily of fixed sign. It is, however, well-known that, at least in one limit, nontrivial initial conditions lead to an enhancement in the net particle creation. Thus, if $c_{kk'} \equiv 0$ and the initial ρ corresponds to a thermal density matrix (or, indeed, an arbitrary mixture of pure states, each with a definite number of particles in each mode), the net number of created particles is necessarily positive and exceeds that produced from an initial vacuum.[5]

A similar result obtains when H is time-independent ($\dot{\omega}_k = \dot{c}_{kk'} = 0$) but $c_{kk'}$ is nonvanishing and sufficiently small that H is a positive quadratic form. One anticipates then that an initial density matrix $\rho \propto \exp(-\beta \sum_k H^0{}_k)$ will evolve towards a true thermal $\rho \propto \exp(-\beta' H)$, and it is easy to see that such an "equilibrium" corresponds to an increase in both $\langle N_k \rangle$ and S.

References

1. The subject matter discussed here is considered in greater detail in two forthcoming papers, "Entropy Generation in Cosmological Particle Creation," by B. L. Hu, H. E. Kandrup, and D. Pavon, and "Nonequilibrium Entropy in Classical and Quantum Field Theory," by H. E. Kandrup.

2. H. E. Kandrup, J. Math. Phys. 25, 3286 (1984); 26, 2850 (1985).

3. Earlier work on this subject is summarized, e.g., by B. L. Hu in L. Z. Fang and R. Ruffini (eds.),"Cosmology of the Early Universe," (World Science, 1984).

4. This net increase is a very general result. A proof for a slightly different Hamiltonian is provided by H. E. Kandrup, Class. and Quantum Grav. Lett. (in press).

5. L. Parker, Phys. Rev. 183, 1057 (1969).

THE ENTROPY OF HAWKING RADIATION

Diego Pavón
Department of Physics and Astronomy
University of Maryland. College Park,
Maryland 20742
and
Departamento de Termología. Universidad Autónoma de Barcelona. Spain

As is well-known, the Hawking radiation surrounding a black hole is not purely thermal as it is contaminated by vacuum polarization effects which die away only at distances large compared with the hole size. Our purpose in this note is to obtain an expression for the entropy of Hawking radiation emitted by a Schwarzschild black hole, both in the two-dimensional space-time case, and in the four-dimensional one. In both of them the radiation is supposed to be in its thermal state. The condition for the stability of equilibrium between the hole and radiation is that the heat capacity of the whole system be negative |1|.

In the two-dimensional case we assume the black hole and radiation enclosed in a linear impermeable box of length 2R, being the hole located at the center of the box. The energy density and pressure of that radiation are given respectively through the exact expressions |2|

$$\rho = \frac{1}{24\pi} \{ f'' + \frac{1}{f} (\kappa^2 + \frac{1}{4} f'^2)\} \quad , \tag{1}$$

$$P = \frac{1}{24\pi} \frac{1}{f} (\kappa^2 - \frac{1}{4} f'^2) \quad , \tag{2}$$

where $f(r) = 1-2M/r$, $\kappa = 1/4M$, and M stands for the black hole mass. Units have been chosen so that $c = G = \hbar = k_B = 1$. In terms of both, the black hole temperature, $T_{bh} = 1/8\pi M$, and the local temperature of radiation, $T_{loc} = T_{bh}/\sqrt{1-2M/r}$, the above expressions read

$$\rho = \frac{\pi}{6} T_{loc}^2 \{ [1 - (1 - \frac{T_{bh}^2}{T_{loc}^2})^4] - 8 \frac{T_{bh}^2}{T_{loc}^2} (1 - \frac{T_{bh}^2}{T_{loc}^2})^3 \} , \tag{3}$$

$$P = \frac{\pi}{6} T_{loc}^2 \{ 1 - (1 - \frac{T_{bh}^2}{T_{loc}^2})^4 \} , \tag{4}$$

respectively. Using $ds = d\rho/T_{loc}$, the entropy density follows at once:

$$s = \frac{\pi}{3} T_{loc} \{ 1 - (1- \frac{T^2_{bh}}{T^2_{loc}})^4 - 4 \frac{T^2_{bh}}{T^2_{loc}} (1- \frac{T^2_{bh}}{T^2_{loc}})^3 \} \quad . \tag{5}$$

Since the number of particles (real as well as virtual) making up the radiative medium is not defined, the corresponding Gibbs function must vanish. Effectively, $\rho+P-T_{loc}\, s = 0$ such as expected. The total entropy of the radiation confined in the box, $2\int_{2M}^{R}(1/\sqrt{f})dr$, yields

$$S = \frac{1}{6}(z + \ln z) - \frac{1}{4} - \frac{1}{6} z^{-1} + \frac{1}{4} z^{-2} \quad , \tag{6}$$

where $z \equiv R/2M > 1$.

As it can be checked, S has some quite reasonable properties: (a) As the radius of the box tends to the Schwarzschild radius, S vanishes. (b) In the flat-spacetime limit (i.e., $R >> 2M$) S tends to its conventional thermal value $(2\pi/3)RT_{bh}$. (c) For $M \to 0$, with R held fixed, S diverges.

In the four-dimensional case the black hole is supposed to be enclosed together with the radiation in a spherical box of impermeable and perfectly reflecting walls and radius R. We shall use the approximate Page's expression |3| for the renormalized stress-energy tensor of radiation in the Hawking-Hartle-Israel state:

$$T^{\mu}{}_{\nu} = - \frac{\pi^2}{90} (\frac{1}{8\pi M})^4 \{ \frac{1 - (4-6M/r)^2 \; (2M/r)^6}{(1 - 2M/r)^2} (\delta^{\mu}{}_{\nu} - 4\delta^{\mu}_{0}\, \delta^{0}{}_{\nu}) + 24 (\frac{2M}{r})^6 (3\delta^{\mu}{}_{0}\, \delta^{0}{}_{\nu} + \delta^{\mu}{}_{1}\, \delta^{1}{}_{\nu})\} \quad . \tag{7}$$

Because the Gibbs function should vanish, the entropy density may be obtained through $s = (\rho+P)/T_{loc}$, where $P = (1/3)(T^{r}{}_{r} + T^{\theta}{}_{\theta} + T^{\phi}{}_{\phi})$ as we are dealing with an anisotropic medium. In terms of temperatures one has

$$s = \frac{4\pi^2}{90} T^3_{loc} \{1 - (1 + 25 \frac{T^4_{bh}}{T^4_{loc}} + 6 \frac{T^2_{bh}}{T^2_{loc}})(1 - \frac{T^2_{bh}}{T^2_{loc}})^6\} \; . \tag{8}$$

Last expression does not exactly coincide with that derived from $d\rho/T_{loc}$, the discrepancy can be traced back to the approximate character of $T^{\mu}{}_{\nu}$ above. Upon integration throughout the volume of the box the total radiation entropy reads

$$S = 4\pi \int_{2M}^{R} r^2 \, s \, dr = \frac{1}{360} \{ (1-z^{-1})^{1/2} \left[\frac{1}{3} z^3 + \frac{11}{6} z^2 + \frac{3}{2} z - 10 \right]$$

$$+ \frac{191}{48} \ln \left[\frac{1+(1-z^{-1})^{1/2}}{1-(1-z^{-1})^{1/2}} \right] - \frac{214}{3}(1-z^{-1})^{3/2} + \frac{88}{5}(1-z^{-1})^{5/2}$$

$$- \frac{50}{7}(1-z^{-1})^{7/2} \} \ . \tag{9}$$

The right-hand side of last equation has the same (a)-(c) properties as above except that in the present four-dimensional case the flat-spacetime limit of S is $(4\pi^2/90)(4\pi R^3/3)T_{bh}^3$.

References

1. D. Pavón and W. Israel, Gen. Rel. Grav., 16 (1984) 563.
2. P.C.W. Davies, S.A. Fulling, and W.G. Unruh, Phys. Rev. D 13 (1976) 2720.
3. D.N. Page, Phys. Rev. D 25 (1982) 1499.

THERMODYNAMICAL REDUCTION OF THE ANISOTROPY OF TIME BY INTRODUCING IRREVERSIBILITY ON MICROSCOPICAL SCALE.

G.J.M. Verstraeten
Hoger Instituut voor Wijsbegeerte
Kardinaal Mercierplein 2, B-3000 Leuven, Belgium.

Physical as well as philosophical research have described the reduction of time anisotropy to thermodynamic or statistical thermodynamic evolution of many-particle systems. These topics can be understood in a more general frame enclosing the reconciliation of reversible mechanics and irreversible statistical thermodynamics. Prigogine's work " From Being to Becoming " tries to reconcile physical reversibility and irreversibility, introducing a new complementarity, linking the mechanical state to the thermodynamical one.[1]
We search for a connection between Prigogine's new complementarity and the Markov postulate which guarantees the irreversibility of a thermodynamical system.
The topology of time will be extremely important. That is why we control Prigogine's initial basic assumptions about time order, and time anisotropy, using Grünbaum's definition of symmetric time order.[2] Before discovering the role of the determined initial point of time and the tacit assumption of the asymmetric time order in Prigogine's theory, we first explain the role of time order in Boltzmann's H-theorem.
The third section explains the connection between the new complementarity and the initial configuration of the considered system in the phase space.

1. THE H-THEOREM AND THE ANISOTROPY OF TIME.

For non-equilibrium states, the distribution is defined by $f_{it} = \frac{\langle N_i \rangle t}{v_i}$, representing the distribution in the ith cell of volume v_i, in the 6th dimensional μ-space, and $\langle N_i \rangle t$ the actual current average amount of particles in the ith cell.
The well-known Boltzmann equation is expressed by: $f_i(t+1) - f_i(t) =$

$$v_i^{-1} \sum_j \sum_k \sum_l \left[B_{ij,kl} f_k(t) f_l(t) - B_{kl,ij} f_i(t) f_j(t) \right] \qquad (1)$$

The terms $B_{ij,kl}$ describe the collisions of two particles. All observable mechanical information is contained in the $B_{ij,kl}$ and the necessary time for the collision is smaller than the observable unit-time translation.
The Boltzmann equation yields in a non-static form of the H-function:

$$H(t) = \sum_i v_i f_i(t) \ln f_i(t). \qquad (2)$$

If A is an observable quantity, and if $t_2 > t_1$ then:

$$H(A(t_1)) \geq H(A(t_2)). \qquad (3)$$

With $S = -H$, we obtain an entropy function based on statistical methods and obeying the second law of thermodynamics. The applied average variation of

$N_i, \langle \Delta N_i \rangle_t$ is not the expected but the actual current variation of N_i. This approximation is known as the Stosszahlansatz. The results of the H-theorem are contradictory to the reversible character of the micro-interactions. This is called the reversibility paradox. The completely defined and observed variation at a particular point of time divides time in past and future, and introduces an anisotropic time concept. Indeed, formula (3) is only valid for $t \geqslant t_1$; for $t < t_1$ the Boltzmann equation as formulated in (1) does not exist. This anisotropic time concept means the tacit introduction of the Markov postulate. The latter cannot be explained as an effect of insufficient information, but should be understood on a microscopic scale.

2. PRIGOGINE'S NEW COMPLEMENTARITY.

For Prigogine " becoming " is not an accidental phenomenon; it is part of the essence of matter. Time is considered as an operator T and the new complementarity is established between the latter and the Liouville operator with a continuous spectrum of eigenvalues apart from one particular discrete eigenvalue: the zero. Mixing and K -flow are such kinds of systems.

Starting from a Gibbs entropy of a thermodynamic isolated system

$$S(t) = k_B \int d\alpha D_t(\alpha) \ln D_t(\alpha) + k_B \ln c, \tag{4}$$

he investigates a Lyapounov-function

$$\Omega_t = \int D_t M D_t \, dp \, dq \geqslant 0 \quad ; \quad \Omega_t \cdot \frac{d\Omega_t}{dt} \leqslant 0 \tag{5}$$

$$M = T^{\dagger} T. \tag{6}$$

The Poincaré-Misra theorem shows that Ω_t can be a Lyapounov function if and only if M is a functional of an operator and not of ordinary phase variables.[1]

Prigogine defines a micro-canonical ensemble with the distribution D_0 in the phase space. The evolution from this distribution to weak stability is caused by the action of T on the initial situation. If $\{e_n\}$ and $\{E_n\}$, n a countable number, are respectively the unity vectors, and the respective projection operators on these unity vectors, in the ensemble of observable points of time. So we find:

$$T = \sum_{-\infty}^{+\infty} n E_n \quad (7) \qquad\qquad M = \sum_{n=-\infty}^{n=+\infty} \lambda_n^2 E_n + P_0. \tag{8}$$

The projection operator P_0 corresponds to the micro-canonical ensemble and the time shift operator U implies:

$$U e_n = e_{n+1} \quad (9) \text{ and } [T, U] = U. \quad (10)$$

Prigogine defines this shift operator by the Baker-transformation:

$$U \phi(\omega) = \phi(B\omega) \qquad \omega = (p, q) \tag{11}$$

and

$$B(\omega) = (2p, \tfrac{q}{2}) \quad \text{if } 0 \leqslant p < 1/2 \tag{12a}$$

$$B(\omega) = (2p-1, \tfrac{q+1}{2}) \quad \text{if } 1/2 \leqslant p < 1. \tag{12b}$$

If τ symbolizes the minimum observable time translation we know that

$D_\tau(U_\tau(\omega)) = D_0(\omega)$ (τ sufficiently small).

So all states are observably equivalent, though the microscopic states in phase space are not identical.

Prigogine pretends from (8) that:

$\lim_{n\to\infty} \lambda_n^t = 0$ (13) and from Stone's theorem[1] he finds:

$U_\tau = e^{iL\tau}$ (14) from (10) and (6): $[L,T] = i$ (15), $[M,L] = i$. (16)

Prigogine identifies the microscopic entropy density to $-M$ and from (13) he discovers a microscopic form of the thermodynamic second law. Formula (16) represents the new complementarity between the mechanical state and thermodynamical one.

We summarize our critical remarks:

1) The order of the event represented by the operator $\{E_n\}$, which corresponds to the important order of the ensemble $\{\lambda_n^t\}$, cannot be defined without supposing a time concept already asymmetric. In order to identify the events $\{e_n\}$, and the ordered points of time, Prigogine needs Reichenbach's Mark method[3] which presupposes irreversibility of thermodynamical processes.

2) Grünbaum's symmetric time order[2] cannot be applied because there will be no n-chain of events between the events represented by E_n and E_{n+1}. Indeed, $\tau = t_{n+1} - t_n$ does not permit other points of time between the two extremities of τ.

3) Time does not appear as a real variable, at its most as a rational variable. So Stone's theorem is not valid.

Prigogine's concept of time presupposes already irreversible behaviour and the link between the determined initial distribution D_0, and the anisotropy of time caused by its determination is not made. Moreover, the tacit introduction of the Markov postulate in the chain E_n, E_{n+1} ..... is not put in evidence.

3. INITIAL BOUNDARY CONDITIONS AND IRREVERSIBILITY.

In order to avoid the critical remarks formulated above, we apply a general relation between a unitary and a hermitian operator. So we find: $U_\tau = e^{iA(\tau)}$ (17)

(U_τ unitary and defined as above, $A(\tau)$ hermitian):

$A(\tau) = -A(-\tau)$ (18); $U_\tau^n = U_{n\tau}$ (19); $A(n\tau) = nA(\tau)$. (20)

The hermitian operator A varies the distribution of states in the phase space which are all observably equivalent (see (11) and (12)). The number of applications corresponds to a particular point of time. The ensemble of these numbers, at its most a countable amount, is isomorphic to the ensemble of events $\{e_n\}$ and so the order in both ensembles is. The determination of the boundary conditions implies a time asymmetrical order. Moreover, $A(\tau)$ and $A(-\tau)$ divide the phase space anisotropically. Transition from regime + to - is impossible, due to the validity of a determined Liouville-equation. For $\tau \ll \Delta\tau$, the instrument's tolerance, it can be shown: with $\mathcal{L}_\tau = -iL_\tau A(\tau)$ and $U_{t_n} = e^{i\mathcal{L}_\tau t_n}$, $\mathcal{A}(n\tau) = \mathcal{L}_\tau t_n$, we find $[\mathcal{A}(n\tau), T] = i$ or $n\tau[\mathcal{L}_\tau, T] = i$. The initial boundary conditions in the phase space determine the complementarity relation and represent the Markov-condition on a microscopical scale.

1. Prigogine, I. (1979), From Being to Becoming
2. Grünbaum, A. (1973), Philosophical Problems of Space and Time
3. Reichenbach, H. (1956), The Direction of Time.

STEEPEST ENTROPY ASCENT IN QUANTUM THERMODYNAMICS

Gian Paolo Beretta
Department of Mechanical Engineering
Massachusetts Institute of Technology
and Dipartimento di Energetica
Politecnico di Milano, Italy

Quantum Thermodynamics [1] is a unified quantum theory that includes within a single uncontradictory nonstatistical structure the whole of Quantum Mechanics and Classical Equilibrium Thermodynamics, as well as a general description of nonequilibrium states, their entropy, and their irreversible motion towards stable equilibrium. Quantum Thermodynamics postulates that a system has access to a much broader set of states than contemplated in Quantum Mechanics. Specifically, for a system that is strictly uncorrelated from any other system, namely, a system for which Quantum Mechanics contemplates only states that are described by a state vector $|\Psi\rangle$, Quantum Thermodynamics postulates that in addition to the quantum mechanical states there exist many other states that cannot be described by a vector $|\Psi\rangle$ but must be described by a self-adjoint, unit-trace, nonnegative-definite linear operator ρ that we call the state operator.

In contrast with the density operators used in Statistical Mechanics to characterize either a heterogeneous ensemble of identical uncorrelated systems that are generally distributed over a range of different states, or a homogeneous ensemble of identical systems that are correlated with some other system such as a heat bath or a reservoir, we emphasize that the state operators in Quantum Thermodynamics are used to characterize a homogeneous ensemble of identical uncorrelated systems each of which is exactly in the same state as all the others. In other words, state operators ρ in Quantum Thermodynamics describe the uncorrelated states of a system, in the same sense as the state vectors $|\psi\rangle$ in Quantum Mechanics describe the uncorrelated states of a system. Clearly, all the states of Quantum Mechanics form a subset of the states contemplated in Quantum Thermodynamics, namely, the subset of idempotent states operators ρ such that $\rho^2 = \rho = |\psi\rangle\langle\psi|$.

Postulating the augmentation just cited of the set of conceivable states of an uncorrelated system allows a unification of mechanics and thermodynamics at the fundamental microscopic level [2]. Entropy emerges as a state property much in the same way in which energy is understood to be a state property. Energy is represented by the state functional $\mathrm{Tr}H\rho$ where H is the Hamiltonian of the system. Entropy is represented by the state functional $-k\mathrm{Tr}\rho\ln\rho$ where k is the Boltzmann constant [2]. Among all the different states ρ with a given value $\langle E\rangle$ of the energy, i.e., such that $\mathrm{Tr}H\rho = \langle E\rangle$, the value of the entropy spans from zero for the idempotent quantum mechanical states to a maximum value for the classical thermodynamical state $\rho = \exp(-\beta H)/\mathrm{Tr}\exp(-\beta H)$, where β is uniquely determined by $\langle E\rangle$. State operators with values of the entropy between zero and the maximum value represent states that, in general, are nonequilibrium and are not contemplated either in Quantum Mechanics or in Classical Thermodynamics.

Because the nonidempotent state operators represent nonmechanical states, the description of their time evolution cannot be derived from the laws of mechanics. Specifically, it cannot be derived from the unitary evolution generated by the Schroedinger equation $d|\psi\rangle/dt = -iH|\psi\rangle/\hbar$. Quantum Thermodynamics postulates that the time evolution of the state operator ρ is given in general by the solution of a general equation of motion [1] which for a single material constituent has the form

$$\frac{d\rho}{dt} = -\frac{i}{\hbar}[H,\rho] - \frac{1}{\tau}F(\rho), \tag{1}$$

where the operator $F(\rho)$ is given explicitly in [1] and can be visualized geometrically [3] as the projection of the gradient of the entropy functional $-kTr\rho\ln\rho$ onto the hyperplane generated by the normalization functional $Tr\rho$, the energy functional $TrH\rho$ and, for a field with number operator N (with $[N,H] = 0$), the number-of-particle functional $TrN\rho$.

The coefficient τ in Equation 1 cannot be inferred other than from experiments on the relaxation of nonequilibrium states. Mathematically, all the general results that we summarize below unfold identically whether τ is a universal constant, a constant that depends on the system, or any positive functional of ρ. At present, we have not found a way to estimate τ on the basis of available experimental data. However, we have discussed specific implications of Equation 1 which should in principle be experimentally verifiable [4].

The operator $F(\rho)$ has many interesting features. It reduces to the null operator whenever $\rho^2 = \rho$, namely, for each quantum mechanical state. Equation 1 maintains idempotent any initially idempotent state operator and, therefore, all the unitary evolutions of mechanical states generated by the Schroedinger equation are also solutions of Equation 1. Thus, we conclude that Quantum Thermodynamics contains the whole of Quantum Mechanics. But it is more general, because for the nonmechanical states, i.e., for $\rho^2 \neq \rho$, $F(\rho)$ does indeed contribute to the time evolution. The two terms in Equation 1 compete with each other in the sense that $-i[H,\rho]/\hbar$ tends to "pull" ρ in a direction tangent to the local constant entropy hypersurface whereas $-F(\rho)/\tau$ tends to "pull" ρ in the local direction of steepest entropy ascent while maintaining it on a constant energy and constant number of particles hyperplane.

The term $-i[H,\rho]/\hbar$ maintains invariant the entropy functional by maintaining invariant each of the eigenvalues of the state operator ρ. If H is time-dependent, then this term describes an adiabatic exchange of energy between the system and some other external systems during which the two systems remain uncorrelated. The adiabatic rate of energy change, $Tr(dH/dt)\rho$, depends on the rate of change of the Hamiltonian operator H. Even if H is time-dependent, the term $-F(\rho)/\tau$ does not contribute to changing the value of the energy functional $TrH\rho$, but for most nonidempotent states it increases the value of the entropy functional $-Tr\rho\ln\rho$. Interestingly, the rate of entropy increase is independent of the rate of change dH/dt of the Hamiltonian operator and, therefore, adiabatic energy exchanges can be made to approach reversibility, i.e., vanishing entropy production, in the limit of very fast changes of the Hamiltonian.

The magnitude of the rate of entropy increase is a nonlinear function of ρ which goes to zero smoothly at many states, including the idempotent states, the equilibrium states, and the limit cycles. It is therefore interesting to note that if a state is very close to, say, an idempotent state, then the term $-F(\rho)/\tau$ may be so small compared to the term $-i[H,\rho]/\hbar$ that its effect may be negligible for a long time, during which the evolution may seem dominated by the unitary term $-i[H,\rho]/\hbar$. According to Equation 1, however, all the idempotent states, the limit cycles, and the less-than-maximum-entropy equilibrium states are unstable in the sense of Lyapunov, i.e., arbitrarily close to each one of them there is a trajectory that after some finite time (perhaps very long) carries the state to a finite distance. The only equilibrium states that are stable in the sense of Lyapunov are the maximum entropy states $\rho = \exp(-\beta H)/Tr\exp(-\beta H)$, i.e., the equilibrium states of classical thermodynamics.

From the results just summarized it follows that if we postulate that the uncorrelated states of a system are described by state operators ρ that are not necessarily idempotent and evolve in time according to Equation 1, then we obtain a quantum theory that when restricted to the idempotent state operators reduces to the whole of Quantum Mechanics, and when restricted to the maximum entropy state operators, which turn out to be the only stable equilibrium states -- a conclusion that is equivalent to the second law of thermodynamics [2], reduces to the whole of Equilibrium Thermodynamics.

Moreover, the theory implies general conclusions on the nature of the nonequilibrium states and their irreversible, energy conserving but entropy increasing, motion towards stable equilibrium. A general state operator ρ can always be written as

$$\rho = B \exp(-\sum_j f_j X_j), \tag{2}$$

where the self-adjoint linear operators X_1, X_2, ..., X_j, ... form a fixed set spanning the real space of all self-adjoint linear operators, the coefficients f_1, f_2, ..., f_j, ... are real numbers, and B is an idempotent, self-adjoint linear operator. In terms of this expression, the entropy functional

$$S = -k\mathrm{Tr}\rho\ln\rho = k\sum_j f_j \mathrm{Tr}X_j\rho = k\sum_j f_j \langle X_j\rangle \tag{3}$$

so that

$$kf_j = \partial S/\partial\langle X_j\rangle \big|_{\langle X_{i\neq j}\rangle}, \tag{4}$$

and the coefficient f_i can be interpreted as a generalized affinity associated with the observable represented by operator X_i.

The particular structure of the dissipative term $-F(\rho)/\tau$ in the equation of motion of Quantum Thermodynamics is such that its contribution to the rate of change $d\mathrm{Tr}X_j\rho/dt$ of the mean value of observable X_j can be written as $-\mathrm{Tr}X_jF(\rho)/\tau = \sum_i f_i L_{ji}(\rho)$, thus implying the existence of linear interrelations between the dissipative contribution to the rate of change of $\langle X_j\rangle$ and the generalized affinities f_i. The coefficients $L_{ji}(\rho)$ may be interpreted as generalized conductivities. The explicit structure of the functionals $L_{ji}(\rho)$ is such that $L_{ji}(\rho) = L_{ij}(\rho)$ and the symmetric matrix $[L_{ij}(\rho)]$ is nonnegative definite. Thus, we conclude that the particular structure of the dissipative term in Equation 1, when coupled with the general expression for nonequilibrium states given by Equation 2, implies a general result that we may identify with an extension of Onsager's reciprocity principle to all nonequilibrium states. We emphasize that this result follows directly with no further assumptions from the specific structure of the operator $-F(\rho)/\tau$, which implies the existence of an endogenous irreversible dynamical tendency of nonequilibrium states to follow the local direction of steepest entropy ascent. The complete dynamical behavior results from the competition between the endogenous irreversible tendency towards the direction of steepest entropy ascent, contributed by the term $-F(\rho)/\tau$ in the equation of motion, and the reversible Hamiltonian tendency towards a unitary evolution, contributed by the term $-i[H,\rho]/\hbar$.

1. G.P. Beretta, E.P. Gyftopoulos, J.L. Park and G.N. Hatsopoulos, Nuovo Cimento B, Vol. 82, 169 (1984). G.P. Beretta, E.P. Gyftopoulos and J.L. Park, Nuovo Cimento B, Vol. 87, 77 (1985).
2. G.N. Hatsopoulos and E.P. Gyftopoulos, Found. Phys., Vol. 6, 15, 127, 439, 561 (1976).
3. G.P. Beretta, articles in Frontiers of Nonequilibrium Statistical Physics, G.T. Moore and M.O. Scully, Editors, Plenum Press, 1986.
4. G.P. Beretta, Int. J. Theor. Phys., Vol. 24, 119, 1249 (1985).

PARTICIPANTS

Ajanapon, Pimon, Dept. of Mathematics, Northeast Missouri State Univ., Kirksville, MO 63501

Albano, A. M.,Dept. of Physics, Bryn Mawr College, Bryn Mawr, PA 19010

Aldaya, Victor, Dept. of Theoretical Physics, Univ. of Valencia, Burjasot (Valencia), SPAIN

Alhassid, Yoram, Dept. of Physics, Yale Univ., New Haven, CT 06511

Ali, S. Twareque, Dept. of Mathematics, Concordia Univ., Montreal, QUE, H4B 1R6, CANADA

Alley, Carroll O., Dept. of Physics, Univ. of Maryland, College Park, MD 20742

Allshouse, Dennis, Dept. of Physics, Drexel Univ., Philadelphia, PA 19104

Alphenaar, Bruce, Dept. of Applied Physics, Yale Univ., New Haven, CT 06520

Amrolia, Z. J., The Mathematical Institute, Oxford Univ., Oxford, OX1, ENGLAND

Andrews, Thomas B., Dept. of Radio and Satellite Systems, Bell Communications Research, Red Bank, NJ 07701

Anninos, Peter, Dept. of Physics, Drexel Univ., Philadelphia, PA 19104

Bakas, Ioannis, Dept. of Physics, Univ. of Utah, Salt Lake City, UT 84112

Balazs, N. L., Dept. of Physics, State Univ. of New York, Stony Brook, NY 11794

Ballentine, L. E., Dept. of Physics, Simon Fraser Univ., Burnaby, BC, V5A 1S6, CANADA

Barocchi, Fabrizio, Dept. of Physics, Univ. of Florence, Largo E. Fermi 2, Florence, ITALY

Barrett, Terence W., Naval Air Systems Command HQ, Code Air-931P, Washington, DC 20361

Barrow, John, Astronomy Centre, Univ. of Sussex, Falmer, Brighton BN1 9QH, ENGLAND

Barut, A. O., Dept. of Physics, Univ. of Colorado, Boulder, CO 80309

Battelino, Peter, Dept. of Plasma Physics, Univ. of Maryland, College Park, MD 20742

Bayfield, James E., Dept. of Physics and Astronomy, Univ. of Pittsburgh, Pittsburgh, PA 15260

Beretta, Gian-Paolo, Dept. of Mechanical Engineering, M.I.T., Room 3-339, Cambridge, MA 02139

Bergeman, Tom, Dept. of Physics, State Univ. of New York, Stony Brook, NY 11794

Bertrand, Jacqueline, Dept. of Theoretical Physics, Univ. of Paris VII, LPTM-Tour 33-43, 75251 Paris Cedex 05, FRANCE

Bertrand, Pierre, Dept. of Aerospace Research, O.N.E.R.A., 29 Avenue de la Division Leclerc, 92 320 Chatillon, FRANCE

Bathia, Nam P., Dept. of Mathematics, Univ. of Maryland - B.C., Catonsville, MD 21228

Bhattacharya, Aniket, Dept. of Physics, Univ. of Maryland, College Park, MD 20742

Bleher, Siegfried, Dept. of Physics and Astro., Univ. of Maryland, College Park, MD 20742

Bombelli, Luca, Dept. of Physics , Syracuse Univ., Syracuse, NY 13244

Brandt, Howard, Strategic Defense Initiative, OSD/SDIO-IST, Washington, DC 20301-7100

Brill, Dieter, Dept. of Physics, Univ. of Maryland, College Park, MD 20742

Brooke, James A., Dept. of Mathematics, Univ. of Saskatchewan, Saskatoon, SAS, 57N OWO, CANADA

Brown, Reggie, Dept. of Physics, Univ. of Maryland, College Park, MD 20742

Buot, Felix A., Naval Research Lab., Code 6811, Washington, D.C. 20375

Busch, Paul, Dept. of Mathematics, Florida Atlantic Univ., Boca Raton, FL 33431

Buske, Norm, Search Technical Services, HCR 11, Box 17, Davenport, WA 99122-9404

Calzetta, Esteban, Dept. of Physics, Univ. of Maryland, College Park, MD 20742

Carey, Frank C., Dept. of Physics, Oakland Univ., Rochester, MI 48063

Carrington, Margaret, Dept. of Physics, State Univ. of New York Stony Brook, NY 11794

Carter, Patricia, Naval Surface Weapons Center, R-41, Silver Spring, MD 20903

Cary, John, Dept. of Astrophysical Sciences, Univ. of Colorado, Box 391, Boulder, CO 80309

Cawley, Robert, Nuclear Brach, R-41, Naval Surface Weapons Center, Silver Spring, MD 20903-5000

Catto, Sultan, Dept. of Physics, Baruch College, Box 502, 17 Lexington Avenue, New York, NY 10010

Chandler, Elaine A., Dept. of Chemistry, D5. Univ. of Pennsylvania, Philadelphia, PA 19104

Chang, Shau-Jin, Dept. of Physics, Univ. of Illinois, Urbana, IL 61801

Chen, Dun, Dept. of Physics, Yale Univ., New Haven, CT 06511

Chen, Jun Liang, Dept. of Physics and Astro., Univ. of Maryland, College Park, MD 20742

Choi, J. B., Dept. of Physics and Astronomy, Univ. of Maryland, College Park, MD 20742

Cohen, Leon, Dept. of Physics, Hunter College of C.U.N.Y., New York, NY 10021

Cohen, Thomas D., Dept. of Physics, Univ. of Maryland, College Park, MD 20742

Cole, James, Dept. of Physics and Astronomy, Univ. of Maryland, College Park, MD 20742

Dahl, Jens Peder, Dept. of Chemistry, Dept. B, The Technical Univ. of Denmark, DTH 301, DK-2800 Lyngby, DENMARK

Dana, Itzhak, Dept. of Chemistry, D5, Univ. of Pennsylvania, Philadelphia, PA 19104

Das, A, Dept. of Mathematics, Simon Fraser Univ., Burnaby, B.C., V5A 1S6, CANADA

Dehn, James, Ballistic Research Laboratory, Aberdeen Proving Ground, Aberdeen, MD 21005

Del Olmo, Mariano A., Center for Mathematica Research, Univ. of Montreal, C.P. 6128, Succ. A, Montreal, QUE, H3C 3J7, CANADA

Delos, John, Dept. of Physics, College of William and Mary, Williamsburg, VA 23185

Devine, David, Dept. of Physics and Astronomy, Univ. of Maryland, College Park, MD 20742

Dirl, Rainer, Dept. of Theoretical Physics, Technische Univ. Vienna, A-1040 Vienna, AUSTRIA

Doebner, H. D., Inst. fur Theoretische Physik, Tech. Univ. Clausthal, 3392 Clausthal-Zellerfeld, W. Germany

Dong, Qiqi, Dept. of Physics, Baylor Univ., Waco, TX 76798

Du, Mengli, Dept. of Physics, College of William and Mary, Williamsburg, VA 23185

Durand, Mireille, Inst. de Sciences Nucleaires, Univ. de Grenoble, 38026 Grenoble, France

Dworzecka, Maria, Dept. of Physics, George Mason Univ., Fairfax, VA 22030

Ehrman, Joachim, Dept. of Applied Mathematics, Univ. of Western Ontario, London, ONT, N6A 5B9, CANADA

Eschenazi, Elia, Dept. of Physics, Drexel Univ., Philadelphia, PA 19104

Feinsilver, Philip, Dept. of Mathematics, Southern Illinois Univ., Carbondale, IL 62901

Fernandez-Fera, Ramon, Dept. of Mechanical Engineering, Yale Univ., New Haven, CT 06520

Ferrell, Richard A., Dept. of Physics and Astro., Univ. of Maryland, College Park, MD 20742

Finn, John M., Lab. for Plasma and Fusion Energy Studies, Univ. of Maryland, College Park, MD 20742

Fishman, Louis, Dept. of Civil Engineering, Catholic Univ. of America, Washington, DC 20064

Floyd, Edward R., Naval Ocean Systems Center, Arctic Submarine Lab., San Diego, CA 92152

Ford, Joseph, School of Physics, Georgia Institute of Technology, Atlanta, GA 30332

Freidkin, Eugen, Dept. of Physics, Polytechnic Univ. of New York, Brooklyn, NY 11201

Fulling, Stephen A., Dept. of Mathematics, Texas A&M Univ., College Station, TX 77843

Gilmore, Robert, Dept. of Physics, Drexel Univ., Philadelphia, PA 19104

Gill, Tepper, Dept. of Mathematics, Howard Univ., Washington, DC 20059

Glick, Arnold, Dept. of Physics, Univ. of Maryland, College Park, MD 20742

Goldin, Gerald A., Dept. of Mathematics, Rutgers Univ., New Brunswick, NJ 08903

Goldman, Joseph, Dept. of Physics, American Univ., Washington, DC 20016

Gotay, Mark J., Dept. of Mathematics, U.S. Naval Academy, Annapolis, MD 21402

Gracia-Bondia, Jose M., Dept. of Mathematics, Univ. of Costa Rica, San Pedro, Montes de Oca, COSTA RICA

Grebogi, Celso, Lab. for Plasma and Fusion Energy Studies, Univ. of Maryland, College Park, MD 20742

Grelland, Hans H., Norsk Kybernetikk A/S, P.O.Box 128, N-1362 Billingstad, NORWAY

Griffin, James, Dept. of Physics, Univ. of Maryland, College Park, MD 20742

Habib, Salman, Dept. of Physics and Astronomy, Univ. of Maryland, College Park, MD 20742
Han, Daesoo, NASA - Goddard Space Flight Center, Greenbelt, MD 20771
Hannay, John H., H.H. Wills Physics Lab., Univ. of Bristol, Bristol BS8 1TL, ENGLAND
Hauser, Avi, Dept. of Physics, Univ. of Maryland, College Park, MD 20742
Hawkins, Raymond J., Bell Communications Research, 331 Newman Springs Rd., 3X-230, Red Bank, NJ 07701
Hazak, Gioka, Dept. of Applied Physics, Yale Univ., New Haven, CT 06511
Heagy, James F., Dept. of Physics, Drexel Univ., Philadelphia, PA 19104
Herdekopf, E., Dept. of Physics, Univ. of Maryland, College Park, MD 20742
Heller, Eric, Dept. of Chemistry, Univ. of Washington, Seattle, WA 98195
Henyey, Frank, Ctr. for Studies of Nonlinear, Dynamics, La Jolla Inst., 3252 Holiday Ct., #208, La Jolla, CA 92037
Hershfield, Selman, Dept. of Physics, Cornell Univ., Ithaca, NY 14853
Hioe, F. T., Dept. of Physics, St. John Fisher College, Rochester, NY 14618
Hizandis, Kyriakos, Dept. of Physics and Astro., Univ. of Maryland, College Park, MD 20742
Horiuchi, Hisashi, Dept. of Physics, Kyoto Univ., Kyoto 606, JAPAN
Hu, Bei-Lok, Dept. of Physics and Astronomy, Univ. of Maryland, College Park, MD 20742
Huang, Yuanshi, Dept. of Mathematics, Fuzhou Univ., Fuzhou, Fujian, P.R. OF CHINA
Huggins, Steven R., Dept. of Theoretical Physics, The Univ., Newcastle-upon-Tyne, UNITED KINGDOM
Hui, Yukkei, Dept. of Physics, Univ. of California, Berkeley, CA 94720
Hwang, Fu-Kwun, Dept. of Physics and Astro., Univ. of Maryland, College Park, MD 20742
Ibort, Luis A., Dept. of Mathematics, Univ. of California, Berkeley, CA 94720
Iida, Shinji, Research Ctr. for Nuclear Physics, Osaka Univ., Osaka 567, JAPAN
Irie, Haruyuki, Dept. of Physics, Nihon Univ., Tokyo 101, JAPAN
Jackson, Eric M., Dept. of Physics and Astronomy, Univ. of Maryland, College park, MD 20742
Jacob, Verne L., Naval Research Laboratory, Code 4174, Washington, DC 20375
Jalabert, Rodolfo, Dept. of Physics and Astronomy, Univ. of Maryland, College Park, MD 20742
Jensen, Roderick V., Dept. of Applied Physics, Yale Univ., New Haven, CT 06520
John, Sarah, Dept. of Physics, College of William and Mary, Williamsburg, VA 23185
Jellinek, Julius, Dept. of Chemistry, Univ. of Chicago, Chicago, IL 60637
Kadtke, James, Los Alamos National Lab., MS-B258, Los Alamos, NM 87545
Kandrup, Henry E., Dept. of Physics, Syracuse Univ., Syracuse, NY 13244
Karimabadi, Homayoun, Astronomy Program, Univ. of Maryland, College Park, MD 20742
Kasperkovitz, Peter, Institute for Theoretical Physics, Technical Univ. of Vienna, A-1040 Vienna, AUSTRIA
Kim, Shoon K., Dept. of Chemistry, Temple Univ. Philadelphia, PA 19122
Kim, Y. S., Dept. of Physics and Astro., Univ. of Maryland, College Park, MD 20742
Kincanon, Eric A., Dept. of Physics and Astro., Univ. of Missouri, Columbia, MO 65211
Klauder, John, Dept. of Theoretical Physics, AT&T Bell Labs., Murray Hill, NJ 07974
Koch, Peter M., Dept. of Physics, State Univ. of New York, Stony Brook, NY 11974-3800
Kocik, Jerzy, Dept. of Mathematics, Southern Illinois Univ., Carbondale, IL 62901
Koiller, Jair, Dept. of Mathematics, Yale Univ., New Haven, CT 06520
Koul, Rabinder K., Dept. of Physics, Syracuse Univ., Syracuse, NY 13244
Kubitscheck, Ulrich, Dept. of Physics and Astronomy, Univ. of Maryland, College Park, MD 20742
Kummer, Martin, Dept. of Mathematics, Univ. of Toledo, Toledo, OH 43606
Kunzle, H. P., Dept. of Mathematics, Univ. of Alberta, Edmonton, ABT, T6G 2G1, CANADA
Lee, Chongmoon, Dept. of Physics, Hunter College of C.U.N.Y., New York, NY 10021
Lee, C. T., Dept. of Physics, Alabama A&M Univ., Normal, AL 35762
Lee, Hai-Woong, Dept. of Physics, Oakland Univ., Rochester, MI 48063
Lee, Joohan, Dept. of Physics, Syracuse Univ., Syracuse, NY 13244
Li, Qiang, Dept. of Physics and Astronomy, Univ. of Maryland, College Park, MD 20472

Licht, A. Lewis, Dept. of Physics, Univ. of Illinois at Chicago, Chicago, IL 60680
Lill, J. V., Dept. of Condensed Matter/Radiation Science, Naval Research Lab., Washington, DC 20375
Lin, En-Bing, Dept. of Mathematics, Univ. of California, Riverside, CA 92521
Littlejohn, Robert G., Dept. of Physics, Univ. of California, Berkeley, CA 94720
Liu, Chuan Sheng, Dept. of Physics, Univ. of Maryland, College Park, MD 20742
Lochner, James, Dept. of Physics and Astro., Univ. of Maryland, College Park, MD 20742
Louck, James D., Theory Div., MS B-284, Los Alamos National Lab., Los Alamos, NM 87545
MacDonald, William, Dept. of Physics and Astro., Univ. of Maryland, College Park, MD 20742
Malek-Madani, Reza, Dept. of Mathematics, U.S. Naval Academy, Annapolis, MD 21402
Malta, Coraci P., Dept. of Physics/Mathematics, Univ. of Sao Paulo, CP 20516, 01000 Sao Paulo, BRAZIL
Marlino, Anthony, Dept. of Physics, Drexel Univ., Philadelphia, PA 19104
Mayer-Kress, G., Center for Nonlinear Studies, Los Alamos National Laboratory, Los Alamos, NM 87545
McMullan, David, Dept. of Physics, Univ. of Utah, Salt Lake City, UT 84112
McWilliams, Roger, Dept. of Physics, Univ. of California, Irvine, CA 92717
Menikoff, Ralph, Dept. of Theor. Physics, MS B214, Los Alamos National Lab., Los Alamos, NM 87545
Michelotti, Leo P., Accelerator Division, MS 345, Fermi National Accel. Lab., Batavia, IL 60510
Miller, H., Theoretical Physics Division, CSIR, P.O. Box 395, Pretoria 0001, SOUTH AFRICA
Millonas, Mark, Dept. of Physics, Yale Univ., New Haven, CT 06511
Molzahn, Frank H., Dept. of Physics, Univ. of Manitoba, Winnipeg, MB, R3T 2N2, CANADA
Monchick, Louis, Applied Physics Lab., Johns Hopkins Univ., Laurel, MD 20707
Morris, Randall D., MSR DIvision, RCA, Morristown, NJ 08057
Moshinsky, Marcos, Dept. of Theor. Physics, U.N.A.M., Apdo. 20-364, 01000 Mexico, D.F., MEXICO
Nakazawa, Hiroshi, Dept. of Physics - I, Kyoto Univ., Kyoto 606, JAPAN
Narcowich, Francis J., Dept. of Mathematics, Texas A&M Univ., College Station, TX 77843
Namyslowski, Jozef, Dept. of Physics and Astro., Univ. of Maryland, College Park, MD 20742
Needels, Jeffrey T., Dept. of Chemistry, Univ. of Kansas, Lawrence, KS 66045
Nessmann, Charlotte, Dept. of Chemistry, Univ. of Pennsylvania, Philadelphia, PA 19104
Newman, Murray N., Dept. of Physics, Univ. of Manitoba, Winnipeg, MB, R3T 2N2, CANADA
Newhouse, S., Dept. of Mathematics, Univ. of Maryland, College Park, MD 20742
Nietendel, Jadwiga, Dept. of Math./Physics/Chem., Pedagogical Univ., 25-509 Kielce, Lesna 16, POLAND
Noz, Marilyn E., Dept. of Radiology, New York Univ., New York, NY 10016
O'Connell, Robert F., Dept. of Physics & Astronomy, Louisiana State Univ., Baton Rouge, LA 70803
Osborn, Thomas, Dept. of Physics, Univ. of Manitoba, Winnipeg, MB, R3T-2N2 CANADA
Ott, Edward, Dept. of Physics & Astronomy, Univ. of Maryland, College Park, MD 20742
Park, Bae-Sig, Dept. of Physics & Astronomy, Univ. of Maryland, College Park, MD 20742
Park, Shim C., Dept. of Physics, Baylor Univ., Waco, TX 76798
Parmentier, Serge, Dept. of Physics, Penn State Univ., University Park, PA 16802
Pascolini, Allessandro, Dept. of Physics, Univ. of Padua, 35131 Padua, Italy
Patton, Charles M., H.C.C.O., Hewlett-Packard Co., 1000 N.E. Circle Blvd., Corvallis, OR 97330
Pavon, Diego, Dept. de Termologia, Facultad de Ciencias, Unv. Autonoma de Barcelona, Bellaterra (Barcelona), SPAIN
Pechukas, Philip, Dept. of Chemistry, Columbia Univ., New York, NY 10027
Pecora, Louis M., Metal Physics Branch, Naval Research Lab., Code 4631, Washington, DC 20375-5000
Peng, Huei, Dept. of Physics, Univ. of Alabama, Huntsville, AL 35899
Pilloff, Hersch, Div. of Physics, Office of Naval Research, Code 1112LO, Arlington, VA 22217

Ploszajczak, M. J., Dept. of Theor. Phys., Inst. of Nuclear Physics, 31-342 Krakow, POLAND
Prakash, Manju, School of Mines, Columbia Univ., New York, NY 10027
Radons, Gunter, Dept. of Theor. Physics, Regensburg Univ., 8400 Regensburg, W. GERMANY
Rajagopal, A. K., Naval Research Laboratory, Code 6830R, Washington, DC 20375
Ralston, John P., Dept. of Physics & Astronomy, Univ. of Kansas, Lawrence, KS 66045
Rangarajan, Gobindan, Dept. of Physics and Astro., Univ. of Maryland, College Park, MD 20742
Rankin, Linda, Dept. of Mathematics, Univ. of Maryland, College Park, MD 20742
Reifler, Frank, MSR Division, RCA, Morrestown, NJ 08057
Reinhardt, W. P., Dept. of Chemistry, Univ. of Pennsylvania, Philadelphia, PA 19104
Remler, Edward G., Dept. of Physics, College of William and Mary, Williamsburg, VA 23185
Rhoades-Brown, Mark, Dept. of Physics, S.U.N.Y. at Stony Brook, Stony Brook, NY 11794
Riesco-Chueca, Pascual. Dept. of Mechanical Engineering, Yale Univ., New Haven, CT 06520
Robbins, Jonathan, Dept. of Physics, Univ. of California, Berkeley, CA 94720
Robbins, Mark O., Dept. of Physics, Johns Hopkins Univ., Baltimore, MD 21218
Robinson, Sam, Dept. of Mathematics, Virginia Poytechnic Inst., Blacksburg, VA 24061
Romeiras, Filipe J., Lab. for Plasma and F. E. Studies, Univ. of Maryland, College Park, MD 20742
Rose, Michael, Dept. of Physics, Penn State Univ., University Park, PA 16802
Rossetto, Bruno, Inst. Univ. de Techn., Univ. of Toulon, F83 130 La Garde, FRANCE
Rossler, Otto E., Dept. of Physical & Theor. Chemistry, Univ. of Tubingen, 7400 Tubingen, W. GERMANY
Royer, Antoine, Center for Mathematical Research, Univ. of Montreal, Montreal, QUE, H3C 3J7, CANADA
Rubin, Morton, Dept. of Physics, Univ. of Maryland at B.C., Baltimore, MD 21228
Rule, Donald W., Nuclear Branch, R-41, Naval Surface Weapons Ctr., Silver Spring, MD 20903
Sáenz, Albert W., Dept. of Condensed Matter & Radiation Sciences Division Naval Research Lab., Washington, DC 20375
Salamin, Yousef, Dept. of Physics, Univ. of Colorado, Boulder, CO 80309
Sanders, Malcom, Dept. of Applied Physics, Yale Univ., New Haven, CT 06520
Sauer, Benjamin, Dept. of Physics, S.U.N.Y. at Stony Brook, Stony Brook, NY 11794-3800
Scarpetta, Gaetano, Dept. of Theoretical Physics, Univ. of Salerno, 84100 Salerno, ITALY
Scharph, William, Dept. of Physics, Drexel Univ., Philadelphia, PA 19104
Schempp, Walter, Lehrstuhl for Mathematics I, Univ. of Siegen, D-5900 Siegen, W. GERMANY
Schleich, Wolfgang P., Dept. of Physics, Ctr. for Theor. Physics, Univ. of Texas, Austin, TX 78712
Schroeck, Franklin E., Jr., Dept. of Mathematics, Florida Atlantic Univ., Boca Raton, FL 33431
Scully, Marlan O., Dept. of Physics, Univ. of New Mexico, Albuquerque, NM 87131
Seligman, Thomas H., Inst. of Physics, Univ. Nac. Autonoma de Mexico, AP 20-364, 01000 Mexico, D.F., MEXICO
Seo, Yoonho, Dept. of Physics & Astronomy, Univ. of Maryland, College Park, MD 20742
Shaarawi, Amr, Dept. of Electrical Engineering, Virginia Polytechnic Institute and State Univ., Balcksburg, VA 24060
Shi, Kangjie, Dept. of Physics, Univ. of Illinois, Urbana, IL 61801
Shi, Yanhua, Dept. of Physics and Astro., Univ. of Maryland, College Park, MD 20742
Shlesinger, Michael F., Physics Division, Office of Naval Research, 800 N. Quincy St., Arlington, VA 22217
Shrauner, Barbara Abraham, Dept. of Electrical Engineering, Washington Univ., St. Louis, MO 63130
Sica, Louis, Applied Optics, Naval Research Lab., Code 6530, Washington, DC 20375
Simms, D. J., Dept. of Mathematics, Trinity College, Dublin 2, IRELAND
Sinha, Sukanya, Dept. of Physics, Univ. of Maryland, College Park, MD 20742
Sinha, Supurna, Dept. of Physics , Syracuse Univ., Syracuse, NY 13244

Snider, R. F., Dept. of Chemistry, Univ. of British Columbia, Vancouver, BC, V6T 1Y6, CANADA

Solari, Hernan G., Dept. of Physics and Atmospheric Science, Drexel Univ., Philadelphia, PA 19104

Sreenivasan, K. R., Dept. of Applied Mechanics, Yale Univ., New Haven, CT 06520

Sternberg, Shlomo, Dept. of Mathematics, Harvard Univ., Cambridge, MA 02138

Stevens, Mark, Dept. of Physics, Johns Hopkins Univ., Baltimore, MD 21218

Sun, G., Lab. of Plasma and F.E. Study, Univ. of Maryland, College Park, MD 20742

Sun, Yuemin, Dept. of Physics, Penn. State Univ., University Park, PA 16802

Sundaram, Bala, Dept. of Physics, Johns Hopkins Univ., Baltimore, MD 21218

Susskind, Silvio M., Dept. of Physics, Yale Univ., New Haven, CT 06520

Svendsen, E. C., Dept. of Mathematical Sciences, George Mason Univ., Fairfax, VA 22030

Swank, Jean H., Goddard Space Flight Center, Bldg. 2, Greenbelt, MD 20771

Swank, L. James, Dept. of Physics & Astronomy, Univ. of Maryland, College Park, MD 20742

Szu, Harold, Naval Research Lab., Code 5709, Washington, DC 20375

Tannenbaum, David, Dept. of Launcher and Missile, Naval Underwater Systems Ctr., Newport, RI 02840

Thurber, James K., Dept. of Mathematics, Purdue Univ., West Lafayette, IN 47907

Toll, John S., Office of the President, Univ. of Maryland, College Park, MD 20742

Toms, David, Dept. of Theoretical Physics, Univ. of Newcastle, Newcastle, NE1 7RU, UNITED KINGDOM

Tufillaro, Nick, Dept. of Physics, Bryn Mawr College, Bryn Mawr, PA 19010

Uhm, Han S., Nuclear Branch, R-41, Naval Surface Weapons Center, Silver Spring, MD 20903

Unterberger, Andre, Dept. of Mathematics, Univ. of Reims, 51062 Reims Cedex, FRANCE

Varvoglis, H., Dept. of Astronomy, Univ. of Thessaloniki, Thessaloniki, GREECE

Verstraeten, Guido, Hoger Institute Wijsbegeerte, Katholieke Univ. Leuven, 3000 Leuven, BELGIUM

Vogt, Andrew, Dept. of Mathematics, Georgetown Univ., Washington, DC 20057

Warnock, Robert L., Theoretical Physics, Room 3115, Lawrence Berkeley Lab., Bldg. 50A, Berkeley, CA 94720

Warschall, Henry, Dept. of Mathematics, Univ. of Rochester, Rochester, NY 14627

Waterland, R. L., Dept. of Physics, College of William and Mary, Williamsburg, VA 23185

Weening, Richard, Dept. of Physics, College of William and Mary, Williamsburg, VA 23185

White, John, Dept. of Physics, American Univ., Washington, DC 20016

Wigner, Eugene P., Dept. of Physics, Princeton Univ., Princeton, NJ 08544

Yabana, Kazuhiro, Dept. of Physics, Kyoto Univ., Kyoto 606, JAPAN

Yoon, Jong H., Dept. of Physics and Astronomy, Univ. of Maryland, College Park, MD 20742

Yorke, James, Inst. for Physical Sciences and Tech., Univ. of Maryland, College Park, MD 20742

Zachariasen, F., Dept. of Physics, California Inst. of Tech., Pasadena, CA 91125

Zachary, W. W., Naval Research Lab., Code 4603-S, Washington, DC 20375

Zhou, Huan-Xiang, Dept. of Physics, Drexel Univ., Philadelphia, PA 19104

Zuckerman, Gregg J., Dept. of Mathematics, Yale Univ., New Haven, CT 06520

Zwanzig, Robert, Inst. for Physical Science and Technology, Univ. of Maryland, College Park, MD 20742

Lecture Notes in Physics

Vol. 257: Statistical Mechanics and Field Theory: Mathematical Aspects. Proceedings, 1985. Edited by T.C. Dorlas, N.M. Hugenholtz and M. Winnink. VII, 328 pages. 1986.

Vol. 258: Wm. G. Hoover, Molecular Dynamics. VI, 138 pages. 1986.

Vol. 259: R.F. Alvarez-Estrada, F. Fernández, J.L. Sánchez-Gómez, V. Vento, Models of Hadron Structure Based on Quantum Chromodynamics. VI, 294 pages. 1986.

Vol. 260: The Three-Body Force in the Three-Nucleon System. Proceedings, 1986. Edited by B.L. Berman and B.F. Gibson. XI, 530 pages. 1986.

Vol. 261: Conformal Groups and Related Symmetries – Physical Results and Mathematical Background. Proceedings, 1985. Edited by A.O. Barut and H.-D. Doebner. VI, 443 pages. 1986.

Vol. 262: Stochastic Processes in Classical and Quantum Systems. Proceedings, 1985. Edited by S. Albeverio, G. Casati and D. Merlini. XI, 551 pages. 1986.

Vol. 263: Quantum Chaos and Statistical Nuclear Physics. Proceedings, 1986. Edited by T.H. Seligman and H. Nishioka. IX, 382 pages. 1986.

Vol. 264: Tenth International Conference on Numerical Methods in Fluid Dynamics. Proceedings, 1986. Edited by F.G. Zhuang and Y.L. Zhu. XII, 724 pages. 1986.

Vol. 265: N. Straumann, Thermodymamik. VI, 140 Seiten. 1986.

Vol. 266: The Physics of Accretion onto Compact Objects. Proceedings, 1986. Edited by K.O. Mason, M.G. Watson and N.E. White. VIII, 421 pages. 1986.

Vol. 267: The Use of Supercomputers in Stellar Dynamics. Proceedings, 1986. Edited by P. Hut and S. McMillan. VI, 240 pages. 1986.

Vol. 268: Fluctuations and Stochastic Phenomena in Condensed Matter. Proceedings, 1986. Edited by L. Garrido. VIII, 413 pages. 1987.

Vol. 269: PDMS and Clusters. Proceedings, 1986. Edited by E.R. Hilf, F. Kammer and K. Wien. VIII, 261 pages. 1987.

Vol. 270: B.G. Konopelchenko, Nonlinear Integrable Equations. VIII, 361 pages. 1987.

Vol. 271: Nonlinear Hydrodynamic Modeling: A Mathematical Introduction. Edited by Hampton N. Shirer. XVI, 546 pages. 1987.

Vol. 272: Homogenization Techniques for Composite Media. Proceedings, 1985. Edited by E. Sanchez-Palencia and A. Zaoui. IX, 397 pages. 1987.

Vol. 273: Models and Methods in Few-Body Physics. Proceedings, 1986. Edited by L.S. Ferreira, A.C. Fonseca and L. Streit. XIX, 674 pages. 1987.

Vol. 274: Stellar Pulsation. Proceedings, 1986. Edited by A.N. Cox, W.M. Sparks and S.G. Starrfield. XIV, 422 pages. 1987.

Vol. 275: Heidelberg Colloquium on Glassy Dynamics. Proceedings, 1986. Edited by J.L. van Hemmen and I. Morgenstern. VIII, 577 pages. 1987.

Vol. 276: R.Kh. Zeytounian, Les Modèles Asymptotiques de la Mécanique des Fluides II. XII, 315 pages. 1987

Vol. 277: Molecular Dynamics and Relaxation Phenomena in Glasses. Proceedings, 1985. Edited by Th. Dorfmüller and G. Williams. VII, 218 pages. 1987.

Vol. 278: The Physics of Phase Space. Proceedings, 1986. Edited by Y.S. Kim and W.W. Zachary. IX, 449 pages. 1987.